Lecture Notes in Computer Science   10095

Commenced Publication in 1973
Founding and Former Series Editors:
Gerhard Goos, Juris Hartmanis, and Jan van Leeuwen

More information about this series at http://www.springer.com/series/7410

Orr Dunkelman · Somitra Kumar Sanadhya (Eds.)

Progress in Cryptology – INDOCRYPT 2016

17th International Conference on Cryptology in India
Kolkata, India, December 11–14, 2016
Proceedings

 Springer

Editors
Orr Dunkelman
University of Haifa
Haifa
Israel

Somitra Kumar Sanadhya
Indraprashtha Institute of Information
    Technology (IIIT-D)
New Delhi
India

ISSN 0302-9743   ISSN 1611-3349   (electronic)
Lecture Notes in Computer Science
ISBN 978-3-319-49889-8   ISBN 978-3-319-49890-4   (eBook)
DOI 10.1007/978-3-319-49890-4

Library of Congress Control Number: 2016957382

LNCS Sublibrary: SL4 – Security and Cryptology

Printed on acid-free paper

This Springer imprint is published by Springer Nature
The registered company is Springer International Publishing AG
The registered company address is: Gewerbestrasse 11, 6330 Cham, Switzerland

Preface

Since its introduction in 2000, INDOCRYPT has been widely acknowledged as the leading Indian venue for cryptography. As part of this tradition, INDOCRYPT 2016 was held during December 11–14, in Kolkata. This was the fourth time the conference was hosted Kolkata since its introduction by Prof. Bimal Roy. Past venues were held throughout India: Kolkata (2000, 2006, 2012, 2016), Chennai (2001, 2004, 2007, 2011), Hyderabad (2002, 2010), New Delhi (2003, 2009, 2014), Bangalore (2005, 2015), Kharagpur (2008), and Mumbai (2013).

INDOCRYPT 2016 attracted 84 submissions from 20 different countries, out of which 23 were selected at the end of a long review process: Most papers were reviewed by at least three committee members, whereas papers co-authored by Program Committee members were reviewed by at least five reviewers. In addition to the 283 reviews (produced with the aid of 91 additional reviewers), the Program Committee generated 223 comments during the discussion phase. We would like to express our sincere gratitude to all the members of the Program Committee, as well as all the external reviewers who helped in the challenging reviewing process.

The submission and review process was done using the iChair software package. We wish to express our sincere gratitude to Thomas Baignères and Matthieu Finiasz for the iChair software, which facilitated a smooth and easy submission and review process.

In addition to the 23 presentations of accepted papers, the attendees of INDOCRYPT also enjoyed three invited talks given by leading experts. Claudio Orlandi (Denmark) spoke about "Faster Zero-Knowledge Protocols for General Circuits and Applications"; the talk by François-Xavier Standaert (Belgium) covered "Leakage-Resilient Symmetric Cryptography"; and Tetsu Iwata (Japan) discussed "Breaking and Repairing Security Proofs of Authenticated Encryption Schemes."

Finally, we would like to thank the general chair, Prof. Bimal Roy, and the local organizing team comprising members from the Applied Statistics Unit, the R.C. Bose Center for Cryptology and Security at ISI Kolkata, and the Cryptology Research Society of India.

December 2016   Orr Dunkelman
                                                                     Somitra Sanadhya

Organization

General Chair

Bimal Roy   Indian Statistical Institute Kolkata, India

Program Chairs

Orr Dunkelman   University of Haifa, Israel
Somitra Sanadhya   Indraprastha Institute of Information Technology
                                          Delhi, India

Program Committee

Diego Aranha	University of Campinas, Brazil
Jean-Philippe Aumasson	Kudelski Security, Switzerland
Steve Babbage	Vodafone Group, UK
Begül Bilgin	KU Leuven, Belgium
Rishiraj Bhattacharya	Indian Statistical Institute Kolkata, India
Céline Blondeau	Aalto University, Finland
Andrey Bogdanov	Technical University of Denmark, Denmark
Itai Dinur	Ben-Gurion University of the Negev, Israel
Helena Handschuh	Cryptography Research, USA and KU Leuven, Belgium
Carmit Hazay	Bar-Ilan University, Israel
Takanori Isobe	Sony Corporation, Japan
Nathan Keller	Bar-Ilan University, Israel
Tanja Lange	Technische Universiteit Eindhoven, The Netherlands
Gaëtan Leurent	Inria, France
Atefeh Mashatan	Ryerson University, Canada
Florian Mendel	Graz University of Technology, Austria
Katerina Mitrokotsa	Chalmers University of Technology, Sweden
Amir Moradi	Ruhr-Universität Bochum, Germany
Debdeep Mukhopadhyay	IIT Kharagpur, India
David Naccache	ENS, France
Michael Naehrig	Microsoft Research, USA
Elisabeth Oswald	University of Bristol, UK
Arpita Patra	Indian Institute of Science, Bangalore
Thomas Peyrin	Nanyang Technological University, Singapore
Axel Poschmann	NXP Semiconductors, Germany
Vanishree Rao	PARC, USA

Francisco Rodríguez-Henríquez	CINVESTAV-IPN, Mexico
Bimal Roy	Indian Statistical Institute Kolkata, India
Santanu Sarkar	IIT Madras, India
Jean-Pierre Seifert	Technische Universität Berlin, Germany
Sourav Sen Gupta	Indian Statistical Institute Kolkata, India
François-Xavier Standaert	UCL, Belgium
Muthuramakrishnan Venkitasubramaniam	University of Rochester, USA
Xiaoyun Wang	Tsinghua University, China

Additional Reviewers

Gora Adj
Shashank Agarwal
Gilad Asharov
Josep Balasch
Subhadeep Banik
Paulo S.L.M. Barreto
Rana Barua
Srimanta Bhattacharya
Johannes Blömer
Debrup Chakraborty
Suvradip Chakraborty
Ayantika Chatterjee
Amit Kumar Chauhan
Chien-Ning Chen
Ran Cohen
Deirdre Connolly
Somindu C.R.
Abhijit Das
Poulami Das
Thomas De Cnudde
David Derler
Sandra Díaz-Santiago
Ning Ding
Christoph Dobraunig
Luis J. Dominguez Perez
Tuyet Duong
Ratna Dutta
Romain Gay
Satrajit Ghosh
Siyao Gou
Lorenzo Grassi

Hannes Gross
Mike Hamburg
Shoichi Hirose
Harunaga Hiwatari
Mike Hutter
Dirmanto Jap
Mahabir Jhawar
Bhavana Kanukurthi
Mikko Kiviharju
Ilya Kizhvatov
François Koeune
Kim Laine
Bei Liang
Patrick Longa
Atul Luykx
Monosij Maitra
Subhamoy Maitra
Daniel Malinowski
Mark Marson
Takahiro Matsuda
Siang Meng Sim
Santos Merino del Pozo
Guillermo Morales-Luna
Pratyay Mukherjee
Sayantan Mukherjee
Mridul Nandi
Khoa Nguyen
Ruben Niederhagen
Eduardo Ochoa-Jiménez
Tobias Oder
Claudio Orlandi

Elena Pagnin
Sumit Kumar Pandey
Tapas Pandit
Sikhar Patranabis
Oxana Poburinnaya
Antigoni Polychroniadou
Somindu Ramanna
Guillaume Rambaud
Shantanu Rane
Joost Renes
Bastian Richter
Lil Rodríguez-Henríquez
Sushmita Ruj
Debapriya Basu Roy
Vishal Saraswat
Pascal Sasdrich
Tobias Schneider
Kyoji Shibutani
Igor Shparlinski
Danilo Šijačić
Deng Tang
Mehdi Tibouchi
Ayineedi Venkateswarlu
Vincent Verneuil
Qingju Wang
Benjamin Wesolowski
Alexander Wild
Bo-Yin Yang
Hồng-Sheng Zhou

Invited Talks

Leakage-Resilient Symmetric Cryptography - Overview of the ERC Project CRASH, Part II

François-Xavier Standaert

ICTEAM Institute, Crypto Group, Université catholique de Louvain,
Ottignies-Louvain-la-Neuve, Belgium
fstandae@uclouvain.be

Abstract. Side-channel analysis is an important concern for the security of cryptographic implementations, and may lead to powerful key recovery attacks if no countermeasures are deployed. Therefore, various types of protection mechanisms have been proposed over the last 20 year. The first solutions in this direction were typically aiming at reducing the amount of information leakage directly at the hardware level, and independent of the algorithm implemented. Over the years, a complementary approach (next denoted as leakage-resilience) emerged, trying to exploit the formalism of modern cryptography in order to design new constructions and security models in which the guarantees of provable security can be extended from mathematical objects towards physical ones. This naturally raises the question whether the formal results obtained in these models are practically relevant (both in terms of performance and security)?

The development of sound connections between the formal models of leakage-resilient (symmetric) cryptography and the practice of side-channel attacks was one of the main objectives of the CRASH project funded by the European Research Council. In this talk, I will survey a number of results we obtained in this direction. For this purpose, I will start with a separation result for the security of stateful and stateless primitives. I will then follow with a discussion of (*i*) pseudorandom building blocks together with the theoretical challenges they raise, and (*ii*) authentication, encryption and authenticated encryption schemes together with the practical challenges they raise. I will finally conclude by discussing emerging trends in the field of physically secure implementations.

The extended version of this abstract is available from [1].

Reference

1. http://perso.uclouvain.be/fstandae/PUBLIS/184.pdf

Faster Zero-Knowledge Protocols for General Circuits and Applications

Claudio Orlandi

Aarhus University, Aarhus, Denmark

Abstract. *Zero-knowledge protocols (ZKP)* [GMR85] are one of the corner-stones of modern cryptography. In a nutshell, a ZKP allows a prover P (with a secret input x) to persuade a verifier V that $f(x) = 1$ for some public function f, without the V learning any other information about x.

A large body of literature has investigated the efficiency of ZKP for statements with a rich algebraic structure, starting from Schnorr's classic ZKP for discrete logarithm [Sch89]. However, the lack of efficient ZKP for interesting, non-algebraic statements (such as "*I know x such that SHA - 256 $(x) = y$*" for a public y), has arguably prevented the application of ZKPs to real-world applications.

In this talk I will describe two recent ZKPs for arbitrary circuits, ZKGC [JKO13] and ZKBoo [GMO16], together with their applications.

The first protocol (ZKGC), leveraging on the impressive advances in the field of practically efficient secure two-party computation (2PC), proposes to perform *zero-knowledge from garbled Boolean circuits*. As opposed to general 2PC (where many copies of the circuit must be garbled to achieve active security), when constructing ZKP it is enough to garble and evaluate *a single circuit*. Moreover, due to the nature of the application (since the verifier has no secret input), more efficient special purpose *privacy-free garbling schemes* [FNO15] can be used instead.

The second protocol instead (ZKBoo) follows a more classic "commit-challenge-response" structure (i.e., is a Σ-protocol). In ZKBoo the prover decomposes the computation of the function f in such a way that subsets of the computation can be checked by the verifier without revealing any information about the input to the computation, following the approach proposed by [IKOS07].

ZKGC and ZKBoo both have interesting properties: ZKGC leads to *smaller proof sizes* and, since it is based on garbled circuits, it can be combined very naturally with pre-existing secure computation tools towards building interesting applications such as: enforcing input validity in secure two-party computation [Bau16, KMW16], attributed-based key exchange with general policies [KKL+16], privacy-preserving credentials [CGM16], ZKPs for RAM programs [HMR15], etc.

ZKBoo on the other hand is *faster* and can be used for both Boolean and arithmetic circuits. Perhaps most importantly, ZKBoo can be made *non-interactive* using the Fiat-Shamir [FS86] heuristic. This qualitative advantage allows to use ZKBoo in applications such as (post-quantum) signature schemes from symmetric-key primitives [DOR+16], blind certificate authorities [WPaR16], etc.

It is exciting to see the growing number of applications which are enabled (or benefit) by the advances in the realm of ZKPs, and it seems likely that future research will make use of these tools in designing cryptographic solutions to interesting problems.

From a technical point of view, the main bottleneck in ZKGC and ZKBoo is their communication complexity, which in both cases is proportional to the number of non-linear gates in f times the security parameter (resulting in proof sizes in the order of hundreds of kylobytes for functions like SHA-1/256). Whether and how we can overcome this is a major and very exciting research question.

Acknowledgements. Research supported by: the Danish National Research Foundation and The National Science Foundation of China (grant 61361136003) for the Sino-Danish Center for the Theory of Interactive Computation; the European Union Seventh Framework Programme ([FP7/2007-2013]) under grant agreement number ICT-609611 (PRACTICE).

References

[Bau16]   Baum, C.: On garbling schemes with and without privacy. In: Zikas, V., De Prisco, R. (eds.) Security and Cryptography for Networks - 10th International Conference, SCN 2016, Amalfi, Italy, 31 August – 2 September 2016, Proceedings, pp. 468–485. Springer, Switzerland (2016)

[CGM16] Chase, M., Ganesh, C., Mohassel, P.: Efficient zero-knowledge proof of algebraic and non-algebraic statements with applications to privacy preserving credentials. In: Robshaw, M., Katz, J. (eds.) Advances in Cryptology - CRYPTO 2016 - 36th Annual International Cryptology Conference, Santa Barbara, CA, USA, 14–18 August 2016, Proceedings, Part III, pp. 499–530. Springer, Heidelberg (2016)

[DOR⁺16] Derler, D., Orlandi, C., Ramacher, S., Rechberger, C., Slamanig, D.: Digital signatures from symmetric-key primitives. In: Manuscript (2016)

[FNO15] Frederiksen, T.K., Nielsen, J.B., Orlandi, C.: Privacy-free garbled circuits with applications to efficient zero-knowledge. In: Oswald, E., Fischlin, M. (eds.) Advances in Cryptology - EUROCRYPT 2015 - 34th Annual International Conference on the Theory and Applications of Cryptographic Techniques, Sofia, Bulgaria, 26–30 April 2015, Proceedings, Part II, pp. 191–219.Springer, Heidelberg (2015)

[FS86]   Fiat, A., Shamir, A.: How to prove yourself: practical solutions to identification and signature problems. In: Odlyzko, A.M. (ed.) Advances in Cryptology — CRYPTO 1986, pp. 186–194. Springer, Heidelberg (1986)

[GMO16] Giacomelli, I., Madsen, J., Orlandi, C.: Zkboo: faster zero-knowledge for boolean circuits. In: 25th USENIX Security Symposium, USENIX Security 16, Austin, TX, USA, 10–12 August 2016, pp. 1069–1083 (2016)

[GMR85] Goldwasser, S., Micali, S., Rackoff, C.: The knowledge complexity of interactive proof-systems (extended abstract). In: Proceedings of the 17th Annual ACM Symposium on Theory of Computing, 6–8 May 1985, Providence, Rhode Island, USA, pp. 291–304 (1985)

[HMR15]   Hu, Z., Mohassel, P., Rosulek, M.: Efficient zero-knowledge proofs of non-algebraic statements with sublinear amortized cost. In: Gennaro, R., Robshaw M. (eds.) Advances in Cryptology - CRYPTO 2015 - 35th Annual Cryptology Conference, Santa Barbara, CA, USA, 16–20 August 2015, Proceedings, Part II, pp. 150–169. Springer, Heidelberg (2015)

[IKOS07]   Ishai, Y., Kushilevitz, E., Ostrovsky, R., Sahai, A.: Zero-knowledge from secure multiparty computation. In: Proceedings of the Thirty-ninth Annual ACM Symposium on Theory of Computing, STOC 2007, pp. 21–30. ACM (2007)

[JKO13]   Jawurek, M., Kerschbaum, F., Orlandi, C.: Zero-knowledge using garbled circuits: how to prove non-algebraic statements efficiently. In: 2013 ACM SIGSAC Conference on Computer and Communications Security, CCS 2013, Berlin, Germany, 4–8 November 2013, pp. 955–966 (2013)

[KKL+16]   Kolesnikov, V., Krawczyk, H., Lindell, Y., Malozemoff, A.J., Rabin, T.: Attribute-based key exchange with general policies. CCS 2016 (2016). http://eprint. iacr.org/2016/518

[KMW16]   Katz, J., Malozemoff, A.J., Wang, X.: Efficiently enforcing input validity in secure two-party computation. Cryptology ePrint Archive, Report 2016/184 (2016). http:// eprint.iacr.org/2016/184

[Sch89]   Schnorr, C.-P.: Efficient identification and signatures for smart cards. In: CRYPTO, pp. 239–252 (1989)

[WPaR16]   Wang, L., Pass, R., Shelat, A., Ristenpart, T.: Secure channel injection and anonymous proofs of account ownership. Cryptology ePrint Archive, Report 2016/925 (2016) http://eprint.iacr.org/2016/925

Contents

Functional Encryption

Symmetric-Key Cryptanalysis

Foundations

New Cryptographic Constructions

Public-Key Cryptography

Blending FHE-NTRU Keys – The Excalibur Property

Louis Goubin and Francisco José Vial Prado[(✉)]

Laboratoire de Mathématiques de Versailles, UVSQ, CNRS,
Université Paris-Saclay, 78035 Versailles, France
Francisco.vial-prado@uvsq.fr

Abstract. Can Bob give Alice his decryption secret and be convinced
that she will not give it to someone else? This is achieved by a proxy
re-encryption scheme where Alice does not have Bob's secret but instead
she can transform ciphertexts in order to decrypt them with her own key.
In this article, we answer this question in a different perspective, relying on a property that can be found in the well-known modified NTRU
encryption scheme. We show how parties can collaborate to *one-way-glue*
their secret-keys together, giving Alice's secret-key the additional ability
to decrypt Bob's ciphertexts. The main advantage is that the protocols we propose can be plugged directly to the modified NTRU scheme
with no post-key-generation space or time costs, nor any modification
of ciphertexts. In addition, this property translates to the NTRU-based
multikey homomorphic scheme, allowing to equip a hierarchic chain of
users with automatic re-encryption of messages and supporting homomorphic operations of ciphertexts. To achieve this, we propose two-party
computation protocols in cyclotomic polynomial rings. We base the security in presence of various types of adversaries on the RLWE and DSPR
assumptions, and on two new problems in the modified NTRU ring.

1 Introduction

Is it possible to avoid betrayal in a hierarchic scenario? Imagine a chain of
users equipped with a public-key encryption scheme, where high level users can
decrypt ciphertexts intended to all lower level users in the chain. This is trivial
to construct using any public-key cryptosystem $\mathcal{E}$: just transfer low-level secret-
keys to upper levels following the hierarchy. The evident drawback is that high-
level users can betray their children and distribute their secrets to other parties.
Using a proxy re-encryption procedure or multiple trapdoors is hence preferred,
because parents do not have direct knowledge of their children's secrets. A proxy
re-encryption scheme is a cryptosystem that allows a public transformation of
ciphertexts such that they become decryptable to an authorized party. This is a
particular case of a cryptosystem allowing delegation of decryption, which finds
applications in mail redirection, for instance. In this article, we give a solution
to the betrayal issue in another perspective, relying on a new property we found
in the well-known modified NTRU encryption scheme, and which we refer to as

© Springer International Publishing AG 2016
O. Dunkelman and S.K. Sanadhya (Eds.): INDOCRYPT 2016, LNCS 10095, pp. 3–24, 2016.
DOI: 10.1007/978-3-319-49890-4_1

"Excalibur". Basically, this feature allows to generate a secret-key that decrypts encryptions under multiple public-keys and behaves like a regular key of the cryptosystem.

1.1   The Excalibur Property

A public-key encryption scheme $\mathcal{E} = (\mathsf{Keygen}, \mathsf{Enc}, \mathsf{Dec})$ with plaintext space $\mathcal{M}$ has the Excalibur property if there is an algorithm that allows two users Alice and Bob with key-pairs $(\mathsf{sk}_A^{old}, \mathsf{pk}_A^{old})$ and $(\mathsf{sk}_B, \mathsf{pk}_B)$ respectively to forge a new key-pair for Alice $(\mathsf{sk}_A, \mathsf{pk}_A)$ such that

- Alice's key sk_A can decrypt ciphertexts in $\mathsf{Enc}(\mathsf{pk}_A, \mathcal{M}) \cup \mathsf{Enc}(\mathsf{pk}_B, \mathcal{M})$.
- Bob cannot decrypt ciphertexts in $\mathsf{Enc}(\mathsf{pk}_A, \mathcal{M})$.
- Alice cannot generate a secret-key sk_B' that is able to decrypt ciphertexts in $\mathsf{Enc}(\mathsf{pk}_B, \mathcal{M})$ but is not able to decrypt ciphertexts in $\mathsf{Enc}(\mathsf{pk}_A, \mathcal{M})$ (i.e. she cannot give away access to Bob's secret without leaking her own).

The intuition is that sk_A is a one-way expression of $(\mathsf{sk}_A^{old}, \mathsf{sk}_B)$. As Alice owns decryption rights over Bob's ciphertexts, this can be seen as *automatic* proxy re-encryption, in the sense that the re-encryption procedure is the identity. The idea is to "glue" Alice and Bob secret-keys together, resulting on a master key given to Alice. This Excalibur master key can be separated into factors only by Bob, hence the name of the feature: Bob plays the role of young Arthur, who is the only man in the kingdom able to separate Excalibur from the stone. Moreover, Alice can glue her key to an upper user's key, who inherits decryption over Bob's ciphertexts, and so forth, and if we suppose that no user is willing to give away own secrets, this achieves automatic N–hop re-encryption and sets a hierarchic chain.

We therefore have a scheme in which a single private-key can decrypt messages under multiple public-keys, and we will see that if a group of low-level users cheated in the joint key generation of this private-key (in order to sabotage or harden decryption), the secret-key holder may be able to trace it back to the wrongdoers, by simply testing decryptions and looking at the private-key's coefficients. In a sense, this is the inverse setting of a public-key traitor tracing scheme, where there are multiple secret-keys associated with a single public-key, and such that if a group of users collude in creating a new private-key achieving decryption with the public-key, it is possible to trace it to its creators, see for instance [4].

Three main advantages of this property over the trivial transfer of keys, over re-encryption schemes and over multiple trapdoor schemes are (i) there are no extra space or time costs: as soon as the keys are blended, the resulting key-pair acts as a fresh one and no ciphertext modification is necessary, (ii) our key generation procedure can be plugged directly into the (multikey) NTRU-based fully homomorphic encryption scheme, supporting homomorphic operations and automatic N–hop re-encryption and (iii) a user with a powerful key does not need to handle a "key ring" of secret-keys of her children; her key-pair $(\mathsf{sk}, \mathsf{pk})$ acts as a regular NTRU key. In contrast, the classical proxy re-encryption scenario is more flexible; a user can agree a decryption delegation at any moment to any user, whereas in our proposal once the keys are blended, modifications in hierarchy involve new key generations. This is why our proposal is more suitable to a rigid pre-defined hierarchic scenario.

1.2   Modified NTRU

The NTRUEncrypt cryptosystem is a public-key encryption scheme whose security is based on short vector problems on lattices. Keys and ciphertexts are elements of the polynomial ring $\mathbb{Z}[X]/\langle\phi(x)\rangle$ where $\phi(x) = x^n - 1$, and coefficients are considered modulo a large prime q. This scheme was defined in 1996 by Hoffstein, Pipher, Silverman and gained much attention since its proposal because of its efficiency and hardness reductions. In [25], Stehlé and Steinfeld provided modifications to the scheme in order to give formal statistic proofs, which ultimately led to support homomorphic operations with an additional assumption in [23]. Among these modifications, we highlight the change of ring and parameters restrictions: $R = \mathbb{Z}[x]/\langle\phi(x)\rangle$ where now $\phi(x) = x^n + 1$, n is a power of 2 (hence ϕ is the $2n$-th cyclotomic polynomial), and the large prime modulus is such that $x^n + 1$ splits into n different factors over $\mathbb{F}_q$ (namely, $q = 1$ mod $2n$). We will consider the modified NTRU scheme, but we believe that, possibly via a stretching of parameters, the original NTRU may also exhibit the Excalibur property.

1.3   Excalibur Key Generation

The way to glue two secret-keys is very simple: just multiply them together! Indeed, the modified NTRU scheme offers a fruitful property: *If one replaces a secret-key with a small polynomial multiple of it, decryption still works.* If this polynomial multiple is itself a secret-key, then by symmetry decryption with the resulting key will be correct in the union of ciphersets decryptable by one key or another. However, addressing the main point of this article, parties must multiply the involved polynomials using multiparty protocols, since they do not want to trust individual secrets to each other. To achieve this joint key generation, we rely on multiparty protocols in the polynomial ring $R_q = \mathbb{F}_q[x]/(x^n + 1)$ in both the secret and shared setting. To this end, we describe two multiplication protocols between mutually distrusting Alice and Bob:

1. **Secret Inputs Setting:** Alice and Bob hold $f, g \in R_q$ respectively. They exchange random polynomials and at the end Alice learns $fg + r \in R_q$ where r is a random polynomial known by Bob, and Bob learns nothing.
2. **Additively Shared Inputs Setting:** Alice and Bob hold $f_A, g_A \in R_q$ and $f_B, g_B \in R_q$ respectively such that $f = f_A + f_B$ and $g = g_A + g_B$. They exchange some random polynomials, and at the end Alice and Bob learn π_A, π_B respectively such that $\pi_A + \pi_B = fg \in R_q$. Revealing π_A or π_B to each other does not leak information about the input shares.

Let us illustrate how to use these protocols in Alice's key generation. Suppose that Bob keys were previously generated. Generating Alice's secret-key is fairly easy: Informally, if $\beta \in R_q$ is Bob's secret-key, let Alice and Bob sample random $\alpha_A, \alpha_B \in R_q$ respectively, with small coefficients. They perform the first protocol on inputs $f = \alpha_A$ and $g = \beta$, and Bob chooses $r = \alpha_B \beta$. At the end, Alice learns $\gamma = \alpha_A \beta + \alpha_B \beta = \alpha \beta \in R_q$, and Bob learns nothing. One may stop here and let Alice compute her public-key $\mathsf{pk}_A = 2h\gamma^{-1} \in R_q$ for suitable $h \in R_q$, but she may cheat and generate other NTRU fresh keys $(\mathsf{sk}_A,' \mathsf{pk}'_A)$ and then distribute freely Bob's secret γ. This is why the public-key is also generated jointly, and moreover, the public-key will be generated before the secret-key, this way Alice must first commit to a public-key pk_A.

1.4   Fully Homomorphic Encryption

Fully Homomorphic Encryption schemes allow public processing of encrypted data. Since Gentry's breakthrough in [10–12], there has been considerable effort to propose FHE schemes that are efficient [1,2,7,14–18,20], secure [2,6,8,9,13], and having other properties [7,9,13,19]. We highlight the existence of Multi-key FHE schemes, in which some ciphertexts can only be decrypted with the collaboration of multiple key-holders. This was first constructed in [23], and it reduces the general multiparty computation problem to a particular instance. We encourage the reader to see the latest version of this article.

All of the above schemes have a PPT encryption algorithm that adds random "noise" to the ciphertext, and propose methods to add and multiply two ciphertexts. With these methods they give an (homomorphic) evaluation algorithm of circuits. The noise in ciphertexts grows with homomorphic operations (especially with multiplication gates) and after it reaches a threshold, the ciphertext can no longer be decrypted. Thus, only circuits of bounded multiplicative degree can be evaluated: these schemes are referred to as leveled FHE schemes. Gentry proposed a technique called "bootstrapping" that transform a ciphertext into one of smaller noise that encrypts the same message, therefore allowing more homomorphic computations. This (algorithmically expensive) technique remains the only known way to achieve pure FHE scheme from a leveled FHE scheme. In order to do this, the decryption circuit of the leveled scheme must be of permitted depth and the new scheme relies on non-standard assumptions.

Nevertheless, leveled FHE schemes with good *a priori* bounds on the multiplicative depth do satisfy most applications requirements, see [22,27]. We suggest that the use of our protocols in the LATV scheme use the leveled version, but as pointed out in [23], the scheme can be transformed into a fully homomorphic scheme by boostrapping and modulus reduction techniques, both adaptable to the use of Excalibur keys.

1.5   FHE and Bidirectional Multi-hop Re-encryption Paradigm

It has been widely mentioned (for instance in the seminal work [11]) that a fully homomorphic encryption scheme allows bidirectional multi-hop proxy re-encryption. The argument is similar to the celebrated bootstrapping procedure: let c be an encryption of m using Bob's secret-key s_B. First publish τ, an encryption of s_B under Alice's public-key, then homomorphically run the decryption circuit on c and τ, the result is an encryption of m decryptable by Alice's secret-key. However, we point out that this is *pure* re-encryption only if Alice never gets access to τ, since she can decrypt and learn s_B directly. This restriction tackles the pure re-encryption definition, and in light of this the NTRU-based FHE scheme with the Excalibur property may be a starting point to clear out this paradigm (as it satisfies the pure definition, but fails to be bidirectional).

1.6   Our Contributions

In this article, we propose a key generation protocol that allows to glue NTRU secret-keys together in order to equip a hierarchic chain of users, such that a given user has the ability to decrypt all ciphertexts intended to all lower users in the chain, and she cannot give away secrets without exposing her own secret-key. This procedure can be plugged directly into the (multikey) FHE-scheme by Lopez-Alt et al., it is compatible with homomorphic operations and has no space costs or ciphertext transformations, and important users do not have to handle key rings. To achieve this, we describe two-party computations protocols in cyclotomic polynomial rings that may be of independent interest. We base the semantic security on the hardness of RLWE and DSPR problems, and the semi-honest and malicious security in a new hardness assumption which we call "Small Factors Assumption". In this assumption we define the "Small GCD Problem" and we show that any algorithm solving this problem can be used to break the semantic security of the modified NTRU scheme.

2   Preliminaries

2.1   Notation

Let q be a large prime. We let the set $\{-\lfloor q/2 \rfloor, \ldots, \lfloor q/2 \rfloor\}$ represent the equivalence classes of $\mathbb{Z}/q\mathbb{Z}$, and both notations $[x]_q$ or $x \bmod q$ represent modular reduction of x into this set. For a ring A, $A^{\times}$ stands for the group of units (or invertible elements) of A, $\langle a \rangle$ or (a) is the ideal generated by $a \in A$. Also, we denote by $\mathbb{F}_k$ the finite field of k elements, for $k = q^l \in \mathbb{Z}$. The notation $e \leftarrow \xi$ indicates that the element e is sampled according to the distribution ξ, and $e \xleftarrow{R} S$ means that e was sampled from the set S using the uniform distribution. Similarly, $A \xleftarrow{R} S$ means that each $a \in A$ was sampled uniformly at random on S. Finally, let $R \overset{def}{=} \mathbb{Z}[x]/(x^n + 1)$, we identify an element of R with its coefficient vector in $\mathbb{Z}^n$, and for $v(x) = v_0 + v_1 x + \cdots + v_{n-1} x^{n-1}$ in R, we denote by $||v||_{\infty}, ||v||_2$ its l_{∞}, l_2 norm respectively.

2.2   The Quotient Ring R_q

Operations in the modified NTRU scheme are between elements of $R_q \overset{def}{=} \mathbb{F}_q[x]/(x^n + 1)$, the ring of polynomials modulo $\Phi_{2n}(x) = x^n + 1$ (i.e. Φ_{2n} is the $2n$–th cyclotomic polynomial) and coefficients in $\mathbb{F}_q$, where n is a power of 2 and q is a large prime. Addition and multiplication of polynomials are performed modulo $\Phi_{2n}(x)$ and modulo q. The ring R_q is not a unique factorization domain, in fact, small units of this ring serve as NTRU secret-keys. The Chinese remainder theorem shows that the group of units is large, and thus $y = ru \in R_q$ where $r \in R_q$ is a random element and u is a unit is a good masking of u: it is unfeasible to recover u from y for large n. Let us collect some lemmas related to the set of invertible elements of R_q.

Lemma 2.2.1. *Let $q \geq 3$ be a prime number and $\Phi_n(x) \in \mathbb{Z}[x]$ be the n–th cyclotomic polynomial. Then $\Phi_n(x)$ is irreducible over $\mathbb{F}_q$ if and only if q is a generator of the group $(\mathbb{Z}/n\mathbb{Z})^\times$.*

Lemma 2.2.2. *If $n > 2$ is a power of 2, then $(\mathbb{Z}/2n\mathbb{Z})^\times$ is not cyclic and therefore $\Phi_{2n}(x) = x^n + 1$ is not irreducible over $\mathbb{F}_q$. In addition, $x^n + 1$ decomposes into l distinct irreducible factors over $\mathbb{F}_q$ for prime $q \geq 3$: Let $(\phi_i)_{i=1}^l \subset \mathbb{F}_q[x]$ respectively such that $x^n + 1 = \prod_{i=1}^l \phi_i(x)$ over $\mathbb{F}_q$. Then we have a ring isomorphism*

$$\pi : \frac{\mathbb{F}_q[x]}{(x^n + 1)} \longrightarrow \prod_{i=1}^l \frac{\mathbb{F}_q[x]}{(\phi_i(x))} \ where \ \frac{\mathbb{F}_q[x]}{(\phi_i(x))} \simeq \mathbb{F}_{q^{\deg \phi_i}}.$$

Corollary 2.2.3. $\mathrm{Card}(R_q^\times) = \prod_{i=1}^l \left(q^{\deg \phi_i} - 1 \right).$

The proofs are straightforward. In the original modifications in [25], $q = 1 \bmod 2n$ and hence $x^n + 1$ splits into n distinct linear factors, yielding $\mathrm{Card}(R_q)^\times = (q - 1)^n$.

2.3   Bounded Gaussian Samplings on $\mathbb{Z}[x]/(x^n + 1)$

Let n be a power of 2 and q a prime number, $R = R_0 \overset{def}{=} \frac{\mathbb{Z}[x]}{(x^n + 1)}$ and as before $R_q \overset{def}{=} \frac{\mathbb{F}_q[x]}{(x^n + 1)}$. The modified NTRU scheme uses a particular distribution in R_q, which we refer to as K-bounded by rejection Gaussian, serving to sample both message noises and secret-keys. Definitions follow.

Definition 2.3.1. *Let $\mathcal{G}_r$ be the Gaussian distribution over R, centered about 0 and of standard deviation r.*

Sampling from $\mathcal{G}_r$ can be done in polynomial time, for instance approximating with Irwin-Hall distributions. Consider the following definitions from [23]:

Definition 2.3.2. *A polynomial $e \in R$ is called K-bounded if $||e||_\infty < K$.*

Definition 2.3.3. *A distribution is called K-bounded over R if it outputs a K-bounded polynomial.*

Definition 2.3.4 *[K-bounded by rejection Gaussian]. Let $\bar{\mathcal{G}}_K$ be the distribution $\mathcal{G}_{K/\sqrt{n}}$ that repeats sampling if the output is not K-bounded.*

Lemma 2.3.5 (Expansion factors for $\phi(x) = x^n + 1$, from [23]). *For any polynomials $s, t \in R$,*

$$\|s \cdot t \mod \phi(x)\|_2 \leq \sqrt{n} \cdot \|s\|_2 \cdot \|t\|_2,$$
$$\|s \cdot t \mod \phi(x)\|_\infty \leq n \cdot \|s\|_\infty \cdot \|t\|_\infty.$$

Corollary 2.3.6. *Let χ be a K-bounded distribution over R and let $s_1, \ldots, s_l \leftarrow \chi$. Then $\prod_{i=1}^{l} s_i$ is $(n^{l-1}K^l)$-bounded.*

3   Modified NTRU Encryption

We review the modified NTRU encryption scheme as presented in [23], and we insist on the multi-key property. The message space is $\{0, 1\}$ and the ciphertext space is $R_q = \frac{\mathbb{F}_q[x]}{(x^n+1)}$. Let q be a large prime, $0 < K \ll q$, n be a power of 2 and $\bar{\mathcal{G}}_K$ be the K-bounded by rejection discrete Gaussian. A key-pair $(\mathsf{sk}, \mathsf{pk})$ is a tuple of polynomials in R_q, the secret-key being K-bounded.

  Keygen(1^κ):

Step 1. Sample a polynomial $f \leftarrow \bar{\mathcal{G}}_K$. Set $\mathsf{sk} = 2f + 1$, if sk is not invertible in R_q start again.
Step 2. Sample a polynomial $g \leftarrow \bar{\mathcal{G}}_K$ and set $\mathsf{pk} = 2g \cdot \mathsf{sk}^{-1} \in R_q$.
Step 3. Output $(\mathsf{sk}, \mathsf{pk})$.

  Enc(pk, m): Sample polynomials $s, e \leftarrow \bar{\mathcal{G}}_K$. For message $m \in \{0, 1\}$, output $c = m + 2e + s \cdot \mathsf{pk} \mod q$.
  Dec(sk, c): For a ciphertext $c \in R_q$, compute $\mu = c \cdot \mathsf{sk} \in R_q$ and output $m = \mu \mod 2$.

3.1   The Multikey Property

We describe a decryption property that states that one can decrypt a ciphertext with the secret-key required for decryption, or a small polynomial multiple of it.

Lemma 3.1.1. *Let $(f, h) \leftarrow$ Keygen(1^κ), $m \in \{0, 1\}$ and let $c \leftarrow$ Enc(h, m). Let $\theta \in R$ be a M-bounded polynomial satisfying $\theta \mod 2 = 1$. If $M < (1/72)(q/n^2 K^2)$, then*

$$\mathsf{Dec}(f, c) = \mathsf{Dec}(\theta \cdot f, c) = m.$$

Proof. There exist K-bounded polynomials s, e such that $c = m + hs + 2e$. Decryption works since

$$[fc]_q = [fm + fhs + 2fe]_q = [fm + 2gs + 2fe]_q$$

and supposing there is no wrap-around modulo q in the latter expression, we
have $[fc]_q \mod 2 = fc \mod 2 = m$. If we replace f by $\theta \cdot f$ and try to decrypt,
we have $\theta f c = \theta f m + 2\theta g s + 2\theta f e$, and then again, if there is no wrap-around
modulo q (i.e. if M is small enough), $\theta f c \mod 2 = m$ is verified. To ensure
that there is no wrap-around modulo q, one has to give an *a priori* relation
between K, n and M. In fact, using Corollary 2.3.6, we have $||gs||_\infty < nK^2$ and
$||fe||_\infty < n(2K+1)K$, and thus

$$||fc||_\infty < 2nK^2 + 2n(2K+1)K + K.$$

Decryption using f is correct if $2nK^2 + 2n(2K+1)K + K < q/2$, and decryption using θf is correct if $nM(2nK^2 + 2n(2K+1)K + K) < q/2$. Therefore,
decryption using f is ensured by $36nK^2 < q/2$, decryption using θf is ensured
by $36n^2MK^2 < q/2$.   □

Corollary 3.1.2 [*The multikey property*]. *Let (f_1, h_1) and (f_2, h_2) be valid keys,
$m_1, m_2 \in \{0, 1\}$ and let $c_1 \leftarrow \mathsf{Enc}(h_1, m_1)$, $c_2 \leftarrow \mathsf{Enc}(h_2, m_2)$. Let $\tilde{f} \leftarrow f_1 \cdot f_2 \in
R_q$. Then*

$$\mathsf{Dec}(\tilde{f}, c_1) = m_1, \quad \mathsf{Dec}(\tilde{f}, c_2) = m_2$$

provided that K is small enough,

Proof. Apply Lemma 3.1.1 with $f = f_1$ and $\theta = f_2$ for the first equation and
$f = f_2, \theta = f_1$ for the second.   □

We can of course extend this facts to show that a highly composite key of
the form $\tilde{f} = \prod_{i=1}^{l} f_i \in R_q$ can decrypt all messages decryptable by any of
f_i: Just apply Lemma 3.1.1 with $f = f_i$ and $\theta = \tilde{f}/f_i$, provided good *a priori*
bounds: In fact $||\tilde{f}||_\infty \leq n^{l-1}K^l$, therefore decryption with this key is ensured
by $n^{l-1}K^l \ll q$.

4   Hardness Assumptions

The modified NTRU-FHE scheme semantic security is based on the celebrated
Ring Learning With Errors problem (RLWE) and the new *Small Polynomial
Ratio problem* (SPR). For the original modified NTRU parameters, the decisional SPR problem reduces to RLWE, but not a single homomorphic operation
can be assured. A stretch of parameters is needed to overcome this, though it
severely harms the statistic proofs of Stehlé and Steinfeld. The *DSPR assumption*
states that the decisional SPR problem with stretched parameters is computationally hard. We adopt this same assumption, and in addition, we base the
security of the honest-but-curious model on two problems that involve decomposing a polynomial into bounded factors. In the first, one wants to factorize a
polynomial in R_q into two K-bounded polynomials, given the information that
this is possible. In the second, one wants to extract a common factor of two
polynomials such that the remaining factors are K-bounded. We first describe
the DSPR assumption and then our *"Small Factors"* assumption.

4.1 Small Polynomial Ratio Problem, from [23]

In [25] Stehlé and Steinfeld based the security of the modified NTRU encryption scheme on the Ring Learning With Errors (RLWE) problem [24]. They showed that the public-key $\mathsf{pk} = 2g \cdot \mathsf{sk}^{-1} \in R_q$ is *statistically close to uniform* over R_q, given that g and $f' = (\mathsf{sk} - 1)/2$ were sampled using discrete Gaussians. Their results holds if (a) n is a power of 2, (b) $x^n + 1$ splits over n distinct factors over R_q (i.e. $q = 1 \mod 2n$) and (c) the Gaussian error distribution has standard deviation of at least $\mathrm{poly}(n)\sqrt{q}$. However, these distributions seem too wide to support homomorphic operations in the NTRU-FHE scheme. To overcome this, authors in [23] defined an additional assumption which states that if the Gaussian is contracted, it is still hard to distinguish between a public-key and a random element of R_q (even if the statistic-closeness result does not hold).

Definition 4.1.1 *[DSPR Assumption]. Let $q \in \mathbb{Z}$ be a prime integer and $\bar{\mathcal{G}}_K$ denote the K-bounded discrete Gaussian distribution over $R_0 = \mathbb{Z}[X]/(x^n + 1)$ as defined in Definition 2.3.4. The decisional small polynomial ratio assumption says that it is hard to distinguish the following two distributions on R_q: (1) A polynomial $h = [2gf^{-1}]_q \in R_q$ where f', g were sampled with $\bar{\mathcal{G}}_K$ and $f = 2f' + 1$ is invertible over R_q, and (2) a polynomial $u \xleftarrow{R} R_q$ sampled uniformly at random.*

Finally, in a work by Bos et al. [5], authors achieved to base the security on RLWE alone, alas achieving multikey FHE for a constant number of keys, a property inherent to any FHE scheme (as proved in the latest version of [23]).

4.2 Small Factorizations in the Quotient Ring

In addition to the RLWE and DSPR assumptions, we rely the semi-honest security on the hardness of the following problems. Let us define the distribution $\bar{\mathcal{G}}_K^{\times}$ which samples repeatedly from $\bar{\mathcal{G}}_K$ until the output is invertible over R_q.

Small Factors Problem: *Let $a, b \leftarrow \bar{\mathcal{G}}_K^{\times}$ and let $c(x) = a(x) \cdot b(x) \in R_q$. Find $a(x)$ and $b(x)$, given $c(x)$ and a test routine $T : R_q \rightarrow \{0, 1\}$ that outputs 1 if the input is in $\{a, b\}$ and 0 otherwise.*

$\bar{\mathcal{G}}_K^{\times}$–GCD Problem: *Let $a, b \leftarrow \bar{\mathcal{G}}_K^{\times}$, and $y \xleftarrow{R} R_q$. Let $u(x) = a(x) \cdot y(x) \in R_q$ and $v(x) = b(x) \cdot y(x) \in R_q$. Find $a(x), b(x)$ and $y(x)$, given $u(x), v(x)$ and a test routine $T : R_q \rightarrow \{0, 1\}$ that outputs 1 if the input is in $\{a, b, y\}$ and 0 otherwise.*

Proposition 4.2.1. *An algorithm solving the $\bar{\mathcal{G}}_K^{\times}$–GCD problem can be used to break the semantic security of the NTRU scheme.*

Proof. Given only a public-key of the form $\mathsf{pk} = [2ab^{-1}]_q$ where a is the secret-key, sample $p \xleftarrow{R} R_q$ and define $(u', v') = (ab^{-1}p, p)$. Define also $T : R_q \rightarrow \{0, 1\}$

that for input $\alpha \in R_q$, samples random $r \xleftarrow{R} \{0,1\}$, checks if $\mathsf{Dec}(\alpha, \mathsf{Enc}(\mathsf{pk}, r)) \overset{?}{=} r$ and outputs 1 if α pass several such tests. Note that $u' = ay'$ and $v' = by'$ for $y = b'^{-1}p$, therefore seeding u', v', T to such algorithm outputs a, b, y'.   □

Small Factors Assumption: *For the modified NTRU parameters, it is unfeasible to solve the small factors problem.*

In the absence of a formal proof, let us motivate the hardness of the small factors problem. The SF problem is equivalent to solve a quadratic system of equations over $\mathbb{F}_q$ with additional restrictions on the unknowns. Indeed, each coefficient of $c(x)$ is a quadratic form on coefficients of $a(x), b(x)$:

$$c_k = \sum_{i=0}^{n-1} a_i b_{k-i \bmod n} \cdot \sigma_k(i) \quad \bmod q,$$

where $\sigma_k(i) = +1$ if $i \leq k$ and -1 otherwise, the unknowns a_i, b_j follow a Gaussian distribution about 0 and are bounded in magnitude by K. As $K \ll q$, one can consider the equations over the integers. This results in a Diophantine quadratic system of n equations in $2n$ variables. Quadratic systems of m equations with n unknowns can be the Achilles heel for strong cryptographic primitives, as they can be attacked in the very overdetermined ($m \geq n(n-1)/2$) or very underdetermined ($n \geq m(m+1)$) cases in fields with even characteristic. In [26], authors adapt an algorithm of Kipnis-Patarin-Goubin [21] to odd characteristic fields and show a gradual change between the determined case ($m = n$, $\exp(m)$ runtime) and the massively underdetermined case ($n \geq m(m + 1)$, $\mathsf{poly}(n)$ runtime). According to their analysis, our system ($n = 2m$) escapes the polynomial-time scope. Let us write the system in clear:

$$\forall i \in \{0, \ldots, n-1\}, \|a_i\|_\infty < K \text{ and } \|b_i\|_\infty < K,$$

$$\begin{cases} c_0 = a_0b_0 - a_1b_{n-1} - a_2b_{n-2} - \ldots - a_{n-1}b_1, \\ c_1 = a_0b_1 + a_1b_0 - a_2b_{n-1} - \ldots - a_{n-1}b_2, \\ c_2 = a_0b_2 + a_1b_1 + a_2b_0 - \ldots - a_{n-1}b_3, \\ \vdots \\ c_{n-1} = a_0b_{n-1} + a_1b_{n-2} + a_2b_{n-3} + \ldots + a_{n-1}b_0. \end{cases}$$

As this is an underdetermined system, the linearization $Z_{i,j} = a_i b_j$ results in a linear system with too many degrees of freedom to select the correct solution; this is not better than guessing in the initial quadratic system. On the other hand, this system presents cyclic anti-symmetry, which one could exploit to find a solution. However, it is not clear how to use the additional symmetry to make progress in finding solutions (this is also the case when trying to solve lattice problems in the particular case of ideal lattices).

From another point of view, we are given an element c in the intersection $\langle a \rangle \cap \langle b \rangle$ of ideals of $R = \mathbb{Z}[x]/(x^n + 1)$, and a test routine $T : R \to \{0,1\}$ that outputs 1 if the input is b and 0 otherwise (in our scenario, the test routine is to simply try out the extracted key $\beta = b$ via decryptions). An algorithmic issue arises again: There is a degree of freedom of one ring unit in the small factors problem, and an algorithm must exclude trivial factorizations of c: for instance, if nothing was required for a and b other than invertibility, the size of the candidates list for (a, b) is at least the number of units of R_q, since it contains all pairs $(a, b) = (cu, cu^{-1})$ for invertible $u \in R_q$. Using the K-boundedness of a, b, the list is to be reduced rejecting all incorrect pairs. To optimize up the rejection, we suggest a study of the distribution $\chi \stackrel{def}{=} (\bar{\mathcal{G}}_K)^{-1}$, which samples e according to $\bar{\mathcal{G}}_K$ and then outputs $e^{-1} \in R_q$ if e is invertible.

5   Two-Party Multiplication Protocols in R_q

In this section we introduce two protocols to jointly achieve multiplication in the quotient ring between two mutually distrusting parties. We distinguish two settings, the "secret inputs" (which is the classical MPC scenario) and the "shared inputs" which supposes that both parties have additive shares of some elements. The latter setting, however, can be regarded as a classical MPC computing a quadratic expression of the inputs.

5.1   Secret Inputs Setting

Alice and Bob hold $f \in R_q$ and $g \in R_q$ respectively. The following protocol allows them to multiply these elements: Alice will learn $fg + r \in R_q$ where r is a polynomial chosen by Bob. The reason of this is that if Alice learns fg, she can compute $g = fg/f$. The utility of this protocol may seem questionable, in the sense that it transfers Alice's obliviousness from g to r, nevertheless we will see that careful selection of r will allow the two parties to generate Alice's NTRU keys. This protocol is inspired on [3], where authors propose a protocol to compute scalar products as a building block to perform much more complex functionalities. It is detailed in Algorithm 1.

Algorithm 1. TMP

Require: Alice holds $f \in R_q$, Bob holds $g \in R_q$. Let p, m be public integers.
Ensure: Alice learns $fg + r \in R_q$ where Bob knows $r \in R_q$

1: Alice generates m random polynomials $\{f_1, \ldots, f_m\} \stackrel{R}{\subset} R_q$ such that $\sum_{i=1}^{m} f_i = f$.
2: Bob generates m random polynomials $\{r_1, \ldots, r_m\} \stackrel{R}{\subset} R_q$ and $r \stackrel{def}{=} \sum_{i=1}^{m} r_i$.
3: **for** $i = 1, \ldots, m$ **do**
4:     Alice generates a secret random number $k, 1 \leq k \leq p$.
5:     Alice generates random polynomials $v_1, \cdots, v_p$, sets $v_k = f_i$, and send all these polynomials to Bob.
6:     Bob computes the products and masks them: For all $j = 1, \ldots, p$ $z_{i,j} = v_j g + r_i$.
7:     Alice extracts $z_{i,k} = f_i g + r_i$ from Bob with a 1–out–of–p OT protocol.
8: **end for**
9: Alice computes $\sum_{i=1}^{m} z_{i,k} = fg + r$.

Note that throughout the protocol, Bob always computed products of random polynomials, and to guess the value of f he has to perform $\approx p^m$ additions.

Lemma 5.1.1. *If it is not feasible to compute $O(p^m)$ additions in R_q, and if the RLWE assumption holds for $q, \phi(x) = x^n + 1$ and uniform χ over R_q, TMP securely outputs $fg + r$ to Alice and r to Bob in the presence of semi-honest parties.*

Proof. In this model, both parties follow exactly the protocol but try to learn as much information as possible from their transcript of the protocol. Let $view_A$, $view_B$ be the collection of learned elements by Alice and Bob respectively. We have that $view_B$ contains only polynomials $v_j^{(i)}$ indistinguishable from uniform (since they were sampled by semi-honest Alice), and these elements are independent from Bob's input, samplings, and computations. Therefore, to learn f, he needs to perform $\approx p^m$ additions. On the other hand Alice wants to learn g or r and she only has m pairs of the form $(f_i, f_i g + r_i)$ (and the output which is the component-wise sum of these), which by the RLWE assumption are indistinguishable from (f_i, u_i) for uniform u_i. In other words, the view of each adversary contains her input, her output, and a list of polynomials indistinguishable from random by construction. We can construct simulators $\mathcal{S}_A, \mathcal{S}_B$ of protocol TMP for both parties, and it follows immediately that the views of Alice and Bob are indistinguishable from the simulators.   □

Remark. If both parties are malicious but they do not want to leak their own inputs, at the end of the protocol they learn nothing about the other party's input.

This holds because Alice may deviate from the samplings, but she sends pm random elements computationally hiding f, Bob will process these pm elements (deviating as much as he wants from the actual required computation) and send m elements computationally hiding g and r to Alice via the OT protocol, thus Bob learns nothing. In this case, deviations from the protocol may cause the output to be incorrect. We do not worry much about this as soon as Bob's input is safe, since we will see that it will result in invalid keys for Alice and the honest party will know that the other is malicious.

5.2   Shared Inputs Setting

In this setting, two parties share two elements of R_q additively, and they want to compute shares of the product of these elements. Let Alice and Bob hold x_A, y_A and x_B, y_B respectively such that

$$x = x_A + x_B \quad \text{and} \quad y = y_A + y_B.$$

We propose a protocol SharedTMP, at the end of which Alice and Bob will learn additive shares π_A, π_B respectively of the product:

$$\pi_A + \pi_B = xy \in R_q.$$

Algorithm 2. SharedTMP

Require: Alice holds $(x_A, y_A) \in R_q^2$, Bob holds $(x_B, y_B) \in R_q^2$ such that $x = x_A + x_B$, $y = y_A + y_B$

Ensure: Alice learns $\pi_A \in R_q$, Bob learns $\pi_B \in R_q$ such that $\pi_A + \pi_B = xy$

1: Alice samples $r_A \xleftarrow{R} R_q$, Bob samples $r_B \xleftarrow{R} R_q$
2: Alice and Bob perform $\mathsf{TMP}(x_A, y_B)$ using Bob's randomness r_B, thus Alice learns $u_A = x_A y_B + r_B$ and Bob learns nothing.
3: Bob and Alice perform $\mathsf{TMP}(x_B, y_A)$ using Alice's randomness r_A, thus Bob learns $u_B = x_B y_A + r_A$ and Alice learns nothing.
4: Alice computes the share $\pi_A = x_A y_A + u_A - r_A \in R_q$
5: Bob computes the share $\pi_B = x_B y_B + u_B - r_B \in R_q$

Note that $\pi_A + \pi_B = (x_A + x_B)(y_A + y_B) = xy$. Since they only communicate in steps 2 and 3, security is reduced to two independent instances of the TMP protocol. We also have the following observation:

Lemma 5.2.1. *Let Alice and Bob perform* SharedTMP *on some non-trivial inputs, learning at the end π_A and π_B respectively. Even if Alice reveals π_A to Bob, he cannot deduce Alice's inputs.*

Proof. This follows directly from the randomness of $u_A - r_A$.   □

6   Excalibur Key Generation

We present our main contribution, three protocols $\mathsf{Keygen}_{\mathsf{pk}}, \mathsf{Keygen}_{\mathsf{sk}}$ and a validation protocol, to be performed by Alice and Bob that will generate the public and the (blended) private-key of Alice, in that order. Let us first give an informal outline of the protocol. Bob has already generated his key-pair $(\beta, 2h\beta^{-1}) \in R_q \times R_q$. They want to compute a new key-pair $(\mathsf{sk}_A, \mathsf{pk}_A) = (\alpha\beta, 2g(\alpha\beta)^{-1}) \in R_q \times R_q$ for Alice, which correctly decrypts encryptions under pk_B since it contains the factor β.

– Excalibur generation of pk_A
   1. They share polynomials α, g, r of R_q additively, such that $\alpha = 1 \mod 2$.
   2. They perform SharedTMP to obtain shares of $\alpha r, gr$. Alice reveals her shares to Bob.
   3. Bob computes $2(gr) \cdot (\alpha r)^{-1} \cdot \beta^{-1} = 2g(\alpha\beta)^{-1}$ in R_q and broadcasts the result.
– Excalibur generation of sk_A (to be performed after publication of pk_A)
   1. Let $\alpha_A + \alpha_B = \alpha$ denote the same additive sharing of α than in the previous steps, where Alice holds α_A and Bob holds α_B. Alice and Bob perform TMP on entries α_A, β respectively, and Bob chooses $r = \alpha_B\beta$ as the randomness in the protocol.
   2. At the end of the protocol, Alice learns $\alpha_A\beta + r = \alpha\beta = \mathsf{sk}_A \in R_q$, and Bob learns nothing.

– Validation protocol: Alice and Bob run tests to be convinced that the keys are well formed and behave as claimed.

The protocols are described formally in Algorithms 3, 4 and 6.

Algorithm 3. Excalibur Keygen$_\mathsf{pk}$

Require: Bob already has his own key-pair $(\mathsf{sk}_B, \mathsf{pk}_B) = (\beta, 2h\beta^{-1}) \in R_q \times R_q$.
Ensure: A public-key for Alice pk_A
1: Alice and Bob sample random shares of elements in R_q:
   – Alice samples $s_A \leftarrow \bar{\mathcal{G}}_K, r_A \xleftarrow{R} R_q, g_A \leftarrow \bar{\mathcal{G}}_K$
   – Bob samples $s_B \leftarrow \bar{\mathcal{G}}_K, r_B \xleftarrow{R} R_q, g_B \leftarrow \bar{\mathcal{G}}_K$
   Let $\alpha = 2(s_A + s_B) + 1, r = r_A + r_B, g = g_A + g_B$ denote the shared elements.
2: Alice and Bob perform SharedTMP twice to obtain shares of $z = \alpha \cdot r$ and $w = g \cdot r$. Alice reveals her shares, thus Bob learns z, w.
3: Bob checks: If z is not invertible in R_q, restart the protocol.
4: Bob computes $2w(z\beta)^{-1} = 2g(\alpha\beta)^{-1}$ and publishes it as pk_A, along with a NIZK proof showing that z, w come from step 2 and that pk_A is well-formed.
5: Alice verifies Bob's proof. If it is not correct, abort the protocol.

If protocol Algorithm 3 was carried out properly, a ciphertext encrypted with pk_A is correctly decrypted by any secret-key having the factor $\alpha\beta$ and reasonable coefficient size. Remark that in step 2, Bob received the element $z = \alpha \cdot r$: this does not allow to deduce a functional equivalent of the secret-key $\alpha\beta$, since r has large coefficients. Also, chances are overwhelmingly high that this element is in fact invertible in view of Sect. 2.2.

Algorithm 4. Excalibur Keygen$_\mathsf{sk}$

Require: Bob's secret-key β and the same sharing of $\alpha = 2(s_A + s_B) + 1$ than in protocol 3.
Ensure: A secret-key for Alice $\mathsf{sk}_A = \alpha\beta$
1: Bob computes $r := (2s_B + 1)\beta \in R_q$
2: Alice and Bob perform the protocol TMP$(2s_A, \beta)$, and Bob uses r as the random polynomial. At the end Alice knows $2s_A\beta + r = \alpha\beta \in R_q$.

Once the keys are generated, they must pass a series of decryption and a well-formedness test. This is described in Algorithms 5 and 6. First, Alice checks if her new secret-key works as expected, and then she convinces Bob, via a game of decryptions that she is indeed capable of decrypting ciphertexts encrypted under pk_A and underpk_B. As we will see, this validation protocol avoids malicious activity.

Algorithm 5. Validation function (performed by Alice)

1: **function** VALIDATE($\mathsf{sk}_A, \mathsf{pk}_A, \mathsf{pk}_B$)
2:    **for** i from 1 to k **do**
3:       $\mu \xleftarrow{R} \{0,1\}$
4:       $\mu_1 \leftarrow \mathsf{Dec}(\mathsf{sk}_A, \mathsf{Enc}(\mathsf{pk}_A, \mu))$
5:       $\mu_2 \leftarrow \mathsf{Dec}(\mathsf{sk}_A, \mathsf{Enc}(\mathsf{pk}_B, \mu))$
6:       **if** $\mu_1 \neq \mu$ or $\mu_2 \neq \mu$ **then** output **reject**
7:       **end if**
8:    **end for**
9:    **if** $||\mathsf{sk}_A||_\infty > n(2K+1)^2$ or $||\mathsf{sk}_A \cdot \mathsf{pk}_B||_\infty > 2(2K+1)$ **then** output **size warning**
10:    **end if**
11:    output **accept**
12: **end function**

Algorithm 6. Validation protocol (performed by Alice and Bob)

Require: Alice holds $(\mathsf{sk}_A, \mathsf{pk}_A)$ and Bob holds $(\mathsf{sk}_B, \mathsf{pk}_B)$
1: Alice runs VALIDATE($\mathsf{sk}_A, \mathsf{pk}_A, \mathsf{pk}_B$). If the output is **reject**, abort.
2: Bob picks $2k$ random messages $(m_1^{(A)}, \ldots, m_k^{(A)})$ and $(m_1^{(B)}, \ldots, m_k^{(B)})$, and for each $i = 1, \ldots, k$ he computes ciphertexts $c_i^{(A)} = \mathsf{Enc}(\mathsf{pk}_A, m_i^{(A)}), c_i^{(B)} = \mathsf{Enc}(\mathsf{pk}_B, m_i^{(B)})$. He send all ciphertexts to Alice.
3: For each $i = 1, \ldots, k$, Alice compute $\mu_i^{(A)} = \mathsf{Dec}(\mathsf{sk}_A, c_i^{(A)}), \mu_i^{(B)} = \mathsf{Dec}(\mathsf{sk}_A, c_i^{(B)})$. She sends all plaintexts to Bob.
4: For each $i = 1, \ldots, k$, Bob checks if $\mu_i^{(A)} = m_i^{(A)}$ and $\mu_i^{(B)} = m_i^{(B)}$.

---

7   Security

We first discuss the honest-but-curious model, where the protocol is strictly followed but parties try to learn secrets. Then we look at the malicious model, where one party does not follow the protocol properly, in order to steal secrets or to sabotage the key generation.

7.1   Honest-But-Curious Model

In this model, we suppose that Alice and Bob follow exactly the instructions in Algorithms 3, 4, 5 and 6 but they try to learn about each other's secret with all collected information.

Proposition. *If Alice is able to extract Bob's key from the protocol, she can solve the Small Factors Problem or the $\bar{\mathcal{G}}_K^\times$-GCD Problem. If Bob is able to extract Alice's key, he can solve the $\bar{\mathcal{G}}_K^\times$-GCD Problem.*

Proof. Let us focus first in Bob's chances on learning α (or a functional equivalent of the form $\theta \cdot \alpha$ for small $\theta \in R_q$). Recall that

$$\begin{cases} \mathsf{pk}_B = 2h\beta^{-1}, \\ \mathsf{pk}_A = 2g(\alpha\beta)^{-1}, \\ z = \alpha \cdot r, \\ w = g \cdot r, \\ \alpha = 2(s_A + s_B) + 1. \end{cases}$$

Let us focus on *Bob's view of the protocol:*

$$V = \{(s_B, r_B, g_B, z_B, w_B, z_A, w_A, \beta, \mathsf{pk}_B), \mathsf{pk}_A, \alpha \cdot r, g \cdot r\} \subset R_q.$$

What Bob is curious about: Any element of the set

$$U = \{\alpha, g, r, s_A, g_A, r_A\} \subset R_q.$$

The parentheses in V indicate that he sampled or received the elements contained, and the rest are results of joint computation. The knowledge of any element in U allows Bob to deduce Alice's secret-key α, and only the last three elements $\mathsf{pk}_A, \alpha \cdot r, g \cdot r$ of V depend on elements in U. Thus, extracting α is equivalent to solve the following system of equations in the unknowns $(X, Y, Z) = (\alpha, r, g)$:

$$\begin{cases} b_1 = XY, \\ b_2 = ZY, \\ b_3 = ZX^{-1}, \end{cases}$$

where $b_1 = \alpha \cdot r, b_2 = g \cdot r, b_3 = \beta \mathsf{pk}_A/2$. We can eliminate the third equation noting that $b_1 b_3 = b_2$, and thus Bob faces the Small GCD Problem of Sect. 4.2.

Let us now focus in Alice chances of learning Bob's secret.

Alice's view of the protocol: $W = \{(s_A, r_A, g_A, z_A, w_A), \mathsf{pk}_B, \mathsf{pk}_A, \alpha\beta\} \subset R_q.$
What Alice is curious about: Any element of the set
$$Q = \{\alpha, \beta, h, s_B, \{w_B, z_B\}, \{w, z\}\}.$$

First, extracting α or β directly from $\alpha \cdot \beta$ is exactly the small factors Problem. Using the only three sensitive elements of W, she faces the following system of equations in $(X, Y, Z) = (h, \beta, \alpha)$

$$\begin{cases} a_1 = XY^{-1}, \\ a_2 = (ZY)^{-1}, \\ a_3 = ZY, \end{cases}$$

where $a_1 = \mathsf{pk}_B/2, a_2 = \mathsf{pk}_A/2g, a_3 = \alpha\beta$. After elimination of the third equation since $a_3 = a_2^{-1}$, Alice also faces the small GCD problem (actually, mapping $Y \mapsto Y^{-1}, Z \mapsto Z^{-1}$ yields to the same gcd problem faced by Bob). $\square$

7.2   Security Against One Malicious Party

We consider the presence of one malicious adversary, a party that deviates as much as she wants from the protocol, but has a list of paramount objectives which she is not willing to sacrifice. We suppose that one of the two parties strictly follows the protocol and the other one is malicious, given the objectives below. We consider the presence of only one somewhat malicious adversary, given that both parties have concurrent objectives (for instance, Bob is trying to protect his key, and Alice to extract it from the protocols). In other words,

What curious Alice wants:

(A1) A functional secret-key sk_A associated with pk_A,
(A2) such that sk_A decrypts encryptions under pk_B,
(A3) protecting elements of $U = \{g_A, r_A, s_A\}$ from Bob and
(A4) to learn β.

What curious Bob wants:

(B1) To give Alice a functional secret-key sk_A associated with pk_A with decryption rights on $\mathsf{Enc}(\mathsf{pk}_B, \mathcal{M})$.
(B2) to protect elements of $Q = \{\beta, h, s_B, \{w_B, z_B\}\}$ from Alice,
(B3) (if malicious) overloading Alice's secret-key sk_A to have large coefficients, and
(B4) to learn α.

We will show that either the keys will be correctly generated or one party will not fulfill all of her objectives.

Malicious Alice, Semi-honest Bob. Suppose that Bob is strictly following the protocol and Alice may deviate from the protocol but wants to fulfill (A1) to (A4). Let us summarize Alice's participation in the key generation:

1. *Samples $s_A \leftarrow \bar{\mathcal{G}}_K, r_A \xleftarrow{R} R_q, g_A \leftarrow \bar{\mathcal{G}}_K$.*
   Trivial samplings of these elements may ultimately leak α to Bob. For instance, if $s_A = 0$, $\alpha = 2s_B + 1$, if $r_B = 0$, $z/r_B = \alpha$, if $g_A = 0$ $g = w/g_B$. Also, if s_A or g_A have large coefficients, there is risk of mod q wrap-around in the decryption procedure with sk_A. As she is sampling only shares of elements, she cannot force algebraic relations with them: regardless of her samples, α, g, r will remain indistinguishable from random.

2. *Participates in $\mathsf{SharedTMP}((2s_A, r_A), (2s_B + 1, r_B))$ and learns z_A, participates in $\mathsf{SharedTMP}((g_A, r_A), (g_B, r_B))$ and learns w_A, then sends z_A, w_A to Bob.*
   As discussed in Sect. 5, TMP and SharedTMP are secure if Bob is honest, in the sense that either Alice learns the correct output, or either she learns indistinguishable from random elements, but she learns nothing about Bob's input. She is limited to alter the inputs of both instances of SharedTMP and then giving wrong z_A or w_A to Bob. Nevertheless if she inputs different r_A's

in both protocols or if she changes the values of z_A, w_A before sending them to Bob, from the linearity of shares and the randomness of Bob's entries it follows that this sabotages the relation $wz^{-1} = g\alpha^{-1}$, needed for correctness of decryption. In other words, in order to ensure (A1) and (A2), she is forced to maintain the input r_A for both instances of SharedTMPand send the correct output to Bob.

3. *Participates in* $\mathsf{TMP}(\{2s_A\}, \{2s_B + 1\})$ *and learns* $\alpha\beta$.
   If she uses the correct value of $2s_A$ (i.e. the same as in step 2), she learns the correct output $\alpha\beta$. If she inputs another value $x \neq 2s_A$, she does learn a functional equivalent of Bob secret (namely, $(x + 2s_B + 1)\beta$), but she is not able to decrypt encryptions under the already published pk_A, failing the verification procedure.

Malicious Bob, Semi-honest Alice. Now suppose the inverse case, where Alice follows the protocol strictly and Bob is protecting β and guessing α, deviating as much as he wants from the protocol but fulfilling (B1) to (B4). We begin by saying that (B3) is unavoidable (unless the presence of a zero-knowledge proof that Bob's polynomials are of the right size), but Alice can tell if Bob overloaded the secret-key $\alpha\beta$ simply looking at the coefficients. Let us now summarize Bob's participation in the key generation:

1. *Samples* $s_B \leftarrow \bar{\mathcal{G}}_K, r_B \xleftarrow{R} R_q, g_B \leftarrow \bar{\mathcal{G}}_K$.
   Trivial sampling may compromise sensible elements as before. He must ensure the randomness of α, r if he wants to protect these elements, and on the other hand Alice will know if he deviates from a K-bounded sampling (just looking at the coefficients in $\alpha \cdot \beta$. Therefore, he gains nothing in deviating from a K-bounded sampling.

2. *Participates in* $\mathsf{SharedTMP}((2s_A, r_A), (2s_B + 1, r_B))$ *and learns* z_B, *participates in* $\mathsf{SharedTMP}((g_A, r_A), (g_B, r_B))$ *and learns* w_B.
   As noted in Sect. 5, because of Alice's randomness in SharedTMP, either Bob obeys the protocol an receive the correct outputs, either he deviates and receives random outputs, from which he cannot deduce secret values and which sabotage key generation. Also, if he uses different r_B's in both instances, Alice will not be able to decrypt since the decryption relation $wz^{-1} = g\alpha^{-1}$ is not fulfilled (and he remains oblivious of r_A, not being able to force this relation). Hence, he is forced to follow SharedTMP and use the same r_B in both instances if he wants to fulfill (B1).

3. *Receives* z_A, w_A, *learning* z, w. *Checks if* z *is invertible and publishes* $\mathsf{pk}_A = 2w(z\beta)^{-1}$. *He then participates in* $\mathsf{TMP}(\{2s_A\}, \{2s_B + 1\})$, *chooses* $R = (2s_B + 1)\beta$ *and learns nothing.*
   Suppose that he published pk'_A as Alice's public-key and participated in the TMP instance with generic values, indicated by an apostrophe. At the end, Alice knows pk'_A and sk'_A. She will run the validate function of Algorithm 5 to check (i) if sk'_A has the expected coefficient size, (ii) if $\mathsf{sk}'_A \cdot \mathsf{pk}'_A = 2g'$ for a vector $g' \in \chi$ and (iii) if she is able to decrypt encryptions of messages under

pk'_A and pk_B. If she is indeed able to decrypt encryptions under pk_B, then sk'_A contains the factor β, thus by randomness of s_A, $\beta' = \theta\beta$ and $R' = \omega\beta$ for small θ and ω. Also, as long as the polynomials sk'_A and $\mathsf{sk}'_A \cdot \mathsf{pk}'_A$ are of the right form, she does not care about how Bob computed pk'_A, as decryption of encryptions under pk'_A work as claimed. If on the contrary a single decryption fails or if sk'_A or $\mathsf{sk}'_A \cdot \mathsf{pk}'_A$ have large coefficients, she can claim one of the following Bob's wrongdoings: Either he did not include β, either he included $\theta\beta$ for too large θ, either he sabotaged entirely the key generation in a change of input or inside a multiparty multiplication protocol. This allows to conclude that if Bob fails to give what is expected, the output keys will be rejected by Alice, who discovers Bob's maliciousness after the validation protocol Algorithm 6.

We should point out another strategy that Bob could maliciously try. When generating Alice's secret-key, he could simply ignore Alice's input share $2s_A$, and thus the protocol gives Alice the key $\mathsf{sk}'_A = \alpha'\beta$, for an $\alpha' \in R_q^\times$ of Bob's choice. Bob, who received no output from the protocol, can reconstruct this key, thus gaining Alice's secret. However, this key will be rejected by Alice since it cannot be associated with the previously generated $\mathsf{pk}_A = 2g(\alpha\beta)^{-1}$. To avoid this rejection, Bob should have published $\mathsf{pk}'_A = 2g'(\alpha'\beta)^{-1}$ instead, but it is easy to see that this publication would contradict the NIZK proof of step 4 of Algorithm 3: Because of the way $\mathsf{SharedTMP}$ works, Bob has no way of choosing α' of his choice in the expression $z = \alpha r$. In view of this, passing the validation protocol with such a key is overwhelmingly unlikely.

8 Extensions

8.1 Chains of Keys

Suppose Alice and Bob perform the latter protocols, such that Alice has now a private-key of the form $\mathsf{sk}_A = \alpha\beta$ where β is Bob's secret-key. Alice can repeat the protocol with a third user Charlie (with slight coefficients size modifications at the validation protocol), who at the end receives a pair of keys of the form $(\mathsf{sk}_C, \mathsf{pk}_C) = (\alpha\beta\gamma, 2g_C(\alpha\beta\gamma)^{-1})$. As his secret-key contains the factors β and $\alpha\beta$, he can decrypt both Bob's and Alice's ciphertexts. This shows that easy modifications to the protocol allows to generate a chain of users, each one inheriting the previous user decryption rights. From Corollary 3.1.2, it is easy to see that the length of such a chain is at most $\approx \log(q/nK)$ to ensure decryptions (this matches the maximum number of keys on the multikey LATV FHE scheme for the same parameters). We point out that intersecting chains are also possible, meaning that a user can glue her secret-keys to two or more upper-level users and even if they collude they are not able to extract his key. This comes from an easy generalization of our $\bar{\mathcal{G}}_K^\times$–GCD problem.

8.2 Plugging in LATV-FHE

Because of the form of an Excalibur key, i.e. $(\mathsf{sk}, \mathsf{pk}) = (\prod_{i=1}^{r} \alpha_i, 2g \prod_{i=1}^{r} \alpha_i^{-1})$, the inclusion of our protocols into the Multikey FHE scheme from [23] is immediate. The only missing element are the evaluation keys, which can be generated easily by the secret-key holder after the (Excalibur) key generation: they are "pseudo-encryptions" of the secret-key sk under the public-key pk. This achieves a somewhat homomorphic encryption scheme in the chain of users, where in addition they can combine ciphertexts generated by any public-key.

9 Conclusion

In this article, we proposed a new protocol to generate NTRU keys with additional decryption rights, allowing to form a hierarchic chain of users. We motivated such a procedure because it avoids betrayal naturally, and since it applies to the FHE-NTRU scheme, it may contribute to clear the bootstrapping-like re-encryption paradigm, since it is to our knowledge the first FHE scenario featuring (the pure definition of) proxy re-encryption. In this light, it concurs with other proxy re-encryption schemes, as, while being rigid, ciphertext transformation is no necessary at all, since decryption rights are defined in key-generation time. We used two-party computation protocols as building blocks, and relied the semantic security on the well-known RLWE and DSPR assumptions, and security in presence of semi-honest parties on a hardness assumption in cyclotomic polynomial rings.

Acknowledgments. We would like to thank Pablo Schinke Gross for suggesting the term Excalibur and the INDOCRYPT 2016 anonymous reviewers for their helpful comments. This work has been supported in part by the FUI CRYPTOCOMP project.

References

1. Alperin-Sheriff, J., Peikert, C.: Practical bootstrapping in quasilinear time. In: Canetti, R., Garay, J.A. (eds.) CRYPTO 2013. LNCS, vol. 8042, pp. 1–20. Springer, Heidelberg (2013). doi:10.1007/978-3-642-40041-4_1
2. Alperin-Sheriff, J., Peikert, C.: Faster bootstrapping with polynomial error. In: Garay, J.A., Gennaro, R. (eds.) CRYPTO 2014. LNCS, vol. 8616, pp. 297–314. Springer, Heidelberg (2014). doi:10.1007/978-3-662-44371-2_17
3. Atallah, M.J., Du, W.: Secure multi-party computational geometry. In: Dehne, F., Sack, J.-R., Tamassia, R. (eds.) WADS 2001. LNCS, vol. 2125, pp. 165–179. Springer, Heidelberg (2001). doi:10.1007/3-540-44634-6_16
4. Boneh, D., Franklin, M.: An efficient public key traitor tracing scheme. In: Wiener, M. (ed.) CRYPTO 1999. LNCS, vol. 1666, pp. 338–353. Springer, Heidelberg (1999). doi:10.1007/3-540-48405-1_22
5. Bos, J.W., Lauter, K., Loftus, J., Naehrig, M.: Improved security for a ring-based fully homomorphic encryption scheme. In: Stam, M. (ed.) IMACC 2013. LNCS, vol. 8308, pp. 45–64. Springer, Heidelberg (2013). doi:10.1007/978-3-642-45239-0_4

6. Brakerski, Z.: Fully homomorphic encryption without modulus switching from classical GapSVP. In: Safavi-Naini, R., Canetti, R. (eds.) CRYPTO 2012. LNCS, vol. 7417, pp. 868–886. Springer, Heidelberg (2012). doi:10.1007/978-3-642-32009-5_50

7. Brakerski, Z., Gentry, C., Halevi, S.: Packed ciphertexts in LWE-based homomorphic encryption. In: Kurosawa, K., Hanaoka, G. (eds.) PKC 2013. LNCS, vol. 7778, pp. 1–13. Springer, Heidelberg (2013). doi:10.1007/978-3-642-36362-7_1

8. Brakerski, Z., Vaikuntanathan, V.: Efficient fully homomorphic encryption from (standard) LWE. In: Proceedings of the 2011 IEEE 52nd Annual Symposium on Foundations of Computer Science. FOCS 2011, pp. 97–106. IEEE Computer Society, Washington, DC (2011)

9. Brakerski, Z., Vaikuntanathan, V.: Fully homomorphic encryption from ring-LWE and security for key dependent messages. In: Rogaway, P. (ed.) CRYPTO 2011. LNCS, vol. 6841, pp. 505–524. Springer, Heidelberg (2011). doi:10.1007/978-3-642-22792-9_29

10. Gentry, C.: Computing on encrypted data. In: Garay, J.A., Miyaji, A., Otsuka, A. (eds.) CANS 2009. LNCS, vol. 5888, p. 477. Springer, Heidelberg (2009). doi:10.1007/978-3-642-10433-6_32

11. Gentry, C.: A fully homomorphic encryption scheme. Ph.D. thesis, Stanford, CA, USA, aAI3382729 (2009)

12. Gentry, C.: Fully homomorphic encryption using ideal lattices. In: Proceedings of the Forty-First Annual ACM Symposium on Theory of Computing. STOC 2009, pp. 169–178. ACM, New York (2009). http://doi.acm.org/10.1145/1536414.1536440

13. Gentry, C., Halevi, S.: Fully homomorphic encryption without squashing using depth-3 arithmetic circuits. In: Proceedings of the 2011 IEEE 52nd Annual Symposium on Foundations of Computer Science, pp. 107–109. FOCS 2011. IEEE Computer Society, Washington, DC (2011)

14. Gentry, C., Halevi, S.: Implementing gentry's fully-homomorphic encryption scheme. In: Paterson, K.G. (ed.) EUROCRYPT 2011. LNCS, vol. 6632, pp. 129–148. Springer, Heidelberg (2011). doi:10.1007/978-3-642-20465-4_9

15. Gentry, C., Halevi, S., Peikert, C., Smart, N.P.: Ring switching in BGV-style homomorphic encryption. In: Visconti, I., Prisco, R. (eds.) SCN 2012. LNCS, vol. 7485, pp. 19–37. Springer, Heidelberg (2012). doi:10.1007/978-3-642-32928-9_2

16. Gentry, C., Halevi, S., Smart, N.P.: Better bootstrapping in fully homomorphic encryption. In: Fischlin, M., Buchmann, J., Manulis, M. (eds.) PKC 2012. LNCS, vol. 7293, pp. 1–16. Springer, Heidelberg (2012). doi:10.1007/978-3-642-30057-8_1

17. Gentry, C., Halevi, S., Smart, N.P.: Fully homomorphic encryption with polylog overhead. In: Pointcheval, D., Johansson, T. (eds.) EUROCRYPT 2012. LNCS, vol. 7237, pp. 465–482. Springer, Heidelberg (2012). doi:10.1007/978-3-642-29011-4_28

18. Gentry, C., Halevi, S., Smart, N.P.: Homomorphic evaluation of the AES circuit. In: Safavi-Naini, R., Canetti, R. (eds.) CRYPTO 2012. LNCS, vol. 7417, pp. 850–867. Springer, Heidelberg (2012). doi:10.1007/978-3-642-32009-5_49

19. Gentry, C., Sahai, A., Waters, B.: Homomorphic encryption from learning with errors: conceptually-simpler, asymptotically-faster, attribute-based. In: Canetti, R., Garay, J.A. (eds.) CRYPTO 2013. LNCS, vol. 8042, pp. 75–92. Springer, Heidelberg (2013). doi:10.1007/978-3-642-40041-4_5

20. Halevi, S., Shoup, V.: Algorithms in HElib. In: Garay, J.A., Gennaro, R. (eds.) CRYPTO 2014. LNCS, vol. 8616, pp. 554–571. Springer, Heidelberg (2014). doi:10.1007/978-3-662-44371-2_31

21. Kipnis, A., Patarin, J., Goubin, L.: Unbalanced oil and vinegar signature schemes. In: Stern, J. (ed.) EUROCRYPT 1999. LNCS, vol. 1592, pp. 206–222. Springer, Heidelberg (1999). doi:10.1007/3-540-48910-X_15

22. Lauter, K., Lopez-Alt, A., Naehrig, M.: Private computation on encrypted genomic data. Techical report (2014). http://research.microsoft.com/apps/pubs/default. aspx?id=219979

23. López-Alt, A., Tromer, E., Vaikuntanathan, V.: On-the-fly multiparty computation on the cloud via multikey fully homomorphic encryption. In: STOC, pp. 1219–1234 (2012)

24. Lyubashevsky, V., Peikert, C., Regev, O.: On ideal lattices and learning with errors over rings. In: Gilbert, H. (ed.) EUROCRYPT 2010. LNCS, vol. 6110, pp. 1–23. Springer, Heidelberg (2010). doi:10.1007/978-3-642-13190-5_1

25. Stehlé, D., Steinfeld, R.: Making NTRU as secure as worst-case problems over ideal lattices. In: Paterson, K.G. (ed.) EUROCRYPT 2011. LNCS, vol. 6632, pp. 27–47. Springer, Heidelberg (2011). doi:10.1007/978-3-642-20465-4_4

26. Thomae, E., Wolf, C.: Solving underdetermined systems of multivariate quadratic equations revisited. In: Fischlin, M., Buchmann, J., Manulis, M. (eds.) PKC 2012. LNCS, vol. 7293, pp. 156–171. Springer, Heidelberg (2012)

27. Yasuda, M., Shimoyama, T., Kogure, J., Yokoyama, K., Koshiba, T.: Secure pattern matching using somewhat homomorphic encryption. In: Proceedings of the 2013 ACM Workshop on Cloud Computing Security Workshop. CCSW 2013, pp. 65–76. ACM, New York (2013). http://doi.acm.org/10.1145/2517488.2517497

Approximate-Deterministic Public Key Encryption from Hard Learning Problems

Yamin Liu[1,2], Xianhui Lu[1,2,3(✉)], Bao Li[1,2,3], Wenpan Jing[1,2],
and Fuyang Fang[1,2,3]

[1] Data Assurance and Communication Security Research Center,
Chinese Academy of Sciences, Beijing, China
[2] State Key Laboratory of Information Security,
Institute of Information Engineering, Chinese Academy of Sciences, Beijing, China
{liuyamin,luxianhui,libao,jingwenpan,fangfuyang}@iie.ac.cn
[3] University of Chinese Academy of Sciences, Beijing, China

Abstract. We introduce the notion of approximate-deterministic public key encryption (A-DPKE), which extends the notion of deterministic public key encryption (DPKE) by allowing the encryption algorithm to be "slightly" randomized. However, a ciphertext convergence property is required for A-DPKE such that the ciphertexts of a message are gathering in a small metric space, while ciphertexts of different messages can be distinguished easily. Thus, A-DPKE maintains the convenience of DPKE in fast search and de-duplication on encrypted data, and encompasses new constructions. We present two simple constructions of A-DPKE, respectively from the learning parity with noise and the learning with errors assumptions.

Keywords: Deterministic public key encryption · Learning parity with noise · Learning with errors

1 Introduction

Deterministic Public Key Encryption. The provable security of deterministic public key encryption (DPKE) was initiated by Bellare, Boldyreva and O'Neill in 2007 [4]. Different from the widely accepted notion of probabilistic encryption [19], the encryption algorithm of DPKE does not require a fresh randomness; consequently, given a plaintext, its ciphertext is unique. Hence DPKE can serve as a candidate for efficiently searchable encryption, and supports de-duplication over encrypted data.

Though DPKE can not satisfy most security requirements of randomized public key encryption due to the deterministic encryption algorithm, Bellare, Boldyreva and O'Neill defined an as-strong-as-possible security notion for DPKE, called PRIV, over plaintext distributions with high min-entropy independent of the public key. More security definitions and constructions of DPKE were discussed in [5,6,13,17,33,35]. Currently DPKE can be instantiated from various intractability assumptions, including lattice-related ones such as the learning with errors assumption (LWE).

O. Dunkelman and S.K. Sanadhya (Eds.): INDOCRYPT 2016, LNCS 10095, pp. 25–42, 2016.
DOI: 10.1007/978-3-319-49890-4_2

Hard Learning Problems. Generally, hard learning problems, such as LWE and LPN (learning parity with noise), refer to learning a secret from several noisy linear equations. LWE was introduced by Regev in [32]. It states that recovering a secret $\mathbf{s}$ giving $(\mathbf{A}, \mathbf{b} = \mathbf{As} + \mathbf{e})$ is intractable, wherein $\mathbf{A} \in \mathbb{Z}_q^{m \times n}$ and $\mathbf{s} \in \mathbb{Z}_q^n$ are chosen at random, and $\mathbf{e}$ is picked from an error distribution, for appropriate secret dimension n, number of samples m, and modulus q. The decisional version of LWE states that $(\mathbf{A}, \mathbf{b} = \mathbf{As} + \mathbf{e})$ is computationally indistinguishable from the uniform pair $(\mathbf{A}, \mathbf{u})$, and is equivalent to the search version. Syntactically, LPN is LWE in the case of $q = 2$ with the errors being picked from the Bernoulli distribution, however, LPN and LWE are different in many aspects.

The hardness of LWE is guaranteed by worst-case hard problems over lattices, as shown in a series of literatures [11,29,32], and LPN is essentially the hard problem of decoding a random binary linear code. In addition, both are believed to be intractable even for quantum algorithms. Thus, it is desirable to instantiate cryptographic primitives from them.

LWE is useful in the construction of various public key cryptographic primitives, such as chosen-ciphertext secure encryption [31], identity-based encryption [20], password-based authenticated key exchange [24], and in our interest, DPKE [35]. The low-noise version of LPN has also been used to construct secure public key encryption [1,16,21,23,34]. And recently, Yu and Zhang showed how to obtain several public key cryptographic primitives from constant-noise LPN [38], such as CPA/CCA secure encryption and oblivious transfer.

However, the syntax of hard learning problems seems incompatible with the definition of DPKE since it involves a randomly sampled error item, which is important to the intractability of the problems while causes a kind of nondeterminacy. Hence, current constructions of DPKE from LWE either take a detour from lossy trapdoor functions (LTDF) [6,35], or use a deterministic variant of LWE called learning with rounding (LWR) [12]. The construction of LTDF from LWE are somewhat complicated [31]. The LWR-based constructions of DPKE [3,37] are very simple, but to ensure the hardness of LWR, the modulus q should be large enough [3,7]. Besides, as far as we know, currently there is no construction of DPKE or even LTDF from LPN, and it is believed that there is no "rounding version" of LPN [2].

Nevertheless, we try to address the problem in another way. Remember that using the LWE assumption to instantiate another important cryptographic primitive, the smooth projective hashing (SPH) [14], is also an open problem as stated in [30]. In 2009, Katz and Vaikuntanathan defined a variant of SPH called **approximate smooth projective hashing**, instantiated it with LWE, and obtained the first password-based authenticated key exchange protocol from a lattice-related assumption [24]. Thus, we believe that a similar solution should work for the case of DPKE.

1.1   Our Contributions

Approximate-DPKE. We extend the definition of DPKE to allow some sort of nondeterminacy while maintaining its advantages, by introducing the notion of **approximate-deterministic public key encryption** (A-DPKE). Compared with DPKE, in A-DPKE the encryption algorithm is "slightly" randomized, thus there will be many ciphertexts corresponding to one message. However, these ciphertexts are not scattered in the ciphertext space, instead they are gathering in a small metric space. Moreover, ciphertexts of different plaintexts are distributed "far enough" that they will not mix up. We call the property *ciphertext convergence*, and with it A-DPKE preserves the advantages of DPKE in encrypted search and de-duplication, since the ciphertexts of a given message can be easily recognized without decryption.

A-DPKE can achieve the same security level of DPKE, namely, the PRIV-series of security definitions. However, though the encryption algorithm of A-DPKE is randomized, it cannot be as secure as probabilistic encryption due to the ciphertext convergence property: encryptions of the same message can be easily recognized while encryptions of different messages can be easily distinguished. It is a tradeoff between security and functionality just as in the case of DPKE.

Then we can bring out simple and natural instantiations of A-DPKE from hard learning problems.

A-DPKE from LPN. To the best of our knowledge, there is no constructions of DPKE from the LPN assumption (neither low-noise nor constant-noise) so far. However, by relaxing DPKE to A-DPKE, immediately we obtain a simple construction of A-DPKE from low-noise LPN, using the trapdoor generation techniques as in [23]. The secret key is a matrix $\mathbf{T} \in \mathbb{Z}_2^{m \times m}$, and the public key is a pair of matrices $(\mathbf{A}, \mathbf{B} = \mathbf{TA}) \in (\mathbb{Z}_2^{m \times n})^2$. To encrypt a message $\mathbf{m} \in \{0,1\}^n$, two ciphertext components are computed as $\mathbf{c}_1 = \mathbf{Am} + \mathbf{e}$, $\mathbf{c}_2 = (\mathbf{B} + \mathbf{G})\mathbf{m} + \bar{\mathbf{T}}\mathbf{e}$, where $\mathbf{e}$, $\bar{\mathbf{T}}$ are small errors, and $\mathbf{G}$ is the generator matrix of an efficiently decodeable binary linear code. We can see that though the encryption is randomized, the Hamming distance of two ciphertexts of a message could be small if the error items are small. By choosing proper parameters, both the ciphertext convergence property and the security will hold.

A-DPKE from LWE. Further, we show a natural A-DPKE construction from LWE. The public key is simply a matrix $\mathbf{A} \in \mathbb{Z}_q^{m \times n}$ generated with the trapdoor generation techniques from [28], and the secret key is the corresponding trapdoor $\mathbf{R}$. Then the encryption and decryption are simply the evaluation and inversion of the LWE function. Note that a message $\mathbf{m} \in \{0,1\}^n$ is encrypted as $\mathbf{c} = \mathbf{Am} + \mathbf{e}$. By choosing the size of the error item $\mathbf{e}$ properly, the ciphertext convergence property will hold. And the security is ensured by the hardness of LWE for high min-entropy secrets [3,18]. Compared with the LWE-based DPKE via LTDF, our A-DPKE is simpler in structure; and compared with the LWR-based DPKE scheme, our LWE-based A-DPKE scheme can use smaller modulus q.

Organization. The rest of the paper is organized as follows. In Sect. 2 some notations and preliminaries about lattice is introduced. In Sect. 3 the definition of A-DPKE is given. In Sect. 4 and Sect. 5 the A-DPKE schemes from LPN and LWE are constructed respectively. Finally, Sect. 6 is the conclusion.

2   Preliminaries

2.1   Notations

We use bold lower-case characters to denote vectors, such as $\mathbf{x}$, and use bold upper-case letters to denote matrices, such as $\mathbf{X}$. If X is a set, then $x \xleftarrow{\$} X$ denotes that x is chosen from X uniformly at random. If X is a distribution, then $x \xleftarrow{\$} X$ denotes that x is randomly sampled according to X.

For a randomized algorithm A, $x \xleftarrow{\$} \mathsf{A}(\cdot)$ denotes that x is assigned the output of A. An algorithm is efficient if it runs in polynomial time in its input length. A function $f(\lambda)$ is negligible if it decreases faster than any polynomial of the security parameter λ, and is denoted as $f(\lambda) \leq \mathsf{negl}(\lambda)$.

The min-entropy of a random variable X is denoted as $H_\infty(X) = -\log\left(\max\limits_{x} P_X(x)\right)$, wherein $P_X(x) = \Pr[X = x]$. X is called a k-source if $H_\infty(X) \geq k$.

The statistical distance between two random variables X and Y is $\Delta(X, Y) = \frac{1}{2} \sum\limits_{x} |P_X(x) - P_Y(x)|$, X and Y are statistically close if $\Delta(X, Y) \leq \epsilon(\lambda)$, and is denoted as $X \overset{s}{\approx} Y$. X and Y are computationally indistinguishable if no efficient algorithm can tell them apart given only oracle access, and is denoted as $X \overset{c}{\approx} Y$.

2.2   Lattices

A full-rank m-dimensional lattice Λ generated by a basis $\mathbf{B} = \{\mathbf{b}_1, ..., \mathbf{b}_m\} \in \mathbb{Z}^{m \times m}$ is defined as

$$\Lambda = \mathcal{L}(\mathbf{B}) = \{\mathbf{B}\mathbf{x} : \mathbf{x} \in \mathbb{Z}^m\},$$

where $\mathbf{b}_1, ..., \mathbf{b}_m$ are linearly independent.

The length of lattice vectors is measured with norms. By default the Euclidean norm is used, i.e., $\|\mathbf{x}\|_2 = \sqrt{\sum x_i^2}$, or solely denoted as $\|\mathbf{x}\|$. In some occasions in this work, the infinity norm is also used, i.e., $\|\mathbf{x}\|_\infty = \max x_i$. Obviously, for an m-dimensional vector x, if $\|\mathbf{x}\|_\infty \leq a$, then $\|\mathbf{x}\|_2 \leq \sqrt{m}a$; and if $\|\mathbf{x}\|_\infty \geq a$, then $\|\mathbf{x}\|_2 \geq a$.

The length of the shortest nonzero vector in a lattice Λ is denoted by $\lambda_1(\Lambda)$. Since lattice points are periodically arranged in every dimension, then $\lambda_1(\Lambda)$ is the distance of two lattice points in the most "compact" dimension.

The LWE problem is essentially the bounded-distance decoding problem over a full-rank m-dimensional q-ary integer lattice $\Lambda_q(\mathbf{A})$ generated by a random matrix $\mathbf{A} \in \mathbb{Z}_q^{m \times n}$:

$$\Lambda_q(\mathbf{A}) = \{y \in \mathbb{Z}^m : \exists \mathbf{s} \in \mathbb{Z}_q^n \text{ s.t. } y = \mathbf{A}\mathbf{s} \mod q\}.$$

3   Approximate-DPKE: Definition and Security

Here we define approximate-DPKE. Compared with the original DPKE definition, the main difference is that the encryption algorithm of A-DPKE is randomized. And compared with the definition of randomized PKE, A-DPKE has the additional property of ciphertext convergence, i.e., the ciphertexts of a message are distributed in a small metric space.

Definition 1. *An approximate-deterministic public key encryption scheme* A-DPKE = (KG, ENC, DEC) *consists of the following three algorithms:*

- *The probabilistic key generation algorithm:* $(pk, sk) \xleftarrow{\$} \mathsf{KG}(1^\lambda)$.
- *The probabilistic encryption algorithm:* $c \xleftarrow{\$} \mathsf{ENC}(pk, m; r)$.
- *The deterministic decryption algorithm:* $m \leftarrow \mathsf{DEC}(sk, c)$.

And we further require that the encryption scheme should satisfy a "ciphertext convergence" property, i.e., there are a function dis *measuring the "distance" of ciphertexts, and a distance parameter t, fulfilling the following requirements:*

- *For arbitrary two ciphertexts c_1, c_2 of a given plaintext m, there is* $\mathsf{dis}(c_1, c_2) \leq t$.
- *For arbitrary two ciphertext-plaintext pairs (c, m) and (c', m'), there is* $\mathsf{dis}(c, c') > t$ *if* $m \neq m'$.

In the definition we explicitly contain the randomness r in the encryption algorithm. In the following we sometimes omit it in occasions that the choice of randomness is unimportant and just use $\mathsf{Enc}(pk, m)$.

The correctness requirement of A-DPKE involves two aspect. One is trivially the decryption correctness, i.e., there should be $\mathsf{DEC}(sk, \mathsf{ENC}(pk, m; r)) = m$. The other is the ciphertext convergence property, wherein the choices of the metric function dis and parameter t depend on specific instantiations.

The definition of A-DPKE is a generalization of that of DPKE. Consider the metric function dis to be the Hamming distance, e.g., the numbers of bit-wise differences between two ciphertexts, and set the parameter $t = 0$, then a DPKE certainly satisfies the ciphertext convergence property.

As to the security requirement, it is clear that A-DPKE can achieve existing security requirements of DPKE, e.g., the PRIV security. The question is whether it can be semantically secure [19]. The answer is NO, but the consequence is not necessarily negative. On one side, though the encryption algorithm of A-DPKE is randomized, it still can not achieve semantic security due to the ciphertext convergence property. On the other side, with this property, A-DPKE preserves the advantages of DPKE in searchable encryption and de-duplication, since the ciphertexts of a certain message can be efficiently recognized without decryption, given dis and t.

In the following we give the definition of PRIV-IND security [5,6] for A-DPKE, which requires that the encryptions of messages from two different high min-entropy distributions are indistinguishable.

Definition 2 (PRIV-IND security for A-DPKE). *An approximate-deterministic public-key encryption scheme* Π = (KG, Enc, Dec) *is PRIV-IND secure if for any probabilistic polynomial time adversary* A, *for any efficiently sampleable distributions* $\{M_\lambda^0\}_{\lambda \in \mathbb{N}}$ *and* $\{M_\lambda^1\}_{\lambda \in \mathbb{N}}$ *with sufficient min-entropy*

$H_\infty(M_\lambda^0) \geq k$ and $H_\infty(M_\lambda^1) \geq k$, there is $(pk, \mathsf{Enc}(pk, \mathbf{m}_0)) \overset{c}{\approx} (pk, \mathsf{Enc}(pk, \mathbf{m}_1))$, where $(pk, sk) \overset{\$}{\leftarrow} \mathsf{Gen}(1^\lambda)$, $\mathbf{m}_0 \overset{\$}{\leftarrow} M_\lambda^0$ and $\mathbf{m}_1 \overset{\$}{\leftarrow} M_\lambda^1$.

In [37] Xie et al. defined a PRIV-INDr security for DPKE, which requires that the encryption is indistinguishable from uniform. It is clear that PRIV-INDr implies PRIV-IND. We also define the PRIV-INDr security for A-DPKE.

Definition 3 (PRIV-INDr security for A-DPKE). *An approximate-deterministic public-key encryption scheme* $\Pi = (\mathsf{KG}, \mathsf{Enc}, \mathsf{Dec})$ *is PRIV-INDr secure if for any probabilistic polynomial time adversary* A, *for any efficiently sampleable distributions* $\{M_\lambda\}_{\lambda \in \mathbb{N}}$ *with sufficient min-entropy* $H_\infty(M_\lambda) \geq k$, *there is* $(pk, \mathsf{Enc}(pk, \mathbf{m})) \overset{c}{\approx} (pk, \mathbf{u})$, *where* $(pk, sk) \overset{\$}{\leftarrow} \mathsf{Gen}(1^\lambda)$, $\mathbf{m} \overset{\$}{\leftarrow} M_\lambda$ *and* $\mathbf{u} \overset{\$}{\leftarrow} C_\lambda$, *where* C_λ *is the ciphertext space.*

Note that other forms of security definitions for DPKE can also be extended to the A-DPKE case naturally, such as PRIV with respect to hard-to-invert auxiliary information [13].

In the definition of PRIV security, the message blocks $\mathbf{m}_0$ and $\mathbf{m}_1$ contain several (possibly correlated) messages. If the block size is restricted to be one, then the security is called PRIV1 [4–6]. Full PRIV security in the standard model is considered to be elusive [36], and currently the only known approach to achieve it is the one proposed by Bellare and Hoang in [8], with the help of a newly introduced strong assumption UCE (universal computational extractor) [9]. Thus, in this work, we are satisfied with just the PRIV1 security.

4   A-DPKE from LPN

So far, there is no known constructions of DPKE from the learning parity with noise assumption. Now we propose an A-DPKE scheme under the LPN assumption, which depicts that the relaxation from deterministic to approximate-deterministic is worthwhile.

4.1   Coding Theory

In the LPN based A-DPKE construction, we will use a linear code as a building block. Thus some preliminaries about the coding theory are recalled below. The notations and definitions mainly come from [26] by Meurer.

For $x \in [0, 1]$, the q-ary entropy function is defined as $H_q(x) = x \log_q(q - 1) - x \log_q x - (1 - x) \log_q(1 - x)$. In particular, when $q = 2$, $H(x) = x \log x - (1 - x) \log(1 - x)$.

Definition 4 (Linear Code). *A linear code* $\mathcal{C}$ *in a finite field* $\mathbb{Z}_q$ *is a linear subspace of the linear space* $\mathbb{Z}_q^m$. *If the dimension of* $\mathcal{C}$ *is* n, *then* $\mathcal{C}$ *is called an* $[m, n]$-code. *The ratio* $R = \frac{n}{m}$ *is called the information rate of* $\mathcal{C}$.

In this work, we use linear codes in $\mathbb{Z}_2^m$. Given a generator matrix $\mathbf{A} \in \mathbb{Z}_2^{m \times n}$, a code $\mathcal{C}(\mathbf{A}) = \{\mathbf{c} = \mathbf{As} : \mathbf{s} \in \mathbb{Z}_2^n\}$ is specified. An important parameter of a code $\mathcal{C}$ is the minimum distance $d(\mathcal{C})$, which is the minimum Hamming distance between two distinct codewords, i.e., $d(\mathcal{C}) = \min_{\mathbf{c}_1 \neq \mathbf{c}_2 \in \mathcal{C}} |\mathbf{c}_1 - \mathbf{c}_2| = \min_{\mathbf{c} \in \mathcal{C} \setminus \{0\}} |\mathbf{c}|$. With the relative Gilbert-Varshamov distance, a lower bound of $d(\mathcal{C})$ can be estimated.

Definition 5 (Relative Gilbert-Varshamov distance). *Let $0 < R < 1$. The relative Gilbert-Varshamov distance $\mathsf{D}_{\mathsf{GV}}(R) \in \mathbb{R}$ is the unique solution in $0 \leq x \leq 1 - \frac{1}{q}$ of the equation $H_q(x) = 1 - R$.*

The following lemma from [26] shows the lower bound of a linear code $\mathcal{C}$.

Lemma 1. *Almost all linear codes meet the relative Gilbert-Varshamov distance, i.e., for almost all linear codes $\mathcal{C}$ of rate R it holds $d(\mathcal{C}) \geq \lfloor \mathsf{D}_{\mathsf{GV}}(R)m \rfloor$.*

4.2 The LPN Assumption

Then we recall the LPN assumption, wherein the error item is sampled from the Bernoulli distribution $\mathcal{B}_\rho$ with $0 < \rho < 1/2$, i.e., $\Pr[x = 1 : x \xleftarrow{\$} \mathcal{B}_\rho] = \rho$.

Definition 6 (Learning Parity with Noise). *Let λ be the security parameter, $n = n(\lambda), m = m(\lambda)$ be integers, and $\rho \in (0, 1/2)$ be the Bernoulli parameter. The $\mathsf{LPN}_{n,m,\rho}$ assumption states that, if we choose $\mathbf{A} \xleftarrow{\$} \mathbb{Z}_2^{m \times n}, \mathbf{s} \xleftarrow{\$} \mathbb{Z}_2^n, \mathbf{e} \leftarrow \mathcal{B}_\rho^m, \mathbf{u} \xleftarrow{\$} \mathbb{Z}_2^m$, then the following distributions are computationally indistinguishable:*

$$(\mathbf{A}, \mathbf{As} + \mathbf{e}) \overset{c}{\approx} (\mathbf{A}, \mathbf{u}).$$

In standard LPN, the Bernoulli parameter ρ is a constant such as $1/10$. However, for the purpose of constructing PKE schemes, we mainly use a low-noise variant of LPN contributed by Alekhnovich [1], in which $\rho = \Theta(1/\sqrt{n})$.

And we still need another variant of LPN called Knapsack LPN (KLPN), which is defined below.

Definition 7 (Knapsack LPN). *Let λ be the security parameter, $n = n(\lambda), m = m(\lambda)$ be integers, and $\rho \in (0, 1/2)$ be the Bernoulli parameter. The $\mathsf{KLPN}_{n,m,\rho}^m$ assumption states that, if we choose $\mathbf{A} \xleftarrow{\$} \mathbb{Z}_2^{m \times (m-n)}, \mathbf{T} \xleftarrow{\$} \mathcal{B}_\rho^{m \times m}, \mathbf{u} \xleftarrow{\$} \mathbb{Z}_2^{m \times (m-n)}$, then the following distributions are computationally indistinguishable:*

$$(\mathbf{A}, \mathbf{TA}) \overset{c}{\approx} (\mathbf{A}, \mathbf{u}).$$

The equivalence of LPN and KLPN assumptions was stated in [23,27] with the following lemma:

Lemma 2 [23]. *For all algorithms B there exists an algorithm A that runs in roughly the same time as B and $\mathsf{Adv}_{\mathsf{LPN}_{n,m,\rho}}(\mathsf{A}) \geq \frac{1}{m}\mathsf{Adv}_{\mathsf{KLPN}_{n,m,\rho}^m}(\mathsf{B})$.*

4.3 Construction 1: A-DPKE from LPN

Now we will describe the A-DPKE construction from LPN. Firstly we set the parameters used in the construction. Some choices of parameters are similar to those in [23].

- λ is the security parameter, $n(\lambda), m(\lambda)$ are integers, where $n = \Theta(\lambda^2)$ and $m \geq 2n$. Besides, $m > 1400$. For reasonable choices of the security parameter, say $\lambda \geq 80$, $m > 1400$ can be trivially met.
- $0 < c < 1/4$ is a constant. And the Bernoulli parameter is $\rho = \sqrt{c/m}$. $\beta = 2\sqrt{cm}$ is a parameter used in the correctness proof of the construction.
- $\mathbf{G} \in \mathbb{Z}_2^{m \times n}$ is the generator-matrix of a binary linear error-correcting code $\mathcal{C} : \mathbb{Z}_2^n \to \mathbb{Z}_2^m$ with an efficient decoding algorithm $\mathsf{Decode_G}$ which corrects up to αm errors with $4c < \alpha \leq 0.05$.

Now the A-DPKE based on LPN is given below:

- $\mathsf{KG}(1^\lambda)$: Choose $\mathbf{T} \xleftarrow{\$} \mathcal{B}_\rho^{m \times m}$, $\mathbf{A} \xleftarrow{\$} \mathbb{Z}_2^{m \times n}$, and compute $\mathbf{B} = \mathbf{TA}$. Set $sk = \mathbf{T}$ and $pk = (\mathbf{A}, \mathbf{B})$.
- $\mathsf{Enc}(pk, \mathbf{m}; (\mathbf{e}, \bar{\mathbf{T}}))$: To encrypt a message $\mathbf{m} \in \{0, 1\}^n$, choose $\mathbf{e} \xleftarrow{\$} \mathcal{B}_\rho^m$, $\bar{\mathbf{T}} \xleftarrow{\$} \mathcal{B}_\rho^{m \times m}$, and compute

$$\mathbf{c}_1 = \mathbf{Am} + \mathbf{e}, \mathbf{c}_2 = (\mathbf{B} + \mathbf{G})\mathbf{m} + \bar{\mathbf{T}}\mathbf{e},$$

  where $\mathbf{G}$ is the generator matrix of a binary linear code defined above as part of the public parameter. Finally, set the ciphertext $\mathbf{C} = (\mathbf{c}_1, \mathbf{c}_2)$.
- $\mathsf{Dec}(sk, \mathbf{C})$: Parse $\mathbf{C}$ as $(\mathbf{c}_1, \mathbf{c}_2)$. Compute $\mathbf{y} = \mathbf{c}_2 - \mathbf{Tc}_1$ and set $\mathbf{m} = \mathsf{Decode_G}(\mathbf{y})$.

4.4 Correctness

To establish the correctness of **Construction 1** over the specified parameter settings, two lemmata from literatures are required. The first one is the Chernoff Bound for bounding the Hamming weight of a vector constituted by independent Bernoulli random variables, e.g., the weight of the error item in the first component of the ciphertext.

Lemma 3 (Chernoff Bound). *For* $\mathbf{d} \xleftarrow{\$} \mathcal{B}_\rho^m$ *and* $\delta > 0$,

$$\Pr[|\mathbf{d}| > (1 + \delta)\rho m] < e^{-\frac{\min(\delta, \delta^2)}{3}\rho m}.$$

In our case, $\delta = 1$. The other lemma is essentially from [23], bounding the Hamming weight of the error item in the second component of the ciphertext.

Lemma 4 [23]. *For* $\mathbf{e} \xleftarrow{\$} \mathcal{B}_\rho^m$ *with* $|\mathbf{e}| \leq 2\rho m$, $\mathbf{T} \xleftarrow{\$} \mathcal{B}_\rho^{m \times m}$, *and* $4c < \alpha < 1$, *there is*

$$\Pr_{\mathbf{T}}[|\mathbf{Te}| > \frac{\alpha}{2}m] < \mathsf{negl}(\lambda).$$

Then we can establish the correctness of **Construction 1** as an A-DPKE.

Theorem 1. *Let λ be the security parameter, $n, m, c, \rho, \beta, \alpha$ and the error correcting code $\mathbf{G}$ be defined above. Choose the distance function* dis *such that* $\mathsf{dis}(\mathbf{C}_1, \mathbf{C}_2) = |\mathbf{C}_1 - \mathbf{C}_2| = (|\mathbf{c}_{1,1} - \mathbf{c}_{1,2}|, |\mathbf{c}_{2,1} - \mathbf{c}_{2,2}|)$ *denotes the Hamming distance of two ciphertexts* $\mathbf{C}_1 = (\mathbf{c}_{1,1}, \mathbf{c}_{1,2}), \mathbf{C}_2 = (\mathbf{c}_{2,1}, \mathbf{c}_{2,2})$, *and set the parameter* $t = (t_1, t_2) = (2\beta, \alpha m)$. *Then the above construction is a correct A-DPKE scheme.*

Proof. The correctness includes the decryption correctness and the ciphertext convergence.

- Decryption correctness.
  Given a ciphertext $\mathbf{C} = (\mathbf{c}_1, \mathbf{c}_2)$, the decryption algorithm computes

$$y = \mathbf{c}_2 - \mathbf{T}\mathbf{c}_1 = \mathbf{G}m + (\bar{\mathbf{T}} - \mathbf{T})e.$$

To ensure that $\mathsf{Decode}_\mathbf{G}(y)$ correctly recovers m, the Hamming weight of the error term $(\bar{\mathbf{T}} - \mathbf{T})e$ should be small, i.e., $|(\bar{\mathbf{T}} - \mathbf{T})e| \leq \alpha m$.
With the parameters $\rho = \sqrt{c/m}, \beta = 2\sqrt{cm} = 2\rho m$, and the Chernoff Bound for $\delta = 1$, the Hamming weight of e is bounded by β with overwhelming probability. That is,

$$\Pr_{e \xleftarrow{\$} B_\rho^m} [|e| > \beta] \leq e^{-\rho m/3} = 2^{-\Theta(\sqrt{m})}.$$

Then with the triangular inequality and Lemma 4 from [23], there is

$$|(\bar{\mathbf{T}} - \mathbf{T})e| \leq |\bar{\mathbf{T}}e| + |\mathbf{T}e| \leq \alpha m,$$

with overwhelming probability. Thus, $\mathsf{Decode}_\mathbf{G}(y)$ will recover m with overwhelming probability.
- Ciphertext convergence.
  - Given a message m, and its arbitrary two ciphertexts:

$$\mathbf{C}_1 = (\mathbf{c}_{1,1} = \mathbf{A}m + e_1, \mathbf{c}_{1,2} = (\mathbf{B} + \mathbf{G})m + \bar{\mathbf{T}}_1 e_1),$$

$$\mathbf{C}_2 = (\mathbf{c}_{2,1} = \mathbf{A}m + e_2, \mathbf{c}_{2,2} = (\mathbf{B} + \mathbf{G})m + \bar{\mathbf{T}}_2 e_2).$$

According to Lemmas 3 and 4, there is

$$\begin{aligned} \mathsf{dist}(\mathbf{C}_1, \mathbf{C}_2) &= (|\mathbf{c}_{1,1} - \mathbf{c}_{1,2}|, |\mathbf{c}_{2,1} - \mathbf{c}_{2,2}|) \\ &= (|e_1 - e_2|, |\bar{\mathbf{T}}_1 e_1 - \bar{\mathbf{T}}_2 e_2|) \\ &\leq (2\beta, \alpha m), \end{aligned}$$

with overwhelming probability, i.e., the ciphertexts of the same message are close in Hamming distance.

- Given two different messages $\mathbf{m}$ and $\mathbf{m}'$, and two ciphertexts of them:

$$\mathbf{C} = (\mathbf{c}_1 = \mathbf{Am} + \mathbf{e}, \mathbf{c}_2 = (\mathbf{B} + \mathbf{G})\mathbf{m} + \bar{\mathbf{T}}\mathbf{e}),$$

$$\mathbf{C}' = (\mathbf{c}_1' = \mathbf{Am}' + \mathbf{e}', \mathbf{c}_2' = (\mathbf{B} + \mathbf{G})\mathbf{m}' + \bar{\mathbf{T}}'\mathbf{e}'),$$

there is $\text{dist}(\mathbf{C}, \mathbf{C}') = (|\mathbf{c}_1 - \mathbf{c}_1'|, |\mathbf{c}_2 - \mathbf{c}_2'|)$.

Let us analyze the two components separately. Consider $\mathbf{A}$ as the generator matrix of a linear code $\mathcal{C}(\mathbf{A})$, then $|\mathbf{A}(\mathbf{m} - \mathbf{m}')|$ is not less than the minimum distance of $\mathcal{C}(\mathbf{A})$. With the triangular inequality and Lemma 1, there is

$$
\begin{aligned}
|\mathbf{c}_1 - \mathbf{c}_1'| &= |\mathbf{A}(\mathbf{m} - \mathbf{m}') + (\mathbf{e} - \mathbf{e}')| \\
&\geq |\mathbf{A}(\mathbf{m} - \mathbf{m}')| - |\mathbf{e} - \mathbf{e}'| \\
&\geq d(\mathcal{C}(\mathbf{A})) - 2\beta \\
&\geq \lfloor D_{\mathsf{GV}}(\frac{n}{m})m \rfloor - 2\beta \\
&\geq D_{\mathsf{GV}}(\frac{1}{2})m - 1 - 2\beta.
\end{aligned}
$$

By a routine calculation based on Definition 5 there is $D_{\mathsf{GV}}(\frac{1}{2}) > 0.11$. Since $m > 1400$ and $c < 0.25$, that is, $c < 0.000179m$, then $2\beta = 4\sqrt{cm} < 0.0536m$. Hence there is

$$
\begin{aligned}
|\mathbf{c}_1 - \mathbf{c}_1'| &\geq 0.11m - 1 - 0.0536m \\
&= 0.0564m - 1 \\
&> 2\beta = t_1.
\end{aligned}
$$

Similarly, view $\mathbf{U} = \mathbf{B} + \mathbf{G}$ as the generator matrix of a linear code $\mathcal{C}(\mathbf{U})$, then there is

$$
\begin{aligned}
|\mathbf{c}_2 - \mathbf{c}_2'| &= |\mathbf{U}(\mathbf{m} - \mathbf{m}') + (\bar{\mathbf{T}}\mathbf{e} - \bar{\mathbf{T}}'\mathbf{e}')| \\
&\geq |\mathbf{U}(\mathbf{m} - \mathbf{m}')| - |\bar{\mathbf{T}}\mathbf{e} - \bar{\mathbf{T}}'\mathbf{e}'| \\
&\geq d(\mathcal{C}(\mathbf{U})) - \alpha m \\
&\geq \lfloor D_{\mathsf{GV}}(\frac{n}{m})m \rfloor - \alpha m \\
&\geq D_{\mathsf{GV}}(\frac{1}{2})m - 1 - \alpha m \\
&> 0.11m - 1 - 0.05m \\
&= 0.06m - 1 \\
&> 0.05m \geq \alpha m = t_2.
\end{aligned}
$$

Hence it holds that $\text{dis}(\mathbf{C}, \mathbf{C}') > (2\beta, \alpha m)$, i.e., ciphertexts of different messages are far enough in Hamming distance. □

4.5   Security

Now we can show the PRIV1-INDr security of **Construction 1**.

Theorem 2. *Let λ be the security parameter, $n, m, c, \rho, \beta, \alpha$ and the error correcting code $\mathbf{G}$ be defined above. If the LPN assumption holds, then the above construction is PRIV1-INDr secure for uniformly distributed messages.*

Proof. Since for $(pk, sk) \xleftarrow{\$} \mathsf{KG}(1^\lambda)$, $\mathbf{m} \xleftarrow{\$} \{0,1\}^n$, $\mathbf{e} \xleftarrow{\$} \mathcal{B}_\rho^m$, $\bar{\mathbf{T}} \xleftarrow{\$} \mathcal{B}_\rho^{m \times m}$, there is

$$(pk, \mathsf{Enc}(pk, \mathbf{m})) = ((\mathbf{A}, \mathbf{B}), (\mathbf{Am} + \mathbf{e}, (\mathbf{B} + \mathbf{G})\mathbf{m} + \bar{\mathbf{T}}\mathbf{e})) \tag{1}$$

$$\stackrel{c}{\approx} ((\mathbf{A}, \mathbf{B}'), (\mathbf{Am} + \mathbf{e}, (\mathbf{B}' + \mathbf{G})\mathbf{m} + \bar{\mathbf{T}}\mathbf{e})) \tag{2}$$

$$\stackrel{c}{\approx} ((\mathbf{A}, \mathbf{B}'), (\mathbf{Am} + \mathbf{e}, \mathbf{Um} + \bar{\mathbf{T}}\mathbf{e})) \tag{3}$$

$$\stackrel{c}{\approx} ((\mathbf{A}, \mathbf{B}'), (\mathbf{u}_1, \mathbf{u}_2)), \tag{4}$$

where $\mathbf{B}', \mathbf{U} \xleftarrow{\$} \mathbb{Z}_2^{m \times n}, \mathbf{u}_1, \mathbf{u}_2 \xleftarrow{\$} \mathbb{Z}_2^m$. Step 1 is straightforward. Step 2 follows from the KLPN assumption. Step 3 is also natural since $\mathbf{B}'$ is uniform. And step 4 follows from the LPN assumption. $\square$

5   A-DPKE from LWE

5.1   The LWE Assumption

Next we will show a natural construction of A-DPKE from the learning with errors assumption. Firstly we recall the definition of (decisional) LWE.

Definition 8 (Learning with Errors [32]). *Let λ be the security parameter, $n = n(\lambda), m = m(\lambda), q = q(\lambda)$ be integers, and $\chi = \chi(\lambda)$ be a distribution over $\mathbb{Z}_q$. The $\mathsf{LWE}_{n,m,q,\chi}$ assumption states that, if we choose $\mathbf{A} \xleftarrow{\$} \mathbb{Z}_q^{m \times n}, \mathbf{s} \xleftarrow{\$} \mathbb{Z}_q^n, \mathbf{e} \leftarrow \chi^m, \mathbf{u} \xleftarrow{\$} \mathbb{Z}_q^m$, then the following distributions are computationally indistinguishable:*

$$(\mathbf{A}, \mathbf{As} + \mathbf{e}) \stackrel{c}{\approx} (\mathbf{A}, \mathbf{u}).$$

Typically, the error distribution is the discrete Gaussian distribution over $\mathbb{Z}_q$ with appropriate variance, or the uniform distribution over a small interval [15].

The equivalent computational version of LWE states that getting the secret $\mathbf{s}$ from $(\mathbf{A}, \mathbf{b} = \mathbf{As} + \mathbf{e})$ is hard. However, with the trapdoor generation technique from [28], the secret $\mathbf{s}$ can be efficiently recovered.

Lemma 5 [28]. *There is an efficient randomized algorithm $\mathsf{GenTrap}(1^n, 1^m, q)$ that, given any integers $n \geq 1, q \geq 2$, and sufficiently large $m = O(n \log q)$, outputs a parity-check matrix $\mathbf{A} \in \mathbb{Z}_q^{m \times n}$ and a 'trapdoor' $\mathbf{R}$ such that the distribution of $\mathbf{A}$ is $\mathsf{negl}(n)$-far from uniform. Moreover, there are an efficient algorithm Invert that with overwhelming probability over all random choices, does the following:*

- *For* $\mathbf{b} = \mathbf{A}\mathbf{s} + \mathbf{e}$, *where* $\mathbf{s} \in \mathbb{Z}_q^n$ *is arbitrary and either* $\|\mathbf{e}\| < q/O(\sqrt{n \log q})$ *or* $\mathbf{e} \leftarrow D_{\mathbb{Z}^m, \alpha q}$ *for* $1/\alpha \geq \sqrt{n \log q} \cdot \omega(\sqrt{\log n})$, *the deterministic algorithm* $\mathsf{Invert}(\mathbf{R}, \mathbf{A}, \mathbf{b})$ *outputs* $\mathbf{s}$ *and* $\mathbf{e}$.

Goldwasser et al. proved that LWE is hard even for non-uniform secret $\mathbf{s}$ with hard-to invert auxiliary information $f(\mathbf{s})$, provided that $\mathbf{s}$ has high min-entropy and the modulus q is super-polynomial [18]. The size of the modulus q was improved to be polynomial by Alwen et al. in [3] (in the appendix of its full version) with the following definition and lemma.

Definition 9 (LWE with Weak and Leaky Secrets [3]). *Let* λ *be the security parameter,* $n = n(\lambda), m = m(\lambda), q = q(\lambda)$ *be integer parameters, and* χ *be a distribution over* $\mathbb{Z}_q$. *Let* $\gamma = \gamma(\lambda) \in (0, q/2)$ *be an integer and* $k = k(\lambda)$ *be a real. The* $\mathrm{LWE}_{n,m,q,\chi}^{\mathrm{WL}(\gamma,k)}$ *problem says that for any efficiently samplable correlated random variables* $(\mathbf{s}, \mathsf{aux})$, *where the support of* $\mathbf{s}$ *is the integer interval* $[-\gamma, \gamma]^n$ *and* $H_\infty(\mathbf{s}|\mathsf{aux}) \geq k$, *the following distributions are computationally indistinguishable:*

$$(\mathsf{aux}, \mathbf{A}, \mathbf{A}\mathbf{s} + \mathbf{e}) \overset{c}{\approx} (\mathsf{aux}, \mathbf{A}, \mathbf{u}),$$

where $\mathbf{A} \overset{\$}{\leftarrow} \mathbb{Z}_q^{m \times n}, \mathbf{u} \overset{\$}{\leftarrow} \mathbb{Z}_q^m, \mathbf{e} \overset{\$}{\leftarrow} \chi^m$ *are chosen randomly and independently of* $(\mathbf{s}, \mathsf{aux})$.

The lemma below states that the hardness of LWE for weak and leaky sources follows from that of the standard LWE.

Lemma 6 [3]. *Let* $k, l, m, n, \beta, \gamma, \sigma, q$ *be integer parameters and* χ *a distribution (all parameterized by* λ) *such that* $\Pr_{x \overset{\$}{\leftarrow} \chi} [|x| \geq \beta] \leq \mathsf{negl}(\lambda)$ *and* $\sigma \geq \beta \gamma n m$. *Let* Ψ_σ *be either:*

- *The discrete Gaussian distribution with standard deviation* σ, *or*
- *The uniform distribution over the integer interval* $[-\sigma, \sigma]$.

Assuming that the $\mathrm{LWE}_{l,m,q,\chi}$ *assumption holds, the weak and leaky* $\mathrm{LWE}_{n,m,q,\Psi_\sigma}^{\mathrm{WL}(\gamma,k)}$ *-assumption holds if* $k \geq (l + \Omega(\lambda)) \log q$.

In our construction, we choose $\gamma = 1$, and use binary secrets $\mathbf{s} \in \{0, 1\}^n$.

5.2   A-DPKE from LWE

5.3   Construction 2: A-DPKE from LWE

The A-DPKE construction from LWE is shown below. Firstly we list the parameter settings:

- λ is the security parameter, and $n(\lambda), m(\lambda), q(\lambda)$ are integers, with $m \geq 2n \log q$. For simplicity, we let q be prime.

- Let Ψ_σ be a suitable error distribution with $\beta nm < \sigma < \min(\frac{q}{16}, \frac{q}{O(\sqrt{n \log q})})$, where β is the parameter for another error distribution χ_β with which the LWE assumption holds.

Then the encryption and decryption of A-DPKE are simply the evaluation and inversion of the LWE function, using the trapdoor generation technique in Lemma 5.

- KG(1^λ). Run $(\mathbf{A}, \mathbf{R}) \xleftarrow{\$} \mathsf{GenTrap}(1^n, 1^m, q)$. Set $pk = \mathbf{A}$ and $sk = \mathbf{R}$.
- Enc(pk, $\mathbf{m}$; $\mathbf{e}$). To encrypt a message $\mathbf{m} \in \{0, 1\}^n$, compute $\mathbf{c} = \mathbf{Am} + \mathbf{e}$, where $\mathbf{e} \in \mathbb{Z}^m$ is randomly sampled according to the distribution Ψ_σ.
- Dec(sk, $\mathbf{c}$). Run $(\mathbf{m}, \mathbf{e}) \leftarrow \mathsf{Invert}(\mathbf{R}, \mathbf{A}, \mathbf{c})$, and output $\mathbf{m}$.

5.4   Correctness

Intuitively, the encryption algorithm of **Construction 2** encodes a message $\mathbf{m}$ to a point near the lattice point $\mathbf{As}$ in the q-ary lattice $\Lambda_q(\mathbf{A})$, and the offset is the error size. Thus, to prove ciphertext convergence property, we need the following lemma bounding the length of the shortest nonzero vector of $\Lambda_q(\mathbf{A})$, in the form of infinity norm.

Lemma 7 [20]. *Let n and q be positive integers with q prime, and let $m \geq 2n \log q$. Then for all but at most q^{-n} fraction of $\mathbf{A} \in \mathbb{Z}_q^{m \times n}$, we have $\lambda_1^\infty(\Lambda_q(\mathbf{A})) \geq q/4$.*

An immediate corollary explains the bound in the form of Euclidean norm.

Corollary 1. *Let n and q be positive integers with q prime, and let $m \geq 2n \log q$. Then for all but at most q^{-n} fraction of $\mathbf{A} \in \mathbb{Z}_q^{m \times n}$, we have $\lambda_1^2(\Lambda_q(\mathbf{A})) \geq q/4$.*

Now we can show the correctness of **Construction 2**.

Theorem 3. *Let λ be the security parameter, $n = n(\lambda), m = m(\lambda), q = q(\lambda)$ be integers, with $m \geq 2n \log q$ and q being prime. Let Ψ_σ be the error distribution with $\beta nm < \sigma < \min(\frac{q}{16}, \frac{q}{O(\sqrt{n \log q})})$, thus $\|\mathbf{e}\| \leq \sigma$. Choose the distance function dis such that $\mathsf{dis}(\mathbf{c}_1, \mathbf{c}_2) = \|\mathbf{c}_1 - \mathbf{c}_2\|$ denotes the distance of the two vectors $\mathbf{c}_1, \mathbf{c}_2 \in \mathbb{Z}_q^m$, and set the parameter $t = 2\sigma$. Then the above construction is a correct A-DPKE scheme.*

Proof.

- The decryption correctness follows from Lemma 5.
- Ciphertext convergence.
  - Given a message $\mathbf{m}$, and its arbitrary two ciphertexts $\mathbf{c}_1 = \mathbf{Am} + \mathbf{e}_1, \mathbf{c}_2 = \mathbf{Am} + \mathbf{e}_2$, then there is

$$\begin{aligned} \mathsf{dis}(\mathbf{c}_1, \mathbf{c}_2) &= \|\mathbf{c}_1 - \mathbf{c}_2\| \\ &= \|\mathbf{e}_1 - \mathbf{e}_2\| \\ &\leq \|\mathbf{e}_1\| + \|\mathbf{e}_2\| \\ &\leq 2\sigma. \end{aligned}$$

It means that the ciphertexts of the same message are close in Euclidean distance.

- Given two different messages $\mathbf{m}, \mathbf{m}'$, and two ciphertexts of them, $\mathbf{c} = \mathbf{Am} + \mathbf{e}, \mathbf{c}' = \mathbf{Am}' + \mathbf{e}'$. With Lemma 7 and Corollary 1 there is

$$
\begin{aligned}
\mathsf{dis}(\mathbf{c}, \mathbf{c}') &= \|\mathbf{c} - \mathbf{c}'\| \\
&= \|(\mathbf{Am} - \mathbf{Am}') + (\mathbf{e} - \mathbf{e}')\| \\
&\geq \lambda_1^2(\Lambda_q(\mathbf{A})) - 2\sigma \\
&> q/4 - q/8 \\
&= 8/q > 2\sigma.
\end{aligned}
$$

It means that the ciphertexts of different messages are far enough in Euclidean distance. □

5.5   Security

Now we show the PRIV1-IND security of **Construction 2**.

Theorem 4. *Let λ be the security parameter, $n = n(\lambda) \geq \lambda, l = l(\lambda), m = m(\lambda), q = q(\lambda)$ be integers, and χ be an efficiently sampleable distribution such that $\Pr_{x \overset{\$}{\leftarrow} \chi} [|x| \geq \beta] \leq \mathsf{negl}(\lambda)$ and $\sigma \geq \beta nm$. Define Ψ_σ as in Lemma 6 and choose $\mathbf{e}$ according to Ψ_σ. If the $\mathsf{LWE}_{l,m,q,\chi}$ assumption holds, then the above construction is PRIV1-IND secure for all k-sources where $k \geq (l + \Omega(\lambda)) \log q$.*

Proof. The parameters are chosen such that the $\mathsf{LWE}^{\mathsf{WL}(1,k)}_{n,m,q,\Psi_\sigma}$-assumption holds. Hence for any distributions M_λ^0, M_λ^1 over $\{0,1\}^n$ with $H_\infty(M_\lambda^0) \geq k$ and $H_\infty(M_\lambda^1) \geq k$, there is

$$(pk, \mathsf{Enc}(pk, \mathbf{m}_0; \mathbf{e}_0)) \overset{s}{\approx} (\mathbf{B}, \mathbf{Bm}_0 + \mathbf{e}_0) \qquad (1)$$

$$\overset{c}{\approx} (\mathbf{B}, \mathbf{u}) \qquad (2)$$

$$\overset{c}{\approx} (\mathbf{B}, \mathbf{Bm}_1 + \mathbf{e}_1) \qquad (3)$$

$$\overset{s}{\approx} (pk, \mathsf{Enc}(pk, \mathbf{m}_1; \mathbf{e}_1)), \qquad (4)$$

wherein $\mathbf{m}_0 \overset{\$}{\leftarrow} M_\lambda^0$, $\mathbf{m}_1 \overset{\$}{\leftarrow} M_\lambda^1$, $(pk, sk) \overset{\$}{\leftarrow} \mathsf{Gen}(1^\lambda)$, $\mathbf{e}_0, \mathbf{e}_1 \overset{\$}{\leftarrow} \Psi_\sigma$, $\mathbf{B} \leftarrow \mathbb{Z}_q^{m \times n}$, and $\mathbf{u} \overset{\$}{\leftarrow} \mathbb{Z}_q^m$. Step 1 and Step 4 follow with Lemma 5, i.e., the trapdoor generation technique. Step 2 and Step 3 follow with Lemma 6, i.e., the LWE assumption with weak secret. □

Remark 1. Xie et al. proposed a very simple DPKE scheme which is basically the evaluation of inversion of the LWR function, by encrypting $\mathbf{m}$ as $\lfloor \mathbf{Am} \rfloor_p$ where $p \ll q$, but the security analysis requires the modulus q to be super-polynomial [37]. Later, Alwen et al. improved the size of the modulus q to be polynomial, and the size of q is roughly $q \geq 2\beta nm^2$. In our A-DPKE scheme, there is roughly $q \geq \beta nm^{\frac{3}{2}}$, i.e., the modulus can be smaller.

Remark 2. Bellare et al. proved that with the trapdoor techniques in [28] the LWE function is a lossy trapdoor function for uniform input distributions, but they did not mention whether it is a secure DPKE [10] for high min-entropy message distributions.

Remark 3. In fact we can prove the construction is PRIV1-IND secure with respect to hard-to-invert auxiliary input [13], as long as the LWE with weak and leaky secrets assumption holds. We only show the "weak" secret aspect for simplicity.

6 Conclusion

In this work we proposed the notion of approximate-deterministic public key encryption by generalizing the original definition of DPKE. A-DPKE maintains the advantages of DPKE in applications such as searchable encryption and data de-duplication, while allows new constructions from quantum-resistant assumptions. We presented two simple constructions of A-DPKE from hard learning problems, e.g., LPN and LWE. The LWE based A-DPKE is as simple as the DPKE scheme from the LWR assumption, with smaller modulus. And we believe that the relaxation from deterministic to approximate-deterministic is meaningful since previously there is no construction of DPKE from LPN.

To make the new concept practical, it is desirable to instantiate A-DPKE with ring-based assumptions, such as ring-LPN [22] and ring-LWE [25]. However, in the current work we have not addressed the problem, and leave it for future work.

Acknowledgments. We are grateful to anonymous reviewers for their inspiring comments. Besides, we thank Yuanyuan Gao and Jingnan He for helpful discussions. Yamin Liu is supported by the National Natural Science Foundation of China (No. 61502480). Xianhui Lu is supported the by National Natural Science Foundation of China (No. 61572495, No. 61272534). Bao Li and Fuyang Fang are supported by the National Natural Science Foundation of China (No. 61379137) and the National Basic Research Program of China (973 project) (No. 2013CB338002).

References

1. Alekhnovich, M.: More on average case vs approximation complexity. In: FOCS 2003, pp. 298–307. IEEE (2003)
2. Akavia, A., Bogdanov, A., Guo, S., Kamath, A., Rosen, A.: Candidate weak pseudorandom functions in AC0 ∘ MOD2. In: ITCS 2014, pp. 251–259. ACM (2014)
3. Alwen, J., Krenn, S., Pietrzak, K., Wichs, D.: Learning with rounding, revisited - new reductions, properties and applications. In: Canetti, R., Garay, J.A. (eds.) CRYPTO 2013. LNCS, vol. 8042, pp. 57–74. Springer, Heidelberg (2013). doi:10.1007/978-3-642-40041-4_4
4. Bellare, M., Boldyreva, A., O'Neill, A.: Deterministic and efficiently searchable encryption. In: Menezes, A. (ed.) CRYPTO 2007. LNCS, vol. 4622, pp. 535–552. Springer, Heidelberg (2007). doi:10.1007/978-3-540-74143-5_30

5. Bellare, M., Fischlin, M., O'Neill, A., Ristenpart, T.: Deterministic encryption: definitional equivalences and constructions without random oracles. In: Wagner, D. (ed.) CRYPTO 2008. LNCS, vol. 5157, pp. 360–378. Springer, Heidelberg (2008). doi:10.1007/978-3-540-85174-5_20

6. Boldyreva, A., Fehr, S., O'Neill, A.: On notions of security for deterministic encryption, and efficient constructions without random oracles. In: Wagner, D. (ed.) CRYPTO 2008. LNCS, vol. 5157, pp. 335–359. Springer, Heidelberg (2008). doi:10.1007/978-3-540-85174-5_19

7. Bogdanov, A., Guo, S., Masny, D., Richelson, S., Rosen, A.: On the hardness of learning with rounding over small modulus. ePrint Archive 2015/769 (2015)

8. Bellare, M., Hoang, V.T.: Resisting randomness subversion: fast deterministic and hedged public-key encryption in the standard model. In: Oswald, E., Fischlin, M. (eds.) EUROCRYPT 2015. LNCS, vol. 9057, pp. 627–656. Springer, Heidelberg (2015). doi:10.1007/978-3-662-46803-6_21

9. Bellare, M., Hoang, V.T., Keelveedhi, S.: Instantiating random oracles via UCEs. In: Canetti, R., Garay, J.A. (eds.) CRYPTO 2013. LNCS, vol. 8043, pp. 398–415. Springer, Heidelberg (2013). doi:10.1007/978-3-642-40084-1_23

10. Bellare, M., Kiltz, E., Peikert, C., Waters, B.: Identity-based (lossy) trapdoor functions and applications. In: Pointcheval, D., Johansson, T. (eds.) EUROCRYPT 2012. LNCS, vol. 7237, pp. 228–245. Springer, Heidelberg (2012). doi:10.1007/978-3-642-29011-4_15

11. Brakerski, Z., Langlois, A., Peikert, C., Regev, O., Stehlé, D.: Classical hardness of learning with errors. In: STOC 2013, pp. 575–584. ACM (2013)

12. Banerjee, A., Peikert, C., Rosen, A.: Pseudorandom functions and lattices. In: Pointcheval, D., Johansson, T. (eds.) EUROCRYPT 2012. LNCS, vol. 7237, pp. 719–737. Springer, Heidelberg (2012). doi:10.1007/978-3-642-29011-4_42

13. Brakerski, Z., Segev, G.: Better security for deterministic public-key encryption: the auxiliary-input setting. In: Rogaway, P. (ed.) CRYPTO 2011. LNCS, vol. 6841, pp. 543–560. Springer, Heidelberg (2011). doi:10.1007/978-3-642-22792-9_31

14. Cramer, R., Shoup, V.: Universal hash proofs and a paradigm for adaptive chosen ciphertext secure public-key encryption. In: Knudsen, L.R. (ed.) EUROCRYPT 2002. LNCS, vol. 2332, pp. 45–64. Springer, Heidelberg (2002). doi:10.1007/3-540-46035-7_4

15. Döttling, N., Müller-Quade, J.: Lossy codes and a new variant of the learning-with-errors problem. In: Johansson, T., Nguyen, P.Q. (eds.) EUROCRYPT 2013. LNCS, vol. 7881, pp. 18–34. Springer, Heidelberg (2013). doi:10.1007/978-3-642-38348-9_2

16. Döttling, N., Müller-Quade, J., Nascimento, A.C.A.: IND-CCA secure cryptography based on a variant of the LPN problem. In: Wang, X., Sako, K. (eds.) ASIACRYPT 2012. LNCS, vol. 7658, pp. 485–503. Springer, Heidelberg (2012). doi:10.1007/978-3-642-34961-4_30

17. Fuller, B., O'Neill, A., Reyzin, L.: A unified approach to deterministic encryption: new constructions and a connection to computational entropy. In: Cramer, R. (ed.) TCC 2012. LNCS, vol. 7194, pp. 582–599. Springer, Heidelberg (2012). doi:10.1007/978-3-642-28914-9_33

18. Goldwasser, S., Kalai, Y.T., Peikert, C., Vaikuntanathan, V.: Robustness of the learning with errors assumption. In: ICS 2010, pp. 230–240. Tsinghua University Press (2010)

19. Goldwasser, S., Micali, S.: Probabilistic encryption. J. Comput. Syst. Sci. **28**(2), 270–299 (1984)

20. Gentry, C., Peikert, C., Vaikuntanathan, V.: How to use a short basis: trapdoors for hard lattices and new cryptographic constructions. In: STOC 2008, pp. 197–206. ACM (2008)

21. Damgård, I., Park, S.: How practical is public-key encryption based on LPN? ePrint Archive, 2012/699 (2012)

22. Heyse, S., Kiltz, E., Lyubashevsky, V., Paar, C., Pietrzak, K.: Lapin: an efficient authentication protocol based on Ring-LPN. In: Canteaut, A. (ed.) FSE 2012. LNCS, vol. 7549, pp. 346–365. Springer, Heidelberg (2012). doi:10.1007/978-3-642-34047-5_20

23. Kiltz, E., Masny, D., Pietrzak, K.: Simple chosen-ciphertext security from low-noise LPN. In: Krawczyk, H. (ed.) PKC 2014. LNCS, vol. 8383, pp. 1–18. Springer, Heidelberg (2014). doi:10.1007/978-3-642-54631-0_1

24. Katz, J., Vaikuntanathan, V.: Smooth projective hashing and password-based authenticated key exchange from lattices. In: Matsui, M. (ed.) ASIACRYPT 2009. LNCS, vol. 5912, pp. 636–652. Springer, Heidelberg (2009). doi:10.1007/978-3-642-10366-7_37

25. Lyubashevsky, V., Peikert, C., Regev, O.: On ideal lattices and learning with errors over rings. In: Gilbert, H. (ed.) EUROCRYPT 2010. LNCS, vol. 6110, pp. 1–23. Springer, Heidelberg (2010). doi:10.1007/978-3-642-13190-5_1

26. Meurer, A.: A coding-theoretic approach to cryptanalysis. Ph.D. dissertation thesis (2012)

27. Micciancio, D., Mol, P.: Pseudorandom knapsacks and the sample complexity of LWE search-to-decision reductions. In: Rogaway, P. (ed.) CRYPTO 2011. LNCS, vol. 6841, pp. 465–484. Springer, Heidelberg (2011). doi:10.1007/978-3-642-22792-9_26

28. Micciancio, D., Peikert, C.: Trapdoors for lattices: simpler, tighter, faster, smaller. In: Pointcheval, D., Johansson, T. (eds.) EUROCRYPT 2012. LNCS, vol. 7237, pp. 700–718. Springer, Heidelberg (2012). doi:10.1007/978-3-642-29011-4_41

29. Peikert, C.: Public-key cryptosystems from the worst-case shortest vector problem: extended abstract. In STOC 2009, pp. 333–342. ACM (2009)

30. Peikert, C., Vaikuntanathan, V., Waters, B.: A framework for efficient and composable oblivious transfer. In: Wagner, D. (ed.) CRYPTO 2008. LNCS, vol. 5157, pp. 554–571. Springer, Heidelberg (2008). doi:10.1007/978-3-540-85174-5_31

31. Peikert, C., Waters, B.: Lossy trapdoor functions and their applications. In: STOC 2008, pp. 187–196. ACM (2008)

32. Regev, O.: On lattices, learning with errors, random linear codes, and cryptography. In: STOC 2005, pp. 84–93. ACM (2005)

33. Raghunathan, A., Segev, G., Vadhan, S.: Deterministic public-key encryption for adaptively chosen plaintext distributions. In: Johansson, T., Nguyen, P.Q. (eds.) EUROCRYPT 2013. LNCS, vol. 7881, pp. 93–110. Springer, Heidelberg (2013). doi:10.1007/978-3-642-38348-9_6

34. Sun, X., Li, B., Lu, X.: Cramer-shoup like chosen ciphertext security from LPN. In: Lopez, J., Wu, Y. (eds.) ISPEC 2015. LNCS, vol. 9065, pp. 79–95. Springer, Heidelberg (2015). doi:10.1007/978-3-319-17533-1_6

35. Wee, H.: Dual projective hashing and its applications — lossy trapdoor functions and more. In: Pointcheval, D., Johansson, T. (eds.) EUROCRYPT 2012. LNCS, vol. 7237, pp. 246–262. Springer, Heidelberg (2012). doi:10.1007/978-3-642-29011-4_16

36. Wichs, D.: Barriers in cryptography with weak, correlated and leaky sources. In: ITCS 2013, pp. 111–126. ACM (2013)

37. Xie, X., Xue, R., Zhang, R.: Deterministic public key encryption and identity-based encryption from lattices in the auxiliary-input setting. In: Visconti, I., Prisco, R. (eds.) SCN 2012. LNCS, vol. 7485, pp. 1–18. Springer, Heidelberg (2012). doi:10.1007/978-3-642-32928-9_1

38. Yu, Y., Zhang, J.: Cryptography with auxiliary input and trapdoor from constant-noise LPN. In: Robshaw, M., Katz, J. (eds.) CRYPTO 2016. LNCS, vol. 9814, pp. 214–243. Springer, Heidelberg (2016). doi:10.1007/978-3-662-53018-4_9. ePrint Archive, 2016/514

Adaptively Secure Strong Designated Signature

Neetu Sharma[1], Rajeev Anand Sahu[2], Vishal Saraswat[2(✉)],
and Birendra Kumar Sharma[1]

[1] PRS University, Raipur, India
neetus.crypto@gmail.com, sharmabk07@gmail.com
[2] CRRao AIMSCS, Hyderabad, India
rajeevs.crypto@gmail.com, vishal.saraswat@gmail.com

Abstract. Almost all the available strong designated verifier signature (SDVS) schemes are either insecure or inefficient for practical implementation. Hence, an efficient and secure SDVS algorithm is desired. In this paper, we propose an efficient strong designated verifier signature on identity-based setting, we call it ID-SDVS scheme. The proposed scheme is strong existentially unforgeable against adaptive chosen message and adaptive chosen identity attack under standard assumptions, the hardness of the decisional and computational Bilinear Diffie-Hellman Problem (BDHP). Though the unverifiability by a non-designated verifier and the strongness are essential security properties of a SDVS, the proofs for these properties are not provided in most of the literature on SDVS we reviewed. We provide the proofs of unverifiability and of strongness of the proposed scheme. Moreover, we show that the proposed scheme is significantly more efficient in the view of computation and operation time than the existing similar schemes.

Keywords: Strong designated verifier signature · Identity-based cryptography · Bilinear Diffie-Hellman problem · Provable security

1 Introduction

Digital signature is a widely accepted tool for authentication in cryptography. The general definition of digital signature in public key cryptography allows any user in public to verify the authentication of the signature. However, in many situations, like proposal of construction bidding, licensing software, electronic voting etc., the signers may desire to sign a document for a particular receiver with control over the verification of their signatures. In these applications, the signed message may include crucial information between the signer and the verifier.

For such scenarios, Chaum et al. [3] introduced the undeniable signature which allows a signer to have a control over the signature with the property that verification of a signature requires the participation of the signer. But a practical issue with such a signature is that the signer's presence for verification requires the signer to be online all the time. To overcome this complication, Jakobsson et al. [7] proposed the concept of designated verifier signature (DVS),

© Springer International Publishing AG 2016
O. Dunkelman and S.K. Sanadhya (Eds.): INDOCRYPT 2016, LNCS 10095, pp. 43–60, 2016.
DOI: 10.1007/978-3-319-49890-4_3

that transforms Chaum's scheme [2] into non-interactive verification using a designated verifier proof. Their scheme allows the signer to convince the validity of a statement to a particular verifier without allowing any third party to verify the validity of that signature.

Saeednia et al. [14] pointed out that given a DVS, anybody can make sure that there are only two potential signers. Hence, if the signatures may be captured on the line before arriving at the designated verifier, then one can identify the signer, since it is now sure that the verifier did not produce the signature. To overcome this issue, they extended the notion of DVS with a property of *strongness* which requires that to a third party, who is none of the signer or designated verifier, the DVS from a signer A to a designated verifier B, is indistinguishable from a DVS from any other signer C to some other verifier D. They call such a signature *strong* designated verifier signature (SDVS).

1.1   Related Work

In 2004, Susilo et al. [15] proposed the first identity-based strong designated verifier signature (ID-SDVS). Unforgeability of their scheme is based on the Bilinear Diffie-Hellman (BDH) assumption. In 2006, Huang et al. [6] proposed a short ID-SDVS scheme based on Diffie-Hellman key exchange protocol. The security of their scheme relies on the Gap Bilinear Diffie-Hellman (GBDH) assumption. Computation cost of the former scheme is more than double of the latter one. However, the scheme in [6] is not strongly unforgeable since the signature of a message always remains the same and a replay attack is always possible and cannot be trivially avoided. Later, in 2008, Zhang et al. [16] proposed another ID-SDVS scheme that is claimed to be non-delegatable, but in 2009, Kang et al. [9] pointed out security flaws in [16] against the strongness property of SDVS scheme. They observed that in [16], an outsider who eavesdrops an old signature and can obtain some information that is to be used for the verification of subsequent signatures. It has also been explained in [9], that how the property of strongness in [16] does not fulfil their claim. In [9], they also proposed another ID-SDVS scheme and an identity-based designated verifier proxy signature (ID-DVPS). However, in 2010, Lee et al. [11] showed that the scheme in [9] is universally forgeable. In 2009, Kang et al. [10] proposed another ID-SDVS scheme which is more efficient than that in [9]. However, this construction was also shown to be universally forgeable in [5].

1.2   Applications

The strong designated verifier signature has crucial applications in various real world scenarios including the following:

1. *Licensing software*: Software companies use digitally signed keys as their software license so that these keys can only be used by the person who has purchased the product. The strong designated verifier signature on keys protects illegal distribution of the software.

2. *Electronic voting*: In electronic voting schemes, a voting center is required to ensure that a vote has been counted in the final tally or not. The verification of the center's signature on the receipt is one way of doing so. But it should also be taken in account that the voters must not have the ability to convince a third party the nature of their votes they have casted. This may cause some gain or threats by the parties depending upon the nature of the vote. To fulfil this requirement in electronic voting schemes, the center's signature should be a strong designated verifier signature.

1.3 Our Contribution

In this paper, we propose an efficient identity-based strong designated verifier signature (ID-SDVS) scheme using bilinear pairing. Proposed scheme is existentially unforgeable (resp. unverifiable) against adaptive chosen message and adaptive chosen identity attack under the computational (resp. decisional) Bilinear Diffie-Hellman (BDH) assumption in the random oracle model. We also provide a proof for the strongness property of the proposed scheme. We reviewed the existing ID-SDVS schemes including [5,6,8–11,15,16] and noticed that most of the papers on ID-SDVS were missing the full proofs of security which we tabulate in Table 1. Moreover, we show that the proposed scheme is upto 120 % more efficient in the sense of computation and operation time than these schemes.

Table 1. Security proofs

Scheme	Proof of unforgeability	Proof of unverifiability	Proof of strongness
Susilo et al. [15]	✓	✗	✗
Huang et al. [6]	✗	✗	✗
Kancharla et al. [8]	✓	✗	✗
Du et al. [5]	✓	✗	✗
Zhang et al. [16]	✗	✗	✗
Kang et al. [9]	✗	✗	✗
Kang et al. [10]	✗	✗	✗
Lee et al. [11]	✓	✗	✗
Our scheme	✓	✓	✓

1.4 Outline of the Paper

The rest of this paper is organized as follows. In Sect. 2, we introduce some related mathematical definitions, problems and assumptions. In Sect. 3, we present the formal definition of an identity-based strong designated verifier signature scheme and a formal security model for it. The proposed signature scheme is presented in Sect. 4. In Sect. 5 we analyze the security of the proposed scheme and in Sect. 6 we do an efficiency comparison with the state-of-art. Finally, in Sect. 7 we conclude our work.

2   Preliminaries

A probabilistic polynomial time (PPT) algorithm is a probabilistic/randomized algorithm that runs in time polynomial in the length of input. We denote by $y \leftarrow A(x)$ the operation of running a randomized or deterministic algorithm $A(x)$ and storing the output to the variable y. If X is a set, then $v \xleftarrow{\$} X$ denotes the operation of choosing an element v of X according to the uniform random distribution on X. We say that a given function $f : N \to [0,1]$ is *negligible in* n if $f(n) < 1/p(n)$ for any polynomial p for sufficiently large n. For a group G and $g \in G$, we write $G = \langle g \rangle$ if g is a generator of G.

Definition 1 (Bilinear Map). Let G_1 be an additive cyclic group with generator P and G_2 be a multiplicative cyclic group. Let both the groups are of the same prime order q. Then a map $e : G_1 \times G_1 \to G_2$ is called a *cryptographic bilinear map* if it satisfies the following properties.

Bilinearity: For all $a, b \in \mathbb{Z}_q^*$, $e(aP, bP) = e(P,P)^{ab}$, or equivalently, for all $Q, R, S \in G_1$, $e(Q+R, S) = e(Q,S)e(R,S)$ and $e(Q, R+S) = e(Q,R)e(Q,S)$.

Non-degeneracy: There exists $Q, R \in G_1$ such that $e(Q,R) \neq 1$. Note that since G_1 and G_2 are groups of prime order, this condition is equivalent to the condition $g := e(P,P) \neq 1$, which again is equivalent to the condition that $g := e(P,P)$ is a generator of G_2.

Computability: There exists an efficient algorithm to compute $e(Q,R) \in G_2$ for all $Q, R \in G_1$.

Definition 2. A *bilinear map parameter generator* $\mathfrak{B}$ is a PPT algorithm that takes as input security parameter λ and outputs a tuple

$$\langle q, e : G_1 \times G_1 \to G_2, P, g \rangle \leftarrow \mathfrak{B}(\lambda) \tag{1}$$

where q, G_1, G_2, e, P and g are as in Definition 1.

Definition 3 (Bilinear Diffie-Hellman Problem). Given a security parameter λ, let $\langle q, e : G_1 \times G_1 \to G_2, P, g \rangle \leftarrow \mathfrak{B}(\lambda)$. Let $BDH : G_1 \times G_1 \times G_1 \to G_2$ be a map defined by

$$BDH(X, Y, Z) = \omega \text{ where } X = xP, Y = yP, Z = zP \text{ and } \omega = e(P,P)^{xyz} .$$

The *bilinear Diffie-Hellman problem* (BDHP) is to evaluate $BDH(X, Y, Z)$ given $X, Y, Z \xleftarrow{\$} G_1$. (Without the knowledge of $x, y, z \in \mathbb{Z}_q$ — obtaining $x \in \mathbb{Z}_q$, given $P, X \in G_1$ is solving the discrete logarithm problem (DLP)).

Definition 4. A *BDHP parameter generator* $\mathfrak{C}$ is a PPT algorithm that takes as input security parameter λ and outputs a tuple

$$\langle q, e : G_1 \times G_1 \to G_2, P, g, X, Y, Z \rangle \leftarrow \mathfrak{C}(\lambda) \tag{2}$$

where q, G_1, G_2, e, P, g, X, Y and Z are as in Definition 3.

Definition 5 (Bilinear Diffie-Hellman Assumption).   Given a security parameter λ, let $\langle q, e : G_1 \times G_1 \rightarrow G_2, P, g, X, Y, Z \rangle \leftarrow \mathfrak{C}(\lambda)$. The *bilinear Diffie-Hellman assumption* (BDHA) states that for any PPT algorithm $\mathcal{A}$ which attempts to solve BDHP, its *advantage*

$$\mathbf{Adv}_{\mathfrak{C}}(\mathcal{A}) := Prob[\mathcal{A}(q, e : G_1 \times G_1 \rightarrow G_2, P, g, X, Y, Z) = BDH(X, Y, Z)]$$

is negligible in λ.

Definition 6 (Decisional Bilinear Diffie-Hellman Problem).   Given a security parameter λ, let $\langle q, e : G_1 \times G_1 \rightarrow G_2, P, g, X, Y, Z \rangle \leftarrow \mathfrak{C}(\lambda)$. Let $\omega \xleftarrow{\$} G_2$. The *decisional bilinear Diffie-Hellman problem* (DBDHP) is to decide if

$$\omega = BDH(X, Y, Z).$$

That is, if $X = xP, Y = yP, Z = zP$, for some $x, y, z \in \mathbb{Z}_q$, then the DBDHP is to decide if
$$\omega = e(P, P)^{xyz}.$$

(Without the knowledge of $x, y, z \in \mathbb{Z}_q$ — obtaining $x \in \mathbb{Z}_q$, given $P, X \in G_1$ is solving the discrete logarithm problem (DLP)).

Definition 7.   A *DBDHP parameter generator* $\mathfrak{D}$ is a PPT algorithm that takes as input security parameter λ and outputs a tuple

$$\langle q, e : G_1 \times G_1 \rightarrow G_2, P, g, X, Y, Z, \omega \rangle \leftarrow \mathfrak{D}(\lambda) \tag{3}$$

where q, G_1, G_2, e, P, g, X, Y, Z and ω are as in Definition 6.

Definition 8 (Decisional Bilinear Diffie-Hellman Assumption).   Given a security parameter λ, let $\langle q, e : G_1 \times G_1 \rightarrow G_2, P, g, X, Y, Z, \omega \rangle \leftarrow \mathfrak{D}(\lambda)$. The *bilinear Diffie-Hellman assumption* (DBDHA) states that, for any PPT algorithm $\mathcal{A}$ which attempts to solve DBDHP, its *advantage*

$$\mathbf{Adv}_{\mathfrak{D}}(\mathcal{A}) := |Prob[\mathcal{A}(q, e : G_1 \times G_1 \rightarrow G_2, P, g, X, Y, Z, \omega) = 1] -$$
$$Prob[\mathcal{A}(q, e : G_1 \times G_1 \rightarrow G_2, P, g, X, Y, Z, BDH(X, Y, Z)) = 1]| \tag{4}$$

is negligible in λ.

3   Identity-Based Strong Designated Verifier Signature

In this section we present the formal definition of an identity-based strong designated verifier signature (ID-SDVS) and formalize a security model for it.

3.1   Identity-Based Strong Designated Verifier Signature

In an ID-SDVS scheme, a signer with identity ID_S intends to send a signed message to a designated verifier with identity ID_V such that no one other than the designated verifier can verify the signature. An ID-SDVS scheme is consists of the following five algorithms:

1. *params* ← *Setup*(λ): An algorithm run by the private key generator (PKG) which takes as input a security parameter λ and outputs the public parameters *params* and a master secret s of the system. In all the algorithms from here onward, *params* will be considered as an implicit input.
2. $(Q_{\mathsf{ID}}, S_{\mathsf{ID}})$ ← *Key Extract*(ID): An algorithm run by the (PKG) which takes input identity ID and outputs its public and private key pair $(Q_{\mathsf{ID}}, S_{\mathsf{ID}})$.
3. σ ← *DVSign*($S_{\mathsf{ID}_S}, Q_{\mathsf{ID}_V}, m$): A probabilistic algorithm run by the signer that takes as input the signer's secret key S_{ID_S}, the designated verifier's public key Q_{ID_V} and a message m to generate a signature σ.
4. b ← *DVVer*($S_{\mathsf{ID}_V}, Q_{\mathsf{ID}_S}, m, \sigma$): A deterministic algorithm run by the verifier that takes the verifier's secret key S_{ID_V}, the signer's public key Q_{ID_S}, a message m and a signature σ, and returns a bit b which is 1 if the signature is valid and 0 if invalid.
5. $\widehat{\sigma}$ ← *DVTrans*($S_{\mathsf{ID}_V}, Q_{\mathsf{ID}_S}, m$): A deterministic algorithm run by the verifier that takes the verifier's secret key S_{ID_V}, and the signer's public key Q_{ID_S} and a message m to generate a signature $\widehat{\sigma}$.

3.2   Security Model for Identity-Based Strong Designated Verifier Signature

An ID-SDVS scheme must satisfy the following security properties.

1. ***Correctness:*** If the signature σ on a message m is correctly computed by a signer ID_S, then the designated verifier ID_V must be able to verify the correctness of the message-signature pair (m, σ). That is,

$$Prob[1 \leftarrow DVVer(S_{\mathsf{ID}_V}, Q_{\mathsf{ID}_S}, m, DVSign(S_{\mathsf{ID}_S}, Q_{\mathsf{ID}_V}, m))] = 1$$

2. ***Unforgeability:*** It is computationally infeasible to construct a valid ID-SDVS signature without the knowledge of the private key of either the signer or the designated verifier. We define below *strong existential unforgeability against an adaptive chosen message and adaptive chosen identities attack.*

Definition 9 (Unforgeability). An ID-SDVS scheme is said to be *strong existential unforgeable against adaptive chosen message and adaptive chosen identities attack* if for any security parameter λ, no probabilistic polynomial time adversary $\mathcal{A}(q_{H_1}, q_{H_2}, q_E, q_S, q_V, \varepsilon_{\mathcal{A}}(\lambda), t)$ which runs in time t has a non-negligible advantage

$$\mathbf{Adv}_{\text{ID-SDVS},\mathcal{A}}^{\text{SEUF-CID2-CMA2}}(\lambda) := \varepsilon_{\mathcal{A}}(\lambda) := Prob[1 \leftarrow DVVer(S_{\mathsf{ID}_V^*}, Q_{\mathsf{ID}_S^*}, m^*, \sigma^*)]$$

against the challenger $\mathcal{B}$ in the following game:

1. *Setup*: The challenger $\mathcal{B}$ generates the system's public parameter *params* for security parameter λ.
2. *Query Phase*:
   - The adversary $\mathcal{A}$ may request upto q_{H_1} hash queries on its adaptively chosen identities and upto q_{H_2} hash queries on its adaptively chosen messages and obtain responses from $\mathcal{B}$ acting as a random oracle.
   - $\mathcal{A}$ may request upto q_E key extraction queries on its adaptively chosen identities and obtain the corresponding private keys.
   - $\mathcal{A}$ may request upto q_S signature queries on its adaptively chosen messages and adaptively chosen identities for the signer and the designated verifier and obtain a valid strong designated verifier signature.
   - $\mathcal{A}$ may request upto q_V verification queries on signatures on its adaptively chosen messages m and adaptively chosen identities for the signer and the designated verifier and obtain the verification result 1 if it is valid and 0 if invalid.
3. *Output*: Finally, $\mathcal{A}$ outputs a (message, signature) pair (m^*, σ^*) for identities $\mathsf{ID}_{\mathcal{S}}^*$ of the signer and $\mathsf{ID}_{\mathcal{V}}^*$ of the designated verifier such that:
   - $\mathcal{A}$ has never submitted $\mathsf{ID}_{\mathcal{S}}^*$ or $\mathsf{ID}_{\mathcal{V}}^*$ during the key extraction queries.
   - σ^* was never given as a response to a signature query on the message m^* with the signer's identity $\mathsf{ID}_{\mathcal{S}}^*$, and the designated verifier's identity $\mathsf{ID}_{\mathcal{V}}^*$;
   - σ^* is a valid signature on the message m^* from a signer with identity $\mathsf{ID}_{\mathcal{S}}^*$, for a designated verifier with identity $\mathsf{ID}_{\mathcal{V}}^*$.

3. **Unverifiability**: It is computationally infeasible to verify the validity of an ID-SDVS without the knowledge of the private key of either the signer or the designated verifier. We define below *existential designated unverifiability against an adaptive chosen message and adaptive chosen identities attack*.

Definition 10 (Unverifiability). An ID-SDVS scheme is said to be *existential designated unverifiable against adaptive chosen message and adaptive chosen identities attack* if for any security parameter λ, no probabilistic polynomial time adversary $\mathcal{A}(q_{H_1}, q_{H_2}, q_E, q_S, q_V, \varepsilon_{\mathcal{A}}(\lambda), t)$ which runs in time t has a non-negligible advantage

$$\mathbf{Adv}_{\mathrm{ID\text{-}SDVS},\mathcal{A}}^{\mathrm{EDV\text{-}CID2\text{-}CMA2}}(\lambda) := \varepsilon_{\mathcal{A}}(\lambda) := |Prob[\mathcal{A}(Q_{\mathsf{ID}_{\mathcal{S}}^*}, Q_{\mathsf{ID}_{\mathcal{V}}^*}, m^*, \sigma^*) = 1]-$$
$$Prob[\mathcal{A}(Q_{\mathsf{ID}_{\mathcal{S}}^*}, Q_{\mathsf{ID}_{\mathcal{V}}^*}, m^*, DVSign(S_{\mathsf{ID}_{\mathcal{S}}^*}, Q_{\mathsf{ID}_{\mathcal{V}}^*}, m^*)) = 1]| \quad (5)$$

against the challenger $\mathcal{B}$'s response σ^* in the following game:

1. *Setup*: Similar to the unforgeability game in Definition 9.
2. *Query Phase 1*: Similar to the unforgeability game in Definition 9.
3. *Challenge*: At some point, $\mathcal{A}$ outputs a message m^* and identities $\mathsf{ID}_{\mathcal{S}}^*$ of the signer and $\mathsf{ID}_{\mathcal{V}}^*$ of the designated verifier on which it wishes to be challenged such that $\mathcal{A}$ has never submitted $\mathsf{ID}_{\mathcal{S}}^*$ or $\mathsf{ID}_{\mathcal{V}}^*$ during the key extraction queries. The challenger $\mathcal{B}$ responds with a "signature" σ^* and challenges $\mathcal{A}$ to verify if it is valid or not.

4. *Query Phase 2*: $\mathcal{A}$ continues its queries as in Query Phase 1 with an additional restriction that now it cannot submit a verification query on σ^*.
5. *Output*: Finally, $\mathcal{A}$ outputs its guessed bit b^* which is 1 if the signature is valid and 0 if invalid.

4. **Non-transferability**: Given a signature σ on message m, it is infeasible for any PPT adversary $\mathcal{A}$ to decide whether σ was produced by the signer or by the designated verifier, even if $\mathcal{A}$ is also given the private keys of the signer and the designated verifier. In other words, it is impossible for the designated verifier to prove to an outsider that the signature is actually generated by the signer.

Definition 11 (Non-transferability). An ID-SDVS scheme is said to be *non-transferable* if the signature generated by the signer is computationally indistinguishable from that generated by the designated verifier, that is,

$$\sigma \leftarrow DVSign(S_{\mathsf{ID}_S}, Q_{\mathsf{ID}_V}, m) \approx \hat{\sigma} \leftarrow DVTrans(S_{\mathsf{ID}_V}, Q_{\mathsf{ID}_S}, m).$$

5. **Strongness**: Let $\sigma \leftarrow DVSign(S_{\mathsf{ID}_S}, Q_{\mathsf{ID}_V}, m)$ be a signature on a message m from a signer $\mathcal{S}$ to a designated verifier $\mathcal{V}$. *Strongness* requires that σ could have been produced by any other third party $\mathcal{S}^*$ other than $\mathcal{S}$ for some designated verifier $\mathcal{V}^*$ other than $\mathcal{V}$.

Definition 12 (Strongness). An ID-SDVS scheme is said to be *strong designated* if given $\sigma \leftarrow DVSign(S_{\mathsf{ID}_S}, Q_{\mathsf{ID}_V}, m)$, anyone, say $\mathcal{V}^*$, other than the designated verifier $\mathcal{V}$ can produce identically distributed transcripts that are indistinguishable from those of σ from someone, say $\mathcal{S}^*$, except the signer $\mathcal{S}$. That is,

$$\sigma \leftarrow DVSign(S_{\mathsf{ID}_S}, Q_{\mathsf{ID}_V}, m) \approx \hat{\sigma} \leftarrow DVTrans(S_{\mathsf{ID}_V^*}, Q_{\mathsf{ID}_S^*}, m).$$

4   Proposed Scheme

We present here our efficient and secure ID-SDVS. As described in Sect. 3, the proposed scheme consists of the following algorithms: Setup, Key Extract, Designated Signature, Designated Verification and Transcript Simulation.

Setup: In the setup phase, PKG on input security parameter λ, generates the system's master secret key $s \in \mathbb{Z}_q^*$ and the system's public parameters *params* $= (1^\lambda, G_1, G_2, q, e, H_1, H_2, P, P_{pub})$, where G_1 is an additive cyclic group of prime order q with generator P, G_2 is a multiplicative cyclic group of prime order q, and $H_1 : \{0,1\}^* \longrightarrow G_1$, $H_2 : \{0,1\}^* \times G_1 \longrightarrow \mathbb{Z}_q^*$ are two cryptographic secure hash functions, and $P_{pub} = sP \in G_1$ is system's public key, $e : G_1 \times G_1 \longrightarrow G_2$ is a bilinear map as defined in Sect. 2.

Key Extract: For a user with identity $\mathsf{ID}_i \in \{0,1\}^*$, the PKG computes its public key as $Q_{\mathsf{ID}_i} = H_1(\mathsf{ID}_i) \in G_1$ and corresponding private key as $S_{\mathsf{ID}_i} = sQ_{\mathsf{ID}_i} \in G_1$.

Designated Signature: To sign a message $m \in \{0,1\}^*$ which can be verified by a designated verifier $\mathcal{V}$, the signer $\mathcal{S}$ chooses a random $r \xleftarrow{\$} \mathbb{Z}_q^*$ and computes

- $U = rP \in G_1$;
- $h = H_2(m, U) \in \mathbb{Z}_q^*$;
- $V = rP_{pub} + hS_{\mathsf{ID}_\mathcal{S}} \in G_1$;
- $\sigma = e(V, Q_{\mathsf{ID}_\mathcal{V}})$.

The strong designated verifier signature on message m is $(U, \sigma) \in G_1 \times G_2$.

Designated Verification: On receiving a message m and a signature (U, σ), a verifier first computes $h = H_2(m, U) \in \mathbb{Z}_q^*$ and accepts the signature if and only if the following equality holds:

$$\sigma = e(U + hQ_{\mathsf{ID}_\mathcal{S}}, S_{\mathsf{ID}_\mathcal{V}}).$$

Transcript Simulation: The designated verifier $\mathcal{V}$ can produce the signature $\widehat{\sigma}$ intended for itself, by performing the following: chooses an integer $\widehat{r} \xleftarrow{\$} \mathbb{Z}_q^*$ and computes

- $\widehat{U} = \widehat{r}P \in G_1$;
- $\widehat{h} = H_2(m, \widehat{U}) \in \mathbb{Z}_q^*$;
- $\widehat{V} = \widehat{r}P + \widehat{h}Q_{\mathsf{ID}_\mathcal{S}} \in G_1$; and
- $\widehat{\sigma} = e(\widehat{V}, S_{\mathsf{ID}_\mathcal{V}})$.

5 Analysis of the Proposed Scheme

5.1 Correctness of the Proposed Scheme

The correctness of the scheme follows since if (U, σ) is a correctly generated signature on a message m from a signer with identity $\mathsf{ID}_\mathcal{S}$ for a designated verifier with identity $\mathsf{ID}_\mathcal{V}$, then

$$\begin{aligned}
e(U + hQ_{\mathsf{ID}_\mathcal{S}}, S_{\mathsf{ID}_\mathcal{V}}) &= e(rP + hQ_{\mathsf{ID}_\mathcal{S}}, sQ_{\mathsf{ID}_\mathcal{V}}) \\
&= e(rP_{pub} + hS_{\mathsf{ID}_\mathcal{S}}, Q_{\mathsf{ID}_\mathcal{V}}) \\
&= e(V, Q_{\mathsf{ID}_\mathcal{V}}) \\
&= \sigma.
\end{aligned}$$

5.2 Unforgeability

We now prove that the proposed ID-SDVS is unforgeable. That is, any third party other than the signer and the designated verifier, cannot forge a valid signature on an adaptively chosen message from an adaptively chosen signer's identity for an adaptively chosen designated verifier's identity with non-negligible probability. We show that if there exists a probabilistic polynomial time (PPT) adaptive chosen message and adaptive chosen identity algorithm which can produce a forgery for the proposed ID-SDVS then there exists another PPT algorithm which can use the forgery to solve the BDHP. In particular, we prove the following theorem:

Theorem 1. *Given a security parameter λ, if there exists a PPT adversary $A(q_{H_1}, q_{H_2}, q_E, q_S, q_V, \varepsilon_A(\lambda), t)$ which breaks the unforgeability of the proposed ID-SDVS scheme in time t with success probability $\varepsilon_A(\lambda)$, then there exists a PPT adversary $B(\varepsilon_B(\lambda), t')$ which solves BDHP with success probability at least*

$$\varepsilon_B(\lambda) \geq \left(1 - \frac{1}{q^2}\right)\left(1 - \frac{2}{q_{H_1}}\right)^{q_E + q_V}\left(1 - \frac{2}{q_{H_1}(q_{H_1} - 1)}\right)^{q_S}\left(\frac{2}{q_{H_1}(q_{H_1} - 1)}\right)\varepsilon_A(\lambda)$$

in time at most

$$t' \leq (q_{H_1} + q_E + 3q_S + q_V)S_{G_1} + (q_S + q_V)P_e + q_S O_{G_1} + O_{G_2} + S_{G_2} + t$$

where S_{G_1} (resp. S_{G_2}) is the time taken for one scalar multiplication in G_1 (resp. G_2), O_{G_1} (resp. O_{G_2}) is the time taken for one group operation in G_1 (resp. G_2), and P_e is the time taken for one pairing computation.

Proof: Let for a security parameter λ, B is challenged to solve the BDHP for

$$\langle q, e, G_1, G_2, P, aP, bP, cP \rangle$$

where G_1 is an additive cyclic group of prime order q with generator P, G_2 is a multiplicative cyclic group of prime order q with generator $e(P, P)$, and $e : G_1 \times G_1 \rightarrow G_2$ is a cryptographic bilinear map as described in Sect. 2. $a, b, c \xleftarrow{\$} \mathbb{Z}_q^*$ are unknown to B. The goal of B is to solve BDHP by computing $e(P, P)^{abc} \in G_2$ using A, the adversary who claims to forge the proposed ID-SDVS scheme. B simulates the security game for unforgeability with A as follows.

Setup: B generates the system's public parameter

$$params = \langle q, e : G_1 \times G_1 \rightarrow G_2, P, P_{pub} := cP, H_1, H_2 \rangle$$

for security parameter λ where the hash functions H_1 and H_2 behave as random oracles and responds to A's queries as below.

H_1-*queries:* To respond to the H_1 queries, B maintains a list

$$L_{H_1} = \{(\mathsf{ID}_i \in \{0,1\}^*, r_i \in \mathbb{Z}_q^*, R_i \in G_1)_{i=1}^{q_{H_1}}\}$$

which is initially empty. B randomly chooses two indices $\alpha, \beta \in [1, q_{H_1}]$ and sets $i = 0$. When A makes an H_1-query for an identity $\mathsf{ID} \in \{0,1\}^*$, B proceeds as follows.

1. If the query ID already appears in L_{H_1} in some tuple $(\mathsf{ID}_i, r_i, R_i)$ then B responds to A with $H_1(\mathsf{ID}) = R_i \in G_1$;
2. otherwise B sets $i = i + 1$ and
   - if $i = \alpha$, B sets $r_i = \perp$ and $R_i = aP$;
   - if $i = \beta$, B sets $r_i = \perp$ and $R_i = bP$;
   - if $i \neq \alpha, \beta$, B chooses $r_i \xleftarrow{\$} \mathbb{Z}_q^*$ and sets $R_i = r_i P$;
3. Finally B adds the tuple $(\mathsf{ID}_i := \mathsf{ID}, r_i, R_i)$ to L_{H_1} and responds to A with $H_1(\mathsf{ID}) = R_i$.

H_2-queries: To respond to the H_2 queries, $\mathcal{B}$ maintains a list

$$L_{H_2} = \{((m, U) \in \{0,1\}^* \times G_1, h \in \mathbb{Z}_q^*)\}$$

which is initially empty. When $\mathcal{A}$ queries the oracle H_2 on (m, U), $\mathcal{B}$ responds as follows.

1. If the query (m, U) already appears in L_{H_2} in some tuple (m, U, h) then $\mathcal{B}$ responds with $H_2(m, U) = h \in \mathbb{Z}_q^*$.
2. Otherwise $\mathcal{B}$ picks a random $h \in \mathbb{Z}_q^*$ and adds the tuple (m, U, h) to L_{H_2} and responds to $\mathcal{A}$ with $H_2(m, U) = h$.

Key extraction queries: When $\mathcal{A}$ makes a private key query on identity ID, $\mathcal{B}$ proceeds as follows.

1. Runs the above algorithm for responding to H_1-query for identity ID and obtains $H_1(\mathsf{ID}) = R_i$.
2. If $i = \alpha$ or β, $\mathcal{B}$ reports failure and halts.
3. If $i \neq \alpha, \beta$, $\mathcal{B}$ responds to $\mathcal{A}$ with the private key $S_{\mathsf{ID}} := r_i P_{pub}$ on the identity ID.

It can be verified that the provided private key $S_{\mathsf{ID}} = r_i P_{pub}$ is a valid private key for the user with identity $\mathsf{ID}_i := \mathsf{ID}$ since

$$r_i P_{pub} = r_i cP = c r_i P = cH_1(\mathsf{ID}).$$

Note that $\mathcal{B}$ aborts the security game during a key extraction query with probability $\frac{2}{q_{H_1}}$.

Signature queries: To respond to the signature queries, $\mathcal{B}$ maintains a list

$$L_S = \{(m_\ell \in \{0,1\}^*, \mathsf{ID}_{S\ell} \in \{0,1\}^*, \mathsf{ID}_{\mathcal{V}\ell} \in \{0,1\}^*, x_\ell \in \mathbb{Z}_q^*, U_\ell \in G_1, \sigma_\ell \in G_2)_{\ell=1}^{q_S}\}$$

which is initially empty with $\ell = 0$. When $\mathcal{A}$ queries the signature on a message m from a signer with identity ID_S for a designated verifier with identity $\mathsf{ID}_{\mathcal{V}}$, $\mathcal{B}$ proceeds as follows.

1. If the query $(m, \mathsf{ID}_S, \mathsf{ID}_{\mathcal{V}})$ already appears in L_S in some tuple $(m_\ell, \mathsf{ID}_{S\ell}, \mathsf{ID}_{\mathcal{V}\ell}, x_\ell, U_\ell, \sigma_\ell)$ then $\mathcal{B}$ responds to $\mathcal{A}$ with the signature (U_ℓ, σ_ℓ).
2. Otherwise $\mathcal{B}$ sets $\ell = \ell + 1$ and runs the above algorithm for responding to H_1-query for identities ID_S and $\mathsf{ID}_{\mathcal{V}}$ and obtains $Q_{\mathsf{ID}_S} = H_1(\mathsf{ID}_S) = R_i$ and $Q_{\mathsf{ID}_{\mathcal{V}}} = H_1(\mathsf{ID}_{\mathcal{V}}) = R_j$.
3. If $\{i, j\} = \{\alpha, \beta\}$, $\mathcal{B}$ reports failure and halts.
4. If $i \neq \alpha, \beta$, $\mathcal{B}$ computes the private key for ID_S, $S_{\mathsf{ID}_S} = r_i P_{pub}$, and proceeds as follows.
   - randomly chooses $x_\ell \in \mathbb{Z}_q^*$;
   - sets $U_\ell = x_\ell P \in G_1$;
   - runs the H_2-query algorithm to obtain $h_\ell = H_2(m, U_\ell) \in \mathbb{Z}_q^*$;
   - sets $V_\ell = x_\ell P_{pub} + h_\ell S_{\mathsf{ID}_S} \in G_1$;
   - computes $\sigma_\ell = e(V_\ell, Q_{\mathsf{ID}_{\mathcal{V}}})$.

5. Otherwise if $j \neq \alpha, \beta$, $\mathcal{B}$ computes the private key for $\mathsf{ID}_{\mathcal{V}}$, $S_{\mathsf{ID}_{\mathcal{V}}} = r_j P_{pub}$, and proceeds as follows.
   - randomly chooses $x_\ell \in \mathbb{Z}_q^*$;
   - sets $U_\ell = x_\ell P \in G_1$;
   - runs the H_2-query algorithm to obtain $h_\ell = H_2(m, U_\ell) \in \mathbb{Z}_q^*$;
   - sets $V_\ell = x_\ell P + h_\ell Q_{\mathsf{ID}_S} \in G_1$;
   - computes $\sigma_\ell = e(V_\ell, S_{\mathsf{ID}_{\mathcal{V}}})$.
6. Finally $\mathcal{B}$ adds the tuple $(m_\ell, \mathsf{ID}_{S\ell}, \mathsf{ID}_{\mathcal{V}\ell}, x_\ell, U_\ell, \sigma_\ell)$ to L_S and responds to $\mathcal{A}$ with the signature (U_ℓ, σ_ℓ).

Note that $\mathcal{B}$ aborts the security game during a signature query with probability $\frac{2}{q_{H_1}(q_{H_1}-1)}$.

Verification queries: When $\mathcal{A}$ makes a verification query on the signature (U, σ) on a message m from a signer with identity ID_S for a designated verifier with identity $\mathsf{ID}_{\mathcal{V}}$, $\mathcal{B}$ proceeds as follows.

1. $\mathcal{B}$ runs the above algorithm for responding to H_1-query for identities ID_S and $\mathsf{ID}_{\mathcal{V}}$ and obtains $H_1(\mathsf{ID}_S) = R_i$ and $H_1(\mathsf{ID}_{\mathcal{V}}) = R_j$.
2. If $j \in \{\alpha, \beta\}$, $\mathcal{B}$ reports failure and halts.
3. If $j \neq \alpha, \beta$, then $\mathcal{B}$ computes $\mathsf{ID}_{\mathcal{V}}$'s private key, $S_{\mathsf{ID}_{\mathcal{V}}} = r_j P_{pub}$, and proceeds as in the verification of the proposed scheme and responds to $\mathcal{A}$ accordingly.

Note that $\mathcal{B}$ aborts the security game during a verification query with probability $\frac{2}{q_{H_1}}$.

Output: After $\mathcal{A}$ has made its queries, it finally outputs a valid signature (U^*, σ^*) on a message m^* from a signer with identity ID_S^* for a designated verifier with identity $\mathsf{ID}_{\mathcal{V}}^*$ with a non-negligible probability $\varepsilon_{\mathcal{A}}(\lambda)$ such that:
   - $\mathcal{A}$ has never submitted ID_S^* or $\mathsf{ID}_{\mathcal{V}}^*$ during the key extraction queries;
   - (U^*, σ^*) was never given as a response to a signature query on the message m^* with the signer's identity ID_S^*, and the designated verifier's identity $\mathsf{ID}_{\mathcal{V}}^*$; and
   - $\sigma^* = e(U^* + h^* Q_{\mathsf{ID}_S}^*, S_{\mathsf{ID}_{\mathcal{V}}}^*)$.

If $\mathcal{A}$ did not make H_1-query for the identities ID_S^* and $\mathsf{ID}_{\mathcal{V}}^*$, then the probability that verification equality holds is less than $1/q^2$. Thus, with probability greater than $1 - 1/q^2$, both the public keys were computed using H_1-oracle and there exist indices $i, j \in [1, q_{H_1}]$ such that $\mathsf{ID}_S^* = \mathsf{ID}_i$ and $\mathsf{ID}_{\mathcal{V}}^* = \mathsf{ID}_j$. If $\{i, j\} \neq \{\alpha, \beta\}$, then $\mathcal{B}$ reports failure and terminates.

Solution to BDHP: Otherwise, as in the forking lemma [13], $\mathcal{B}$ repeats the game with the same random tape for x_ℓ but with different choices of a random set for H_2-queries to obtain another forgery (U^*, σ') on the message m^* with h' such that $h^* \neq h'$ and $\sigma^* \neq \sigma'$. Then,

$$\frac{\sigma^*}{\sigma'} = \frac{e(U^* + h^* Q_{\mathsf{ID}_S}, S_{\mathsf{ID}_{\mathcal{V}}})}{e(U^* + h' Q_{\mathsf{ID}_S}, S_{\mathsf{ID}_{\mathcal{V}}})} = \frac{e(h^* Q_{\mathsf{ID}_S}, S_{\mathsf{ID}_{\mathcal{V}}})}{e(h' Q_{\mathsf{ID}_S}, S_{\mathsf{ID}_{\mathcal{V}}})} = \frac{e(Q_{\mathsf{ID}_S}, S_{\mathsf{ID}_{\mathcal{V}}})^{h^*}}{e(Q_{\mathsf{ID}_S}, S_{\mathsf{ID}_{\mathcal{V}}})^{h'}}$$
$$= e(Q_{\mathsf{ID}_S}, S_{\mathsf{ID}_{\mathcal{V}}})^{(h^* - h')} = e(aP, bcP)^{(h^* - h')} = (e(P, P)^{abc})^{(h^* - h')}. \quad (6)$$

Let $(h^* - h')^{-1} \mod q = \hat{h}$. Then, from the above equation, $\mathcal{B}$ solves the BDHP by computing

$$e(P, P)^{abc} = (\sigma^*/\sigma')^{\hat{h}} \tag{7}$$

Probability calculation: If $\mathcal{B}$ does not abort during the simulation then $\mathcal{A}$'s view is identical to its view in the real attack. The responses to H_1-queries and H_2-queries are as in the real attack, since each response is uniformly and independently distributed in G_1 and $\mathbb{Z}_q^*$ respectively. The key extraction, signature and verification queries are answered as in the real attack.

The probability that $\mathcal{B}$ does not abort during the simulation is

$$\left(1 - \frac{2}{q_{H_1}}\right)^{q_E + q_V} \left(1 - \frac{2}{q_{H_1}(q_{H_1} - 1)}\right)^{q_S}. \tag{8}$$

The probability that $\mathcal{A}$ did H_1-query for the identities $\mathsf{ID}_{\mathcal{S}}^*$ and $\mathsf{ID}_{\mathcal{V}}^*$ and that $\{\mathsf{ID}_{\mathcal{S}}^*, \mathsf{ID}_{\mathcal{V}}^*\} = \{\mathsf{ID}_\alpha, \mathsf{ID}_\beta\}$ is

$$\left(1 - \frac{1}{q^2}\right)\left(\frac{2}{q_{H_1}(q_{H_1} - 1)}\right). \tag{9}$$

Clearly $\mathcal{B}$'s advantage $\varepsilon_{\mathcal{B}}(\lambda)$ for solving the BDHP, that is, the total probability that $\mathcal{B}$ succeeds to solve BDHP, is the product of $\mathcal{A}$'s advantage $\varepsilon_{\mathcal{A}}(\lambda)$ of forging the proposed ID-SDVS and the above two probabilities. Hence

$$\varepsilon_{\mathcal{B}}(\lambda) \geq \left(1 - \frac{1}{q^2}\right)\left(1 - \frac{2}{q_{H_1}}\right)^{q_E + q_V} \left(1 - \frac{2}{q_{H_1}(q_{H_1} - 1)}\right)^{q_S} \left(\frac{2}{q_{H_1}(q_{H_1} - 1)}\right) \varepsilon_{\mathcal{A}}(\lambda).$$

Time calculation: It can be observed that running time of the algorithm $\mathcal{B}$ is same as that of $\mathcal{A}$ plus time taken to respond to the hash queries, key extraction queries, signature queries and verification queries, $q_{H_1} + q_{H_2} + q_E + q_S + q_V$. Hence the maximum running time required by $\mathcal{B}$ to solve the BDHP is

$$t' \leq (q_{H_1} + q_E + 3q_S + q_V)S_{G_1} + (q_S + q_V)P_e + q_S O_{G_1} + O_{G_2} + S_{G_2} + t$$

as $\mathcal{B}$ requires to compute one scalar multiplication in G_1 to respond to H_1 hash query, one scalar multiplication in G_1 to respond to key extraction query, three scalar multiplications in G_1 to respond to signature query, one scalar multiplication in G_1 to respond to verification query; one pairing computation to respond to signature query, one pairing computation to respond to verification query, one group operation in G_1 to respond to signature query, and, one group operation in G_2 and one scalar multiplication in G_2 to output a solution of BDHP.

5.3   Unverifiability

We now prove that the proposed ID-SDVS is strongly designated. That is, any third party other than the signer and the designated verifier, cannot verify the validity of a signature from a signer for a designated verifier with non-negligible probability. We show that if there exists a PPT adaptive chosen message and adaptive chosen identity algorithm which can verify the proposed ID-SDVS, then there exists another PPT algorithm which can use the earlier algorithm to solve the DBDHP. In particular, we prove the following theorem:

Theorem 2. *Given a security parameter λ, if there exists a PPT adversary $\mathcal{A}(q_{H_1}, q_{H_2}, q_E, q_S, q_V, \varepsilon_{\mathcal{A}}(\lambda), t)$ which breaks the designated unverifiability of the proposed ID-SDVS scheme in time t with success probability $\varepsilon_{\mathcal{A}}(\lambda)$, then there exists a PPT adversary $\mathcal{B}(\varepsilon_{\mathcal{B}}(\lambda), t')$ which solves DBDHP with success probability at least*

$$\varepsilon_{\mathcal{B}}(\lambda) \geq \left(1 - \frac{1}{q^2}\right)\left(1 - \frac{2}{q_{H_1}}\right)^{q_E + q_V}\left(1 - \frac{2}{q_{H_1}(q_{H_1} - 1)}\right)^{q_S}\left(\frac{2}{q_{H_1}(q_{H_1} - 1)}\right)\varepsilon_{\mathcal{A}}(\lambda)$$

in time at most

$$t' \leq (q_{H_1} + q_E + 3q_S + q_V)S_{G_1} + (q_S + q_V)P_e + q_S O_{G_1} + S_{G_1} + S_{G_2} + P_e + t$$

where $S_{G_1}, S_{G_2}, O_{G_1}, O_{G_2}$ and P_e are as defined in Theorem 1.

Proof: Let for a security parameter λ, $\mathcal{B}$ is challenged to solve the DBDHP for

$$\langle q, e : G_1 \times G_1 \rightarrow G_2, P, aP, bP, cP, \omega \rangle$$

where G_1 is an additive cyclic group of prime order q with generator P, G_2 is a multiplicative cyclic group of prime order q with generator $e(P, P)$, and $e : G_1 \times G_1 \rightarrow G_2$ is a cryptographic bilinear map as described in Sect. 2 and $\omega \xleftarrow{\$} G_2$. $a, b, c \xleftarrow{\$} \mathbb{Z}_q^*$ are unknown to $\mathcal{B}$. The goal of $\mathcal{B}$ is to solve DBDHP by verifying if $e(P, P)^{abc} = \omega$ using $\mathcal{A}$, the adversary who claims to forge the proposed ID-SDVS scheme.

$\mathcal{B}$ simulates the security game for strongness with $\mathcal{A}$ by doing the *Setup* and by responding the H_1-queries, H_2-queries, Key extraction queries, Signature queries and Verification queries as in the security game for unforgeability.

Output: After $\mathcal{A}$ has made its queries, it finally outputs a message m^*, an identity ID_S^* of a signer and an identity ID_V^* of a designated verifier on which it wishes to be challenged.

If $\mathcal{A}$ did not make H_1-query for the identities ID_S^* and ID_V^*, then the probability that verification equality holds is less than $1/q^2$. Thus, with probability greater than $1 - 1/q^2$, both the public keys were computed using H_1-oracle and there exist indices $i, j \in [1, q_{H_1}]$ such that $\mathsf{ID}_S^* = \mathsf{ID}_i$ and $\mathsf{ID}_V^* = \mathsf{ID}_j$. If $\{i, j\} \neq \{\alpha, \beta\}$, then $\mathcal{B}$ reports failure and terminates.

Solution to DBDHP: Otherwise, $\mathcal{B}$

- chooses a random $r \xleftarrow{\$} \mathbb{Z}_q^*$;
- sets $U = rP$;

- sets $h = H_2(m^*, U)$;
- sets $\sigma = e(bP, cP)^r \omega^h$;

and challenges $\mathcal{A}$ to verify the validity of the signature (U, σ).

Then, the verification holds if and only if each of the following holds

$$\sigma = e(U + hQ_{\mathsf{ID}_\mathcal{S}}, S_{\mathsf{ID}_\mathcal{V}})$$

$$\Longleftrightarrow \quad e(bP, cP)^r \omega^h = e(rP + haP, bP_{pub})$$

$$\Longleftrightarrow \quad e(P, P)^{bcr} \omega^h = e(rP + haP, bcP)$$

$$\Longleftrightarrow \quad e(P, P)^{bcr} \omega^h = e(P, P)^{(r+ha)bc}$$

$$\Longleftrightarrow \quad \omega^h = (e(P, P)^{abc})^h$$

$$\Longleftrightarrow \quad \omega = e(P, P)^{abc}$$

Then, from the above equation, $\mathcal{B}$ solves the DBDHP by simply returning the response of $\mathcal{A}$ to the strongness challenge.

Probability calculation: If $\mathcal{B}$ does not abort during the simulation then $\mathcal{A}$'s view is identical to its view in the real attack. The responses to H_1-queries and H_2-queries are as in the real attack, since each response is uniformly and independently distributed in G_1 and $\mathbb{Z}_q^*$ respectively. The key extraction, signature and verification queries are answered as in the real attack.

The probability that $\mathcal{B}$ does not abort during the simulation is

$$\left(1 - \frac{2}{q_{H_1}}\right)^{q_E + q_V} \left(1 - \frac{2}{q_{H_1}(q_{H_1} - 1)}\right)^{q_S}. \tag{10}$$

The probability that $\mathcal{A}$ did H_1-query for the identities $\mathsf{ID}_\mathcal{S}^*$ and $\mathsf{ID}_\mathcal{V}^*$ and that $\mathsf{ID}_\mathcal{S}^* = \mathsf{ID}_\alpha$ and $\mathsf{ID}_\mathcal{V}^* = \mathsf{ID}_\beta$ is

$$\left(1 - \frac{1}{q^2}\right)\left(\frac{2}{q_{H_1}(q_{H_1} - 1)}\right). \tag{11}$$

Clearly $\mathcal{B}$'s advantage $\varepsilon_\mathcal{B}(\lambda)$ for solving the DBDHP, that is, the total probability that $\mathcal{B}$ succeeds to solve DBDHP, is the product of $\mathcal{A}$'s advantage $\varepsilon_\mathcal{A}(\lambda)$ of breaking the strongness of the proposed ID-SDVS and the above two probabilities. Hence

$$\varepsilon_\mathcal{B}(\lambda) \geq \left(1 - \frac{1}{q^2}\right)\left(1 - \frac{2}{q_{H_1}}\right)^{q_E + q_V}\left(1 - \frac{2}{q_{H_1}(q_{H_1} - 1)}\right)^{q_S}\left(\frac{2}{q_{H_1}(q_{H_1} - 1)}\right)\varepsilon_\mathcal{A}(\lambda).$$

Time calculation: It can be observed that running time of the algorithm $\mathcal{B}$ is same as that of $\mathcal{A}$ plus time taken to respond to the hash queries, key extraction queries, signature queries and verification queries, that is, $q_{H_1} + q_{H_2} + q_E + q_S + q_V$. Hence the maximum running time required by $\mathcal{B}$ to solve the DBDHP is

$$t' \leq (q_{H_1} + q_E + 3q_S + q_V)S_{G_1} + (q_S + q_V)P_e + q_S O_{G_1} + S_{G_1} + S_{G_2} + P_e + t$$

since during the query phase, $\mathcal{B}$ requires to compute the same operations as in the security game for unforgeability and additionally, one scalar multiplication in G_1, one scalar multiplication in G_2 and one pairing computation to output a solution of DBDHP.

5.4   Non-transferability

The proposed scheme achieves the property of non-transferability as defined in Sect. 3. For this, we show that the transcripts simulated by the designated verifier are indistinguishable from the signatures that he receives from the signer. In the proposed scheme it can be observed that it is hard to distinguish the signature (U, σ) on a message m by the signer from the signature $(\widehat{U}, \widehat{\sigma})$ on the message m by the designated verifier, that is, the distributions

$$
\begin{aligned}
U &= rP \in G_1 \\
h &= H_2(m, U) \in \mathbb{Z}_q^* \\
V &= rP_{pub} + hS_{\mathsf{ID}_S} \in G_1 \\
\sigma &= e(V, Q_{\mathsf{ID}_V})
\end{aligned}
\qquad \text{and} \qquad
\begin{aligned}
\widehat{U} &= \widehat{r}P \in G_1 \\
\widehat{h} &= H_2(m, \widehat{U}) \in \mathbb{Z}_q^* \\
\widehat{V} &= \widehat{r}P + \widehat{h}Q_{\mathsf{ID}_S} \in G_1 \\
\widehat{\sigma} &= e(\widehat{V}, S_{\mathsf{ID}_V})
\end{aligned}
$$

are identical.

5.5   Strongness

The proposed scheme also achieves the property of strongness as defined in Sect. 3. Let $\sigma \leftarrow DVSign(S_{\mathsf{ID}_S}, Q_{\mathsf{ID}_V}, m)$. Then $\sigma \leftarrow DVTrans(S_{\mathsf{ID}_V^*}, Q_{\mathsf{ID}_S^*}, m)$ (where $Q_{\mathsf{ID}_S^*}$ and $S_{\mathsf{ID}_V^*}$ are defined as in the following) since

$$
\begin{aligned}
\sigma &= e(rP_{pub} + hS_{\mathsf{ID}_S}, Q_{\mathsf{ID}_V}) \\
&= e(rP_{pub} + hS_{\mathsf{ID}_S}, xQ_{\mathsf{ID}_V^*}) && \text{where } Q_{\mathsf{ID}_V} = xQ_{\mathsf{ID}_V^*} \\
&= e(rxP_{pub} + hxS_{\mathsf{ID}_S}, Q_{\mathsf{ID}_V^*}) \\
&= e(rP_{pub} + r(x-1)P_{pub} + hxS_{\mathsf{ID}_S}, Q_{\mathsf{ID}_V^*}) \\
&= e(rP_{pub} + r(x-1)hY + hxS_{\mathsf{ID}_S}, Q_{\mathsf{ID}_V^*}) && \text{where } Y = h^{-1}P_{pub} \\
&= e(rP_{pub} + h(r(x-1)Y + xS_{\mathsf{ID}_S}), Q_{\mathsf{ID}_V^*}) \\
&= e(rP_{pub} + hS_{\mathsf{ID}_S^*}, Q_{\mathsf{ID}_V^*}) && \text{where } S_{\mathsf{ID}_S^*} = r(x-1)Y + xS_{\mathsf{ID}_S} \, .
\end{aligned}
$$

6   Comparative Analysis

Here, we compare our scheme with similar existing ID-SDVS schemes [8,11,15] and show that our scheme is more efficient in the sense of computation and operation time than these schemes.

For the computation of operation time in pairing-based scheme, to achieve the 1024-bit RSA level security, Tate pairing defined over the supersingular elliptic curve $E = F_p : y^2 = x^3 + x$ with embedding degree 2 was used, where q is a 160-bit Solinas prime $q = 2^{159} + 2^{17} + 1$ and p a 512-bit prime satisfying $p + 1 = 12qr$, using MIRACL [12], a standard cryptographic library, and the hardware platform is a PIV 3 GHZ processor with 512 M bytes memory and the Windows XP operating system. For computation of operation time, we refer to [4] where the operation time for various cryptographic operations have been obtained. The OT(Operation Time) for one scalar multiplication is 6.38 ms, for one exponentiation in G_2 it is 5.31 ms, for one map-to-point hash function it is 3.04 ms and for one pairing computation it is 20.04 ms. Other operations are omitted in the following analysis since their computation cost is trivial, such as the cost of an inverse operation over Z_q^* takes only 0.03 ms and one general hash function takes less than 0.001 ms which are negligible with compare to the time taken by the other operations.

To evaluate the total operation time in the efficiency comparison tables, we use the method from [1,4]. In each of the two phases: signature generation and verification, we compare the total number of scalar multiplications (SM), exponentiations (E), map-to-point hash functions (H), bilinear pairings (P) and the consequent operation time (OT) (Table 2).

Table 2. Efficiency comparision

Scheme	SM	E	H	P	OT(ms)
Susilo et al. [15]	2	1	0	1	38.11
Kancharla et al.[8]	6	0	1	0	61.36
Lee et al. [11]	2	1	0	2	58.15
Our scheme	**3**	**0**	**0**	**1**	**39.18**

Scheme	SM	E	H	P	OT(ms)
Susilo et al. [15]	0	2	0	2	50.70
Kancharla et al.[8]	0	0	1	4	83.20
Lee et al. [11]	1	0	0	2	46.46
Our scheme	**1**	**0**	**0**	**1**	**26.42**

Signature Generation   Verification

Scheme	SM	E	H	P	OT(ms)
Susilo et al. [15]	2	3	0	3	88.81
Kancharla et al.[8]	6	0	2	4	144.56
Lee et al. [11]	3	1	0	4	104.61
Our scheme	**4**	**0**	**0**	**2**	**65.60**

Overall Scheme

7   Conclusion

In this paper, we have proposed a strong designated verifier signature scheme on the identity-based setting. Our scheme is strong existentially unforgeable against adaptive chosen message and adaptive chosen identity attack under standard assumptions, the hardness of the computational and decisional Bilinear Diffie-Hellman problems. We also provide a proof for the strongness property of our scheme. Moreover, we do an efficiency comparison of our scheme with the existing similar schemes. In the view of computational cost and operation time our scheme is significantly more efficient than the existing schemes. The scheme is suitable for the environments in which less computational cost with strong security is required.

References

1. Cao, X., Kou, W., Xiaoni, D.: A pairing-free identity-based authenticated key agreement protocol with minimal message exchanges. Inf. Sci. **180**(15), 2895–2903 (2010)

2. Chaum, D.: Zero-knowledge undeniable signatures (extended abstract). In: Damgård, I.B. (ed.) EUROCRYPT 1990. LNCS, vol. 473, pp. 458–464. Springer, Heidelberg (1991). doi:10.1007/3-540-46877-3_41

3. Chaum, D., Antwerpen, H.: Undeniable signatures. In: Brassard, G. (ed.) CRYPTO 1989. LNCS, vol. 435, pp. 212–216. Springer, Heidelberg (1990). doi:10.1007/0-387-34805-0_20

4. Debiao, H., Jianhua, C., Jin, H.: An identity-based proxy signature schemes without bilinear pairings. Ann. Telecommun. **66**(11–12), 657–662 (2011)

5. Du, H., Wen, Q.: Attack on Kang et al.'s identity-based strong designated verifier signature scheme. IACR Cryptology ePrint Archive, 2008:297 (2008)

6. Huang, X., Susilo, W., Mu, Y., Zhang, F.: Short (identity-based) strong designated verifier signature schemes. In: Chen, K., Deng, R., Lai, X., Zhou, J. (eds.) ISPEC 2006. LNCS, vol. 3903, pp. 214–225. Springer, Heidelberg (2006). doi:10.1007/11689522_20

7. Jakobsson, M., Sako, K., Impagliazzo, R.: Designated verifier proofs and their applications. In: Maurer, U. (ed.) EUROCRYPT 1996. LNCS, vol. 1070, pp. 143–154. Springer, Heidelberg (1996). doi:10.1007/3-540-68339-9_13

8. Kancharla, P.K., Gummadidala, S., Saxena, A.: Identity based strong designated verifier signature scheme. Informatica **18**(2), 239–252 (2007)

9. Kang, B., Boyd, C., Dawson, E.: Identity-based strong designated verifier signature schemes: attacks and new construction. Comput. Electr. Eng. **35**(1), 49–53 (2009)

10. Kang, B., Boyd, C., Dawson, E.D.: A novel identity-based strong designated verifier signature scheme. J. Syst. Softw. **82**(2), 270–273 (2009)

11. Lee, J.-S., Chang, J.H., Lee, D.H.: Forgery attacks on Kang et al.'s identity-based strong designated verifier signature scheme and its improvement with security proof. Comput. Electr. Eng. **36**(5), 948–954 (2010)

12. MIRACL. Multiprecision integer and rational arithmetic cryptographic library. http://certivox.org/display/EXT/MIRACL

13. Pointcheval, D., Stern, J.: Security arguments for digital signatures and blind signatures. J. Cryptol. **13**(3), 361–396 (2000)

14. Saeednia, S., Kremer, S., Markowitch, O.: An efficient strong designated verifier signature scheme. In: Lim, J.-I., Lee, D.-H. (eds.) ICISC 2003. LNCS, vol. 2971, pp. 40–54. Springer, Heidelberg (2004). doi:10.1007/978-3-540-24691-6_4

15. Susilo, W., Zhang, F., Mu, Y.: Identity-based strong designated verifier signature schemes. In: Wang, H., Pieprzyk, J., Varadharajan, V. (eds.) ACISP 2004. LNCS, vol. 3108, pp. 313–324. Springer, Heidelberg (2004). doi:10.1007/978-3-540-27800-9_27

16. Zhang, J., Mao, J.: A novel ID-based designated verifier signature scheme. Inf. Sci. **178**(3), 766–773 (2008)

The Shortest Signatures Ever

Mohamed Saied Emam Mohamed[1]([✉]) and Albrecht Petzoldt[2]

[1] Technische Universität Darmstadt, Darmstadt, Germany
mohamed@cdc.informatik.tu-darmstadt.de
[2] Kyushu University, Fukuoka, Japan
petzoldt@imi.kyushu-u.ac.jp

Abstract. Multivariate Cryptography is one of the main candidates for creating post quantum public key cryptosystems. Especially in the area of digital signatures, there exist many practical and secure multivariate schemes. In this paper we present a general technique to reduce the signature size of multivariate schemes. Our technique enables us to reduce the signature size of nearly all multivariate signature schemes by 10 to 15% without slowing down the scheme significantly. We can prove that the security of the underlying scheme is not weakened by this modification. Furthermore, the technique enables a further reduction of the signature size when accepting a slightly more costly verification process. This trade off between signature size and complexity of the verification process can not be observed for any other class of digital signature schemes. By applying our technique to the Gui signature scheme, we obtain the shortest signatures of all existing digital signature schemes.

Keywords: Post quantum cryptography · Multivariate cryptography · Digital signatures · Signature size

1 Introduction

Cryptographic techniques are an essential tool to guarantee the security of communication in modern society. Today, the security of nearly all of the cryptographic schemes used in practice is based on number theoretic problems such as factoring large integers and solving discrete logarithms. The best known schemes in this area are RSA [18], DSA [10] and ECC. However, schemes like these will become insecure as soon as large enough quantum computers arrive. The reason for this is Shor's algorithm [19], which solves number theoretic problems like integer factorization and discrete logarithms in polynomial time on a quantum computer. Therefore, one needs alternatives to those classical public key schemes, which are based on hard mathematical problems not affected by quantum computer attacks (so called post quantum cryptosystems). The increasing importance of research in this area has recently been emphasized by a number of authorities. For example, the American National Security Agency (NSA) has recommended governmental organizations to change their security infrastructures from schemes like RSA to post quantum schemes [13] and the National

© Springer International Publishing AG 2016
O. Dunkelman and S.K. Sanadhya (Eds.): INDOCRYPT 2016, LNCS 10095, pp. 61–77, 2016.
DOI: 10.1007/978-3-319-49890-4_4

Institute of Standards and Technologies (NIST) is preparing to standardize these schemes [14].

According to [14], one of the main candidates for this standardization is multivariate cryptography. Multivariate schemes are in general very fast and require only modest computational resources, which makes them attractive for the use on low cost devices like smart cards and RFID chips [2,3]. Additionally, at least in the area of digital signatures, there exists a large number of practical multivariate schemes [6,11,17].

In this paper we present a general technique to reduce the signature size of multivariate signature schemes. The key idea of our strategy is to send only a part of the signature to the receiver of a signed message. The verification process consists in solving a multivariate quadratic system in a very small number of variables.

By our technique we can reduce the signature size of nearly every multivariate signature scheme by 10 to 15% without increasing the key sizes or slowing down the scheme significantly. Furthermore, we can prove that the security of the signature schemes is not weakened by our modifications. Moreover, we can achieve a further reduction of the signature length when accepting a small increase of the verification cost. This trade off is, to our knowledge, unique for multivariate schemes and can not be observed for any other class of digital signature schemes. By applying our technique to the Gui signature scheme of Asiacrypt 2015 [17], we can reduce the signature size of this scheme to 110 bit (80 bit security), by which we obtain the shortest signatures of all existing schemes.

Our technique is especially attractive in situations, in which the connection between the sender and the receiver is very slow. An example for this are (wireless) sensor networks. In such systems, the actual messages are often very short and therefore a major part of the communication consists of the signatures itself. Reducing the signature length therefore reduces the communication cost and speeds up the system significantly.

The rest of this paper is organized as follows. In Sect. 2, we give a short overview of (wireless) sensor networks and show how these systems can be sped up by our technique. Section 3 gives an overview of the basic concepts of multivariate cryptography and introduces two of the best known and most widely studied multivariate signature schemes, namely the Rainbow and the HFEv-signature schemes. Section 4 describes the most important methods for solving systems of multivariate quadratic equations which can be used in the verification process of our schemes. In Sect. 5 we present our technique in detail and show that our modifications do not weaken the security of the underlying schemes. Section 6 shows, for concrete parameter sets of Rainbow and HFEv-, which actual reductions in terms of the signature size can be achieved by our technique. Furthermore, in this section, we analyze the efficiency of our technique using a large set of experiments. In Sect. 7 we show how to apply our technique to the Gui signature scheme, which allows us to efficiently generate secure signatures of size only 110 bit (80 bit security). Finally, Sect. 8 gives a short discussion of the benefits of our technique and Sect. 9 concludes the paper.

2   (Wireless) Sensor Networks

In this section we describe a possible application scenario for our technique. Since we aim at reducing the signature size, our technique is especially attractive in situations in which the connection between sender and receiver is very slow. An example for this are (wireless) sensor networks.

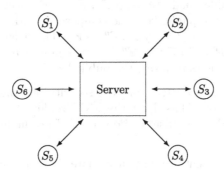

Fig. 1. Sensor network

Let us assume that we have a number of small sensors, which report at regular intervals their status to a server (as shown in Fig. 1). The status messages itself might be very short, but the messages have to be signed in order to prevent adversaries to send false messages to the server. The single sensors have only restricted computing, power and memory capacities. Therefore, multivariate signature schemes are an attractive candidate to generate the above mentioned signatures. To do this, one has to store the private key of the multivariate scheme on the sensor. For current multivariate schemes such as Gui, the size of the private key is only 3 kB. If this is still too big for the sensor, the private key can easily be stored as a random seed.

However, the main problem in our scenario is the slow connection between the sensors and their server. Therefore, our goal is to reduce the communication between the sensors and the server as far as possible. Since, in the upper status messages, the message itself might be very short, the length of the signature plays a major role. Now, multivariate scheme already generate relatively short signatures (usually a few hundred bits). However, as we will show in Sect. 5, we can reduce the length of multivariate signatures further, without weakening the security of the scheme, increasing the key sizes or making the signature generation process more costly. By doing so, we can reduce the amount of communication and therefore speed up the system significantly.

3   Multivariate Cryptography

The basic objects of multivariate cryptography are systems of multivariate quadratic polynomials. The security of multivariate schemes is based on the

MQ Problem: Given m multivariate quadratic polynomials $p^{(1)}(\mathbf{x}), \ldots, p^{(m)}(\mathbf{x})$ in n variables $x_1, \ldots, x_n$, find a vector $\bar{\mathbf{x}} = (\bar{x}_1, \ldots, \bar{x}_n)$ such that $p^{(1)}(\bar{\mathbf{x}}) = \ldots = p^{(m)}(\bar{\mathbf{x}}) = 0$.

The MQ problem (for $m \approx n$) is proven to be NP-hard even for quadratic polynomials over the field GF(2) [8].

To build a public key cryptosystem based on the MQ problem, one starts with an easily invertible quadratic map $\mathcal{F} : \mathbb{F}^n \to \mathbb{F}^m$ (central map). To hide the structure of $\mathcal{F}$ in the public key, one composes it with two invertible affine (or linear) maps $\mathcal{S} : \mathbb{F}^m \to \mathbb{F}^m$ and $\mathcal{T} : \mathbb{F}^n \to \mathbb{F}^n$. The *public key* of the scheme is therefore given by $\mathcal{P} = \mathcal{S} \circ \mathcal{F} \circ \mathcal{T} : \mathbb{F}^n \to \mathbb{F}^m$. The *private key* consists of $\mathcal{S}$, $\mathcal{F}$ and $\mathcal{T}$ and therefore allows to invert the public key.

Note: Due to the above construction, the security of multivariate public key cryptosystems is not only based on the MQ-Problem but also on the EIP-Problem ("Extended Isomorphism of Polynomials") of finding the composition of $\mathcal{P}$ [5].

In this paper we concentrate on multivariate signature schemes. For this we require $n \geq m$, which ensures that every message has a signature. The standard signature generation and verification process of a multivariate signature scheme works as shown in Fig. 2.

Signature Generation

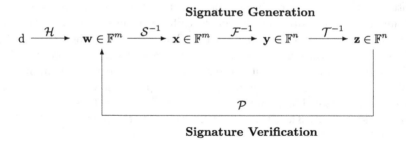

Signature Verification

Fig. 2. General workflow of multivariate signature schemes

Signature Generation: To generate a signature for a document d, we use a hash function $\mathcal{H}$ to compute a hash value $\mathbf{w} = \mathcal{H}(d) \in \mathbb{F}^m$. After that, one computes recursively $\mathbf{x} = \mathcal{S}^{-1}(\mathbf{w}) \in \mathbb{F}^m$, $\mathbf{y} = \mathcal{F}^{-1}(\mathbf{x}) \in \mathbb{F}^n$ and $\mathbf{z} = \mathcal{T}^{-1}(\mathbf{y})$. The signature of the document d is $\mathbf{z} \in \mathbb{F}^n$. Here, $\mathcal{F}^{-1}(\mathbf{x})$ means finding one (of approximately q^{n-m}) pre-image of $\mathbf{x}$ under the central map $\mathcal{F}$.

Verification: To check, if $\mathbf{z} \in \mathbb{F}^n$ is indeed a valid signature for the document d, we compute the hash value $\mathbf{w} = \mathcal{H}(d) \in \mathbb{F}^m$ and $\mathbf{w}' = \mathcal{P}(\mathbf{z}) \in \mathbb{F}^m$. If $\mathbf{w}' = \mathbf{w}$ holds, the signature is accepted, otherwise rejected.

Since the late 1980s, many multivariate schemes both for encryption and digital signatures have been proposed. The first one was the Matsumoto-Imai cryptosystem [12], which was later extended to schemes like Sflash [16] and HFE [15]. A different research direction lead to the development of multivariate schemes such as UOV [11], Rainbow [6], enTTS [21] and SimpleMatrix [20]. Although several of these schemes have been broken due to newly developed attacks, a number of multivariate schemes such as UOV, Rainbow and HFEv- has withstood (for suitable parameter sets) cryptanalysis for more than 20 years now. In the next two subsections, we introduce two of these schemes.

3.1   The Rainbow Signature Scheme

The Rainbow signature scheme [6] is one of the most promising and best studied multivariate schemes. The scheme can be described as follows.

Let $\mathbb{F}$ be a finite field, $n \in \mathbb{N}$ and $v_1 < v_2 < \ldots < v_\ell < v_{\ell+1} = n$ be a sequence of integers. We set $O_i = \{v_i + 1, \ldots, v_{i+1}\}$ and $V_i = \{1, \ldots, v_i\}$ $(i = 1, \ldots \ell)$.

Key Generation: The *private key* of the scheme consists of two invertible affine maps $\mathcal{S} : \mathbb{F}^m \to \mathbb{F}^m$ and $\mathcal{T} : \mathbb{F}^n \to \mathbb{F}^n$ and a quadratic map $\mathcal{F}(\mathbf{x}) = (f^{(v_1+1)}(\mathbf{x}), \ldots, f^{(n)}(\mathbf{x})) : \mathbb{F}^n \to \mathbb{F}^m$. Here, $m = n - v_1$ is the number of components of $\mathcal{F}$. The components of the central map $\mathcal{F}$ are of the form

$$f^{(i)} = \sum_{k,l \in V_j} \alpha_{kl}^{(i)} \cdot x_k \cdot x_l + \sum_{k \in V_j, l \in O_j} \beta_{kl}^{(i)} \cdot x_k \cdot x_l + \sum_{k \in V_j \cup O_j} \gamma_k^{(i)} \cdot x_k + \eta^{(i)}. \quad (1)$$

Here, j is the only integer such that $i \in O_j$. The *public key* is the composed map $\mathcal{P} = \mathcal{S} \circ \mathcal{F} \circ \mathcal{T} : \mathbb{F}^n \to \mathbb{F}^m$.

Signature Generation: To generate a signature for a document d, one uses a hash function $\mathcal{H}$ to compute the hash value $\mathbf{w} = \mathcal{H}(d) \in \mathbb{F}^m$. After that, one computes recursively $\mathbf{x} = \mathcal{S}^{-1}(\mathbf{w})$, $\mathbf{y} = \mathcal{F}^{-1}(\mathbf{x})$ and $\mathbf{z} = \mathcal{T}^{-1}(\mathbf{y})$. Here, $\mathcal{F}^{-1}(\mathbf{x})$ means finding one (of approximately q^{v_1}) pre-image of $\mathbf{x}$ under the central map $\mathcal{F}$. In the case of Rainbow, this is done as follows.

Algorithm 1. Inversion of the Rainbow central map

Input: Rainbow central map $\mathcal{F}$, vector $\mathbf{x} \in \mathbb{F}^m$
Output: vector $\mathbf{y} \in \mathbb{F}^n$ such that $\mathcal{F}(\mathbf{y}) = \mathbf{x}$
1: Choose random values for the variables $y_1, \ldots, y_{v_1}$ and substitute these values into the polynomials $f^{(i)}$ $(i = v_1 + 1, \ldots, n)$.
2: **for** $k = 1$ to ℓ **do**
3:    Perform Gaussian Elimination on the polynomials $f^{(i)}$ $(i \in O_k)$ to get the values of the variables y_i $(i \in O_k)$.
4:    Substitute the values of y_i $(i \in O_k)$ into the polynomials $f^{(i)}$ $(i \in \{v_{k+1} + 1, \ldots n\})$.
5: **end for**

It might happen that one of the linear systems in step 3 of the algorithm does not have a solution. In this case one has to choose other values for $y_1, \ldots, y_{v_1}$ and start again.

The signature of the document d is $\mathbf{z} \in \mathbb{F}^n$.

Signature Verification: To check if $\mathbf{z} \in \mathbb{F}^n$ is indeed a valid signature for the document d, one computes $\mathbf{w} = \mathcal{H}(d)$ and $\mathbf{w}' = \mathcal{P}(\mathbf{z})$. If $\mathbf{w}' = \mathbf{w}$ holds, the signature is accepted, otherwise rejected.

3.2   The HFEv- Signature Scheme

Another widely known construction is the HFEv- signature scheme, which is often used as the basis of more advanced multivariate signature schemes such as QUARTZ and Gui [17] (see Sect. 7).

Let $\mathbb{F} = \mathbb{F}_q$ be a finite field with q elements and $\mathbb{E}$ be a degree n extension field of $\mathbb{F}$. Furthermore, we choose integers D, a and v. Let $\varPhi$ be the canonical isomorphism between $\mathbb{F}^n$ and $\mathbb{E}$, i.e.

$$\varPhi(x_1,\ldots,x_n) = \sum_{i=1}^{n} x_i \cdot X^{i-1}. \tag{2}$$

The central map $\mathcal{F}$ of the HFEv- scheme is a map from $\mathbb{F}^v \times \mathbb{E}$ to $\mathbb{E}$ of the form

$$\mathcal{F}(X) = \sum_{\substack{0 \leq i \leq j}}^{q^i+q^j \leq D} \alpha_{ij} \cdot X^{q^i+q^j} + \sum_{i=0}^{q^i \leq D} \beta_i(v_1,\ldots,v_v) \cdot X^{q^i} + \gamma(v_1,\ldots,v_v), \tag{3}$$

with $\alpha_{ij} \in \mathbb{E}$, $\beta_i : \mathbb{F}^v \to \mathbb{E}$ being linear and $\gamma : \mathbb{F}^v \to \mathbb{E}$ being a quadratic function.

Due to the special form of $\mathcal{F}$, the map $\bar{\mathcal{F}} = \varPhi^{-1} \circ \mathcal{F} \circ \varPhi$ is a multivariate quadratic map from $\mathbb{F}^{n+v}$ to $\mathbb{F}^n$. To hide the structure of $\bar{\mathcal{F}}$ in the public key, one combines it with two affine (or linear) maps $\mathcal{S} : \mathbb{F}^n \to \mathbb{F}^{n-a}$ and $\mathcal{T} : \mathbb{F}^{n+v} \to \mathbb{F}^{n+v}$ of maximal rank.

The *public key* of the scheme is the composed map $\mathcal{P} = \mathcal{S} \circ \bar{\mathcal{F}} \circ \mathcal{T} : \mathbb{F}^{n+v} \to \mathbb{F}^{n-a}$, the *private key* consists of $\mathcal{S}$, $\mathcal{F}$ and $\mathcal{T}$.

Signature Generation: To generate a signature for a document d, we use a hash function $\mathcal{H}$ to compute the hash value $\mathbf{w} = \mathcal{H}(d) \in \mathbb{F}^{n-a}$. After that, the signer performs the following three steps.

1. Compute a preimage $\mathbf{x} \in \mathbb{F}^n$ of $\mathbf{w}$ under the affine map $\mathcal{S}$.
2. Lift $\mathbf{x}$ to the extension field $\mathbb{E}$ (using the isomorphism $\varPhi$). Denote the result by X.
   Choose random values for the Vinegar variables $v_1,\ldots,v_v \in \mathbb{F}$ and compute $\mathcal{F}_V = \mathcal{F}(v_1,\ldots,v_v)$.
   Solve the univariate polynomial equation $\mathcal{F}_V(Y) = X$ by Berlekamp's algorithm and compute $\mathbf{y}' = \varPhi^{-1}(Y) \in \mathbb{F}^n$. Set $\mathbf{y} = (\mathbf{y}'||v_1||\ldots||v_v)$.
3. Compute the signature $\mathbf{z} \in \mathbb{F}^{n+v}$ of the document d by $\mathbf{z} = \mathcal{T}^{-1}(\mathbf{y})$.

Signature verification: To check if $\mathbf{z} \in \mathbb{F}^{n+v}$ is indeed a valid signature for the document d, we compute the hash value $\mathbf{w} = \mathcal{H}(d)$ and $\mathbf{w}' = \mathcal{P}(\mathbf{z}) \in \mathbb{F}^{n-a}$. If $\mathbf{w}' = \mathbf{w}$ holds, the signature is accepted, otherwise rejected.

4 Solving Multivariate Quadratic Systems

In this section we give a short overview of the most important techniques for solving systems of multivariate quadratic equations that might be used in our scheme.

4.1 The Relinearization Technique

In [11], Kipnis and Patarin proposed the Relinearization technique, which allows to solve highly overdetermined multivariate quadratic systems in polynomial time. In particular, a system can be solved by this technique if the number of equations m is greater or equal to

$$m \geq \frac{(n+1) \cdot (n+2)}{2} - 1. \tag{4}$$

The idea can be described as follows:

1. Interpret each of the quadratic monomials $x_i \cdot x_j$ in the system as a new variable x_{ij}.
2. Solve the resulting linear system by Gaussian Elimination.

If Eq. 4 holds, the linear system in step 2 has (in most cases) exactly one solution, which directly yields the solution of the quadratic system.

4.2 Other Techniques

If the number n of variables in the system exceeds the upper bound given by Eq. 4, the Relinearization method produces a huge set of fake solutions, which are solutions to the linear system, but do not solve the original quadratic one. In this case one has to use other methods to solve the quadratic system, for example XL (see Algorithm 2) or a Gröbner Basis algorithm such as F_4 or F_5 [7].

Algorithm 2. XL-Algorithm

Input: Set of polynomials $F = (f^{(1)}, \ldots, f^{(m)})$
Output: vector $\mathbf{x} = (x_1, \ldots, x_n)$ such that $f^{(1)}(\mathbf{x}) = \ldots = f^{(m)}(\mathbf{x}) = 0$
1: **for** $i = 1$ to n **do**
2:     Fix an integer $D > 2$.
3:     Generate all polynomials $h \cdot f^{(j)}$ with $h \in T^n_{D-2}$ and $j = 1, \ldots, m$.
4:     Perform Gaussian elimination on the set of all polynomials generated in the previous step to generate one equation containing only x_i.
5:     If step 4 produced at least one univariate polynomial in x_i, solve this polynomial by e.g. Berlekamp's algorithm.
6:     Simplify the equations by substituting the value of x_i.
7: **end for**
8: **return** $\mathbf{x} = (x_1, \ldots, x_n)$

The XL Algorithm ("eXtended Linearization") was proposed by Courtois et al. in [4]. In order to solve a quadratic system F, the algorithm enlarges the system by multiplying the polynomials $f^{(i)} \in F$ by all monomials of degree $d \leq D - 2$. By doing so, it obtains a large system $\tilde{F}$ of degree D polynomials. The algorithm performs Gaussian Elimination on the system $\tilde{F}$ in order to find a univariate polynomial which then can be solved by Berlekamp's algorithm. In this case, it substitutes the solution into $\tilde{F}$ to simplify the system.

However, if the degree D chosen in step 2 of the algorithm is too small, the enlarged system $\tilde{F}$ will not contain a univariate polynomial. In this case one has to increase the degree D and try again. The smallest degree for which the XL algorithm outputs a solution of the system F is called the *degree of regularity* d_{reg}. This degree of regularity mainly determines the complexity of the algorithm.

For our purposes we want the quadratic system F to be efficiently solvable. In the following, we therefore only consider multivariate systems which can be solved by the XL algorithm at degree 3 or 4. This directly yields an upper bound on the number of variables in our systems. However, for multivariate quadratic systems, it is a hard task to find explicit formulas for this upper bound. In Sect. 6, we therefore try to find these upper bound for concrete instances of Rainbow and HFEv- using a large number of experiments.

5   Reducing the Signature Size of Multivariate Schemes

In this section we present our technique to reduce the signature size of multivariate schemes. Our technique can be applied to nearly all multivariate signature schemes, including UOV [11], Rainbow [6], HFEv- [15] and TTS [21].

However, it is not possible to apply our technique directly to more advanced multivariate signature schemes such as QUARTZ and Gui [17]. In order to reduce the signature size of these schemes, we have to modify our technique slightly (see Sect. 7).

Let $((\mathcal{S}, \mathcal{F}, \mathcal{T}), \mathcal{P})$ be a key pair of a multivariate signature scheme.

Signature Generation: The sender of a message d uses his private key $(\mathcal{S}, \mathcal{F}, \mathcal{T})$ to compute a signature $\mathbf{z} \in \mathbb{F}^n$ for the document d just as in the case of the standard signature scheme. After that, he removes the last k $\mathbb{F}$-elements from the signature $\mathbf{z}$ to obtain a partial signature $\tilde{\mathbf{z}} \in \mathbb{F}^{n-k}$ and sends $\tilde{\mathbf{z}}$ to the verifier.

Signature Verification: The receiver of a signed message checks if $\tilde{\mathbf{z}}$ is indeed part of a valid signature for the document d. To do this, he computes the hash value $\mathbf{w} = \mathcal{H}(d) \in \mathbb{F}^m$ of the document d, substitutes the elements of the partial signature $\tilde{\mathbf{z}}$ into the public key $\mathcal{P}$ and uses one of the techniques described in the previous section to solve the resulting system $\tilde{\mathcal{P}}(z_{n-k+1}, \ldots, z_n) = \mathbf{w}$ of m quadratic equations in k variables. If the system has a solution, the signature $\tilde{\mathbf{z}}$ is accepted, otherwise it is rejected.

Remark: Indeed we do not have to reconstruct the complete signature $\mathbf{z} \in \mathbb{F}^n$ by solving the system $\tilde{\mathcal{P}}(z_{n-k+1}, \ldots, z_n) = \mathbf{w}$. Instead of this, it suffices to check whether the system has a solution. If we use the Relinearization technique for the verification, it therefore suffices to bring the linearized system into row-echelon form and check if the last equations (which contain only zero terms on the left) hold. The reason for this is given by Proposition 1.

Algorithms 3 and 4 show the standard verification process for multivariate schemes and our modified one in algorithmic form.

5.1 How to Choose the Parameter k?

In this section we consider the question how we should choose the number k of $\mathbb{F}$-elements removed from the original signature $\mathbf{z}$. Increasing the number k will lead to shorter signatures but increase the computational effort to check the authenticity of a signature.

If $\frac{(k+1) \cdot (k+2)}{2} - 1 \leq m$, the system $\tilde{\mathcal{P}}(\mathbf{x}) = \mathbf{w}$ from step 3 of Algorithm 3 can be solved by the Relinearization technique (see Sect. 4.1). This means that we can find a solution in polynomial time. Therefore we get

Proposition 1. *Let $\frac{(k+1) \cdot (k+2)}{2} - 1 \leq m$. Then the partial signature $\tilde{\mathbf{z}} \in \mathbb{F}^{n-k}$ can not be found significantly faster than the full signature $\mathbf{z} \in \mathbb{F}^n$.*

Proof. Let us assume that we have a valid partial signature $\tilde{\mathbf{z}} \in \mathbb{F}^{n-k}$ for a document d. By substituting the elements of $\tilde{\mathbf{z}}$ into the public key $\mathcal{P}$ we obtain a system $\tilde{\mathcal{P}}$ of m quadratic equations in k variables. Due to our assumption we can solve this system and therefore recover the whole signature $\mathbf{z}$ in polynomial time using the Relinearization technique (see Sect. 4.1).

Remark: The above proposition states that an attacker who is able to generate a valid partial signature $\tilde{\mathbf{z}} \in \mathbb{F}^{n-k}$ can generate the whole signature $\mathbf{z} \in \mathbb{F}^n$ quasi without additional computational cost. This shows that the security of the underlying signature schemes is not weakened by our modifications.

Algorithm 3. Standard Verification Algorithm for Multivariate Schemes	**Algorithm 4.** Modified Verification Algorithm for Multivariate Schemes
Input: public key $\mathcal{P}$, document d, signature $\mathbf{z} \in \mathbb{F}^n$	**Input:** public key $\mathcal{P}$, document d, partial signature $\tilde{\mathbf{z}} \in \mathbb{F}^{n-k}$
Output: boolean value **TRUE** or **FALSE**	**Output:** boolean value **TRUE** or **FALSE**
1: $\mathbf{w} = \mathcal{H}(d) \in \mathbb{F}^m$	1: $\mathbf{w} = \mathcal{H}(d) \in \mathbb{F}^m$
2: $\mathbf{w}' = \mathcal{P}(\mathbf{z})$	2: $\tilde{\mathcal{P}} = \mathcal{P}(\tilde{\mathbf{z}})$
3: if $\mathbf{w}' = \mathbf{w}$ then	3: if IsConsistent $(\tilde{\mathcal{P}}(\mathbf{x}) = \mathbf{w}) = $ **TRUE** then
4:	4: return **TRUE**
5: return **TRUE**	5: else
6: else	6: return **FALSE**
7: return **FALSE**	7: end if
8: end if	

From Eq. 4, we can derive the maximal number k of elements by which we can reduce the length of the original signature $\mathbf{z}$ such that the partial signature $\tilde{\mathbf{z}}$ can be verified using the Relinearization technique by

$$k = \left\lfloor \frac{1}{2} \cdot (\sqrt{9 + 8 \cdot m} - 3) \right\rfloor. \tag{5}$$

In the following, we slacken this condition a bit. If $\frac{(k+1)\cdot(k+2)}{2} - 1 > m$, we can not solve the system $\tilde{\mathcal{P}}(\mathbf{x}) = \mathbf{w}$ by the Relinearization technique any more and therefore can not recover the whole signature $\mathbf{z}$ in polynomial time. However, if the number k is not too large, we can solve the system $\tilde{\mathcal{P}}(\mathbf{x}) = \mathbf{w}$ by the XL Algorithm at a very low degree (e.g. $d_{\text{reg}} \in \{3, 4\}$). In this case, the computational effort of recovering the whole signature $\mathbf{z}$ is still very small. We therefore come to the conjecture

Conjecture: If the system $\tilde{\mathcal{P}}(z_{n-k+1}, \ldots, z_n) = \mathbf{w}$ of m quadratic equations in k variables can be solved by the XL Algorithm at degree 3 or 4, our technique does not weaken the security of the underlying signature scheme.

Furthermore, in this case, the additional computational effort needed to verify the reduced signature is still acceptable.

However, as already mentioned in Sect. 4.2, it is a hard task to find explicit upper bounds for the parameter k such that the system $\tilde{\mathcal{P}}(\mathbf{x}) = \mathbf{w}$ can be solved by the XL Algorithm at a given degree. In the next section we try to find, for concrete instances of Rainbow and HFEv-, these upper bounds by performing a large set of experiments.

6   Results

In this section we show, for concrete instances of the multivariate schemes Rainbow and HFEv-, the possible reduction of the signature size enabled by our technique. Furthermore we analyze the efficiency of our technique using a straightforward implementation of the schemes.

Table 1 shows, for the multivariate signature schemes Rainbow and HFEv-, possible choices of the parameter k of our technique and the resulting reduction in terms of the signature size of the schemes.

For every scheme and parameter set, the 7-th column of the table gives the maximal values of the parameter k such that the partial signature $\tilde{\mathbf{z}}$ can be verified by the Relinearization technique, the XL Algorithm with $d_{\text{reg}} = 3$ and the XL Algorithm with $d_{\text{reg}} = 4$ respectively. While the first of these numbers can be computed using formula 5, the values of k corresponding to the XL Algorithm were obtained experimentally. In the 8-th column of the table we give first the length of a standard signature (without reduction). The second number shows the length of the shortest partial signature which can be verified using the Relinearization technique, while the third and fourth numbers show the lengths of the shortest signatures that can be verified using the XL Algorithm at degree 3 and 4 respectively. The 9-th column shows the corresponding reduction factors.

Table 1. Possible reduction of the signature length for rainbow and HFEv-

Security level	Scheme	Private key size (kB)	Public key size (kB)	Hash size (bit)	verification by	k	Signature size (bit)	Reduction in %
80	Rainbow(GF(2^8)17,13,13)	19.9	25.1	208	standard	0	344	-
					Relinearization	5	304	12
					XL ($d_{\mathrm{reg}} = 3$)	10	264	23
					XL ($d_{\mathrm{reg}} = 4$)	14	232	33
	HFEv-(GF(7),62,8,2,2)	2.9	47.1	168	standard	0	192	-
					Relinearization	9	165	14
					XL ($d_{\mathrm{reg}} = 3$)	17	135	27
					XL ($d_{\mathrm{reg}} = 4$)	23	126	34
100	Rainbow(GF(2^8),26,16,17)	44.4	59.0	264	standard	0	472	-
					Relinearization	6	424	10
					XL ($d_{\mathrm{reg}} = 3$)	12	384	19
					XL ($d_{\mathrm{reg}} = 4$)	16	344	27
	HFEv-(GF(7),78,8,3,3)	4.5	93.5	210	standard	0	243	-
					Relinearization	11	216	14
					XL ($d_{\mathrm{reg}} = 3$)	17	192	22
					XL ($d_{\mathrm{reg}} = 4$)	24	171	30

In the case of the HFEv- signature scheme, we use GF(7) as the underlying field. We store one element of GF(7) in 3 bits, while 5 elements of GF(7) can be efficiently used to store 14 bits. To store a hash value of length 160 bit, we therefore need 60 elements of GF(7), to store a hash value of 200 bit, we need 75 GF(7) elements.

As can be seen from Table 1 we can, in the case of HFEv-, get signatures which are smaller than the input size of the scheme, even when restricting to verifying the partial signatures with the Relinearization technique. But also for the Rainbow scheme, our technique enables us to reduce the signature length by up to 12%

When verifying the reduced signatures with the XL Algorithm (with $d_{\mathrm{reg}} = 3$), we can obtain a reduction of the signature length by 20–25%. When allowing the XL Algorithm to reach degree 4, we can achieve reductions of up to 34%.

6.1 Efficiency of the Verification Process

To estimate the efficiency of the modified verification process, we created a straightforward implementation of the Rainbow and HFEv- signature schemes in MAGMA code. Our scheme runs on a single core of a server with 24 AMD Opteron processors (2.4 GHz) and 128 GB of RAM. Table 2 shows the time needed for the verification of a (partial) signature (average time of 10,000 verification processes).

As the table shows, there is no significant difference between the running times of the standard verification process and the modified verification process combined with the Relinearization technique. We therefore can achieve a reduction of the signature length by up to 15% at quasi no cost. A further reduction of the signature length is possible if we accept an increase in the verification time. We hence observe a trade off between signature size and efficiency of the verification process, which, to our knowledge, is unique for multivariate signature schemes and can not be observed for any other class of digital signature schemes.

Table 2. Verification times for rainbow and HFEv- schemes with reduced signature length

Security level (bit)	Scheme	Verification by	k	Signature size (bit)	Reduction in %	Verification time (ms)
80	Rainbow(GF(2^8),17,13,13)	standard	0	344	-	0.21
		Relinearization	5	304	12	0.25
		XL Algorithm ($d_{reg} = 3$)	10	264	23	10.8
		XL Algorithm ($d_{reg} = 4$)	14	232	33	425.0
	HFEv-(GF(7),62,8,2,2)	standard	0	192	-	0.86
		Relinearization	9	165	14	0.91
		XL Algorithm ($d_{reg} = 3$)	17	141	27	21.3
		XL Algorithm ($d_{reg} = 4$)	22	126	34	634.6
100	Rainbow(GF(2^8),26,16,17)	standard	0	472	-	0.42
		Relinearization	6	424	10	0.47
		XL Algorithm ($d_{reg} = 3$)	12	376	20	14.3
		XL Algorithm ($d_{reg} = 4$)	16	344	27	534.6
	HFEv-(GF(7),78,8,3,3)	standard	0	243	-	1.25
		Relinearization	10	213	12	1.32
		XL Algorithm ($d_{reg} = 3$)	19	186	23	37.2
		XL Algorithm ($d_{reg} = 4$)	24	171	30	928.6

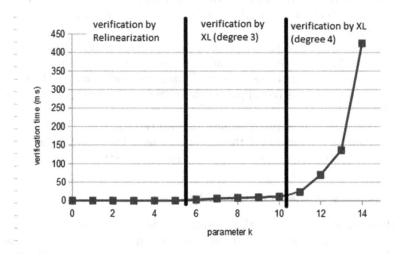

Fig. 3. Verification time (ms) for rainbow(17,13,13) and different values of k

Figure 3 shows this trade off for the example of the Rainbow signature scheme with parameters $(v_1, o_1, o_2) = (17, 13, 13)$. In particular, we note that we can achieve a reduction of the signature size of Rainbow by 5 byte (12%) at quasi no cost. However, for larger values of k (i.e. solution of $\tilde{\mathcal{P}}(\mathbf{x}) = \mathbf{w}$ by the XL Algorithm with $d_{reg} \in \{3, 4\}$) the running time of the verification process increases significantly. More experimental data regarding this trade off between signature size and verification time can be found in Table 2.

7   Application of Our Technique to Gui

The Gui signature scheme as proposed by Petzoldt et al. in [17] is currently the multivariate signature scheme with the shortest signatures. In this section we show how to apply our technique to Gui, by which we obtain the shortest signatures of all currently existing signature schemes. However, due to the special signature generation process of Gui, this can not be done straightforward. In order to show this, we start with a short description of Gui.

The Gui signature scheme [17] is an extension of the HFEv- signature scheme introduced in Sect. 3.2. Indeed, the public and private keys of Gui are just HFEv-keys over GF(2) with specially chosen parameters n, D, a and v. The signature generation process of Gui is very fast [17] and produces very short signatures of size not more than 120 bit. However, due to the parameter choice of Gui, the input length of the HFEv- scheme is only 90 bit. Therefore, it would be possible for an attacker to come up with two messages d_1 and d_2 whose hash values collide in these first 90 bits (birthday attack). To overcome this problem, the authors of [17] developed a specially designed signature generation process for their scheme. Roughly spoken, to generate a signature for a message d, Gui computes $r \in \{3, 4\}$ HFEv- signatures for different hash values of the document d and combines them to a single Gui signature of length $(n - a) + r \cdot (a + v)$ bits. Algorithm 5 shows the signature generation process of Gui in algorithmic form.

In order to verify a Gui signature $\sigma \in GF(2)^{(n-a)+r\cdot(a+v)}$, we have to evaluate the public key of Gui r times (see Algorithm 6).

This repeated evaluation of the public key prevents us from applying our technique to Gui directly. However, we are still able to reduce the signature size of Gui by $a + v$ bit.

Algorithm 5. Signature Generation Process of Gui

Input: Gui private key $(\mathcal{S}, \mathcal{F}, \mathcal{T})$ message d, repetition factor r
Output: signature $\sigma \in GF(2)^{(n-a)+r(a+v)}$
  1: $\mathbf{h} \leftarrow$ SHA-256(d)
  2: $S_0 \leftarrow \mathbf{0} \in GF(2)^{n-a}$
  3: **for** $i = 1$ to r **do**
  4:     $D_i \leftarrow$ first $n - a$ bits of $\mathbf{h}$
  5:     $(S_i, X_i) \leftarrow$ HFEv-$^{-1}(D_i \oplus S_{i-1})$
  6:     $\mathbf{h} \leftarrow$ SHA-256$(\mathbf{h})$
  7: **end for**
  8: $\sigma \leftarrow (S_r || X_r || \ldots || X_1)$
  9: **return** σ

In particular, while the original Gui signature σ is of the form $\sigma = (S_r, X_r, \ldots, X_1)$, we just transmit the partial signature $\tilde{\sigma} = (S_r, X_r, \ldots, X_2)$. The modified verification algorithm of Gui works as shown in Algorithm 7.

Algorithm 6. Standard Verification Algorithm of Gui

Input: Gui public key $\mathcal{P}$, message d, repetition factor r, signature $\sigma \in \mathrm{GF}(2)^{(n-a)+r(a+v)}$

Output: boolean value **TRUE** or **FALSE**

1: $\mathbf{h} \leftarrow \mathrm{SHA\text{-}256}(d)$
2: $(S_r, X_r, \ldots, X_1) \leftarrow \sigma$
3: **for** $i = 1$ to r **do**
4:     $D_i \leftarrow$ first $n - a$ bits of $\mathbf{h}$
5:     $\mathbf{h} \leftarrow \mathrm{SHA\text{-}256}(\mathbf{h})$
6: **end for**
7: **for** $i = r - 1$ to 0 **do**
8:     $S_i \leftarrow \mathcal{P}(S_{i+1} \| X_{i+1}) \oplus D_{i+1}$
9: **end for**
10: **if** $S_0 = 0$ **then**
11:     **return TRUE**
12: **else**
13:     **return FALSE**
14: **end if**

Algorithm 7. Modified Verification Algorithm of Gui

Input: Gui public key $\mathcal{P}$, message d, repetition factor r, partial signature $\tilde{\sigma} \in \mathrm{GF}(2)^{(n-a)+(r-1)\cdot(a+v)}$

Output: boolean value **TRUE** or **FALSE**

1: $\mathbf{h} \leftarrow \mathrm{SHA\text{-}256}(d)$
2: $(S_r, X_{r-1}, \ldots, X_1) \leftarrow \tilde{\sigma}$
3: **for** $i = 1$ to r **do**
4:     $D_i \leftarrow$ first $n - a$ bits of $\mathbf{h}$
5:     $\mathbf{h} \leftarrow \mathrm{SHA\text{-}256}(\mathbf{h})$
6: **end for**
7: **for** $i = r - 1$ to 1 **do**
8:     $S_i \leftarrow \mathcal{P}(S_{i+1} \| X_{i+1}) \oplus D_{i+1}$
9: **end for**
10: $\tilde{\mathcal{P}} \leftarrow \mathcal{P}(S_1)$
11: **if** IsConsistent($\tilde{\mathcal{P}}(\mathbf{x}) = D_1$) **then**
12:     **return TRUE**
13: **else**
14:     **return FALSE**
15: **end if**

In our modified verification process, the hash values $D_1, \ldots, D_r$ are computed just as in the case of the original Gui scheme (line 3 to 6). We follow the original verification process of Gui (see Algorithm 6) further to compute the vectors $S_{r-1}, \ldots, S_1$ (line 7 to 9). After that, we substitute S_1 into the public key $\mathcal{P}$ (line 10) to obtain a system $\tilde{\mathcal{P}}$ of $n - a$ quadratic equations in $a + v$ variables. The signature $\tilde{\sigma}$ is accepted, if and only if the system $\tilde{\mathcal{P}}(\mathbf{x}) = D_1$ has a solution. For the parameters proposed in [17], this step can be performed by the Relinearization technique (see Sect. 4.1) in polynomial time.

By our technique, we can therefore reduce the signature size of Gui from $(n-a) + r \cdot (a+v)$ bit to $(n-a) + (r-1) \cdot (a+v)$ bit. Table 3 shows the possible reduction of the signature size for the four parameter sets proposed in [17]. As the table shows, we can, for the parameters $(n, D, a, v, r) = (95, 9, 5, 5, 3)$ (80 bit

Table 3. Possible reduction of the public key size for Gui

Security level	Scheme	Private key size (kB)	Public key size (kB)	Verification by	k	Signature size (bit)	Reduction in %
80	Gui(GF(2),96,5,6,6,4)	3.1	61.5	standard	0	126	-
				Relinearization	13	113	10
	Gui(GF(2),95,9,5,5,3)	3.0	59.2	standard	0	120	-
				Relinearization	11	109	9
	Gui(GF(2),94,9,4,4,4)	2.9	56.8	standard	0	122	-
				Relinearization	9	113	7
120	Gui(GF(2),127,9,4,6,4)	5.2	139.2	standard	0	163	-
				Relinearization	11	152	7

security), obtain signatures of size 110 bit, which are the shortest signatures of all existing schemes (both in the classical and the post quantum world). Note that by our modifications the security of the scheme is not weakened at all (c.f. Proposition 1), and the performance is not reduced significantly.

8   Discussion

The technique proposed in Sect. 5 of this paper enables us to reduce the signature size of multivariate schemes in a very flexible way. In particular, as Fig. 3 shows, we achieve a reduction of the signature size by 10–15% at quasi no cost and a trade off between further reduction and the efficiency of the verification process. Again we note that our modifications do not weaken the security of the schemes.

To our knowledge, this possibility to reduce the signature length without extra cost is unique for multivariate signature schemes, and no other family of signature schemes allows something similar. While, for some lattice based signature schemes [9], there also exist techniques to reduce the signature size, they are much less flexible than the proposed technique and require significant extra computation.

Our technique can be applied to every standard multivariate signature scheme such as UOV [11], Rainbow [6], HFEv- [15] and TTS [21] and, with some modifications, to more advanced schemes such as QUARTZ and Gui [17], too. Although, for these schemes, we can not reach as high reduction factors as for Rainbow and HFEv-, we obtain, by applying our technique to Gui, the shortest signatures of all existing signature schemes (both in the classical and the post quantum world).

As demonstrated in Sect. 2 of this paper, our technique is especially suitable in situations where the connection between sender and receiver is slow. In particular, if the original messages are very short (e.g. status messages of a node in a sensor network) and therefore the signatures are a major part of the communication, our technique helps to reduce the communication cost and therefore speeds up the system significantly. Furthermore, if the small sensors report their status to a computationally powerful server, a slightly more complicated verification process as implied by our technique should be no major problem.

9   Conclusion

In this paper we proposed a general technique to reduce the signature size of multivariate schemes. Our technique enables us to decrease the signature size of nearly all multivariate signature schemes such as Rainbow and HFEv- by up to 15%, without slowing down the verification process of the schemes significantly. We can prove that the security of the underlying scheme is not weakened by this modification. We can achieve a further reduction of the signature size when accepting a slightly more costly verification process. Therefore, we observe a trade off between signature length and the efficiency of the verification process, which, to our knowledge, is unique for multivariate signature schemes.

By applying our technique to the Gui signature scheme of [17], we obtain signatures of size only 110 bit (80 bit security), which are the shortest signatures of all existing digital signature schemes (both in the classical and the post quantum world).

Acknowledgments. The first and the second authors are supported by the EU-project PQCRYPTO ICT-645622 and JSPS KAKENHI 15F15350 respectively.

References

1. Bernstein, D.J., Buchmann, J., Dahmen, E. (eds.): Post Quantum Cryptography. Springer, Heidelberg (2009)
2. Bogdanov, A., Eisenbarth, T., Rupp, A., Wolf, C.: Time-area optimized public-key engines: $\mathcal{MQ}$-cryptosystems as replacement for elliptic curves? In: Oswald, E., Rohatgi, P. (eds.) CHES 2008. LNCS, vol. 5154, pp. 45–61. Springer, Heidelberg (2008). doi:10.1007/978-3-540-85053-3_4
3. Chen, A.I.-T., Chen, M.-S., Chen, T.-R., Cheng, C.-M., Ding, J., Kuo, E.L.-H., Lee, F.Y.-S., Yang, B.-Y.: SSE implementation of multivariate PKCs on modern x86 CPUs. In: Clavier, C., Gaj, K. (eds.) CHES 2009. LNCS, vol. 5747, pp. 33–48. Springer, Heidelberg (2009). doi:10.1007/978-3-642-04138-9_3
4. Courtois, N., Klimov, A., Patarin, J., Shamir, A.: Efficient algorithms for solving overdefined systems of multivariate polynomial equations. In: Preneel, B. (ed.) EUROCRYPT 2000. LNCS, vol. 1807, pp. 392–407. Springer, Heidelberg (2000). doi:10.1007/3-540-45539-6_27
5. Ding, J., Gower, J.E., Schmidt, D.S.: Multivariate Public Key Cryptosystems. Advances in Information Security, vol. 25. Springer, New York (2006). doi:10.1007/978-0-387-36946-4
6. Ding, J., Schmidt, D.: Rainbow, a new multivariable polynomial signature scheme. In: Ioannidis, J., Keromytis, A., Yung, M. (eds.) ACNS 2005. LNCS, vol. 3531, pp. 164–175. Springer, Heidelberg (2005). doi:10.1007/11496137_12
7. Faugère, J.C.: A new efficient algorithm for computing Gröbner bases (F4). J. Pure Appl. Algebra **139**, 61–88 (1999)
8. Garey, M.R., Johnson, D.S.: Computers and Intractability: A Guide to the Theory of NP-Completeness. W.H. Freeman and Company, San Francisco (1979)
9. Güneysu, T., Lyubashevsky, V., Pöppelmann, T.: Practical lattice-based cryptography: a signature scheme for embedded systems. In: Prouff, E., Schaumont, P. (eds.) CHES 2012. LNCS, vol. 7428, pp. 530–547. Springer, Heidelberg (2012). doi:10.1007/978-3-642-33027-8_31
10. Kravitz, D.: Digital signature algorithm. US Patent 5231668, July 1991
11. Kipnis, A., Patarin, J., Goubin, L.: Unbalanced oil and vinegar signature schemes. In: Stern, J. (ed.) EUROCRYPT 1999. LNCS, vol. 1592, pp. 206–222. Springer, Heidelberg (1999). doi:10.1007/3-540-48910-X_15
12. Matsumoto, T., Imai, H.: Public quadratic polynomial-tuples for efficient signature-verification and message-encryption. In: Barstow, D., Brauer, W., Brinch Hansen, P., Gries, D., Luckham, D., Moler, C., Pnueli, A., Seegmüller, G., Stoer, J., Wirth, N., Günther, C.G. (eds.) EUROCRYPT 1988. LNCS, vol. 330, pp. 419–453. Springer, Heidelberg (1988). doi:10.1007/3-540-45961-8_39
13. Goodin, D.: NSA preps quantum-resistant algorithms to head off cryptoapocalypse. http://arstechnica.com/security/2015/08/nsa-preps-quantumresistant-algorithms-to-head-o-crypto-apocolypse/

14. National Institute of Standards and Technology: Report on Post Quantum Cryptography. NISTIR draft 8105, http://csrc.nist.gov/publications/drafts/nistir-8105/nistir_8105_draft.pdf

15. Patarin, J.: Hidden fields equations (HFE) and isomorphisms of polynomials (IP): two new families of asymmetric algorithms. In: Maurer, U. (ed.) EUROCRYPT 1996. LNCS, vol. 1070, pp. 33–48. Springer, Heidelberg (1996). doi:10. 1007/3-540-68339-9_4

16. Patarin, J., Courtois, N., Goubin, L.: FLASH, a fast multivariate signature algorithm. In: Naccache, D. (ed.) CT-RSA 2001. LNCS, vol. 2020, pp. 298–307. Springer, Heidelberg (2001). doi:10.1007/3-540-45353-9_22

17. Petzoldt, A., Chen, M.-S., Yang, B.-Y., Tao, C., Ding, J.: Design principles for HFEv- based multivariate signature schemes. In: Iwata, T., Cheon, J.H. (eds.) ASIACRYPT 2015. LNCS, vol. 9452, pp. 311–334. Springer, Heidelberg (2015). doi:10.1007/978-3-662-48797-6_14

18. Rivest, R.L., Shamir, A., Adleman, L.: A method for obtaining digital signatures and public-key cryptosystems. Commun. ACM 21(2), 120–126 (1978)

19. Shor, P.: Polynomial-time algorithms for prime factorization and discrete logarithms on a quantum computer. SIAM J. Comput. 26(5), 1484–1509 (1997)

20. Tao, C., Diene, A., Tang, S., Ding, J.: Simple matrix scheme for encryption. In: Gaborit, P. (ed.) PQCrypto 2013. LNCS, vol. 7932, pp. 231–242. Springer, Heidelberg (2013). doi:10.1007/978-3-642-38616-9_16

21. Yang, B.-Y., Chen, J.-M.: Building secure tame-like multivariate public-key cryptosystems: the new TTS. In: Boyd, C., González Nieto, J.M. (eds.) ACISP 2005. LNCS, vol. 3574, pp. 518–531. Springer, Heidelberg (2005). doi:10.1007/11506157_43

Cryptographic Protocols

CRT-Based Outsourcing Algorithms
for Modular Exponentiations

Lakshmi Kuppusamy[✉] and Jothi Rangasamy

Society for Electronic Transactions and Security (SETS), Chennai, India
{lakshdev,jothiram}@setsindia.net

Abstract. The problem of securely outsourcing cryptographic computations to the untrusted servers was formally addressed first by Hohenberger and Lysyanskaya in TCC 2005. They presented an algorithm
which outsources computation of modular exponentiations securely to
two non-interacting third-party servers but the checkability of third-
party computations has probability 1/2. Chen *et al.* improved this algorithm for two non-colluding servers by increasing the checkability probability to 2/3. For real-world cryptographic applications it is desirable
that the checkability probability is $1 - \epsilon$, where ϵ becomes negligible for
appropriate parameter choices. Towards a more practical use, we present
an algorithm(s) for secure outsourcing of (simultaneous) modular expo-
nentiation(s) which can be seen as another application of the Chinese
remainder theorem (CRT). Interestingly the checkability probability of
our algorithm is 1 in the presence of two non colluding servers. Our algorithm is superior in both efficiency and checkability compared to that of
the previously known schemes of the same kind. Finally we discuss the
potential practical applications for our outsourcing schemes, for example
computing the final exponentiation in pairings.

1 Introduction

Secure transmission of sensitive information is vital not only for government and
business sectors but also for individuals. In this information age, we perform
many operations such as banking transactions that need to send some sensitive
data over the Internet. Also the usage of mobile devices such as smart phones,
tablets and PDAs for executing critical transactions and communications brings
new challenges related to security and performance. Though the contemporary
computation had become pervasive, the need for customary security services such
as authentication and confidentiality remains the same. Adapting the public-key
and private-key cryptographic algorithms in combination could make the smart
devices useful in diverse applications. However the indispensable fundamental
operations such as pairing and modular exponentiation in public-key cryptosystems could hinder the performance of such smart devices as they are usually
battery-powered and generally built with less space and/or less computational
power or both. Note that their battery life is inversely proportional to the amount
of computational work they get engaged.

© Springer International Publishing AG 2016
O. Dunkelman and S.K. Sanadhya (Eds.): INDOCRYPT 2016, LNCS 10095, pp. 81–98, 2016.
DOI: 10.1007/978-3-319-49890-4_5

The desire to outsource (cryptographic) computation from a resource-limited device to a vastly powerful server has been growing like never before. This is due to the numerous contributing trends including the upsurge in the usage of mobile devices and the advent of cloud computing. The delegation of expensive computations to a (cloud) server could benefit the smart devices in two ways; first it helps these computationally-poor devices realise all the public-key cryptographic functionalities and second it helps to prolong their battery lifetime. In an outsourcing scenario, one main requirement is that the amount of work needed by the client to delegate and verify the output of the worker(s)[1] should be significantly cheaper than computing the function or value on its own.

1.1 Related Work

Secure outsourcing of expensive computations has been an active area of research in cryptography [1–6]. Gentry's discovery on fully homomorphic encryption (FHE) [7] shows the possibility of outsourcing any computation to cloud by keeping the secrets. However, there is no straightforward guarantee provided by FHE to validate the correctness of the (cloud) server's outputs. FHE has efficacy but lacks efficiency and thus using FHE mechanisms in real-world cryptographic applications would not be immediate. In recent years, verifiable delegation of modular exponentiation, occurring predominantly in public-key cryptography has been receiving considerable attention [8–12]. The work of Matsumoto *et al.* [13] for outsourcing RSA computations with untrusted servers was shown to be insecure by Nguyen and Shparlinski [14]. Motivated by this, Hohenberger and Lysyanskaya [8] proposed a formal security definition of outsourcing that aims to capture two important security properties, namely *secrecy* and *checkability*. Often the computation tasks outsourced to these third party servers contains secret inputs and/or outputs and thus, secrecy is the first major challenge to be dealt with. Moreover a third-party server which we call a cloud server can be potentially malicious. Hence the foremost requirement for an outsourcing scheme is that it should be computationally infeasible for the cloud server to learn anything about the secrets other than performing the computation task assigned to it. Another requirement is checkability as the untrusted server could try to cause failures by returning an invalid result. Therefore, there must be a way for the outsourcing device to check the validity of the output returned by the server. Note that this check must be efficient for the device than performing the computation task by itself. For instance if computing modular exponentiation is outsourced then the outsourcing device should validate server's output by just performing modular multiplications, additions and subtractions.

Two-Server Algorithms. Hohenberger and Lysyanskaya also gave the first secure outsourcing algorithm for modular exponentiation using two untrusted servers that are non colluding after deciding on an initial strategy. The running time of the algorithm is $O(log^2 n)$ which is an asymptotic improvement over the

---

[1] We use the terms worker, program and oracle interchangeably to refer to the untrusted server.

$1.5n$ multiplications required to compute an exponentiation (without outsourcing) using the square-and-multiply algorithm. The probability of detecting an error in the output of the oracles is $1/2$. In ESORICS 2012, Chen *et al.* [10] improved the Hohenberger-Lysyanskaya algorithm in terms of both efficiency (reducing the number of oracle queries) and checkability (i.e., error detection probability is $2/3$). Chen *et al.* also proposed secure outsourcing algorithm for *simultaneous* modular exponentiations in the two server model.

Single-Server Algorithms. Wang *et al.* [11] were the first to present a generic outsourcing algorithm for the single untrusted program model with the error-detection probability is $1/2$ only. Chevalier *et al.* [15] found an attack in Wang *et al.*'s scheme and claimed that fixing the attack makes the algorithm inefficient. Further, Chevalier *et al.* showed that an optimal single server based outsourcing algorithms would require a non-constant number of group operations. The algorithms proposed in [15] are proven unconditionally secure only in the presence of honest-but-curious adversary and the checkability property is not considered for security analysis in the untrusted server setting.

Kiraz and Uzunkol [16] proposed an algorithm to outsource single modular exponentiation to a single untrusted server with adjustable checkablity probability (close to 1) for the appropriate pre-determined parameter choices. However the algorithm incurs significant communication overhead in terms of transmission of group elements to the server from the resource constrained device.

1.2 Our Motivation and Contributions

On one hand we have two-server based algorithms [8,10] to outsource modular exponentiation that achieve error-detection probability of $1/2$ and $2/3$ respectively. On the other hand Chevalier *et al.* [15] showed that the single-server algorithm due to Wang *et al.* [11] violates the secrecy property. The new single server based algorithms proposed in [15] does not provide error checkability property and the security is analysed only in the presence of honest-but-curious adversary. Though the algorithm by Kiraz and Uzonkol [16] enables the outsourcer to check the correctness of the result returned by the server, it suffers from a significant communication overhead compared to the existing schemes.

If the checkability probability is less than 1 it means that the client is not able to validate certain number of outputs corresponding to its queries to the server and hence there is a possibility that the client device may not end up computing the intended task. Thus the problem that remains to be solved is to design an algorithm for secure outsourcing of modular exponentiations using untrusted servers achieving both privacy, high checkability and minimal communication between the resource constrained device and the server. Note that Dijk *et al.*'s scheme [9] has checkability of 1 in a single untrusted server model, but it fails to maintain the privacy of the inputs in queries. This motivates us to examine how to securely outsource modular exponentiations while ensuring privacy and high checkability. Our contributions in this paper are as follows.

- First we present an algorithm to securely outsource variable-base variable-exponent modular exponentiation for the two untrusted server model. Using two non colluding untrusted servers, our algorithm achieves the checkability probability 1, thereby making it error free. We then extend the scheme to the case of simultaneous modular exponentiations while retaining all the advantages of the single exponentiation case. We analyse the proposed algorithms in the one-malicious security model of Hohenberger and Lysyanskaya and prove that our algorithms are secure and checkable.
- Finally we show how to use our secure outsourcing algorithm for (simultaneous) modular exponentiations in various applications. Especially we consider the delegation of the final exponentiation which is arguably the most expensive operation in pairing computations.

OUTLINE. The rest of the paper is organised as follows. In Sect. 2 we recall the security definitions of Hohenberger and Lysyanskaya [8] which is described in Appendix A for secure outsourcing schemes. In Sect. 3 we present our secure outsourcing algorithm for single modular exponentiation using two non-colluding servers and analyse the security of our outsourcing algorithms under the Hohenberger-Lysyanskaya security model in Sect. 3.2. An algorithm to outsource simultaneous modular exponentiations is described in Sect. 4. Section 5 discusses some applications for the proposed algorithms and Sect. 6 concludes the paper.

2   Security Definitions

Hohenberger and Lysyanskaya [8] presented the first formal security definition to the problem of secure outsourcing of computations from a computational resource constrained device to an untrusted server. They gave a framework to quantify the efficiency and checkability of the outsourced task. Chen et al. [10] and Wang et al. [11] followed these security definitions to prove the security properties of their algorithms. We briefly describe the Hohenberger-Lysyanskaya model in this section and refer the reader to [8] for a complete description.

ADVERSARIAL BEHAVIOUR. Let Alg be a cryptographic algorithm executed by two components: (i) a computationally weak and trusted component $\mathcal{C}$ (i.e., a client); and (ii) a computationally powerful and untrusted component $\mathcal{U}$ invoked by $\mathcal{C}$ through oracle queries. We say that $(\mathcal{C}, \mathcal{U})$ forms an outsource-secure implementation of an algorithm Alg $= \mathcal{C}^{\mathcal{U}}$ where $\mathcal{C}$ carries out the tasks by invoking $\mathcal{U}$.

The adversary $\mathcal{A} = (\mathcal{E}, \mathcal{U}')$ is modeled as being categorized into two parts: (i) the adversarial environment $\mathcal{E}$ that submits adversarially chosen inputs to Alg; (ii) a malicious oracle $\mathcal{U}'$ operating instead of $\mathcal{U}$. Note that both U and $\mathcal{U}'$ are invoked in the same manner to mirror the view of U during execution. It is assumed that $\mathcal{E}$ and $\mathcal{U}$ after agreeing on a joint strategy initially, no longer have a direct communication channel once they begin to interact with $\mathcal{C}$. They can only communicate with each other by passing messages through $\mathcal{C}$.

INPUT/OUTPUT SPECIFICATIONS. The inputs/outputs to the algorithm may have the following information:

Secret information available only to $\mathcal{C}$;
Protected information available to both $\mathcal{C}$ and $\mathcal{E}$ but should be kept secret from $\mathcal{U}$. Note that the protected information can be further categorized based on whether the inputs are generated honestly or adversarially;
Unprotected information available to $\mathcal{C}$, $\mathcal{E}$ and $\mathcal{U}$.

An outsource algorithm Alg is defined to accept three inputs (secret, protected, unprotected) and produce three outputs (secret, protected, unprotected). With the above input/output specifications, a formal definition (Definition 1) due to [8] is reproduced in Appendix A.

2.1 Outsource-Security Definitions

Informally, the outsource-security definition must satisfy the following security requirements:

- *Secrecy.* The malicious environment $\mathcal{E}$ should not gain any knowledge about the secret inputs and outputs of the algorithm Alg, even it has a joint strategy agreed with the oracle $\mathcal{U}'$ initially. To achieve this property, the model assumes that there exist a simulator $\mathcal{S}_1$ which simulates the view of $\mathcal{E}$ without access to the secret inputs.
- *Checkability.* Any information that a malicious oracle $\mathcal{U}'$ learns about the inputs to Alg by acting as an oracle to $\mathcal{C}$ instead of $\mathcal{U}$, it can also learn without that. To achieve this property, the model assumes that there exist a simulator $\mathcal{S}_2$ which simulates the view of $\mathcal{U}'$ without access to the secret or protected inputs.

To capture the above requirements formally, the outsource-security definitions have been introduced by Hohenberger-Lysyanskaya [8] and are reproduced in Appendix A through Definitions 2, 3, 4 and 5.

3 Secure Outsourcing of Modular Exponentiations to Two Non-colluding Untrusted Servers

Adaptation of Hohenberger-Lysyanskaya algorithm by Chen *et al.* is the best known algorithm in the related literature to securely outsource variable-base, variable-exponent modular exponentiations in two non-colluding untrusted servers model. The issue with their scheme is that the outsourcing party cannot verify all the outputs of the server due to its design and this makes their algorithm less suitable for practical use. In this section we present an algorithm which preserves privacy of the inputs and achieves the checkability probability 1. We denote our two-server outsourcing algorithm for single exponentiation by **2EXP**.

3.1 2EXP: Secure Outsourcing Algorithm for Single Modular Exponentiation

Let p and q be two primes such that $q|(p-1)$. Then there exists a multiplicative subgroup of $\mathbb{Z}_p^*$ of order q. The task of the client $\mathcal{C}$ is to compute $u^a \bmod p$ where $u \in \mathbb{Z}_p^*$ is a variable base and $a \in \mathbb{Z}_q$ is a variable exponent. The task of oracles $\{\mathcal{U}_1, \mathcal{U}_2\}$ is to return the output $i^j \bmod k$ on input (i, j, k).

In our algorithm, to maintain the secrecy of the values u and a they are computationally masked before being given as input to $\{\mathcal{U}_1, \mathcal{U}_2\}$. Unlike in the previously known schemes, we mask u using the Chinese remainder theorem (CRT). The details are as follows:

Pre-computation. The client $\mathcal{C}$ generates a table (T1) of pairs of the form $(t_i, g_1^{t_i})$ for a fixed base $g_1 \in \mathbb{Z}_p^*$. Whenever a new pair of the form (t, g_1^t) is needed, the client selects a small number of randomly chosen pairs in the table and computes t by adding all the first arguments and g_1^t by multiplying all the second arguments from the selected pairs. Similarly, the client $\mathcal{C}$ generates another table, (T2) of pairs of the form $(\alpha_i, g_2^{\alpha_i})$ for a prime modulus $r_2 \neq p$ and a fixed base $g_2 \in \mathbb{Z}_{r_2}^*$. By setting the parameters appropriately, both the tables enable the client to generate a new pair (α, g_i^α) for $i = 1, 2$ which incurs small number of multiplications rather than the actual cost of performing a modular exponentiation. The previously known related schemes use this pre-processing method extensively despite calling them in different names; RAND algorithm in [8], EBPV algorithm in [10], BPV$^+$ or SMBL in [11]. This technique was invented by Boyko et al. [17] and Nguyen et al. proved that a pair generated using this method will be statistically indistinguishable from a pair computed using a randomly chosen exponent [18]. As in [11] we use the BPV$^+$ technique to avoid online multiplications to produce new pairs. For more details on BPV$^+$, please refer to [11, Sect. 3.1]. Our algorithm **2EXP** proceeds as follows.

Masking u. Let $n = pr_1r_2$ for primes r_1 and r_2. (Note that for a sufficiently large 3-prime modulus n there is no known polynomial time algorithm to factor it into primes even if one of the primes, say p, is known [19].) Then $\phi(n) = (p-1)(r_1-1)(r_2-1)$. Choose $h \in \mathbb{Z}_{r_1}^*$. Compute (θ, g_1^θ) using the BPV$^+$ technique. From CRT, we know that the following system of three simultaneous congruences

$$x \equiv ug_1^\theta \bmod p \quad \text{and} \quad x \equiv h \bmod r_1 \quad \text{and} \quad x \equiv g_2 \bmod r_2$$

has a unique solution $x \bmod n$, which the client computes as below.

$$
\begin{aligned}
x = {} & u \cdot g_1^\theta \cdot r_1 \cdot r_2 \cdot (r_1^{-1}r_2^{-1} \bmod p) + h \cdot r_2 \cdot p \cdot (r_2^{-1}p^{-1} \bmod r_1) \\
& + g_2 \cdot p \cdot r_1 \cdot (p^{-1}r_1^{-1} \bmod r_2).
\end{aligned}
\tag{1}
$$

Masking a. In order to mask a, C invokes the BPV$^+$ algorithm to get the pairs (α, g_2^α) and (β, g_2^β). Now C computes:

1. $a_1 = a - \alpha$;
2. $a_2 = a - \beta$;

Queries to $\mathcal{U}_1$. Now C invokes the BPV$^+$ algorithm to get the pair (t, g_1^t) and sends the following queries to $\mathcal{U}_1$ in random order:

1. $(x, a_1, n) \to X_1$;
2. $(x, \beta, n) \to X_2$;
3. $(g_1^t, -a\theta/t, p) \to X_4$;

Queries to $\mathcal{U}_2$. Now C sends the following queries to $\mathcal{U}_2$ in random order:

1. $(x, a_2, n) \to X_3$;
2. $(g_1^t, -a\theta/t, p) \to X_5$;

Verifying the correctness of $\{\mathcal{U}_1, \mathcal{U}_2\}$**'s output.**

1. C checks whether
$$g_2^\beta \bmod r_2 \stackrel{?}{=} X_2 \bmod r_2. \tag{2}$$

2. C checks whether
$$[X_1 \bmod r_2 \cdot g_2^\alpha] \bmod r_2 \stackrel{?}{=} X_2 \cdot X_3 \bmod r_2. \tag{3}$$

3. C checks whether
$$X_4 \stackrel{?}{=} X_5 \bmod p. \tag{4}$$

Recovering u^a. If Eqs. 2, 3 and 4 hold, then C believes that all the values X_1, X_2, X_3, X_4 and X_5 have been computed correctly. This implies that $X_2 \cdot X_3 \cdot X_4 \equiv u^a \bmod p$ due to simultaneous system of congruences. If the check fails, then C outputs an error message.

Remark 1. (Checkability) Note that all the queries output are checked by the client and then only the output u^a is computed. Hence the client now believes that u^a has been computed correctly. In the above algorithms, both p and n are sent to both the servers $\mathcal{U}_1$ and $\mathcal{U}_2$ and thus reveal $r_1 r_2$, but the servers do not gain any advantage in knowing $r_1 r_2$. Note that the correctness of the outputs X_1 X_2 and X_3 are checked using the client's stored values g_2^α and g_2^β and hence the server trying to send bogus results can be detected.

3.2   Security and Efficiency Analysis

In this section, we analyse the security properties of the proposed algorithm using the one malicious model of Hohenberger and Lysyanskaya [8].

Lemma 1 *(Correctness). In the one malicious model, the algorithms $(C, \mathcal{U}_1, \mathcal{U}_2)$ are correct implementation of* **2EXP***, where the inputs (a, u, p) may be honest, secret; or honest, protected; or adversarial protected.*

Proof. We know that for any integer $m \geq 0$,

$$x^m = \{u \cdot g_1^\theta \cdot r_1 \cdot r_2 \cdot (r_1^{-1} r_2^{-1} \bmod p) + h \cdot r_2 \cdot p \cdot (r_2^{-1} p^{-1} \bmod r_1)$$
$$+ \quad g_2 \cdot p \cdot r_1 \cdot (p^{-1} r_1^{-1} \bmod r_2)\}^m \tag{5}$$

$$= \sum_{k_1+k_2+k_3=m} \frac{m!}{k_1! + k_2! + k_3!} A^{k_1} B^{k_2} C^{k_3}$$

where $A = u \cdot g_1^\theta \cdot r_1 \cdot r_2 \cdot (r_1^{-1} r_2^{-1} \bmod p)$, $B = h \cdot r_2 \cdot p \cdot (r_2^{-1} p^{-1} \bmod r_1)$, $C = g_2 \cdot p \cdot r_1 \cdot (p^{-1} r_1^{-1} \bmod r_2)$ and $k_i, 1 \leq i \leq 3$ are non-negative integers. By taking a reduction modulo r_2 to both sides of Eq. 5, we obtain,

$$x^m \bmod r_2 = g_2^m \bmod r_2, \tag{6}$$

since all the terms except for the case $k_3 = m$ in the above summation vanishes. Similarly, a reduction modulo p to both sides of Eq. 5, we obtain

$$x^m \bmod p = u^m \cdot g_1^{m\theta} \bmod p; \tag{7}$$

If one of $\mathcal{U}_i, i = 1, 2$ performs honestly, then it easy to check whether

$$X_4 \stackrel{?}{=} X_5 \bmod p. \tag{8}$$

Also using Eqs. 6, and 7 we see that

$$X_1 = x^{a_1} \bmod n = \begin{cases} u^{a_1} g_1^{a_1 \theta} \bmod p \\ g_2^{a_1} \bmod r_2 \end{cases} \tag{9}$$

and

$$X_2 = x^\beta \bmod n = \begin{cases} u^\beta g_1^{\beta\theta} \bmod p \\ g_2^\beta \bmod r_2 \end{cases} \tag{10}$$

and

$$X_3 = x^{a_2} \bmod n = \begin{cases} u^{a_2} g_1^{a_2 \theta} \bmod p \\ g_2^{a_2} \bmod r_2 \end{cases} \tag{11}$$

Using the stored g_2^β value, it is easy to see from Eq. 10 that $X_2 \bmod r_2 = g_2^\beta$. Once the value X_2 is checked for correctness, then it is easy to check whether $X_1 \bmod r_2 \cdot g_2^\alpha \bmod r_2 \stackrel{?}{=} X_2 \cdot X_3 \bmod r_2$ using Eqs. 9 and 11. If all the equalities hold, then the desired result is obtained using reduction modulo p.

In the following theorem, we give a proof sketch to show that $(\mathcal{C}, \mathcal{U}_1, \mathcal{U}_2)$ is an outsource-secure implementation of **2EXP**, as per the Hohenberger-Lysyanskaya security model.

Theorem 1 *(Privacy).* *In the one malicious program model, the pair of algorithms $(\mathcal{C}, \mathcal{U})$ is an outsource-secure implementation of* **2EXP***, where the input (a, u, p) may be honest, secret; or honest, protected; or adversarial protected.*

Proof. Assume that $\mathcal{A} = (\mathcal{E}, \mathcal{U}_1', \mathcal{U}_2')$ be a probabilistic polynomial time (PPT) adversary which interacts with the PPT algorithm $\mathcal{C}$ in the one malicious program model.

Pair One: ($\mathcal{E}$ learns nothing) $\text{EVIEW}_{real} \sim \text{EVIEW}_{ideal}$

If the input (a, u, p) is not honest, secret, the simulator $\mathcal{S}_1$ behaves the same way as in the real experiment. If the input is honest, secret, then $\mathcal{S}_1$ behaves as follows: $\mathcal{S}_1$ ignores the input it received in the ith round and chooses three primes p^*, r_1^* and r_2^* of same size as p, r_1 and r_2 respectively and set $n^* = p^* r_1^* r_2^*$. The main task of $\mathcal{S}_1$ in the ith round is to outsource and verify the computation of $(u^*)^{a^*}$, where u^* and a^* are chosen by $\mathcal{S}_1$ from the same distribution used to choose u and a. $\mathcal{S}_1$ mask u^* and a^* the same way as in our algorithm using appropriately chosen α^*, and β^* and computes x^*, a_1^* and a_2^*. Now, s_1 query $\mathcal{U}_1'$ with the inputs (x^*, a_1^*, n^*), (x^*, β^*, n^*) and $(g_1^{*t_1^*}, \frac{-a^*\theta^*}{t_1^*}, p^*)$. Then s_1 query $\mathcal{U}_2'$ with the inputs (x^*, a_2^*, n^*) and $(g_1^{*t_1^*}, \frac{-a^*\theta^*}{t_1^*}, p^*)$. $\mathcal{S}_1$ checks all the outputs as per the algorithm description. If no error is detected, then $\mathcal{S}_1$ outputs $Y_p^i = \emptyset, Y_u^i = \emptyset$, $replace^i = 0$. That is, the output of the ideal process is set to $(estate^i, y_p^i, y_u^i)$; otherwise, $\mathcal{S}_1$ selects a random element r and outputs $Y_p^i = r, Y_u^i = \emptyset, replace^i = 1$. That is the output of the ideal process is set to $(estate^i, r, \emptyset)$. In both the cases, $\mathcal{S}_1$ saves the appropriate states.

In real process, all the components $(x, g_1^{t_1}, a_1, \beta, a_2, \frac{-a\theta}{t_1})$ in the queries made by $\mathcal{C}$ are independently re-randomized to achieve computational indistinguishability. In the ideal process the values $(x^*, g_1^{*t_1^*}, a_1^*, \beta^*, a_2^*, \frac{-a^*\theta^*}{t_1^*})$ are chosen uniformly at random from the same distribution. Hence, both in the real and the ideal process, the input distributions to $\mathcal{U}_1'$ and $\mathcal{U}_2'$ are computationally indistinguishable. Now, consider the following scenarios:

- If either one of $\mathcal{U}_i'$ behaves honestly in the ith round, then it perfectly executes the algorithm 2EXP such that the outputs from $\mathcal{U}_i'$ in the ideal experiment matches. Therefore, the simulator $\mathcal{S}_1$ does not replace the output of 2EXP in the real experiment (i.e., $replace^i = 0$). Hence $\text{EVIEW}_{real}^i \sim \text{EVIEW}_{ideal}^i$.
- If one of $\mathcal{U}_i'$ is dishonest and outputs an incorrect value in round i, then it will be detected by both $\mathcal{C}$ and $\mathcal{S}_1$ with probability 1 as all the ouptuts could be validated. In the real experiment, the outputs from $\mathcal{U}_i'$ are further processed as in Eqs. 9, 10. Similarly, in the ideal experiment, $\mathcal{S}_1$ simulates with the random value r. Hence, $\text{EVIEW}_{real}^i \sim \text{EVIEW}_{ideal}^i$ even if $\mathcal{U}_i'$ is dishonest in the ith round.

By the hybrid argument, it can be shown that $\text{EVIEW}_{real} \sim \text{EVIEW}_{ideal}$.

Pair Two: ($\mathcal{U}_i'$ learns nothing): $\text{UVIEW}_{real} \sim \text{UVIEW}_{ideal}$

Let $\mathcal{S}_2$ be a PPT simulator that behaves in the same manner regardless of whether the input (a, u, p) is honest, secret or honest, protected or adversarial protected. That is, $\mathcal{S}_2$ ignores the input in the ith round, and makes queries of the form (x^*, a_1^*, n^*), (x^*, β^*, n^*), (x^*, a_2^*, n^*), $(g_1^{*t_1^*}, \frac{-a^*\theta^*}{t_1^*}, p^*)$, according to

$\mathcal{U}'_i$. Then $\mathcal{S}_2$ saves both its state and $\mathcal{U}'$'s state. Note that $\mathcal{E}$ can easily distinguish between these two experiments as the inputs to the experiment might be honest, protected and adversarial protected. But it cannot communicate this information to $\mathcal{U}'_i$. This is due to the fact that, in the ideal experiment, the inputs are computationally blinded by $\mathcal{C}$ before being given as input to $\mathcal{U}'_i$. In the ideal experiment, the simulator $\mathcal{S}_2$ always query the components $(x^*, g_1^{*t_1^*}, a_1^*, \beta^*, a_2^*, \frac{-a^*\theta^*}{t_1^*})$ that are selected uniform at random from the same distribution. Hence $\text{UVIEW}^i_{real} \sim \text{UVIEW}^i_{ideal}$ for each round i. By the hybrid argument, it can be shown that $\text{UVIEW}_{real} \sim \text{UVIEW}_{ideal}$.

Theorem 2 *(Checkability). In the one malicious program model, the above algorithms* $(\mathcal{C}, \mathcal{U}_1, \mathcal{U}_2)$ *are an* $(3,1)$*-outsource-secure implementation of* **2EXP***.*

Proof. For an $k-$bit exponent a, the computation of $u^a \bmod p$ requires roughly $1.5k$ modular multiplications (MM) using square and multiply method. Note that the computation of Eq. 12 and t^{-1} can be done offline. The online computations could be as in Eqs. 2, 3 and for the computation of $a\theta t^{-1}$. Thus our algorithm **2EXP** requires 3 Modular Multiplications. Therefore our algorithm $(\mathcal{C}, \mathcal{U})$ is an $3-$efficient implementation of **2EXP**.

Since the third query (X_4 and X_5) is same for both the servers $\mathcal{U}_i$, it is easy to detect if one of the server exhibits malicious behaviour. The other outputs X_1, X_2, X_3 are validated by the client by reducing the values to modulo r_2. If the check passes then by simulataneous system of linear congruences, the client computes the required result by reducing the values to modulo p. However, a malicious server $\mathcal{U}_i$ outputs incorrect values without being detected by $\mathcal{C}$ provided if it can find the modulus r_2. That is, if the malicious $\mathcal{U}_i$ knows r_2 and reduces x to modulo r_2 to obtain g_2 then it can submit the results of X_1, X_2, X_3 for the base g_2 instead of x_i. As the client validates $\mathcal{U}_i$'s output by reducing them to r_2, the validation goes through. But the values X_1, X_2, X_3 will not help the client to find u^a as reducing X_i to modulo p will not lead to the value with base u. Since the server knows p and $r_1 * r_2$ through the input to the queries, it has to solve the factorisation problem for $r_1 * r_2$. That is it is evident from [19] that for a sufficiently large 3-prime modulus $n = p * r_1 * r_2$ there is no known polynomial time algorithm to factor it into primes even if one of the primes, say p, is known.

Hence our algorithm is a 1-checkable implementation of **2EXP**. Combining the above arguments, we prove the theorem.

3.3   Comparison

Table 1 lists the number of operations performed by $\mathcal{C}$ in the previous outsourcing schemes that achieves both privacy and efficiency with two untrusted servers. From the table, it is evident that our algorithm is superior in terms of both the efficiency and checkability parameters. We use MM to denote modular multiplication.

Table 1. Comparison of outsourcing algorithms for single exponentiation

Exp	Algorithm	MMs	Servers	Queries to $\mathcal{U}_1 + \mathcal{U}_2$	Checkability
u^a	Hohenberger-Lysyanskaya [8]	6 $O(\text{RAND})$ +9	2	8	1/2
	Chen *et al.* [10]	5 $O(\text{RAND})$ + 7	2	6	2/3
	Ours	3	2	5	1

4 2GEXP: Algorithm to Outsource Simultaneous Modular Exponentiations

Consider two primes p and q such that $q|(p-1)$ and the multiplicative subgroup of $\mathbb{Z}_p^*$ with order q. The task of $\mathcal{C}$ is to compute $\prod_{i=1}^{s} u_i^{a_i} \bmod p$ where $u_i \in \mathbb{Z}_p^*$ are the variable bases and $a_i \in \mathbb{Z}_q$ are the variable exponents. The task of $\{\mathcal{U}_1, \mathcal{U}_2\}$ is to return the output $i^j \bmod k$ on input (i, j, k). Since $\mathcal{C}$ aims to protect the values u_i and a_i for $1 \leq i \leq s$ from $\{\mathcal{U}_1, \mathcal{U}_2\}$, they are computationally blinded before being given as input to $\{\mathcal{U}_1, \mathcal{U}_2\}$. As seen in Sect. 3.1, the variable-bases u_i are masked using CRT. Our algorithm is described as follows:

Pre-computation. We use the same pre-processing technique, namely the BPV$^+$ algorithm and primes r_1 and r_2 as in Sect. 3.1. The only difference is that $\mathcal{C}$ needs to call it for each pair (a_i, u_i). Choose a base $h \in \mathbb{Z}_{r_1}^*$.

Masking u_i. Compute $(\theta_i, g_1^{\theta_i})$, $i = \{1, \ldots, s\}$ using the BPV$^+$ technique. Using CRT for each $i = \{1, \ldots, s\}$ leads us to the following system of three simultaneous congruences

$$x_i \equiv u_i g_1^{\theta_i} \bmod p \quad \text{and} \quad x_i \equiv h \bmod r_1 \quad \text{and} \quad x_i \equiv g_2 \bmod r_2$$

and its unique solution $x_i \bmod n$ can be computed as

$$\begin{aligned} x_i = {}& u_i \cdot g_1^{\theta_i} \cdot r_1 \cdot r_2 \cdot (r_1^{-1} r_2^{-1} \bmod p) + h \cdot r_2 \cdot p \cdot (r_2^{-1} p^{-1} \bmod r_1) \\ & + g_2 \cdot p \cdot r_1 \cdot (p^{-1} r_1^{-1} \bmod r_2). \end{aligned} \quad (12)$$

Masking a_i. To blind $a_i (1 \leq i \leq s)$, $\mathcal{C}$ runs BPV$^+$ $2s$ times to obtain the pairs $\{(\alpha_i, g_2^{\alpha_i})\}_{i=1}^{s}$ and $\{(\beta_i, g_2^{\beta_i})\}_{i=1}^{s}$. Then $\mathcal{C}$ computes: $a_{i1} = a_i - \alpha_i$ and $a_{i2} = a_i - \beta_i$.

Queries to $\mathcal{U}_1$. Now $\mathcal{C}$ invokes the BPV$^+$ algorithm to get a pair (t, g_1^t) and sends the following queries to $\mathcal{U}_1$ in random order:

1. $(x_i, a_{i1}, n) \rightarrow X_{i1}$;
2. $(x_i, \beta_i, n) \rightarrow X_{i2}$;
3. $(g_1^t, \sum_{i=1}^{s} (-a_i \theta_i)/t, p) \rightarrow X_4$;

Queries to $\mathcal{U}_2$. Now $\mathcal{C}$ makes the following queries to $\mathcal{U}_2$ in random order:

1. $(x_i, a_{i2}, n) \rightarrow X_{i3}$;
2. $(g_1^t, \sum_{i=1}^{s}(-a_i\theta_i)/t, p) \rightarrow X_5$;

Verifying the correctness of $\{\mathcal{U}_1, \mathcal{U}_2\}$ output.

1. $\mathcal{C}$ checks first if

$$\prod_{i=1}^{s} g_2^{\beta_i} \bmod r_2 \stackrel{?}{=} \prod_{i=1}^{s} X_{i2} \bmod r_2. \tag{13}$$

2. $\mathcal{C}$ then checks whether

$$\left[\prod_{i=1}^{s} X_{i1} \bmod r_2 \cdot \prod_{i=1}^{s} g_2^{\alpha_i}\right] \bmod r_2 \stackrel{?}{=} \prod_{i=1}^{s}(X_{i2} \cdot X_{i3}) \bmod r_2. \tag{14}$$

3. and

$$X_4 \stackrel{?}{=} X_5 \bmod p. \tag{15}$$

Recovering $\prod_{i=1}^{s} u_i^{a_i}$. If Eqs. 13, 14 and 15 hold, then $\mathcal{C}$ believes that for $1 \le i \le s$ all the values $X_{i1}, X_{i2}, X_{i3}, X_{i4}$ and X_{i5} have been computed correctly. This implies that $\prod_{i=1}^{s} X_{i2} \cdot X_{i3} \cdot X_{i4} \equiv \prod_{i=1}^{s} u_i^{a_i} \bmod p$ due to simultaneous system of congruences. If the check fails, then $\mathcal{C}$ outputs an error message.

4.1 Security and Efficiency Analysis

The algorithms proposed in Sects. 3 and 4 depend on the same techniques to mask the inputs (a, u). Thus, due to space limitation we do not repeat the correctness lemma as in Lemma 1 and the proof of following theorem for the simulataneous exponentiation case.

Theorem 3. *In the one malicious program model, the pair of algorithms $(\mathcal{C}, \mathcal{U})$ is an outsource-secure implementation of* **2GEXP***, where the input (a_i, u_i, p) may be honest, secret; or honest, protected; or adversarial protected.*

Theorem 4. *In the one malicious program model, the above algorithms $(\mathcal{C}, \mathcal{U}_1, \mathcal{U}_2)$ are an $(3s, 1)$-outsource-secure implementation of* **2GEXP***.*

4.2 Comparison

Table 2 lists the number of operations performed by $\mathcal{C}$ in the previous outsourcing schemes that achieves both privacy and efficiency with two untrusted servers. From the table, it is evident that our algorithm is superior in terms of both the efficiency and checkability parameters. We achieved $1-$checkable property at the cost of little communication overhead compared to the existing one. However we could reduce the communication overhead with lesser checkability probability. We use MM to denote modular multiplication.

Table 2. Comparison of outsourcing algorithms for simultaneous exponentiation

Exp	Algorithm	MMs	Servers	Queries to $\mathcal{U}_1 + \mathcal{U}_2$	Checkablity
$\prod_{i=1}^{s} u_i^{a_i}$	Chen $et\ al.$ [10]	$5O(\text{RAND}) + 3s + 4$	2	$2s + 4$	$2/(s+2)$
	Ours	$3s$	2	$3s + 2$	1

5   Potential Applications of Our Algorithms

In this section we identify and discuss the usefulness of our secure outsourcing algorithms for single and simultaneous modular exponentiation.

5.1   Securely Offloading the Final Exponentiation in Pairings

Pairings have been widely used in cryptography and have become an important and attractive research area. Given the advantages of pairing-based schemes, recent studies focused on deploying them in various scenarios [20–22]. As the usage of hand-held resourced-constrained devices grows exponentially, equipping them with pairings may help the devices cater to variety of security services. A typical pairing computation consists of two steps, namely the Miller's algorithm [23] and the final exponentiation. Research works focused to speed up pairing computation are mainly about improving the complexity of Miller's algorithm and there are only few works related to speeding up the final exponentiation step [24,25]. Now we will show how our secure outsourcing algorithms could benefit the final exponentiation step.

Let us consider the Tate pairing, the most widely used pairing type. Consider an elliptic curve E over a finite field $\mathbb{F}_q$ with k being the embedding degree. For any two points P and Q on $E(\mathbb{F}_{q^k})$, The Tate pairing value at (P, Q) is specified to be $e(P, Q) = f_{r,P}(Q)^{\frac{q^k - 1}{r}}$ where $f_{r,P} \in \mathbb{F}_{q^k}[x, y]$ is called as the Miller function. Thus the final exponentiation in the Tate pairing is obtained by raising an element in $\mathbb{F}_{q^k}$ to the exponent $(q^k - 1)/r$. We note that our outsourcing algorithm for single exponentiation can be adapted to this setting to securely offload the final exponentiation in full.

Another common method is to split the exponent $(q^k - 1)/r$ accordingly so that the the computation of final exponentiation is given by a finite product of the following form $h \prod_{i=1}^{s} g_i^{r_i}$, where h is relatively easy part to compute and the rest of the product is considered to be hard part. Since every g_i being dependent on P and Q is not a fixed element, the hard part of the final exponentiation can be seen as a simultaneous exponentiation and thus our outsourcing algorithm for simultaneous case may be used to securely offload the hard part of the final exponentiation.

5.2 Outsource-Secure Cryptographic Schemes and Primitives

In this section we recall the applications identified in the related literature for outsourcing algorithms for modular exponentiation. Hohenberger and Lysyanskaya [8] showed how to use their outsourcing algorithm for the single exponentiation to obtain outsource-secure Cramer-Shoup encryption [26] and outsource-secure Schnorr signature [27,28] schemes. Chen *et al.* [10] demonstrated the use of their outsourcing algorithm for simultaneous modular exponentiation in cryptographic primitives such as chameleon hashing [29] and trapdoor commitment [30]. Recently Wang *et al.* [11] showed evidence of advantages in using an outsourcing algorithm for the simultaneous case in provable data possession (PDP) [31] schemes for cloud storage. Since our algorithm is generic and superior, it may provide more benefits when used in any of the above mentioned cryptographic protocols.

6 Conclusion

We presented secure outsourcing algorithms for single and simultaneous modular exponentiations. Our algorithms are superior compared to the existing schemes with two servers. We then showed that the proposed algorithms meet the security notions of the Hohenberger-Lysyanskaya security model. Finally we discussed some interesting applications such as outsourcing of final exponentiation in pairings. We leave open the problem of securely outsourcing modular exponentiations using single untrusted server achieving checkability probability 1.

Appendix

A Security Defintions

Definition 1 (Algorithm with IO-outsource). *The outsource algorithm* Alg *obeys the input/output specification if it accepts five inputs and produces three outputs. The honest entity generates the first three inputs and the last two adversarially chosen inputs are generated by the environment $\mathcal{E}$. The first three inputs can be further classified based on the information about them available to the adversary $\mathcal{A} = (\mathcal{E}, \mathcal{U})$. The first input is the honest, secret input which is unknown to both $\mathcal{E}$ and $\mathcal{U}$. The second input is the honest, protected input which may be known by $\mathcal{E}$, but is protected from $\mathcal{U}$. The third input is the honest, unprotected input which may be known by both $\mathcal{E}$ and $\mathcal{U}$. The fourth input is the adversarial, protected input which may be known by $\mathcal{E}$, but is protected from $\mathcal{U}$. The fifth input is the the adversarial, protected input which may be known by $\mathcal{E}$, but is protected from $\mathcal{U}$. Similarly, the first, second and third outputs are called secret, protected and unprotected outputs respectively.*

Definition 2 (Outsource-security). *A pair of algorithms $(\mathcal{C}, \mathcal{U})$ is said to be an outsource-secure implementation of an algorithm* Alg *with IO-outsource if:*

Correctness $\mathcal{C}^{\mathcal{U}}$ is a correct implementation of Alg.

Security *For all probabilistic polynomial-time adversaries* $\mathcal{A} = (\mathcal{E}, \mathcal{U}')$, *there exist probabilistic expected polynomial-time simulators* $(\mathcal{S}_1, \mathcal{S}_2)$ *such that the following pairs of random variables are computationally indistinguishable.*

Pair One *(*$\mathcal{E}$ *learns nothing):* $\text{EVIEW}_{real} \sim \text{EVIEW}_{ideal}$.

The real process: This process proceeds in rounds. Assume that the honestly generated inputs are chosen by a process I. *The view that the adversarial environment obtains by participating in the following process:*

$$\text{EVIEW}^i_{real} = \{\left(istate^i, x^i_{hs}, x^i_{hp}, x^i_{hu}\right) \leftarrow I\left(1^k, istate^{i-1}\right);$$
$$\left(estate^i, j^i, x^i_{ap}, x^i_{au}, stop^i\right) \leftarrow \text{E}\left(1^k, \text{EVIEW}^{i-1}_{real}, x^i_{hp}, x^i_{hu}\right);$$
$$\left(tstate^i, ustate^i, y^i_s, y^i_p, y^i_u\right) \leftarrow \mathcal{C}^{\mathcal{U}'(ustate^{i-1})}\left(tstate^{i-1}, x^{j^i}_{hs}, x^{j^i}_{hp}, x^{j^i}_{hu}, x^i_{ap}, x^i_{au}\right):$$
$$\left(estate^i, y^i_p, y^i_u\right)\}$$

$$\text{EVIEW}_{real} = \text{EVIEW}^i_{real} if stop^i = \text{TRUE}.$$

In round i, *The adversarial environment does not have access to the honest inputs* $(x^i_{hs}, x^i_{hp}, x^i_{hu})$ *that are picked using an honest, stateful process* I. *The environment based on its view from last round, chooses the value of its estate$_i$ variable that is used to recall what it did next time it is invoked. Then, among the previously generated honest inputs, the environment chooses a input vector* $(x^{j^i}_{hs}, x^{j^i}_{hp}, x^{j^i}_{hu})$ *to give it to* $\mathcal{C}^{\mathcal{U}'}$. *Observe that the environment can specify the index* j^i *of the inputs but not the values. The environment also chooses the adversarial protected and unprotected input* x^i_{ap} *and* x^i_{au} *respectively. It also chooses the boolean variable* $stop^i$ *that determines whether round* i *is the last round in this process.*

Then, $\mathcal{C}^{\mathcal{U}'}$ *is run on inputs* $(tstate^{i-1}, x^{j^i}_{hs}, x^{j^i}_{hp}, x^{j^i}_{hu}, x^i_{ap}, x^i_{au})$ *where* $tstate^{i-1}$ *is* $\mathcal{C}$'s *previously saved state. The algorithm produces a new state* $tstate^i$ *for* $\mathcal{C}$ *along with the secret* y^i_s, *protected* y^i_p *and unprotected* y^i_u *outputs. The oracle* $\mathcal{U}'$ *is given* $ustate^{i-1}$ *as input and the current state in saved in* $ustate^i$. *The view of the real process in round* i *consists of* $estate^i$, *and the values* y^i_p *and* y^i_u. *The overall view of the environment in the real process is just its view in the last round.c*

The ideal process:

$$\text{EVIEW}^i_{ideal} = \{\left(istate^i, x^i_{hs}, x^i_{hp}, x^i_{hu}\right) \leftarrow I\left(1^k, istate^{i-1}\right);$$
$$\left(estate^i, j^i, x^i_{ap}, x^i_{au}, stop^i\right) \leftarrow \text{E}\left(1^k, \text{EVIEW}^{i-1}_{ideal}, x^i_{hp}, x^i_{hu}\right);$$
$$\left(astate^i, y^i_s, y^i_p, y^i_u\right) \leftarrow \text{Alg}\left(astate^{i-1}, x^{j^i}_{hs}, x^{j^i}_{hp}, x^{j^i}_{hu}, x^i_{ap}, x^i_{au}\right);$$
$$\left(sstate^i, ustate^i, Y^i_p, Y^i_u, replace^i\right) \leftarrow \mathcal{S}_1^{\mathcal{U}'(ustate^{i-1})}\left(sstate^i, x^{j^i}_{hp}, x^{j^i}_{hu}, x^i_{ap}, x^i_{au}, y^i_p, y^i_u\right);$$
$$\left(z^i_p, z^i_u\right) = replace^i\left(Y^i_p, Y^i_u\right) + \left(1 - replace^i\right)\left(y^i_p, y^i_u\right):$$
$$\left(estate^i, z^i_p, z^i_u\right)\}$$

$$\text{EVIEW}_{ideal} = \text{EVIEW}^i_{ideal} if stop^i = \text{TRUE}.$$

This process also proceeds in rounds. The secret input x^i_{hs} *is hidden from the stateful simulator* $\mathcal{S}_1$. *But, the non-secret inputs produced by the algorithm* *that is run on all inputs of round i is given to $\mathcal{S}_1$. Now, $\mathcal{S}_1$ decides whether to output the values (y_p^i, y_u^i) generated by the algorithm* Alg *or replace them with some other values (Y_p^i, Y_u^i). This replacement is captured using the indicator variable* $replace^i \in \{0,1\}$. *The simulator is allowed to query the oracle* $\mathcal{U}'$ *which saves its state as in the real experiment.*

Pair Two *($\mathcal{U}'$ Learns Nothing):* $\text{UVIEW}_{real} \sim \text{UVIEW}_{ideal}$.

The view that the untrusted entity $\mathcal{U}'$ *obtains by participating in the real process is described in pair one.* $\text{UVIEW}_{real} = ustate^i \, if \, stop^i = \text{TRUE}$. *The ideal process:*

$$\text{UVIEW}_{ideal}^i = \{ \left(istate^i, x_{hs}^i, x_{hp}^i, x_{hu}^i \right) \leftarrow I \left(1^k, istate^{i-1} \right);$$
$$\left(estate^i, j^i, x_{ap}^i, x_{au}^i, stop^i \right) \leftarrow \text{E} \left(1^k, estate^{i-1}, x_{hp}^i, x_{hu}^i, y_p^{i-1}, y_u^{i-1} \right);$$
$$\left(astate^i, y_s^i, y_p^i, y_u^i \right) \leftarrow \text{Alg} \left(astate^{i-1}, x_{hs}^{j^i}, x_{hp}^{j^i}, x_{hu}^{j^i}, x_{ap}^i, x_{au}^i \right);$$
$$\left(sstate^i, ustate^i \right) \leftarrow \mathcal{S}_2^{\mathcal{U}'(ustate^{i-1})} \left(sstate^{i-1}, x_{hu}^{j^i}, x_{au}^i \right);$$
$$\left(ustate^i \right) \}$$

$$\text{UVIEW}_{ideal} = \text{UVIEW}_{ideal}^i \, if \, stop^i = \text{TRUE}.$$

In the ideal process, the stateful simulator $\mathcal{S}_2$ *is given with only the unprotected inputs* (x_{hu}^i, x_{au}^i), *queries* $\mathcal{U}'$. *As before,* $\mathcal{U}'$ *may maintain state.*

Definition 3 (α–efficient, secure outsourcing). *A pair of algorithms $(\mathcal{C}, \mathcal{U})$ is said to be an α–efficient implementation of an algorithm* Alg *if $(\mathcal{C}, \mathcal{U})$ is an outsource secure implementation of algorithm* Alg *and for all inputs x, the running time of $\mathcal{C}$ is $\le$ an α– multiplicative factor of the running time of* Alg(x)

Definition 4 (β–checkable, secure outsourcing). *A pair of algorithms $(\mathcal{C}, \mathcal{U})$ is a β–checkable implementation of an algorithm* Alg *if $(\mathcal{C}, \mathcal{U})$ is an outsource secure implementation of algorithm* Alg *and for all inputs x, if $\mathcal{U}'$ deviates from its advertised functionality during the execution of $\mathcal{C}^{\mathcal{U}'}(x)$, $\mathcal{C}$ will detect the error with probability $\ge \beta$*

Definition 5 ((α, β)–outsource-security). *A pair of algorithms $(\mathcal{C}, \mathcal{U})$ is said to be an (α, β)–outsource-secure implementation of an algorithm* Alg *if they are both α–efficient and β–checkable.*

References

1. Abadi, M., Feigenbaum, J., Kilian, J.: On hiding information from an oracle. In: Proceedings of the Second Annual Conference on Structure in Complexity Theory, pp. 195–203. IEEE Computer Society (1987)
2. Golle, P., Mironov, I.: Uncheatable distributed computations. In: Naccache, D. (ed.) CT-RSA 2001. LNCS, vol. 2020, pp. 425–440. Springer, Heidelberg (2001). doi:10.1007/3-540-45353-9_31

3. Girault, M., Lefranc, D.: Server-aided verification: theory and practice. In: Roy, B. (ed.) ASIACRYPT 2005. LNCS, vol. 3788, pp. 605–623. Springer, Heidelberg (2005). doi:10.1007/11593447_33

4. Wu, W., Mu, Y., Susilo, W., Huang, X.: Server-aided verification signatures: definitions and new constructions. In: Baek, J., Bao, F., Chen, K., Lai, X. (eds.) ProvSec 2008. LNCS, vol. 5324, pp. 141–155. Springer, Heidelberg (2008). doi:10.1007/978-3-540-88733-1_10

5. Gennaro, R., Gentry, C., Parno, B.: Non-interactive verifiable computing: outsourcing computation to untrusted workers. In: Rabin, T. (ed.) CRYPTO 2010. LNCS, vol. 6223, pp. 465–482. Springer, Heidelberg (2010). doi:10.1007/978-3-642-14623-7_25

6. Green, M., Hohenberger, S., Waters, B.: Outsourcing the decryption of ABE ciphertexts. In: USENIX Security Symposium 2011. USENIX Association (2011)

7. Gentry, C.: Fully homomorphic encryption using ideal lattices. In: Mitzenmacher, M. (ed.) Proceedings of the 41st Annual ACM Symposium on Theory of Computing, STOC 2009, pp. 169–178. ACM (2009)

8. Hohenberger, S., Lysyanskaya, A.: How to securely outsource cryptographic computations. In: Kilian, J. (ed.) TCC 2005. LNCS, vol. 3378, pp. 264–282. Springer, Heidelberg (2005). doi:10.1007/978-3-540-30576-7_15

9. van Dijk, M., Clarke, D.E., Gassend, B., Suh, G.E., Devadas, S.: Speeding up exponentiation using an untrusted computational resource. Des. Codes Cryptogr. **39**(2), 253–273 (2006)

10. Chen, X., Li, J., Ma, J., Tang, Q., Lou, W.: New algorithms for secure outsourcing of modular exponentiations. In: Foresti, S., Yung, M., Martinelli, F. (eds.) ESORICS 2012. LNCS, vol. 7459, pp. 541–556. Springer, Heidelberg (2012). doi:10.1007/978-3-642-33167-1_31

11. Wang, Y., Wu, Q., Wong, D.S., Qin, B., Chow, S.S.M., Liu, Z., Tan, X.: Securely outsourcing exponentiations with single untrusted program for cloud storage. In: Kutyłowski, M., Vaidya, J. (eds.) ESORICS 2014. LNCS, vol. 8712, pp. 326–343. Springer, Heidelberg (2014). doi:10.1007/978-3-319-11203-9_19

12. Kiraz, M.S., Uzunkol, O.: Efficient and verifiable algorithms for secure outsourcing of cryptographic computations. Cryptology ePrint Archive, Report 2014/748 (2014). http://eprint.iacr.org/

13. Matsumoto, T., Kato, K., Imai, H.: Speeding up secret computations with insecure auxiliary devices. In: Goldwasser, S. (ed.) CRYPTO 1988. LNCS, vol. 403, pp. 497–506. Springer, Heidelberg (1990). doi:10.1007/0-387-34799-2_35

14. Nguyen, P.Q., Shparlinski, I.E.: On the insecurity of a server-aided RSA protocol. In: Boyd, C. (ed.) ASIACRYPT 2001. LNCS, vol. 2248, pp. 21–35. Springer, Heidelberg (2001). doi:10.1007/3-540-45682-1_2

15. Chevalier, C., Laguillaumie, F., Vergnaud, D.: Privately outsourcing exponentiation to a single server: cryptanalysis and optimal constructions. IACR Cryptology ePrint Archive 2016/309 (2016)

16. Kiraz, M.S., Uzunkol, O.: Efficient and verifiable algorithms for secure outsourcing of cryptographic computations. Int. J. Inf. Secur. **15**(5), 519–537 (2016). doi:10.1007/s10207-015-0308-7

17. Boyko, V., Peinado, M., Venkatesan, R.: Speeding up discrete log and factoring based schemes via precomputations. In: Nyberg, K. (ed.) EUROCRYPT 1998. LNCS, vol. 1403, pp. 221–235. Springer, Heidelberg (1998). doi:10.1007/BFb0054129

18. Nguyen, P., Shparlinski, I., Stern, J.: Distribution of modular sums and the security of the server aided exponentiation. In: Proceedings of the Workshop on Cryptography and Computational Number Theory (CCNT 1999), Singapore, Birkhäuser, pp. 257–268 (2001)

19. Hinek, M.: On the security of multi-prime RSA. http://cacr.uwaterloo.ca/techreports/2006/cacr2006-16.pdf

20. Chevallier-Mames, B., Coron, J.-S., McCullagh, N., Naccache, D., Scott, M.: Secure delegation of elliptic-curve pairing. In: Gollmann, D., Lanet, J.-L., Iguchi-Cartigny, J. (eds.) CARDIS 2010. LNCS, vol. 6035, pp. 24–35. Springer, Heidelberg (2010). doi:10.1007/978-3-642-12510-2_3

21. Guillevic, A.: Comparing the pairing efficiency over composite-order and prime-order elliptic curves. In: Jacobson, M., Locasto, M., Mohassel, P., Safavi-Naini, R. (eds.) ACNS. LNCS, vol. 7954, pp. 357–372. Springer, Heidelberg (2013)

22. Scott, M.: Unbalancing pairing-based key exchange protocols. IACR Cryptology ePrint Archive 2013/688 (2013). http://eprint.iacr.org/2013/688

23. Miller, V.S.: The weil pairing, and its efficient calculation. J. Cryptol. **17**(4), 235–261 (2004)

24. Scott, M., Benger, N., Charlemagne, M., Dominguez Perez, L.J., Kachisa, E.J.: On the final exponentiation for calculating pairings on ordinary elliptic curves. In: Shacham, H., Waters, B. (eds.) Pairing 2009. LNCS, vol. 5671, pp. 78–88. Springer, Heidelberg (2009). doi:10.1007/978-3-642-03298-1_6

25. Kim, T., Kim, S., Cheon, J.H.: On the final exponentiation in tate pairing computations. IEEE Trans. Inf. Theory **59**(6), 4033–4041 (2013)

26. Cramer, R., Shoup, V.: Design and analysis of practical public-key encryption schemes secure against adaptive chosen ciphertext attack. SIAM J. Comput. **33**(1), 167–226 (2003)

27. Schnorr, C.P.: Efficient identification and signatures for smart cards. In: Brassard, G. (ed.) CRYPTO 1989. LNCS, vol. 435, pp. 239–252. Springer, Heidelberg (1990). doi:10.1007/0-387-34805-0_22

28. Schnorr, C.P.: Efficient signature generation by smart cards. J. Cryptol. **4**(3), 161–174 (1991)

29. Ateniese, G., Medeiros, B.: Identity-based chameleon hash and applications. In: Juels, A. (ed.) FC 2004. LNCS, vol. 3110, pp. 164–180. Springer, Heidelberg (2004). doi:10.1007/978-3-540-27809-2_19

30. Fischlin, M., Fischlin, R.: Efficient non-malleable commitment schemes. In: Bellare, M. (ed.) CRYPTO 2000. LNCS, vol. 1880, pp. 413–431. Springer, Heidelberg (2000). doi:10.1007/3-540-44598-6_26

31. Ateniese, G., Burns, R.C., Curtmola, R., Herring, J., Kissner, L., Peterson, Z.N.J., Song, D.X.: Provable data possession at untrusted stores. In: Ning, P., di Vimercati, S.D.C., Syverson, P.F. (eds.) ACM CCS 2007, pp. 598–609. ACM (2007)

Verifiable Computation for Randomized Algorithm

Muhua Liu[1,2], Ying Wu[1,2], and Rui Xue[1,2(✉)]

[1] State Key Laboratory of Information Security,
Institute of Information Engineering, Chinese Academy of Sciences,
Beijing 100093, China
{liumuhua,wuying,xuerui}@iie.ac.cn
[2] University of Chinese Academy of Sciences, Beijing 100049, China

Abstract. Verifiable computation enables a computationally weak client to delegate the difficult computation to a more powerful cloud server. When the client receives the returned result, it can verify the correctness of the result. As a new computing model, it has been widely studied. But, the previous works on verifiable computation was restricted to the case where the target function was deterministic.

In this paper, we present a novel definition for verifiable computation supporting randomized algorithms, which allows a resource constrained client to outsource the computation of a randomized algorithm to an untrusted server. We consider a new setting where the random string, as the random input of the randomized algorithm, is generated by the cloud server. In our security definition, it must guarantee that the server can not tamper with the randomness which means that the generated string is indistinguishable with a truly random string. Our construction is based on indistinguishability obfuscation, constrained verifiable random function and functional pseudorandom function.

Keywords: Verifiable computation · Randomized algorithms · Indistinguishability obfuscation

1 Introduction

Outsourced computation allows a client to outsource difficult computations to a more powerful cloud server. The client can cost less resource to finish an expensive computation. In the past years, outsourced computation has become a very hot research area in cryptography. There are many new issues that we should consider carefully. The first one is computation reliability. When the client receives the result of an outsourced task, how to guarantee the correctness of computation? After all, the server is not reliable or honest. It may make mistakes for all kinds of reasons, such as an incorrect implementation of the algorithm or returning a random result even for saving the computation time. In addition, the key requirement is efficiency, which requires that the total running time of the

© Springer International Publishing AG 2016
O. Dunkelman and S.K. Sanadhya (Eds.): INDOCRYPT 2016, LNCS 10095, pp. 99–118, 2016.
DOI: 10.1007/978-3-319-49890-4_6

client must be less than the time of computing on its own. Therefore, secure outsourced computation has received widespread attention in cryptography.

Gennaro et al. [13] first introduced the notion of non-interactive verifiable computation. They constructed a scheme based on Yao's garbled circuit and fully homomorphic encryption. Subsequently, there are many works [2,4,5,8,20] for verifiable computation. However, the previous works on verifiable computation was restricted to the case the target function was deterministic. In this work, we continue the research to move beyond deterministic algorithms, and consider the case of randomized algorithms. Randomized algorithm has received widespread application in cryptography, such as encryption algorithm, key generation algorithm. In the randomized algorithm, it needs to toss a random string as the input of the algorithm. We consider a new case, and outsource the randomized algorithm to a server. To understand this problem, an illustrative example is given as follows.

We show the necessity for verifiable computation for randomized algorithms by giving an example of assessing some data by random sampling. A research institute wants to assess the quality of teaching by random sampling of database entries from each university. Due to the weak computation ability, the institute delegates the assessment algorithm to a server. The server needs to sample a large database, and then assess the sampled data. Therefore, the server should choose some random strings to sample the assessment data, and then implement the assessment algorithm. In this procedure, the research institute mainly concerns two primaries:

- The institute wants to ensure that the cloud server can not decide the sampled result, i.e. the random string generated by the server must be indistinguishable with a real random string.
- The institute should cost less time than the time required for the institute to execute the randomized algorithm by itself.

Think for another instance of basic algorithm in statistics such as Monte Carlo methods. A client tries to search molecular configuration which satisfies a given condition of energy by Monte Carlo method. In this procedure, it first needs to generate a random molecular configuration. Then, it generates a new molecular configuration by randomly changing the coordinate of the particle, and computes the energy of the new molecular configuration. Finally, it decides whether it accepts this new molecular configuration. As far as we know, the more times the experiment executes, the more accurate the results are. The client needs to interactive with the server if it chooses the random string for each experiment. Therefore, it delegates the execution of the algorithm to a more powerful server. All the random strings of each experiment are chosen by the server.

In this work, we focus on the non-interactive verifiable outsourced computation for randomized algorithms. To construct a scheme which satisfies our setting, we define verifiable outsourced computation for randomized algorithms as follows. A randomized algorithm can be represented as a function $f(x; r)$ with two inputs, an evaluated point x and a random string r. Given an encoding σ_x of the input x, the cloud server can compute an encoding σ_y of the evaluation

$f(x; r)$, where r is generated by the cloud server. After receiving the encoding σ_y of $f(x; r)$, the client can verify the correctness of the evaluation $f(x; r)$. We require that the returned results $\{f(x; r_i)\}_{i=1}^{t}$ are indistinguishable with the results $\{f(x; r_i')\}_{i=1}^{t}$, where r_i' is true randomness, r_i is generated by the cloud server, and t is the number of computation required by the client. Sometimes, the assessment results will be different for the same data due to the different random strings. In order to increase the precision of assessment results, the institute may require the server to implement the sample assessment algorithm many times.

1.1 Our Results

We give a formal definition based on the definition of Gennaro et al. [13]. We then construct a verifiable computation scheme for randomized algorithms which supports arbitrary probabilistic polynomial time algorithms. Selective security is considered in our prove. Adaptive security can be obtained generically by standard complexity leveraging.

We focus on the soundness security, which requires that an adversary can not persuade a client to accept an incorrect output. The detailed definition is given in Sect. 2.

In our construction, the core problem is to handle the secure generation of the random string. In order to solve this problem, we construct an obfuscation of a circuit which has a pseudorandom function's key sk' hardwired in it. On input a value s, it outputs a pseudorandom value r. However, the secret key sk' can be hidden only if the obfuscator is a black box obfuscation [3], and the indistinguishability obfuscator can not ensure the security of the pseudorandom function's key sk'. To overcome this issue, we use a constrained PRF key sk_S to replace the key sk' hardwired in the circuit, and introduce a new value to the circuit, which can guarantee that the obfuscation of circuits implemented identical functions are indistinguishable. By the pseudorandom property of the constrained PRF at the constrained points, the cloud server can not distinguish the pseudorandom value from a true random value.

To verify the correctness of random strings generated by the server, we use the constrained verifiable random function, introduced by Fuchsbauer [11], to replace the constrained random function. We modify the construction of Fuchsbauer [11], and make it satisfy a new pseudorandom definition, where $V.Eval(sk', x)$ should remain pseudorandomness even if an adversary is given the corresponding proof. To verify the evaluation correctness of randomized function, we use functional pseudorandom functions which were introduced by Boyle, Goldwasser and Ivan [6]. In the key generation phase, we generate a functional key sk_f for the function $f(x; r)$, which allows one to evaluate a pseudorandom value on any y for which there exists an pair (x, r) such that $f(x; r) = y$.

1.2 Related Work

Alwen et al. [1] studied functional encryption for randomized function. They gave a construction of functional encryption for randomized functions based on the deterministic functional encryption. In another independent work, Goyal et al. [17] presented the first definition and construction for functional encryption which supported randomized functionalities. In their definitions, they not only considered the dishonest decryptor, but also considered the dishonest encryptor. They added security requirements to ensure that the encryptor can not improperly tamper with the randomness used to the outputs. Komargodski et al. [18] introduced a construction for any family of randomized functionalities in the private key setting.

Choi et al. [7] presented a new notion which was called multi-client verifiable computation. It allows a set of clients to delegate the computation of a function f over a joint inputs $(x_1, \ldots, x_n)$ to an untrusted server without interacting with each other. Recently, Goldwasser et al. [15] proposed an alternative construction based on multi-input functional encryption. Subsequently, Gordon et al. [16] introduced a systematic study of multi-client verifiable computation which satisfies a stronger security requirements. At first glance, the setting we consider may seem similar to multi-client verifiable computation [7,15,16]. All the inputs of the algorithm in the multi-client verifiable algorithm are generated by the clients, while the random string is generated by the server in our setting. Just like the second instance, the client needs not interact with the server in our setting.

Gentry et al. [14] constructed the first outsourcing private random access machine (RAM) computation scheme, which allows a client to privately outsource arbitrary program executing to a remote server. In their construction, every time, the client wants to run a new program execution on input x, he needs to choose a fresh randomness r, and garbles (x, r) under the reusable circuit garbling scheme. Therefore, their scheme does not satisfy our setting.

There are some verifiable computation schemes for some special functions, such as polynomial functions, matrix computation. Fiore and Gennaro [9] proposed two new protocols which supported to delegate large polynomials and matrix computations. Papamanthou et al. [19] introduced a new model for verifying dynamic computations, which was called signatures of correct computation. Fiore et al. [10] proposed highly efficient schemes to delegate various classes of polynomial functions, which included linear combinations, high-degree univariate polynomials and multivariate quadratic polynomials.

1.3   Organization

The rest of this work is organized as follows. We introduce the definition of verifiable computation for randomized algorithms in Sect. 2. Some basic building blocks are given in Sect. 3. We present a construction for randomized algorithms and give its security proof in Sect. 4.

2   Definitions

Throughout the paper, we denote the security parameter by $\lambda \in \mathbb{N}$. A function negl(n) is negligible if for every positive polynomial function $p(\cdot)$ and all sufficiently large n, it holds that $\text{negl}(n) < 1/p(n)$. $x \xleftarrow{R} X$ denotes that the element x is chosen uniformly at random from the set X. For $n \in \mathbb{N}$, let $[n] = \{1, \ldots, n\}$.

2.1 Formal Definition of Verifiable Computation for Randomized Algorithms

In this subsection, we give the formal definitions for verifiable computation for randomized algorithms (or rand − VC).

A randomized algorithm can be implemented as an 2−ary function f that takes a value and a random string as inputs:

$$f : \mathcal{X} \times \mathcal{R} \to \mathcal{Y}.$$

We assume the random string are chosen randomly from the set $\mathcal{R}$. Next, we give the definitions of verifiable computation for randomized algorithms.

Syntax. A verifiable computation for randomized algorithms $\mathcal{F}$ consists of four algorithms rand − VC = (KeyGen, ProbGen, Compute, Verify) :

- KeyGen$(f, \lambda) \to (pk, sk)$. This algorithm is run by the client. On input a security parameter λ and a randomized algorithm which was implemented as a function f, this algorithm generates a public key pk and a matching secret key sk. The public key pk is used by the server to compute with, and the secret key sk is kept private by the client.
- ProbGen$(pk, x, t) \to (\sigma_x, \tau_x)$. This algorithm is run by the client. The problem generation algorithm takes the public key pk, the function input x and the number of computation t as input. It outputs a public value σ_x and a secret value τ_x. The public value σ_x is given to the server to compute with, and the secret value τ_x is kept private by the client to verify the correctness of returned results.
- Compute$(pk, \sigma_x) \to \sigma_y$. This algorithm is run by the server. This algorithm takes as input the client's public key pk and the encoded input σ_x. Then, it chooses strings $s_1, \ldots, s_t$ to generate the random inputs $r_1, \ldots, r_t$, and outputs an encoded version of the function's output.
- Verify$(sk, \tau_x, \sigma_y) \to y$ or $\perp$. This algorithm is run by the client. On input the secret key sk, the secret "decoding" τ_x, and the server's output σ_y, the verification algorithm outputs $y_i = f(x; r_i), i = 1, \ldots, t$, or outputs $\perp$ which indicates that σ_y does not represent the valid outputs of f on x.

A verifiable computation scheme for randomized algorithm is correct if the values produced by the key generation algorithm and the problem generation algorithm allow an honest server to compute values which can verify successfully. In additional, we require that the distribution of the decoded output values is computationally indistinguishable from that obtained by sampling the randomized functionality directly on the inputs. We give the formal definition as follows:

Definition 1 (Correctness). *A verifiable computation for randomized algorithm f is correct if for every $x \in \mathcal{M}$, the following two distributions are computationally indistinguishable:*

- ***Real:*** $\mathsf{Verify}(sk, \tau_x, \sigma_y) = \{y_i = f(x; r_i)\}_{i=1}^{t_x}$, *where:*
    - $(pk, sk) \leftarrow \mathsf{KeyGen}(f, \lambda)$,
    - $(\sigma_x, \tau_x) \leftarrow \mathsf{ProbGen}(sk, x, t_x)$,
    - $\sigma_y \leftarrow \mathsf{Compute}(pk, \sigma_x)$.
- ***Ideal:*** $\{f(x; r_i)\}_{i=1}^{t_x}$, *where* $r_i \xleftarrow{R} \mathcal{R}$.

A verifiable computation scheme for randomized algorithms is secure if a malicious server cannot persuade the verification algorithm to accept an incorrect output. We consider selective security for randomized algorithms, which requires that the adversary outputs the strings $s_1, \ldots, s_t$ before outputting the challenge value σ_x. The secure experiment $\mathsf{Exp}_{\mathcal{A}}^{\mathsf{Verify}}[\mathsf{rand} - \mathsf{VC}, f, \lambda]$ is described as follows:

- The challenger represents the randomized algorithm as a function $f(\cdot; \cdot)$, and runs $(pk, sk) \leftarrow \mathsf{KeyGen}(1^\lambda, f)$. Then he returns pk to the adversary.
- The adversary can generate the encoding of multiple problem instances by himself, and then, returns a challenge $(x, t, s_1, \ldots, s_t)$.
- The challenger computes $(\sigma_x, \tau_x) \leftarrow \mathsf{ProbGen}(pk, x, t)$, and returns the public value σ_x.
- The adversary uses the challenge public value σ_x and the strings $s_1, \ldots, s_t$ to compute an evaluation value $\hat{\sigma}_y$.
- The challenger runs the algorithm $\mathsf{Verify}(sk, \tau_x, \hat{\sigma}_y)$ and gets the output $\hat{y}$. If $\hat{y} \neq \perp^1$, and there exists an index $j \in [t]$, for all $r_j \in \mathcal{R}$, s.t. $\hat{y}_j \neq f(x; r_j)$, or there exists an distinguisher $\mathcal{D}$ which can distinguish the results $\hat{y} = \{\hat{y}_j = f(x; r_j)\}_{j=1}^{t}$ from $\{f(x; r'_j)\}_{j=1}^{t}$, where $r'_j \xleftarrow{R} \mathcal{R}$, the experiment outputs '1', else '0'.

Now, We define the selective security of the verifiable computation based on the adversary's success in the above experiment.

Definition 2 (Selective security). *A verifiable computation scheme for randomized algorithms* $\mathsf{rand} - \mathsf{VC}$ *is selective secure for a function f, if for any probabilistic polynomial time adversary $\mathcal{A}$,*

$$\Pr[\mathsf{Exp}_{\mathcal{A}}^{\mathsf{Verify}}[\mathsf{rand} - \mathsf{VC}, f, \lambda] = 1] \leq \mathsf{negl}(\lambda).$$

Adaptive Security. We give the description of adaptive security for verifiable computation for randomized algorithms. Formally, adaptive security is defined in the same manner as Definition 2, expect that the adversary $\mathcal{A}$ needs not choose strings before the public value σ_x is computed by the challenger. In order to avoid repetition, we omit the formal definition.

The last condition we considered for a randomized verifiable computation scheme is that the verification stage costs less time than the time required to compute the function from scratch.

---

[1] It means that for all $i \in [t]$, $y_i \neq \perp$, where y_i is the ith component of the vector $\hat{y}$.

Definition 3 (Outsourceable). *A* rand − VC *can be outsourced if it permits efficient verification. That is to say, for any x and any* σ_y, *the time required for* Verify(sk, τ_x, σ_y) *is* $o(T)$, *where* T *is the time required to compute* $f(x)$.

Note that we just consider the time to verify the output and do not consider the time to compute the key generation algorithm and the problem generation algorithm (i.e., the encoding of the randomized algorithm itself, and the encoding of the input). It is different with the previous definition of efficiency for deterministic algorithms which requires the time of the encoding of the function itself is in an amortized sense. Our definition also requires the encoding of input itself can be amortized over many different randomized string evaluations.

3 Preliminaries

In this section, we give the definitions of various cryptographic primitives that will be used in our construction. It contains indistinguishability obfuscation, constrained verifiable random functions, and functional pseudorandom functions.

3.1 Indistinguishability Obfuscation

We present the formal definition of indistinguishability obfuscation following the notion of Garg et al. [12].

Definition 4 (Indistinguishability Obfuscation *(iO*)) [12, Definition 4]. *An uniform PPT machine iO is called an indistinguishability obfuscator for a circuit class* $\{\mathcal{C}_\lambda\}$ *if it satisfies the following conditions:*

− **(Correctness:)** *For all security parameters* $\lambda \in \mathbb{N}$, $C \in \mathcal{C}_\lambda$, *and inputs* x, *we have that*
$$\Pr[C'(x) = C(x) : C' \leftarrow i\mathcal{O}(\lambda, C)] = 1.$$

− **(Indistinguishability:)** *For any (not necessarily uniform) PPT distinguisher* (Samp, $\mathcal{D}$), *there exists a negligible function* negl *such that the following holds: if* $\Pr[\forall x, C_0(x) = C_1(x); (C_0, C_1, \sigma) \leftarrow \mathsf{Samp}(1^\lambda)] \geq 1 - \mathrm{negl}(\lambda)$, *then:*
$$| \Pr[\mathcal{D}(\sigma, i\mathcal{O}(\lambda, C_0)) = 1 : (C_0, C_1, \sigma) \leftarrow \mathsf{Samp}(1^\lambda)]$$
$$- \Pr[\mathcal{D}(\sigma, i\mathcal{O}(\lambda, C_1)) = 1 : (C_0, C_1, \sigma) \leftarrow \mathsf{Samp}(1^\lambda)]|$$
$$\leq \mathrm{negl}(\lambda).$$

In a recent work, Garg et al. [12] showed how to construct an indistinguishability obfuscator $i\mathcal{O}$ for the circuit class P/poly.

3.2 Constrained Verifiable Random Functions

In our construction, we will use the constrained verifiable random function (VRF), which is a constrained pseudorandom function (PRF). Below we recall their definition, which was proposed by Fuchsbauer [11]:

Definition 5 (Constrained verifiable random functions). Let $F : \mathcal{K} \times \mathcal{X} \rightarrow \mathcal{Y} \times \mathcal{P}$ be a function which is computed in polynomial time, where $\mathcal{K}$ is the key space, $\mathcal{X}$ is the domain, $\mathcal{Y}$ is the range and $\mathcal{P}$ is the proof space. The function F is said to be a constrained VRF with regard to the sets $\mathcal{S} \subseteq 2^{\mathcal{X}}$, if there exists a constrained-key space $\mathcal{K}'$, and five poly-time algorithms (V.Setup, V.Eval, V.Constrain, V.Prove, V.Verify) :

- V.Setup(1^λ) → (sk', vk') : It takes the security parameter λ as input, and outputs a pair of keys (sk', vk').
- V.Eval(sk', x) → (y, π) : On input the secret key sk' and a point $x \in \mathcal{X}$, this algorithm outputs a evaluation y and a proof π.
- V.Verify(vk', x, y, π) → 0 or 1 : It takes the verifiable key vk', the values x and y, and the proof π as input, and outputs a bit '1' or '0', where '1' indicates that $y = F(sk', x)$.
- V.Constrain(sk', S) → sk_S : On input the secret key sk' and a set $S \in \mathcal{S}$, it outputs a constrained key $sk_S \in \mathcal{K}'$.
- V.Prove(sk_S, x) → (y, π) : On input the constrained key sk_S and a point x, it outputs a pair $(y, \pi) \in \mathcal{Y} \times \mathcal{P} \cup \{(\perp, \perp)\}$, where y is the function value and π is the corresponding proof.

A constrained verifiable random function is required to satisfy the following properties:

- **Provability.** For all $\lambda \in \mathbb{N}$, (sk', vk') ← V.Setup(1^λ), $S \in \mathcal{S}$, sk_S ← V.Constrain(sk', S), $x \in \mathcal{X}$ and (y, π) ← V.Prove(sk_S, x), it holds that
  - If $x \in S$, then $(y, \pi) = $ V.Eval(sk', x) and V.Verify(vk', x, y, π) = 1;
  - If $x \notin S$, then $(y, \pi) = (\perp, \perp)$.
  which implies that for every $x \notin S$, V.Prove(sk_S, x) = V.Eval(sk', x), and for every $x \notin S$, V.Prove(sk_S, x) = $(\perp, \perp)$.
- **Uniqueness.** For all $\lambda \in \mathbb{N}$, all $vk' \in \mathcal{K}$, all $x \in \mathcal{X}$, $y_0, y_1 \in \mathcal{Y}$ and $\pi_0, \pi_1 \in \mathcal{P}$, there is only one of the following condition satisfied:
  - $(y_0, \pi_0) = (y_1, \pi_1)$,
  - V.Verify(vk', x, y_0, π_0) = 1, or
  - V.Verify(vk', x, y_1, π_1) = 1,
  which implies that for every x there is only one pair (y, π) such that $F(sk', x) = (y, \pi)$.
- **Pseudorandomness.** We consider the following experiment $\text{Exp}_\mathcal{A}^{VRF}(1^\lambda, b)$ for $\lambda \in \mathbb{N}$:
  - The challenger generates (sk', vk') by running the algorithm V.Setup(1^λ).
  - The challenger initializes three sets V, T and Q, and sets $V := \emptyset, T := \emptyset, Q := \emptyset$, where V contains all the points that the adversary $\mathcal{A}$ can evaluate, Q contains the points at which the adversary queries the evaluation oracle, and T contains the pairs at which the adversary queries the verifiable oracle.
  - The adversary $\mathcal{A}$ is given the following oracle:
    *∗**Query:** Given an input $x \in \mathcal{X}$, the challenger returns V.Eval(sk', x) → (y, π), and sets $Q := Q \cup \{x\}$;

* **Constrain:** On input a set $S \in \mathcal{S}$, the challenger returns $sk_S \leftarrow$ V.Constrain(sk', S) and sets $V := V \cup S$.
* **Verify:** On input a triple (x, y, π), the challenger returns the result of V.Verify(vk', x, y, π) and sets $T := T \cup \{(x, y, \pi)\}$.
- The adversary $\mathcal{A}$ outputs a challenge point $x^* \in \mathcal{X}$. If $x^* \in V$ or $x^* \in Q$, the challenger returns $\bot$. Else the challenger chooses a bit $b \leftarrow \{0, 1\}$: If $b = 0$ then the challenger returns $(y^*, \pi^*) \leftarrow$ Eval(sk', x^*); else the challenger chooses a random pair (y^*, π^*) from $\mathcal{Y} \times \mathcal{P}$ and returns the pair (y^*, π^*) to the adversary.
- The adversary can also query the oracles of **Query** and **Constrain**. If $x^* \in V$ or $x^* \in Q$, then the oracles return $\bot$. At last, the adversary returns a bit $b' \in \{0, 1\}$. We require that the adversary can not query the verifiable oracle after he receives the challenge pair.

A constrained VRF is pseudorandom if for all PPT adversary $\mathcal{A}$, it holds that:

$$|\Pr[\mathsf{Exp}_{\mathcal{A}}^{VRF}(1^\lambda, b = 0) = 1] - \Pr[\mathsf{Exp}_{\mathcal{A}}^{VRF}(1^\lambda, b = 1) = 1]| \leq \mathrm{negl}(\lambda).$$

Remark 1. Note that our definition of pseudorandom is different to [11]. In our definition, the adversary is given another oracle which is used to verify the correctness of evaluation. We require that the evaluation value keeps pseudorandomness even if the adversary gets the proof of evaluation. We modify the construction of [11] and make it satisfy our definition of pseudorandomness. The detailed construction and proof are given in Appendix A.

3.3 Functional Pseudorandom Functions

In this section, we recall the formal definition of functional pseudorandom functions which was given by Boyle et al. [6].

Definition 6 (Functional PRF) [6, Definition 5.1]. *A functional family of PRFs $\mathcal{F} = \{F_k : D_k \to R_k\}$ consists of three algorithms which are described as follows:*

- *F.Setup$(1^\lambda) \to (k, pp)$: On input a security parameter λ, the algorithm F.Setup outputs a PRF key $k \in \mathcal{K}$: $F(k, \cdot) : D_k \to R_k$, and some public information.*
- *F.KeyGen$(k, f) \to sk_f$: On input a PRF key $k \in \mathcal{K}$ and function description $f : A_f \to D_f$, the algorithm F.KeyGen outputs a key sk_f, where D_f is a subset of D_k.*
- *F.Eval$(sk_f, f, x) \to F(k, f(x))$: On input key sk_f, the function description $f : A_f \to D_f$, and input $x \in A_f$, then F.Eval outputs a PRF evaluation $F(k, f(x))$.*

A functional PRF must satisfy the following properties:

- ***Correctness:*** *For any (efficiently computable) $f : A_f \to D_f, \forall x \in A_f$, it holds that*

$$\Pr\left[\begin{array}{l} \mathsf{F.Eval}(sk_f, f, x) \\ = F(k, f(x)) \end{array} : \begin{array}{l} (k, pp) \leftarrow \mathsf{F.Setup}(1^\lambda), \\ \forall sk_f \leftarrow \mathsf{F.KeyGen}(k, f) \end{array}\right] = 1.$$

– **Pseudorandomness:** *Given a set of keys* $sk_{f_1}, \ldots, sk_{f_\ell}$, *the evaluation of* $F(k, y)$ *should remain pseudorandom on all inputs* y *that are not in the range of any of the function* $f_1, \ldots, f_\ell$. *That is, for any PPT adversary* $\mathcal{A}$, *the advantage of* $\mathcal{A}$ *in the following experiment is negligible (for any polynomial* $\ell = \ell(\lambda)$):
- **Setup:** *The challenger runs* F.Setup(1^λ) *and gets a secret key* k *and the public information* pp.
- **Key query:** *After receiving the public information* pp, *the adversary* $\mathcal{A}$ *can repeatedly make the key generation queries:*
  * *The adversary submits the function description* f_i *to the challenger.*
  * *The challenger runs* $sk_{f_i} \leftarrow$ F.KeyGen(k, f_i) *and returns* sk_{f_i} *to the adversary.*
- **Challenge:** *The challenger randomly picks a bit* $b \in \{0,1\}$, *and returns an oracle* $\mathcal{O}$ *to the adversary* $\mathcal{A}$. *If* $b = 1$, *the challenger sets* $\mathcal{O} = F(k, \cdot)$. *If* $b = 0$, *the challenger chooses a random function* $H : D_k \rightarrow R_k$ *and constructs an oracle*

$$\mathcal{O}_{k,H}^{\{f_i\}} := \begin{cases} F(k, y) & \text{if } \exists i \in [\ell] \text{ and } x \text{ s.t. } f_i(x) = y, \\ H(y) & \text{otherwise.} \end{cases}$$

- **Guess:** $\mathcal{A}$ *outputs a guess* b'. *The advantage of the adversary* $\mathcal{A}$ *in this experiment is defined*

$$\text{Adv}_{\mathcal{A}}^{FPRF}(1^\lambda) := \Pr[b = b'] - 1/2.$$

4   Construction

In this section we present our construction of the verifiable outsourced computation for randomized algorithms. Let (V.Setup, V.Eval, V.Constrain, V.Prove, V.Verify) be a constrained verifiable random function family. Let $i\mathcal{O}$ be an indistinguishability obfuscator for all efficiently computable circuits. Let (F.Setup, F.KeyGen, F.Eval) be a functional PRF family. $\hat{f}(x; r)$ represents the function description of the randomized algorithm. We now proceed to describe our scheme rand $-$ VC = (KeyGen, ProbGen, Compute, Verify).

– KeyGen($\hat{f}, \lambda$) → (pk, sk) :
  1. Define a new function $f(x; r) = x|r|\hat{f}(x; r)$.
  2. Choose a functional PRF key $k \leftarrow$ F.Setup(1^λ).
  3. Run the key generation algorithm of functional PRF $sk_f \leftarrow$ F.KeyGen(k, f).
  4. Set the public key as sk_f and the secret key as the functional PRF key k, i.e. $pk = (f, sk_f), sk = k$.
– ProbGen(x, pk, t) → (σ_x, τ_x) :
  1. Run the setup algorithm of constrained verifiable random function $(sk', vk') \leftarrow$ V.Setup(1^λ).

2. Construct a circuit $PP = i\mathcal{O}(\mathcal{G})$, where the circuit $\mathcal{G}$ is described in Fig. 1. Note that $\mathcal{G}$ has the PRF key sk' hardwired in it.

3. Set the public value $\sigma_x = (x, t, PP)$ and the secret value $\tau_x = (vk', x, t)$.

Size of Circuit $\mathcal{G}$. In order to prove that rand $-$ VC is security, we require the circuit $\mathcal{G}$ to be padded with zeros such that $|\mathcal{G}| = |\mathcal{G}^*|$, where the circuit $\mathcal{G}^*$ is described later in Fig. 2.

- Compute$(pk, \sigma_x) \rightarrow \sigma_{\boldsymbol{y}}$:

  1. Parse the public value $\sigma_x = (x, t, PP)$, and the public key $pk = (sk_f, f)$.

  2. Choose the strings s_i, compute $(s_i, r_i, \pi_i) = PP(s_i)$, $y_{i,1} = $ F.Eval(sk_f, f, x, r_i) and $f(x; r_i) = x|r_i|y_{i,2}$, for $i = 1, \ldots, t$.

  3. Return $\sigma_{\boldsymbol{y}} = ((s_1, r_1, \pi_1, y_{1,1}, y_{1,2}), \ldots, (s_t, r_t, \pi_t, y_{t,1}, y_{t,2}))$.

- Verify$(\sigma_{\boldsymbol{y}}, \tau_x, sk) \rightarrow \boldsymbol{y}$ or $\perp$:

  1. It firstly checks the number of the tuples. if it is not equal to t, it rejects.

  2. Else, for $i = 1, \ldots, t$, it checks V.Verify$(vk', s_i, r_i, \pi_i) \overset{?}{=} 1$, and $F(k, x|r_i|y_{i,2}) \overset{?}{=} y_{i,1}$, if not, it rejects, else it outputs $\boldsymbol{y} = (y_{1,2}, \ldots, y_{t,2})$.

Input: a string s
Constants: constrained VRF key sk'

(a) compute $(r, \pi) \leftarrow$ V.Eval(sk', s)
(b) output (s, r, π)

Fig. 1. Functionality $\mathcal{G}$

Theorem 1. *If* (V.Setup, V.Eval, V.Constrain, V.Prove, V.Verify) *is a constrained VRF, then the proposed construction* rand $-$ VC *satisfies correctness.*

Proof. For a randomized algorithm A, it is represented as a function $\hat{f}(\cdot; \cdot)$, which has two inputs x and r. Consider the distribution **Real**$_1$:{Verify$(\sigma_{\boldsymbol{y}}, \tau_x, sk)$}, where $(pk, sk) \leftarrow$ KeyGen$(1^\lambda, f), (\sigma_x, \tau_x) \leftarrow$ ProbGen$(x, t, pk), \sigma_{\boldsymbol{y}} \leftarrow$ Compute(pk, σ_x). Similarly, consider the **Ideal**$_1$ distribution $\{f(x; r_i)\}_{i=1}^t$, where $r_i \overset{R}{\leftarrow} \mathcal{R}$.

We prove that that if the distribution **Real**$_1$ and **Ideal**$_1$ can be distinguished by an adversary $\mathcal{A}$ with non-negligible advantage. Then, another adversary $\mathcal{B}$ can be constructed to break the constrained VRF security with the same advantage. We construct this algorithm as follows:

1. VRF challenger $\mathcal{C}$ chooses a bit b sampled uniformly at$\{0, 1\}$.
2. For $i = 1$ to t
   (a) $\mathcal{B}$ sends s_i to the challenger $\mathcal{C}$, and receives (r_i, π_i). If $b = 0, (r_i, \pi_i) = $ V.Eval(sk', s_i), else (r_i, π_i) is chosen randomly.
   (b) $\mathcal{B}$ computes $y_i = f(x; r_i)$.
3. $\mathcal{B}$ sends $\boldsymbol{y} = (y_1, \ldots, y_t)$ to $\mathcal{A}$. If $\mathcal{A}$ guesses 0, $\mathcal{B}$ outputs 0; else $\mathcal{B}$ outputs 1.

Note that if $\mathcal{C}$ returns the evaluation of constrained VRF, then $\mathcal{B}$ perfectly simulates the real distribution, else it simulates the ideal distribution. We can get that, if $\mathcal{A}$ can distinguish the two distributions with non-negligible advantage, then the adversary $\mathcal{B}$ can break the security of constrained VRF with the same advantage.

Theorem 2. *If* (V.Setup, V.Eval, V.Constrain, V.Prove, V.Verify) *is a constrained verifiable random function and* (F.Setup, F.KeyGen, F.Eval) *is a functional pseudorandom function, then our construction is a secure outsourced computation scheme.*

Proof. We consider selective verifiable security for rand $-$ VC. Suppose there exists an adversary $\mathcal{A}$ that can break the verifiable security, then there exists an adversary $\mathcal{B}$ that can break the functional PRF security or the constrained VRF security.

We first give the original security experiment which is described in the following:

Game 1. The first game is the original security game instantiated for our construction.

1. Challenger represents the randomized algorithm as a function $\hat{f}(\cdot;\cdot)$, which has inputs x and r, and constructs a new function $f(x,r) = x|r|\hat{f}(x;r)$.
2. Challenger computes $k \leftarrow$ F.Setup(1^λ) and $sk_f \leftarrow$ F.KeyGen(f,k), and returns $pk = (f, sk_f)$, sets $sk = k$.
3. $\mathcal{A}$ sends x^* and the challenge random string $s_1^*, \ldots, s_t^*$ to the challenger.
4. Challenger computes $(sk', vk') \leftarrow$ V.Setup(1^λ), and constructs a circuit $PP = i\mathcal{O}(\mathcal{G})$, which is described in Fig. 1. Then it sends $\sigma_x = (PP, x^*)$ to the adversary.
5. The adversary $\mathcal{A}$ returns a value $\sigma_y = ((s_1^*, r_1, \pi_1, y_{1,1}, y_{1,2}), \ldots, (s_t^*, r_t, \pi_t, y_{t,1}, y_{t,2}))$.
6. The adversary successes if V.Verify$(vk', s_j^*, r_j, \pi_j) = 1$, $F(k_1, x^*|r_j|y_{j,2}) = y_{j,1}$ for $j = 1, \ldots, t$, and there exists an index i such that $y_{i,2} \neq \hat{f}(x^*; r_i)$, or there exists an distinguisher $\mathcal{D}$ that can distinguish the two results $y = (y_{1,2}, \ldots, y_{t,2})$ and $\{f(x; r_i')\}_{i=1}^t$, where $r_i' \xleftarrow{R} \mathcal{R}$.

There are two cases for forging an output $\sigma_y = ((s_1^*, r_1, \pi_1, y_{1,1}, y_{1,2}), \ldots, (s_t^*, r_t, \pi_t, y_{t,1}, y_{t,2}))$ which can be verified successfully:

- **Type 1 forgery:** There exists an index $j \in [t]$, such that V.Verify$(vk', s_j^*, r_j, \pi_j) = 1$, $F(k, x^*|r_j|y_{j,2}) = y_{j,1}$, and $\hat{f}(x^*; r_j) \neq y_{j,2}$;
- **Type 2 forgery:** It satisfies that V.Verify$(vk', s_j^*, r_j, \pi_j) = 1$, $F(k, x^*|r_j|y_{j,2}) = y_{j,1}$ for $j = 1, \ldots, t$, and there exists an distinguisher $\mathcal{D}$ that can distinguish the two results $y = (y_{1,2}, \ldots, y_{t,2})$ and $\{\hat{f}(x^*; r_i')\}_{i=1}^t$, where $r_i' \xleftarrow{R} \mathcal{R}$.

We first prove that the **Type 1 forgery** is not possible.

Lemma 1. *If there exists an adversary $\mathcal{A}$ that can output **Type 1 forgery**, then there exists another adversary $\mathcal{B}$ that breaks the pseudorandomness of functional PRF.*

Proof. Suppose that the adversary outputs a forgery $\sigma_y = ((s_1^*, r_1, \pi_1, y_{1,1}, y_{1,2}), \ldots, (s_t^*, r_t, \pi_t, y_{t,1}, y_{t,2}))$ that satisfies V.Verify$(vk', s_j^*, r_j, \pi_j) = 1$, $F(k, x^*|r_j|y_{j,2}) = y_{j,1}$ for $j = 1\ldots, t$, and there exists an index i, such that $f(x^*; r_i) \neq y_{i,2}$.

First, we show that the adversary can not forge a random string due to the uniqueness of the constrained VRF. Suppose that the adversary $\mathcal{A}$ outputs a forgery $\sigma_y = ((s_1^*, r_1, \pi_1, y_{1,1}, y_{1,2}), \ldots, (s_t^*, r_t, \pi_t, y_{t,1}, y_{t,2}))$ that satisfies V.Verify$(vk', s_j^*, r_j, \pi_j) = 1$, $F(k, x^*|r_j|y_{j,2}) = y_{j,1}$ for $j = 1, \ldots, t$, and there exists an index i, such that $(r_i, \pi_i) \neq$ V.Eval(sk', s_i^*), then we can construct another adversary $\mathcal{B}_1$ that breaks the uniqueness of the constrained VRF. The adversary $\mathcal{B}_1$ computes that $(r_i', \pi_i') =$ V.Eval(sk', s_i^*) which must satisfy that V.Verify$(vk', s_i^*, r_i', \pi_i') = 1$. Because $(r_i, \pi_i) \neq (r_i', \pi_i')$, it is contradictive to the uniqueness of the constrained VRF.

Then, we construct a probabilistic polynomial time reduction algorithm $\mathcal{B}_2$ that attacks the functional PRF security game. $\mathcal{B}_2$ first runs the step 1, and sends the function $f(x; r) = x|r|\hat{f}(x; r)$ to the functional PRF challenger. It receives back a functional key sk_f, runs the steps 3–6, and gets $\sigma_y = ((s_1^*, r_1, \pi_1, y_{1,1}, y_{1,2}), \ldots, (s_t^*, r_t, \pi_t, y_{t,1}, y_{t,2}))$. If the adversary successes, then it must satisfies V.Verify$(vk', s_j^*, r_j, \pi_j) = 1$, $F(k, x^*|r_j|y_{j,2}) = y_{j,1}$ for $j = 1, \ldots, t$, and there exists an index i, such that $\hat{f}(x^*; r_i) \neq y_{i,2}$. We have shown that the string r_i must be computed correctly due to the uniqueness of the constrained VRF. Because $y_{i,2} \neq \hat{f}(x^*; r_i)$, we can get that $x^*|r_i|y_{i,2}$ does not belong the domain of $f(x; r)$. Therefore, $\mathcal{B}_2$ sends $x^*|r_i|y_{i,2}$ to the challenger of the functional PRF security game and receives a challenge values y^*. If $y_{i,1} = y^*$, then $\mathcal{B}_2$ guesses '1' to indicate that $y^* = F(k, x^*|r_i|y_{i,2})$; otherwise, it outputs '0' to indicate that y^* was chosen randomly. Therefore, if $\mathcal{A}$ can forge an output with non-negligible probability, then $\mathcal{B}_2$ must have non-negligible advantage against the security of the functional PRF.

Lemma 2. *If there exists an adversary $\mathcal{A}$ that can output **Type 2 forgery**, then there exists another adversary $\mathcal{B}$ that breaks the pseudorandomness of the constrained VRF.*

Proof. To prove this lemma, we first define a sequence of experiments where the first experiment is the original security experiment. Then we show that any polytime adversary can not distinguish the two neighbour games with non-negligible advantage. The first experiment is given in the above, we describe the sequence of experiments where each experiment is described by its modification from the previous experiment.

Game 2

1. Challenger represents the randomized algorithm as a function $\hat{f}(\cdot; \cdot)$, which has inputs x and r, and constructs a new function $f(x; r) = x|r|\hat{f}(x; r)$.

2. Challenger computes $k \leftarrow$ F.Setup(1^λ) and $sk_f \leftarrow$ F.KeyGen(f, k), and returns $pk = (f, sk_f)$, sets $sk = k$.
3. $\mathcal{A}$ sends x^* and the challenge random string $s_1^*, \ldots, s_t^*$ to the challenger.
4. Challenger sets $S \leftarrow \{0,1\}^* \backslash \{s_1^*, \ldots, s_t^*\}$, computes $(sk', vk') \leftarrow$ V.Setup(1^λ), $sk_S \leftarrow$ V.Constrain(vk', S), $(z_j^*, \pi_j^*) =$ V.Eval(sk', s_j) for $j = 1, \ldots, t$, and constructs a circuit $PP = i\mathcal{O}(\mathcal{G}^*)$, where the circuit $\mathcal{G}^*$ is described in Fig. 2. Note that $\mathcal{G}^*$ has the constrained key sk_S, the challenge random strings $s_1^*, \ldots, s_t^*$, the values $z_1^*, \ldots, z_t^*$ and the corresponding proof $\pi_1^*, \ldots, \pi_t^*$ hardwired in it. Then it sends $\sigma_x = (PP, x^*)$ to the adversary.
5. The adversary $\mathcal{A}$ returns a value $\sigma_y = ((s_1^*, r_1, \pi_1, y_{1,1}, y_{1,2}), \ldots, (s_t^*, r_t, \pi_t, y_{t,1}, y_{t,2}))$.
6. The adversary successes if V.Verify$(vk', s_j^*, r_j, \pi_j) = 1$, $F(k_1, x^*|r_j|y_{j,2}) = y_{j,1}$ for $j = 1, \ldots, t$, and there exists an distinguisher $\mathcal{D}$ that can distinguish the two results $\boldsymbol{y} = (y_{1,2}, \ldots, y_{t,2})$ and $\{\hat{f}(x^*; r_i')\}_{i=1}^t$, where $r_i' \xleftarrow{R} \mathcal{R}$.

Game 3

1. Challenger represents the randomized algorithm as a function $\hat{f}(\cdot; \cdot)$, which has inputs x and r, and constructs a new function $f(x; r) = x|r|\hat{f}(x; r)$.
2. Challenger computes $k \leftarrow$ F.Setup(1^λ) and $sk_f \leftarrow$ F.KeyGen(f, k), and returns $pk = (f, sk_f)$, sets $sk = k$.
3. $\mathcal{A}$ sends x^* and the challenge random string s^* to the challenger.
4. Challenger sets $S \leftarrow \{0,1\}^* \backslash \{s_1^*, \ldots, s_t^*\}$, computes $(sk', vk') \leftarrow$ V.Setup(1^λ), $sk_S \leftarrow$ V.Constrain(vk', S), chooses random strings $z_1^*, \ldots, z_t^* \xleftarrow{R} \mathcal{R}$, and the proofs $\pi_1^*, \ldots, \pi_t^* \xleftarrow{R} \mathcal{P}$, and constructs a circuit $PP = i\mathcal{O}(\mathcal{G}^*)$, where the circuit $\mathcal{G}^*$ is described in Fig. 2. Note that $\mathcal{G}^*$ has the constrained VRF key sk_S, the challenge random strings $s_1^*, \ldots, s_t^*$, the values $z_1^*, \ldots, z_t^*$ and the proofs $\pi_1^*, \ldots, \pi_t^*$ hardwired in it. Then it sends $\sigma_x = (PP, x^*)$ to the adversary.
5. The adversary $\mathcal{A}$ returns a value $\sigma_y = ((s_1^*, r_1, \pi_1, y_{1,1}, y_{1,2}), \ldots, (s_t^*, r_t, \pi_t, y_{t,1}, y_{t,2}))$.
6. The adversary successes if V.Verify$(vk', s_j^*, r_j, \pi_j) = 1$, $F(k_1, x^*|r_j|y_{j,2}) = y_{j,1}$ for $j = 1, \ldots, t$, and there exists an distinguisher $\mathcal{D}$ that can distinguish the two results $\boldsymbol{y} = (y_{1,2}, \ldots, y_{t,2})$ and $\{\hat{f}(x^*; r_i')\}_{i=1}^t$, where $r_i' \xleftarrow{R} \mathcal{R}$.

Lemma 3. *If $i\mathcal{O}$ is an indistinguishability obfuscator, Game 1 and Game 2 are computationally indistinguishable.*

Proof. Note that the only difference in Game 1 and Game 2 is that in the former game, we output $\sigma_x = (PP = i\mathcal{O}(\mathcal{G}), x^*)$, while in the later game, we output $\sigma_x = (PP = i\mathcal{O}(\mathcal{G}^*), x^*)$. In order to prove that for any probabilistic polynomial time adversary, it can not distinguish the two games. We need to show that the two circuits have the identical function. That is to say, the two circuits $i\mathcal{O}(\mathcal{G})$ and $i\mathcal{O}(\mathcal{G}^*)$ output the same value when they take the same value as input. Then, we can get that $i\mathcal{O}(\mathcal{G})$ and $i\mathcal{O}(\mathcal{G}^*)$ are computationally indistinguishable, which would imply Game 1 and Game 2 are computationally indistinguishable.

Input: a string s

Constants: constrained VRF key sk_S, the challenge random strings $s_1^*, \ldots, s_t^*$, the evaluation values $z_1^*, \ldots, z_t^*$ and the corresponding proofs $\pi_1^*, \ldots, \pi_t^*$

(a) if there exists an index i, such that $s = s_i^*$, then outputs (s_i^*, z_i^*, π_i^*);
(b) else, compute $(r, \pi) = \mathsf{V.Prove}(sk_S, s)$;
(c) output (s, r, π)

Fig. 2. Functionality $\mathcal{G}^*$

Now, we prove that the two circuits have identical input-output behaviour. We only discuss the situation $t = 1$. The situation of $t \geq 2$ can be proved via a standard hybrid argument. Considering the two following cases: $s \neq s^*$ and $s = s^*$. First, we can get that $\mathsf{V.Eval}(sk', s) = \mathsf{V.Prove}(sk_S, s) = (r, \pi)$ from the property of constrained VRF. Second, $\mathcal{G}$ computes $(r, \pi) \leftarrow \mathsf{V.Eval}(sk', s^*)$, and $\mathcal{G}^*$ outputs the hard-wired value (z^*, π^*), when $s = s^*$. Note that $(r, \pi) = (z^*, \pi^*)$, therefore $\mathcal{G}(s) = \mathcal{G}^*(s)$.

Lemma 4. *Assuming* $(\mathsf{V.Setup}, \mathsf{V.Eval}, \mathsf{V.Verify}, \mathsf{V.Constrain}, \mathsf{V.Prove})$ *is a constrained VRF, Game 2 and Game 3 are computationally indistinguishable.*

Proof. We give a probabilistic polynomial time reduction algorithm $\mathcal{B}$ that attacks the constrained VRF security game. $\mathcal{B}$ first runs steps 1–2, receives a value x and challenge strings $s_1^*, \ldots, s_t^*$, and submits $S \leftarrow \{0,1\}^* \backslash \{s_1^*, \ldots, s_t^*\}$ to the constrained VRF challenger. It receives a punctured key sk_S and challenge values $((z_1^*, \pi_1^*), \ldots, (z_t^*, \pi_t^*))$. It runs step 4 for $\mathcal{A}$ as in Game 2. If the adversary $\mathcal{A}$ wins, then $\mathcal{B}$ guesses '1' to indicate that $(z_i^*, \pi_i^*) = \mathsf{V.Eval}(sk', s_i^*)$ for $i = 1, \ldots, t$; otherwise, it outputs '0' to indicate that $(z_i^*, \pi_i^*)_{i=1}^t$ are chosen randomly. We observe that when $(z_i^*, \pi_i^*)_{i=1}^t$ is generated as $F(sk', s_i^*)$ for $i = 1, \ldots, t$, then $\mathcal{B}$ gives exactly the view of Game 2 to $\mathcal{A}$. Otherwise if $(z_i^*, \pi_i^*)_{i=1}^t$ are chosen randomly, the view is of Game 3. Therefore, if $\mathcal{A}$ can distinguish the Game 2 and Game 3 with non-negligible advantage, $\mathcal{B}$ must have the same advantage against the security of the constrained VRF.

Lemma 5. *If* $(\mathsf{V.Setup}, \mathsf{V.Eval}, \mathsf{V.Verify}, \mathsf{V.Constrain}, \mathsf{V.Prove})$ *is a constrained verifiable random function and* $(\mathsf{F.Setup}, \mathsf{F.KeyGen}, \mathsf{F.Eval})$ *is a functional pseudorandom function, then the adversary can successfully break the security of this scheme with negligible probability in the Game 3.*

Proof. We observe that the outputs $z_1^*, \ldots, z_t^*$ are chosen randomly in Game 3. Therefore, for any adversary, they can not distinguish the results $\mathbf{y} = (y_{1,2}, \ldots, y_{t,2}) = (\hat{f}(x^*; z_1^*), \ldots, \hat{f}(x^*; z_t^*))$ and $(\hat{f}(x^*; r_1'), \ldots, \hat{f}(x^*; r_t'))$, where $r_j' \xleftarrow{R} \mathcal{R}, j = 1, \ldots, t$.

Efficiency. To be a verifiable computation scheme for randomized algorithm, we require that $\mathsf{rand} - \mathsf{VC}$ must satisfy the property of outsourceable as in Definition 3. Since our Verify algorithm requires one verifiable computation of constrained VRF and a PRF computation, the running time is independent of the algorithm $f(\cdot;\cdot)$.

5   Conclusion

In this work, we introduced a new notion for verifiable computation for randomized algorithms and give a construction from the constrained verifiable random function, functional pseudorandom function and indistinguishability obfuscation.

Our work leaves one open interesting problem. Can we construct a verifiable computation scheme for randomized algorithm without indistinguishability obfuscation? After all, indistinguishability obfuscation is not efficient. If it can be constructed, the problem generation algorithm will not be in the amortized sense.

Acknowledgment. This work is supported by the "Strategic Priority Research Program" of the Chinese Academy of Sciences, Grants No. XDA06010701, National Natural Science Foundation of China (No. 61402471, 61472414, 61170280), and IIE's Cryptography Research Project.

A   Modified Construction of Constrained VRF

A.1   Assumption

Let $\mathcal{G}$ be a group generator, which takes the security parameter 1^λ and $n \in \mathbb{N}$ as input, and outputs a sequence of groups $\mathbb{G} = (\mathbb{G}_1, \ldots, \mathbb{G}_n)$ of prime order p. Suppose that $g_i \in \mathbb{G}_i$ is the generator of the i-th group. For all $i, j \geq 1$ and $i + j \leq n$, there exists a map $e_{i,j} : \mathbb{G}_i \times \mathbb{G}_j \rightarrow \mathbb{G}_{i+j}$ such that $\forall \alpha, \beta \in \mathbb{Z}_p :$ $e_{i,j}(g_i^a, g_j^b) = (g_{i+j}^{ab})$.

Next, we recall the n-MDDH assumption, which is proposed by Fuchsbauer [11].

Definition 7 (n-MDDH Assumption) [11, Assumption 1]. *Let $\mathcal{G}$ be a generator of multilinear groups, and let $(p, g, \mathbb{G} = (\mathbb{G}_1, \ldots, \mathbb{G}_n)) \leftarrow \mathcal{G}(1^\lambda, n)$, where g is the generator of $\mathbb{G}_1$. We say that the n-multilinear Decisional Diffie-Hellman (n-MDDH) assumption is hard for $\mathcal{G}$ if for random $c_1, \ldots, c_{n+1} \leftarrow \mathbb{Z}_p$ and $T \leftarrow \mathbb{Z}_p$ and for every PPT adversary $\mathcal{A}$:*

$$
|\Pr[\mathcal{A}(1^\lambda, \mathbb{G}, g, g^{c_1}, \ldots, g^{c_{n+1}}, g_n^{\prod_{j \in [n+1]} c_j}) = 1] -
$$
$$
\Pr[\mathcal{A}(1^\lambda, \mathbb{G}, g, g^{c_1}, \ldots, g^{c_{n+1}}, g_n^T) = 1]| \leq \mathrm{negl}(\lambda).
$$

A.2   Construction

The construction of constrained VRF is given in [11, p. 8]. We modify it and describe the new construction in the following:

- V.Setup$(1^\lambda, 1^n) \rightarrow (sk', vk')$: On input a security parameter λ and a length n, the setup algorithm runs $\mathcal{G}(1^\lambda, n)$ and gets a sequence of groups $\mathbb{G}_1, \ldots, \mathbb{G}_n, g_1, \ldots, g_n$ of prime order p, where $g_1, \ldots, g_n$ is defined the generators of $\mathbb{G} = (\mathbb{G}_1, \ldots, \mathbb{G}_n)$ respectively. Then, it chooses $\gamma \leftarrow \mathbb{Z}_p$ and $(d_{1,0}, d_{1,1}), \ldots, (d_{n,0}, d_{n,1}) \leftarrow \mathbb{Z}_p^2$ randomly and sets $R = g^\gamma$ and $D_{i,b} = g^{d_{i,b}}$ for $i \in [n]$ and $b \in \{0, 1\}$. The VRF secret key and verifiable key are defined as

$$vk' = (\mathbb{G}, \{D_{i,b}\}_{i \in [n], b \in \{0,1\}}, \gamma)$$

$$sk' = (\mathbb{G}, R, \{D_{i,b}\}_{i \in [n], b \in \{0,1\}}, \{d_{i,b}\}_{i \in [n], b \in \{0,1\}})$$

- V.Eval$(sk', x) \rightarrow (y, \pi)$: On input a value $x \in \mathcal{X} = \{0, 1\}^n$, it outputs a function value y and a proof π as follows:

$$y = g_n^{\gamma \prod_{i \in [n]} d_{i,x_i}}, \qquad \pi = g_n^{\prod_{i \in [n]} d_{i,x_i}}$$

- V.Verify$(vk', x, y, \pi) \rightarrow 0$ or 1 : On input the pair (x, y, π), this algorithm first computes $D(x) = g_n^{\prod_{i \in [n]} d_{i,x_i}}$ by applying the multilinear maps to $(D_{1,x_1}, \ldots, D_{n,x_n})$. Then, it verifies the following equations:

$$D(x) = \pi, \qquad \pi^\gamma = y.$$

If it holds, the algorithm outputs 1, else, outputs 0.

- V.Constrain$(sk', v) \rightarrow sk_v$: Let $V = \{i \in [n] | v_i \neq ?\}$ be the set of indices for which the input bit is fixed to 0 or 1. On input the secret key sk' and a vector $v \in \{0, 1, ?\}^n$, where v is described the constrained domain $S_v = \{x \in \{0, 1\}^n | \forall i \in [n] : x_i = v_i \vee v_i =?\}$, It computes $sk_v = (\mathbb{G}, R, \{D_{i,b}\}_{i \in [n], b \in \{0,1\}}, k_v)$, with k_v defined as follows:
    - If $|V| > 1$, then it computes $k_v = (g_{|V|-1})^{\prod_{i \in V} d_{i,x_i}}$.
    - If $V = \{j\}$, then it sets $k_v = d_{j,v_j}$.
    - If $V = \emptyset$, then it sets $k_v = \{d_{i,b}\}_{i \in [n], b \in \{0,1\}}$, i.e. $sk_v = sk'$.
- V.Prove$(sk_v, x) \rightarrow (y, \pi)$: Let $V = \{i \in [n] | v_i \neq ?\}$ and let $\bar{V} = \{i \in [n] | v_i =?\}$ be its complement. If $x_i \neq v_i$ for some $i \in V$, then it returns $(\bot, \bot)$; else it applies the multilinear maps to $(\{D_{i,x_i}\}_{i \in \bar{V}}, R)$ and computes $D_{\bar{V}}(x) = (g_{|\bar{V}|+1})^{\gamma \prod_{i \in \bar{V}} d_{i,x_i}}$.
    - If $|V| > 1$, then it computes $\pi = g_n^{\prod_{i \in [n]} d_{i,x_i}}$ and

$$y = e(D_{\bar{V}}(x), k_v) = e((g_{|V|-1})^{\prod_{i \in V} d_{i,x_i}}, (g_{|\bar{V}|+1})^{\gamma \prod_{i \in \bar{V}} d_{i,x_i}}) = g_n^{\gamma \prod_{i \in [n]} d_{i,x_i}}.$$

- If $V = \{j\}$, then it computes $\pi = g_n^{\prod_{i \in [n]} d_{i,x_i}}$ and

$$y = D_{\bar{V}}(x)^{k_v} = ((g_{|\bar{V}|+1})^{\gamma \prod_{i \in \bar{V}} d_{i,x_i}})^{d_{j,v_j}} = g_n^{\gamma \prod_{i \in [n]} d_{i,x_i}}.$$

- If $V = \emptyset$, the computation of (y, π) is similar to the evaluation algorithm. Finally, it outputs $\mathsf{V.Prove}(sk_v, x) = (y, \pi)$.

The properties of provability and uniqueness can be verified easily, we omit the detailed procedures.

Theorem 3. *If there exists a PPT adversary $\mathcal{A}$ that can break the pseudorandomness of the modified construction with non-negligible advantage, then there exists another PPT adversary $\mathcal{B}$ that breaks the n-multilinear decisional Diffie-Hellman assumption with non-negligible advantage.*

Here we show the intuition of the proof. In our proof, we just consider the selective security of PRF in which the adversary is required to output a challenge point before he queries the oracles. Suppose that x^* is the challenge point. If the adversary $\mathcal{A}$ can break the pseudorandonness of the modified construction with non-negligible advantage, we construct another adversary $\mathcal{B}$ that breaks the n-MDDH assumption. $\mathcal{B}$ chooses a sequence random values $z_1, \ldots, z_n \xleftarrow{R} \mathbb{Z}_p$ and sets $R = g^{c_{n+1}}, D_{i,x_i^*} = g^{c_i}$ and $D_{i,1-x_i^*} = g^{z_i}$. When $\mathcal{A}$ queries the oracles (query, constrain and verify) on input $x \neq x^*$, there must exists a bit j such that $x_j \neq x_j^*$. Because $\mathcal{B}$ knows the value of z_j, it can answer the oracle queries. Finally, the adversary $\mathcal{B}$ returns $(y = T, \pi = g_n^{\prod_{i \in [n]} c_i})$, where T is either $g_n^{\prod_{i \in [n+1]} c_i}$ or a random element from $\mathbb{G}_n$. If the adversary $\mathcal{A}$ guesses '1', then $\mathcal{B}$ outputs '1', which indicates that $T = g_n^{\prod_{i \in [n+1]} c_i}$. Otherwise, $\mathcal{B}$ outputs '0' to indicate that T is chosen randomly. Therefore, if $\mathcal{A}$ can distinguish the randomness of this experiment with non-negligible advantage, $\mathcal{B}$ can break the security of the MDDH assumption with non-negligible probability.

References

1. Alwen, J., Barbosa, M., Farshim, P., Gennaro, R., Gordon, S.D., Tessaro, S., Wilson, D.A.: On the relationship between functional encryption, obfuscation, and fully homomorphic encryption. In: Proceedings of 14th IMA International Conference on Cryptography and Coding, IMACC 2013, Oxford, UK, 17–19 December 2013, pp. 65–84 (2013)
2. Applebaum, B., Ishai, Y., Kushilevitz, E.: From secrecy to soundness: efficient verification via secure computation. In: Abramsky, S., Gavoille, C., Kirchner, C., Meyer auf der Heide, F., Spirakis, P.G. (eds.) ICALP 2010. LNCS, vol. 6198, pp. 152–163. Springer, Heidelberg (2010). doi:10.1007/978-3-642-14165-2_14
3. Barak, B., Goldreich, O., Impagliazzo, R., Rudich, S., Sahai, A., Vadhan, S., Yang, K.: On the (im)possibility of obfuscating programs. In: Kilian, J. (ed.) CRYPTO 2001. LNCS, vol. 2139, pp. 1–18. Springer, Heidelberg (2001). doi:10.1007/3-540-44647-8_1

4. Barbosa, M., Farshim, P.: Delegatable homomorphic encryption with applications to secure outsourcing of computation. In: Dunkelman, O. (ed.) CT-RSA 2012. LNCS, vol. 7178, pp. 296–312. Springer, Heidelberg (2012). doi:10.1007/978-3-642-27954-6_19

5. Benabbas, S., Gennaro, R., Vahlis, Y.: Verifiable delegation of computation over large datasets. In: Rogaway, P. (ed.) CRYPTO 2011. LNCS, vol. 6841, pp. 111–131. Springer, Heidelberg (2011). doi:10.1007/978-3-642-22792-9_7

6. Boyle, E., Goldwasser, S., Ivan, I.: Functional signatures and pseudorandom functions. In: Krawczyk, H. (ed.) PKC 2014. LNCS, vol. 8383, pp. 501–519. Springer, Heidelberg (2014). doi:10.1007/978-3-642-54631-0_29

7. Choi, S.G., Katz, J., Kumaresan, R., Cid, C.: Multi-client non-interactive verifiable computation. In: Sahai, A. (ed.) TCC 2013. LNCS, vol. 7785, pp. 499–518. Springer, Heidelberg (2013). doi:10.1007/978-3-642-36594-2_28

8. Chung, K.-M., Kalai, Y., Vadhan, S.: Improved delegation of computation using fully homomorphic encryption. In: Rabin, T. (ed.) CRYPTO 2010. LNCS, vol. 6223, pp. 483–501. Springer, Heidelberg (2010). doi:10.1007/978-3-642-14623-7_26

9. Fiore, D., Gennaro, R.: Publicly verifiable delegation of large polynomials and matrix computations, with applications. In: The ACM Conference on Computer and Communications Security, CCS 2012, Raleigh, NC, USA, 16–18 October 2012, pp. 501–512 (2012)

10. Fiore, D., Gennaro, R., Pastro, V.: Efficiently verifiable computation on encrypted data. In: Proceedings of the 2014 ACM SIGSAC Conference on Computer and Communications Security, Scottsdale, AZ, USA, 3–7 November 2014, pp. 844–855 (2014)

11. Fuchsbauer, G.: Constrained verifiable random functions. In: Abdalla, M., Prisco, R. (eds.) SCN 2014. LNCS, vol. 8642, pp. 95–114. Springer, Heidelberg (2014). doi:10.1007/978-3-319-10879-7_7

12. Garg, S., Gentry, C., Halevi, S., Raykova, M., Sahai, A., Waters, B.: Candidate indistinguishability obfuscation and functional encryption for all circuits. In: 54th Annual IEEE Symposium on Foundations of Computer Science, FOCS 2013, 26–29 October 2013, Berkeley, CA, USA, pp. 40–49 (2013)

13. Gennaro, R., Gentry, C., Parno, B.: Non-interactive verifiable computing: outsourcing computation to untrusted workers. In: Rabin, T. (ed.) CRYPTO 2010. LNCS, vol. 6223, pp. 465–482. Springer, Heidelberg (2010). doi:10.1007/978-3-642-14623-7_25

14. Gentry, C., Halevi, S., Raykova, M., Wichs, D.: Outsourcing private RAM computation. In: 55th IEEE Annual Symposium on Foundations of Computer Science, FOCS 2014, Philadelphia, PA, USA, 18–21 October 2014, pp. 404–413 (2014)

15. Goldwasser, S., Gordon, S.D., Goyal, V., Jain, A., Katz, J., Liu, F.-H., Sahai, A., Shi, E., Zhou, H.-S.: Multi-input functional encryption. In: Nguyen, P.Q., Oswald, E. (eds.) EUROCRYPT 2014. LNCS, vol. 8441, pp. 578–602. Springer, Heidelberg (2014). doi:10.1007/978-3-642-55220-5_32

16. Gordon, S.D., Katz, J., Liu, F., Shi, E., Zhou, H.: Multi-client verifiable computation with stronger security guarantees. IACR Cryptol. ePrint Arch. **2015**, 142 (2015)

17. Goyal, V., Jain, A., Koppula, V., Sahai, A.: Functional encryption for randomized functionalities. In: Dodis, Y., Nielsen, J.B. (eds.) TCC 2015. LNCS, vol. 9015, pp. 325–351. Springer, Heidelberg (2015). doi:10.1007/978-3-662-46497-7_13

18. Komargodski, I., Segev, G., Yogev, E.: Functional encryption for randomized functionalities in the private-key setting from minimal assumptions. In: Dodis, Y., Nielsen, J.B. (eds.) TCC 2015. LNCS, vol. 9015, pp. 352–377. Springer, Heidelberg (2015). doi:10.1007/978-3-662-46497-7_14
19. Papamanthou, C., Shi, E., Tamassia, R.: Signatures of correct computation. In: Sahai, A. (ed.) TCC 2013. LNCS, vol. 7785, pp. 222–242. Springer, Heidelberg (2013). doi:10.1007/978-3-642-36594-2_13
20. Parno, B., Raykova, M., Vaikuntanathan, V.: How to delegate and verify in public: verifiable computation from attribute-based encryption. In: Cramer, R. (ed.) TCC 2012. LNCS, vol. 7194, pp. 422–439. Springer, Heidelberg (2012). doi:10.1007/978-3-642-28914-9_24

UC-secure and Contributory Password-Authenticated Group Key Exchange

Lin Zhang and Zhenfeng Zhang[✉]

Trusted Computing and Information Assurance Laboratory,
Institute of Software, Chinese Academy of Sciences, Beijing, China
{zhanglin,zfzhang}@tca.iscas.ac.cn

Abstract. The contributory property allows participants of group key exchange fairly to engage in the generation of the random session key rather than an entity or some part of members solely to determinate it or force it to lie in an undesired distribution. In this paper, we put forth a password-authenticated group key exchange (GPAKE) in which principals cooperate to agree a strong session key just in possession of a short password. The scheme realizes the optimality of contributory property—full-contributiveness—as long as there is one honest party, the uniform distribution of final session keys can be guaranteed. Moreover, it reaches the security definitions in the well-known universal composability (UC) framework under the random oracle model based on the one-more gap Diffie-Hellman assumption. In particular, our scheme that achieves these results with only two-round messages, has better performances on round complexity in comparison with the existing UC-secure schemes.

Keywords: Group key exchange · Password-based protocols · Contributiveness · Universal composability

1 Introduction

In recent decades, as electronic communications and information systems become more and more complicated, applications, such as video- or tele-conferencing involving multiple participants, are widespread throughout the Internet. In order to satisfy the requirement of secure communication channels within the insecure public network, it is necessary to design authenticated key exchange protocols for groups of principals.

To date, a collection of schemes has been designed elegantly. Bresson, et al. [10] is the first to usher in a formal model of security for group key exchange protocols, and the first to give a concrete scheme with a rigorous proof. However, in their protocol, the number of communication rounds depends upon the number of group members, so that this construction is impractical in the situation where the number of players is very large. Fewer rounds generally mean easier implementation and more effective reducing to synchronization problems. Subsequent to their work, several solutions [8,24] to constant-round protocol

© Springer International Publishing AG 2016
O. Dunkelman and S.K. Sanadhya (Eds.): INDOCRYPT 2016, LNCS 10095, pp. 119–134, 2016.
DOI: 10.1007/978-3-319-49890-4_7

for group key exchange are provided and proven secure in formal models. In addition, the desirable security goals of this kind of protocol not just focus on resistance against the outsider adversary who lies outside of the target group and seeks to get any information about the session key with observing and modifying protocol messages. Additionally, a certain degree of security properties against malicious insiders are expected in the designs of protocols. In Katz and Shin's work [23], they define the insider security for group key exchange protocols: one prevents the adversary from determining the session key entirely, unless at least there exists one corrupted party in the group.

Many schemes, including ones mentioned above, relies on possession of shared keys with other peers or authentic public/private pairs. In some scenarios, passwords, the ubiquitous keys to on-line communications, are the proper alternative in the group key exchanges, which benefits from password's convenience and low-cost. Namely, in the password-authenticated group key exchange (GPAKE), members only share a low-entropy password that can be reliably remembered by humans to bootstrap a high-entropy session key. Compared with other group key exchange protocols, GPAKEs, as password-based protocols, bear additional vulnerabilities to off-line dictionary attacks and the inevitable on-line dictionary attacks because of relatively small dictionary space. Thus, how to resist off-line attacks and restrict to adversaries eliminating at most one password per party instance, is also the basic security requirement in designing password-authenticated group key exchange protocols.

The first solution to the GPAKE is proposed by Bresson et al. [9]. Still, their protocol's round complexity is related to the number of group users. Abdalla et al. [1,4] demonstrate the first password-based group key exchange protocols in a constant number of rounds, in the random oracle /ideal cipher models [7,25], and in the standard model, respectively. Later, they give provably secure schemes [2,3] universal composability (UC) framework [15]. Recently, Xu et al. [27] present an one-round scheme in the standard model, using indistinguishability obfuscation as the main tool. Specifically, the works of [2,3] achieve a strong notion of (t, n)-contributiveness which captures that no adversary can bias the key if no more t players in the group of n players have been corrupted.

These schemes have shown important outcomes in GPAKE, yet much work remains to be done to enhance the efficiency and practicality of existing schemes. We focus on how to realize as few rounds as possible for the design of GPAKE scheme, without the expense of desirable security involved in the preceding description.

1.1   Our Contributions

In this literature, we put forward a UC-secure solution for password-authenticated group key exchange protocol against the static adversary in the password-only model, where the players do not have public keys authenticated by a certificate authority, pre-shared symmetric keys or other auxiliary equipments.

In the aspect of security models, this scheme is provably secure in the UC framework by the help of random oracles under the one-more gap Diffie-Hellman assumption. Compared with the game-based security models, such as [6,9], not only does the UC framework inherently provide the secure composition property, but it also has conspicuous advantages in distribution of passwords. Specifically, UC framework brings about strong composability properties: (1) UC-secure protocols remain secure even if many protocol instances (may be various kinds protocols) execute concurrently, and (2) The powerful universal composition theorem guarantees that they can be securely used as sub-routine protocols of other UC protocols. Besides, rather ideal assumptions on passwords independently chosen from pre-determined dictionary space in the game-based models, UC framework designates the environment (i.e. the distinguisher) to provide passwords to parties, which models arbitrary distributions and dependencies between passwords. Thereout, it captures the cases in real-life settings where the honest parties with incorrect but related passwords interacts with others—when a user obliviously makes typos.

Furthermore, we incorporate the full-contributiveness property (or called as (n, n)-contributiveness) into our rewritten ideal functionality for GPAKE, which means that the adversary cannot bias the key if there exists at least one honest player in the group, while our scheme has proven to be capable of realizing it. In fact, the notion of contributiveness brings several advantages in group key exchange protocols. First, it pledges each party equally contributes to the session key, which makes one intuitively feel that key agreement is "fairer" than key distribution. Second, it still results in high quality random keys even though some malicious parties improperly choose their contributions. Third, it deters the case where the insider adversary determinates session keys to specific values known by an outsider adversary in advance, so that the latter can eavesdrop on the later communications without the former's direct divulging to them. To a certain degree, the destructibility of the insider adversary is decreased.

An important measure of a protocol's efficiency is the communication complexity (number of protocol rounds) of the given protocol. Our protocol achieves the properties above with only two rounds. It distinctly has better performance on round complexity than the other UC-secure ones [2,3]. On the minus side, our scheme also has to perform $O(n)$ exponent calculations per member.

Table 1. Comparison of GPAKE schemes

Scheme	Security	Contributeness	Model	Rounds	Computation
BCP [9]	Game	$(1, n)$	RO&IC	n	$O(n)$
ABCP [1]	Game	$(1, n)$	RO&IC	4	$O(1)$
AP [4]	Game	$(1, n)$	Std	5	$O(n)$
ACCP [2]	UC	$(n/2, n)$	RO&IC	5	$O(n)$
ACGP [3]	UC	(n, n)	Std	6	$O(n)$
XHZ [27]	Game	$(1, n)$	Std	1	iO[1]
Ours	UC	(n, n)	RO	2	$O(n)$

[1] The "iO" means a program indistinguishability obfuscator.

However, according to the Moore's laws which declare the computing power grows faster than communication power, it is therefore an acceptable and reasonable compromise trade communication power for computing power in a group key exchange protocol. Comparison of some existing schemes for GPAKE is shown in Table 1.

2   Security Definitions

In this section, we will begin by reviewing the UC framework and the general split functionality. Then a detailed description of the ideal functionality for the password-authenticated group key exchange and related discussion are followed.

2.1   Universal Composability Framework

Universal composability [15] is the definition of secure computation that considers an execution of a protocol in the setting involving an *environment* $\mathcal{Z}$, an adversary and parties. This framework involves two worlds—the real world and the ideal world. $\mathcal{Z}$'s aim is to distinguish two worlds. For it, the environment provides the inputs to the parties and observes their outputs. On one hand, as usual, the real world consists of participants of the protocol and an *adversary* $\mathcal{A}$ that controls the communication channel and potentially attacks protocols. On the other hand, in the ideal world, there exists an entirely trusted entity $\mathcal{F}$ called *ideal functionality*, and dummy participates of the target protocol simply hand their inputs to $\mathcal{F}$. The *ideal adversary* $\mathcal{S}$ directly interacts with $\mathcal{F}$, and their communication essentially models the information it can obtain and its abilities to attack the protocols. Namely, the functionality describes the security goals we expect. Intuitively, the adversary, with a variety of means of attacks, should not learn more information than the functionality's outputs to it. Thus, security requires that, for any adversary $\mathcal{A}$ attacking a protocol ρ, there exists an ideal adversary $\mathcal{S}$ such that no environment $\mathcal{Z}$ can distinguish the case that it is interacting with $\mathcal{A}$ and parties in the real world, and the case which it is interacting with $\mathcal{S}$ and a functionality $\mathcal{F}$ in the ideal world. If so, we say that ρ *UC-realizes* $\mathcal{F}$. From the point of view of the environment, the real-world protocol is at least as secure as the ideal-world one. In particular, the universal composability theorem guarantees that the protocol ρ continues to behave like the ideal functionality $\mathcal{F}$ even if it is executed in an arbitrary network environment. The complete details refer to [15].

2.2   Split Functionalities

In a network, without any authentication mechanism, an adversary of the network can simply "disconnect" the parties completely, and engage in separate executions with the other parties on behalf of the honest ones. Such an attack is inevitable. Players cannot distinguish the case in which they interact with the expected ones from the case where they interact with the adversary. Hence, in

For a given functionality $\mathcal{F}$, the split version $s\mathcal{F}$ proceeds as follows:

Initialization:

- Upon receiving (Init, sid) from P_i, send (Init, sid, P_i) to the adversary $\mathcal{S}$.
- Upon receiving (Init, $sid, P_i, \mathcal{H}, sid_\mathcal{H}$) from $\mathcal{S}$, where $\mathcal{H}$ is the set of party identities, check that P_i has sent (Init, sid) and that for all previous sets $\mathcal{H}'$, either (1) $\mathcal{H} = \mathcal{H}', sid_\mathcal{H} = sid_{\mathcal{H}'}$, or (2) $\mathcal{H} \cap \mathcal{H}' = \varnothing, sid_\mathcal{H} \neq sid_{\mathcal{H}'}$. If so, record $(\mathcal{H}, sid_\mathcal{H})$, send (Init, $sid, sid_\mathcal{H}$) to P_i, and initialize a new instance of $\mathcal{F}$ with $sid_\mathcal{H}$, denoted as $\mathcal{F}^\mathcal{H}$. Otherwise, ignore this message.

Computation:

- Upon receiving (Input, sid, m) from P_i, find the set $\mathcal{H}$ such that $P_i \in \mathcal{H}$, and forward m to $\mathcal{F}^\mathcal{H}$.
- Upon receiving (Input, $sid, \mathcal{H}, P_j, m$) from $\mathcal{S}$, if $\mathcal{F}^\mathcal{H}$ exists and $P_j \notin \mathcal{H}$, then forward m to $\mathcal{F}^\mathcal{H}$ as if coming from P_j. Otherwise, do nothing.
- When a copy $\mathcal{F}^\mathcal{H}$ generates an output m for party $P_i \in \mathcal{H}$, send m to P_i. if m is for a party $P_j \notin \mathcal{H}$ or for $\mathcal{S}$, $s\mathcal{F}$ sends the output to $\mathcal{S}$.

Fig. 1. The split version of ideal functionality $\mathcal{F}$

the work of [5], Barak et al. propose a new model based on split functionalities which guarantees that this attack is the only one available to the adversary.

The split functionality is a generic construction based upon a normal ideal functionality $\mathcal{F}$. Its formal description can be found on Fig. 1. It models security by allowing the adversary to carry out such an "attack" in the ideal world. In the initialization stage, the adversary adaptively chooses subsets of the honest parties' $\mathcal{H}$ under two constraints: (1) these subsets are disjoint; (2) the adversary must choose a unique session identifier $sid_\mathcal{H}$ for each authentication set $\mathcal{H}$. That is, the subsets create a partition of the players. During the computation stage of $s\mathcal{F}$, each subset $\mathcal{H}$ activates a different and independent instance of the ideal functionality $\mathcal{F}$, denoted as $\mathcal{F}^\mathcal{H}$. In each such execution, the parties $P_i \in \mathcal{H}$ provide their own inputs, and the adversary $\mathcal{S}$ provides the inputs for all $P_i \notin \mathcal{H}$. Similarly, the parties $P_i \in \mathcal{H}$ all receive their specified outputs as computed by their copy of $\mathcal{F}$. However, the adversary receives all of its own outputs, as well as the outputs of the parties $P_i \notin \mathcal{H}$ who are controlled by $\mathcal{S}$. It's important to note that there is no interaction between different instances of $\mathcal{F}$ run by $s\mathcal{F}$.

2.3   The Ideal Functionality for GPAKE

The formalized description of GPAKE's ideal functionality $\mathcal{F}_{\mathsf{GPAKE}}$ is presented in Fig. 2. In order to reduce repeated representations, assume that the ideal functionality only takes notice of the first query or input for each sid and party,

The functionality $\mathcal{F}_{\mathsf{GPAKE}}$ parameterized by the security parameter κ, interacts with an adversary $\mathcal{S}$ and a ordered set of parties $\mathcal{H} = \{P_1, \ldots, P_n\}$ (where $n \geq 3$) via the following queries:

- **Initialization.** Upon receiving $(\mathsf{NewSession}, sid, P_i, pw_i)$ from P_i, record (sid, P_i, pw_i), if P_i is honest, mark it **fresh**, and send $(\mathsf{NewSession}, sid, P_i)$ to $\mathcal{S}$. Otherwise, this record is marked as **corrupted** instead. If there exists $n - 1$ recorded tuples (sid, P_j, pw_j) for $P_j \in \mathcal{H} \backslash \{P_i\}$, then record $(sid, \mathcal{H}, \mathsf{ready})$ and send it to $\mathcal{S}$.
- **Key Generation.** Upon receiving $(\mathsf{NewKey}, sid, P_i, \mathcal{H}^*, sk^*)$ from $\mathcal{S}$, abort if there is no record of the form $(sid, \mathcal{H}^*, \mathsf{ready})$ or $\mathcal{H} \neq \mathcal{H}^*$. Otherwise, proceed for record (sid, P_i, pw_i) as follows:
  - If all the records whose identities belong to $\mathcal{H}^*$ are corrupted, then output (sid, sk^*) to player P_i.
  - If this record is **fresh**, and there is a record (sid, P_j, pw') with $pw' = pw_i$, and a key sk' was sent to P_j, then output (sid, sk') to P_i.
  - In any other case, pick a new random key sk' of length κ, and send (sid, sk') to P_i.
  Either way, mark the record (sid, P_i, pw_i) as **completed**.

Fig. 2. The ideal functionality $\mathcal{F}_{\mathsf{GPAKE}}$

and subsequent ones for the same sid and party are straightly ignored. What's more, the session identifier sid are considered to be globally unique so that several sessions running in parallel can be distinguished. Note that we take into account the static corruption—the adversary could selectively designate and corrupt some participants, but only prior to the beginning of a protocol instance. From the corruption on, it not only obtains their inputs resulted from $\mathcal{Z}$, and also fully controls their behaviors in the following executions.

In the ideal world, the functionality $\mathcal{F}_{\mathsf{GPAKE}}$ interacts with an adversary $\mathcal{S}$ (i.e. the simulator), n parties $P_1, \ldots, P_n$ and the environment $\mathcal{Z}$ (through the parties). Before beginning, $\mathcal{Z}$ chooses the passwords pw_i (may be unequal) on its own for participants, which captures the arbitrary password distribution, including the case of making typos. As a bonus, this approach provides forward secrecy for free, which preserve the security of session keys even if the password is used for other purposes.

Though such a query $(\mathsf{NewSession}, sid, P_i, pw_i)$, every party initiates a new session with the expected ones in the group $\mathcal{H}$. Then $\mathcal{F}_{\mathsf{GPAKE}}$ is triggered to create the corresponding records, such as (sid, P_i, pw_i), for them to store their inputs locally, and labeled it as **fresh**. Actually, among these group members, some may be impersonated or corrupted by the adversary $\mathcal{S}$ to take part in the protocol instance with $\mathcal{S}$'s own password attempt. In both cases, the records are marked as **corrupted**. Once all the parties in the group $\mathcal{H}$ have sent NewSession queries, the ideal functionality $\mathcal{F}_{\mathsf{GPAKE}}$ stores a record $(sid, \mathcal{H}, \mathsf{ready})$, and informs the adversary $\mathcal{S}$ with it as a notification. When the adversary $\mathcal{S}$ commences with impersonating a party P_i with the NewSession query, it is allowed temporarily to submit a character $\perp$ instead of the password, and replenish it before sending the corresponding NewKey query. This stipulation contributes to a more smooth simulation in the security proof. In principle, after this phase, the parties basically wait for receiving the session keys.

When receiving $(\mathsf{NewKey}, sid, P_i, \mathcal{H}^*, sk^*)$ query from the adversary $\mathcal{S}$, the ideal functionality $\mathcal{F}_{\mathsf{GPAKE}}$ is instructed to release the session key to P_i. Note that $\mathcal{H}^*$ is the set of participants that is specified by the adversary and may not be equal to $\mathcal{H}$ in the NewSession queries. When $\mathcal{H} = \mathcal{H}^*$, the computation happen within pre-assigned members, while $\mathcal{H} \neq \mathcal{H}^*$ means that the adversary introduces outsiders (may be fictional entities) into the group to replace some honest ones. The latter case is forbidden in our definition. Besides, if there no exists a ready record for $\mathcal{H}^*$, i.e. members of $\mathcal{H}^*$ do not entirely join this session, $\mathcal{F}_{\mathsf{GPAKE}}$ also abort this execution. Unlike previous key exchange functionalities [16,23], in that if one of NewSession records is corrupted the adversary is given the ability to fully determine the resulting session key into sk^*, ours deprives of this ability of $\mathcal{S}$ unless all parties are corrupted simultaneously. By this definition, we integrate the full-contributiveness property in the functionality. Participants shall share the same, uniformly distributed session keys with whom have the matching password. Namely, $\mathcal{F}_{\mathsf{GPAKE}}$ has to traverse the records and the session keys sent to some parties, and return the corresponding ones. If no one is found out, $\mathcal{F}_{\mathsf{GPAKE}}$ chooses a random value from the range of session keys. In consideration of implicit authentication, the protocol will not end up with the case where no key is established for parties unless the inevitable abortions occur.

When $\mathcal{F}_{\mathsf{GPAKE}}$ outputs the session key to the specified party, the corresponding NewSession record is marked as completed to avoid undesired on-line guessing attacks from $\mathcal{S}$ even after the authentication has ended.

2.4   Discussions

For the completeness of the ideal functionality, the adversary $\mathcal{S}$ should be acquiescently allowed to abort the instance at any time to capture some trivial cases. For instance, in the real network, the attackers can always delay, hijack messages or revise them into irregular formats in the communication channel, resulting in a failed session among parties.

In our context of the ideal functionality $\mathcal{F}_{\mathsf{GPAKE}}$, the TestPwd query is completely abandoned, since split functionality has modeled the adversary's active attacks which enable it to take apart in the group. In the view of security analysis, the simulator does have to learn the results whether the passwords are matching, which is totally left to the functionality along with generation of session keys. Moreover, an outsider should not get the final results of protocol executions without the later communications in realistic scenarios.

In the UC framework, as per the formalism of [15], assume that multiple protocol instances are running concurrently. As the case in the real world, numerous execution instances often invoke the same common random strings or random oracles. Roughly, we have to consider the multi-session extension $\mathcal{\bar{F}}$ through the JUC theorem [18]. We refer to [16,18] for more discussions.

3 Our GPAKE Scheme

The basic idea of this protocol is inspired by Jarecki et al's (verifiable) oblivious pseudorandom functions (V-OPRF) in [21,22] and Camenisch et al's constructions for distributed password verification protocol of [12], and then utilized to build our GPAKE scheme. Briefly, each participant $P_i \in \{P_1, \ldots, P_n\}$ has the shared password pw_i, along with its own ephemeral secret key x_i. The session key computed by parties for session and sub-session identifiers sid, $ssid$ is $H_2(sid, ssid, H_1(sid, pw_i)^X)$, where $X = \sum_{i=1}^{n} x_i \bmod q$, and H_1, H_2 are hash functions. In order to get this value, each party chooses $r_i \leftarrow_R \mathbb{Z}_q^*$ to blind the password hash $a_i := (H_1(sid, pw_i))^{r_i}$ and sends the result to the others. When $(b_{i,j} := a_i^{x_j})_{j \in [n], j \neq i}$ are returned, it can compute $v_i := (a_i^{x_i} \cdot \prod_{j=1, j \neq i}^{n} b_{j,i})^{1/r_i} = H_1(sid, pw_i)^X$. Simultaneously, P_i proceed to the similar power operation to $\{a_j\}_{j \neq i}$ from the others with its own secret key x_i. Nevertheless, such simple proposition of GPAKE cannot reach the UC-security in the unauthenticated channel. Hence, we provide other primitives, such as the zero-knowledge proof of knowledge and the extra hash function to ensure the simulator that the participants always use the coincident public/secret keys, and also to help it extract the secret key for simulation. More details are presented as follows.

3.1 Concrete Construction

Let $\mathbb{G}$ be a multiplicative group of prime order q with the generator g generated through an algorithm of parameters generation by the security parameter κ. The hash functions $H_1 : \{0,1\}^* \times \{0,1\}^* \to \mathbb{G}$, $H_2 : \{0,1\}^* \times \mathbb{G} \times \mathbb{G} \to \{0,1\}^{2\kappa}$ and $H_3 : \{0,1\}^* \times \{0,1\}^* \times \mathbb{G} \to \{0,1\}^\kappa$ are modeled as random oracles. The public parameters also consist of the common random strings crs for the zero-knowledge proofs of knowledge. PK denotes the non-interactive proof of knowledge (which is formally defined by Camenisch et al. [11,14]), showing $y_i = g^{x_i} \wedge (b_{i,j} = a_j^{x_i})_{j \in [n], j \neq i}$.

Assume that the actual members are known in advance, and we simply denote them as $P_1, \ldots, P_n$ according to a certain order. In a protocol instance, the parties communicate over an unauthenticated broadcast channel, where messages can be arbitrarily observed, modified, and delayed by the adversary $\mathcal{A}$. Particularly, the adversary can corrupt or impersonate the valid ones to join the group as an insider with its own password attempt.

When a protocol execution begins, each party randomly chooses a blinding factor r_i and ephemeral secret key x_i from $\mathbb{Z}_q^*$, and then generates the blinded password hash $a_i := (H_1(sid, pw_i))^{r_i}$ and the ephemeral public key $y_i := g^{x_i}$, respectively. By the end of this round, it computes a hash to the values a_i and y_i, i.e. $h_i := H_2(sid, a_i, y_i)$. Then each party broadcasts $\langle P_i, a_i, h_i \rangle$ to others.

Shared information: Generator g of group $\mathbb{G}$. Hash functions H_1, H_2, H_3.
　　　　　　　　Common reference strings for proofs of knowledge crs.
Information held by P_i: Password pw_i.
===

Round 1:

　1). Choose $x_i \leftarrow_R \mathbb{Z}_q^*$ and generate $y_i := g^{x_i}$;
　2). Pick $r_i \leftarrow_R \mathbb{Z}_q^*$ and compute $a_i := (H_1(sid, pw_i))^{r_i}$;
　3). Make a hash $h_i := H_2(sid, a_i, y_i)$;
　4). Broadcast $\langle P_i, a_i, h_i \rangle$.

Round 2:

　1). On receiving $\langle P_j, a_j, h_j \rangle$ from all $P_j \in \mathcal{SP} \backslash \{P_i\}$, compute $b_{i,j} := a_j^{x_i}$,
　　and set $ssid := (a_1, h_1) \| \ldots \| (a_n, h_n)$;
　2). Produce the non-interactive proof of knowledge π_i;
　3). Broadcast $\langle P_i, y_i, (b_{i,j})_{j \in [n], j \neq i}, \pi_i \rangle$.

Key Generation:

　1). Upon receiving $\langle P_j, (b_{j,k})_{k \in [n], k \neq j}, \pi_j \rangle$ from all $P_j \in \mathcal{SP} \backslash \{P_i\}$, check
　　$h_j = H_2(sid, a_j, y_j)$ and continue. If not, abort this instance;
　2). Verify $(\pi_j)_{j \in [n], j \neq i}$, and abort if one of them is invalid;
　3). Compute $v_i := (a_i^{x_i} \cdot \prod_{j=1, j \neq i}^n b_{j,i})^{1/r_i}$, and then output the session
　　key $sk_i := H_3(sid, ssid, pw_i, v_i)$;

Fig. 3. The description of our GPAKE protocol for each party P_i

Note that the sub-session identifier is defined as messages received in this round
$ssid := a_1, h_1 \| \ldots \| a_n, h_n$, which be included in the subsequent hash evaluations.
Specifically, it means that parties are partitioned by the shared messages among
them. The purpose of usage of H_2 is, during the formal security proof, to embed
the one-more gap Diffie-Hellman problem in the next round rather than this
round when the group has not partitioned by the adversary yet (Fig. 3).

　In the second round, each party computes blinded responses $b_{i,j} := a_j^{x_i}$ for
the others in this group using its ephemeral secret key x_i. Moreover, it is required
to generate a non-interactive zero-knowledge proof of knowledge PK that $b_{i,j}$ is
generated correctly using y_i's discrete logarithm. That is,

$$\pi_i := \mathsf{PK}\{x_i : y_i = g^{x_i} \wedge (b_{i,j} = a_j^{x_i})_{j \in [n], j \neq i}\}$$

Note that these zero-knowledge proofs should be on-line extractable, since the
simulator $\mathcal{S}$ needs to extract the adversary's ephemeral secret keys to obtain
the solutions to the one-more gap Diffie-Hellman problem in the simulation of

random oracle H_3. It ends this round communication with broadcasting the message $\langle P_i, y_i, (b_{i,j})_{j\in[n], j\neq i}, \pi_i \rangle$ to the other participants.

In the end, the parties check the hash values and the proofs of knowledge. As soon as a value received by P_i doesn't be verified correctly, it aborts this instance and outputs nothing. Otherwise, it computes the key material $v_i := (a_i^{x_i} \cdot \prod_{j=1, j\neq i}^n b_{j,i})^{1/r_i}$ and the session key $sk_i := H_3(sid, ssid, pw_i, v_i)$, outputs $(sid, ssid, sk)$, and terminates this session.

Throughout this scheme, the hash values in the first round and the proofs of knowledge play important roles in ensuring the full-contributory property. For the existence of hash values and proofs of knowledge, it is impossible for a malicious party P_i adaptively to choose its ephemeral secret key x_i after it gets $\prod_{j=1, j\neq i}^n H_1(sid, pw_i)^{x_j}$. Namely, even if there is only one honest party, the remaining $n-1$ ones still cannot have the sum of secret keys depend on its.

Remarkably, this scheme achieves the implicit authentication by only two rounds communications among the participants. Using general techniques, such as the hash values of session key materials along with new tags, it is easy to get explicit authentication at the cost of one more round messages.

In the scheme, PK is a non-interactive transformation of a proof of knowledge with the Fiat-Shamir heuristic [19] in the random oracle model. It can be extended to be online-extractable, by verifiably encrypting the witness with a public key defined in the common reference string. The witness can be extracted from the CRS by the simulator without rewinding by decrypting the ciphertext. A practical instantiation is given by the CPA-secure version of Camenisch and Shoup's encryption scheme [13], which is secure under the DCR assumption [26].

4 Security Analysis

In this section, we prove the security of our scheme utilizing the (N, Q) one-more gap Diffie-Hellman assumption, which states that for the group $\mathbb{G}$ there no exists polynomial-time adversary $\mathcal{A}$ so that the following probability is non-negligible:

$$\mathsf{Prob}[\mathcal{A}^{(\cdot)^k, \mathsf{DDH}(\cdot)}(g, g^k, g_1, \ldots, g_N) = \{(g_{j_s}, g_{j_s}^k)|s = 1, \ldots, Q+1\}]$$

where $k \in \mathbb{Z}_q^*$ and Q is the number of the queries $\mathcal{A}$ makes to the $(\cdot)^k$ oracle. Moreover, $\mathcal{A}$'s other inputs $g_1, \ldots, g_N$ are assumed to be sampled from $\mathbb{G}$.

We can draw a conclusion for the GPAKE scheme in this theorem below:

Theorem 1. *Under the one-more gap Diffie-Hellman assumption in $\mathbb{G}$, if the zero-knowledge proofs involved are online extractable, then the password-authenticated group key exchange presented in Sect. 3 securely realizes $s\widetilde{\mathcal{F}}_{\mathsf{GPAKE}}$ under static corruptions in the $(\mathcal{F}_{\mathsf{CRS}}, \mathcal{F}_{\mathsf{RO}})$-hybrid model.*

In order to prove this theorem, it is an ideal-world adversary (i.e. simulator) $\mathcal{S}$ that needs to be constructed such that an arbitrary environment $\mathcal{Z}$ cannot distinguish between protocol executions in the ideal world and ones in the real world, which is described in Sect. 4.1. Then, in Sect. 4.2, we demonstrate the indistinguishability between two worlds through a sequence of games.

4.1 Description of Simulator

The simulator $\mathcal{S}$ not only interacts with the functionality $\mathcal{F}_{\mathsf{GPAKE}}$ in the ideal world, but also acts as honest parties and environment $\mathcal{Z}$ against a copy of the real-world adversary $\mathcal{A}$ invoked by $\mathcal{S}$ internally, and provide it a simulated real world. Moreover, $\mathcal{S}$ faithfully forwards all messages between $\mathcal{A}$ and $\mathcal{Z}$.

Simulating Random Oracles and Common Random Strings. When the simulator receives the queries to random oracles H_1, H_2 and H_3, it chooses random values from appropriate ranges to provide answers, and then records inputs and outputs. Here, $\mathcal{S}$ is allowed to maintain a list Λ to store them, which is also helpful to ensure the consistency of simulated random oracles. The simulator answers $\mathcal{A}$'s queries and updates the lists according to the rules which are analogous to the description in Fig. 4. Particularly, the random oracle H_3 is answered by the help of NewKey queries in some points.

Furthermore, the simulated common reference string is chosen by $\mathcal{S}$ for the adversary $\mathcal{A}$ as $\mathcal{F}_{\mathsf{CRS}}$ presented in the Appendix A.2. $\mathcal{S}$ runs the initial simulator for proofs of knowledge and gets (crs, τ). $\mathcal{S}$ sets the common reference string to crs and locally stores τ as the trapdoor for generating simulated PKs and extracting the adversary's witnesses.

Simulating the Party P_i. Once receiving $(\mathsf{Init}, sid, P_i)$ and $(\mathsf{NewSession}, sid, P_i)$ from the functionality, the simulator $\mathcal{S}$ randomly samples an element g_i from the group $\mathbb{G}$, and then sets $a_i := g_i$, due to the fact that it has no access to the correct password for the honest party P_i. And it randomly chooses the value h_i from $\{0,1\}^*_{H_2}$. Such messages from honest participants are delivered to the adversary $\mathcal{A}$ in the simulated real world.

The adversary $\mathcal{A}$ can make its decision about the subgroups participants belong to, on account of lack of strong authentication assumptions. It sends the first flow to target parties on behalf of ones it wants to impersonate (or they have been corrupted since the beginning of the session). These subgroups $\mathcal{H}$s are defined according to the messages $(a_i, h_i)_{i \in [n]}$ in the first round. $\mathcal{S}$ forwards these $\mathcal{H}$s, which make a partition of all parties, to the split functionality. Namely, the players in the same session receive and share the same $(a_i, h_i)_{i \in [n]}$. The simulator $\mathcal{S}$ also sends NewSession queries for the corrupted parties or ones in disguise correspondingly. Note that the simulator might have no knowledge about the latter's passwords in this moment, and thus it has to pass $(\mathsf{NewSession}, sid, P_j, \bot)$ for the dishonest party P_j to the ideal functionality instead. Moreover, we assume that the simulator is permitted to fill in the blanks later.

During the second round, on the behalf of honest parties, $\mathcal{S}$ just follows the protocol description to send $\langle P_i, y_i, (b_{i,j})_{j \in [n], j \neq i}, \pi_i \rangle$ back to $\mathcal{A}$ in the broadcast channel, where π_i is a simulated one, and H_2 is programmed as $h_i := H_2(sid, a_i, y_i)$. Finally, $\mathcal{S}$ makes a call $(\mathsf{NewKey}, sid, P_i, \bot)$ to $\widehat{\mathcal{F}}_{\mathsf{GPAKE}}$.

To make the session keys indistinguishable in the view of $\mathcal{Z}$, the simulator deals with $\mathcal{A}$'s queries $H_3(sid, ssid, pw_j, v_j)$ for some party P_j as follows. If $v_j \neq H_1(sid, pw_j)^{\Sigma x_i^* + \Sigma x_k}$, where x_k results from an honest party P_k, while x_l^*

is extracted from the proofs of knowledge provided by the dishonest one P_l, $\mathcal{S}$ just return a random value. Otherwise, $\mathcal{S}$ fills the blank $\perp$ in P_j's NewSession record with pw_j, and sends (NewKey, sid, P_j, $\perp$) to $\widehat{\mathcal{F}}_{\text{GPAKE}}$, and then obtain sk. Finally, it sets $H_3(sid, ssid, pw_j, v_j) := sk$ and output it to $\mathcal{A}$.

4.2   Sequence of Games

Here, via a sequence of games $\mathbf{G}_i$, we will prove that the real world with the arbitrary $\mathcal{A}$ and the ideal world with $\widehat{\mathcal{F}}_{\text{GPAKE}}$ and $\mathcal{S}$ as defined above are indistinguishable in the view of environment $\mathcal{Z}$. This needs to be stressed that, the simulator $\mathcal{S}$ "magically" obtains the inputs of honest parties provided by $\mathcal{Z}$ in the intermediate games, but they are no longer needed at the end of simulation. Following is the sequence of concrete games:

Game $\mathbf{G}_0$: Let $\mathbf{G}_0$ be the real-world game. As we noted above, the simulator $\mathcal{S}$ "magically" receives inputs from $\mathcal{Z}$, and just simply runs the real-world protocol executions for all the honest parties.

Game $\mathbf{G}_1$: It is identical to $\mathbf{G}_0$, except that we change the generation of crs and proofs of knowledge in the protocol. More specifically, on one hand, the common random strings are replaced with the simulated ones, and $\mathcal{S}$ knows the secret keys. On the other hand, whenever the honest parties perform the proofs of knowledge, $\mathcal{S}$ provides the simulated ones instead. The indistinguishability between them follows from the zero-knowledge properties of the proof system.

Game $\mathbf{G}_2$: Since $\mathbf{G}_2$, $\mathcal{S}$ begins to simulate the hash functions H_1, H_2 and H_3 instead of the real ones. It is distinguishable with the previous game in the view of the environment $\mathcal{Z}$, if there happen collisions that multiple inputs of oracles correspond to an output. Obviously, this case occurs with negligible probability, due to the birthday paradox.

Game $\mathbf{G}_3$: Let $\mathbf{G}_3$ be the modification of $\mathbf{G}_2$, where the honest party P_i replaces a normal $a_i := H_1(sid, pw_i)^{r_i}$ with a random element g_i from $\mathbb{G}$ in the first round. Actually, in the previous game, r_i is randomly chosen from $\mathbb{Z}_q^*$ by P_i locally without leakage to the adversary $\mathcal{A}$. Therefore, $H_1(sid, pw_i)^{r_i}$ cannot be distinguished from the random g_i, from $\mathcal{Z}$'s view.

Game $\mathbf{G}_4$: Compared with $\mathbf{G}_3$, the simulator $\mathcal{S}$ makes the party P_i output sk which $\widehat{\mathcal{F}}_{\text{GPAKE}}$ forwards to it after the NewKey query, rather than the normal value $H_3(sid, ssid, pw_i, v_i)$. Only by querying the $H_3(\cdot)$ oracle can the environment distinguish between these two outputs. Concretely, When the adversary $\mathcal{A}$ makes a query $(sid, ssid, pw_j, H_1(sid, pw_j)^{\Sigma x_i^* + \Sigma x_k})$, $\mathcal{S}$ interacts with the functionality $\widehat{\mathcal{F}}_{\text{GPAKE}}$, and obtains the proper output value sk as the response. Besides, We define an event Γ that the adversary $\mathcal{A}$ queries the oracle $H_3(\cdot)$ on the input $(sid, ssid, pw_j, H_1(sid, pw_j)^{\Sigma x_i^* + \Sigma x_k})$ without communicating with some honest parties of $\mathcal{H}$, which gives rise to an abort in $\mathbf{G}_4$, since $\mathcal{S}$ has to send (NewKey, sid, P_j, $\mathcal{H}^*$, $\perp$) to the functionality, where $\mathcal{H}^* \neq \mathcal{H}$. It is observed that, provided that the event Γ does not occur, the environment is not able to distinguish between two cases.

Here, by the help of reduction from the one-more gap Diffie-Hellman problem, we conclude that the event Γ just happens with a negligible probability. Given an instance of the one-more gap Diffie-Hellman problem $(Q, g, y = g^k, g_1, \ldots, g_N)$, we revise the simulator's behaviors as follows. Before the beginning, the simulator initializes a counter $c(k) := 0$ modeling the number of times that the $(\cdot)^k$ oracle is invoked and a set of pairs of the form (z, z^k), where z is in $\{g_1, \ldots, g_N\}$, denoted as $\mathcal{T}(k) := \varnothing$. It uses the challenges $g_1, \ldots, g_N$ as the responses to $H_1(\cdot)$ queries instead of random values from $\mathbb{G}$, and as the value a_j for the honest party P_j to the other members in the first round.

In the second round, without loss of generality, assume that there exists s ($s \geq 1$) honest parities in the target group $\mathcal{H}$, the simulator randomly samples $s - 1$ values from $\mathbb{Z}_q^*$ and implicitly sets $k_t = k - \sum_{i=1, i \neq t}^{s} k_i$ for a honest party P_t. Note that $\mathcal{S}$ has no access to k_t, but it still can provide the P_t's ephemeral public key $y_t := y / (\prod_{i=1, i \neq t}^{s} g^{k_i})$. If a_j is the first round message sent to P_t, the simulator calls the $(\cdot)^k$ oracle to compute $b_{t,j} := a_j^k / (\prod_{i=1, i \neq t}^{s} a_j^{k_i})$ and simulates the proof of knowledge π_t using the trapdoor τ corresponding to the specified CRS. Moreover, once the oracle $(\cdot)^k$ is invoked, it increases the counter $c(k)$. For the other honest ones, the simulations proceeds as before. When $\mathcal{A}$ later queries $H_2(sid, ssid, pw_i, v_i)$, the simulator invokes the DDH oracle to check whether it satisfies $v_i = H_1(sid, pw_i)^{k + \Sigma x_l^*}$. If so, it adds $(H_1(sid, pw_i), v_i / H_1(sid, pw_i)^{\Sigma x_l^*})$ to the set $\mathcal{T}(k)$. Once the event $c(k) < |\mathcal{T}(k)|$, which the adversary never communicates with P_t since the end of the first round but submits an appropriate H_3 query corresponding the solution, occurs, the simulator $\mathcal{S}$ addresses the one-more gap DH problem by returning the set $\mathcal{T}(k)$.

Now, the ideal-world is identical to $\mathbf{G_4}$, except that $\mathcal{S}$ no longer owns the specified inputs from $\mathcal{Z}$. Thus, the proof of theorem is completed.

Acknowledgement. We would like to thank the anonymous reviewers for their beneficial comments. This work is supported by the National Natural Science Foundation of China (No. U1536205, 61170278) and the National Basic Research Program of China (No.2013CB338003).

A   Auxiliary Ideal Functionalities

In this section, we list the formal ideal functionalities of random oracles and common random strings used as setup assumptions in our work.

A.1   Random Oracles

The random oracle model (e.g. [7]) captures an idealization of a hash function. In particular, it allows only black-box access and cannot be "predicted" without explicitly evaluating it. The outputs are uniformly selected random strings of specified size. We present the random oracle functionality $\mathcal{F}_{\mathsf{RO}}$ that has been defined by Hofheinz and Müller-Quade [20] in Fig. 4.

$\mathcal{F}_{\mathsf{RO}}$ proceeds as follows, running on security parameter κ, with parties $P_1, \ldots, P_n$ and an adversary $\mathcal{S}$:

- $\mathcal{F}_{\mathsf{RO}}$ keeps a list L (which is initially empty) of pairs of bitstrings.
- Upon receiving a value (RO, sid, m), where $m \in \{0,1\}^*$ from some party P_i or from $\mathcal{S}$, do:
  - If there is a pair $(m, \tilde{h})$ for some $\tilde{h} \in \{0,1\}^\kappa$ in the list L, set $h := \tilde{h}$;
  - If there is no such pair, choose uniformly $h \in \{0,1\}^\kappa$ and store the pair (m, h) in L.
  Once h is set, reply to the activating machine (i. e., either P_i or $\mathcal{S}$) with (RO, sid, h).

Fig. 4. The ideal functionality $\mathcal{F}_{\mathsf{RO}}$

A.2   Common Reference Strings

The common reference string functionality $\mathcal{F}_{\mathsf{CRS}}$ [15,17] captures that a common string drawn from a pre-specified distribution D can be accessible by all parties in the system, including the adversary. Furthermore, it guarantees that no party can be aware of the information related to the process of generating this string. The functionality illustrated in Fig. 5 results from the 2005 version of [15].

The functionality $\mathcal{F}_{\mathsf{CRS}}$ running on distribution D proceeds as follows:

- When receiving input (CRS, sid) from party P, first verify that $sid = (\mathcal{P}, sid')$ where $\mathcal{P}$ is a set of identities, and that $P \in \mathcal{P}$; else ignore the input. Next, if there is no value r recorded then choose and record $r \xleftarrow{R} D$. Finally, send a public delayed output (CRS, sid, r) to P.

Fig. 5. The ideal functionality $\mathcal{F}_{\mathsf{CRS}}$

References

1. Abdalla, M., Bresson, E., Chevassut, O., Pointcheval, D.: Password-based group key exchange in a constant number of rounds. In: Yung, M., Dodis, Y., Kiayias, A., Malkin, T. (eds.) PKC 2006. LNCS, vol. 3958, pp. 427–442. Springer, Heidelberg (2006). doi:10.1007/11745853_28
2. Abdalla, M., Catalano, D., Chevalier, C., Pointcheval, D.: Password-authenticated group key agreement with adaptive security and contributiveness. In: Preneel, B. (ed.) AFRICACRYPT 2009. LNCS, vol. 5580, pp. 254–271. Springer, Heidelberg (2009). doi:10.1007/978-3-642-02384-2_16

3. Abdalla, M., Chevalier, C., Granboulan, L., Pointcheval, D.: Contributory password-authenticated group key exchange with join capability. In: Kiayias, A. (ed.) CT-RSA 2011. LNCS, vol. 6558, pp. 142–160. Springer, Heidelberg (2011). doi:10.1007/978-3-642-19074-2_11

4. Abdalla, M., Pointcheval, D.: A scalable password-based group key exchange protocol in the standard model. In: Lai, X., Chen, K. (eds.) ASIACRYPT 2006. LNCS, vol. 4284, pp. 332–347. Springer, Heidelberg (2006). doi:10.1007/11935230_22

5. Barak, B., Canetti, R., Lindell, Y., Pass, R., Rabin, T.: Secure computation without authentication. In: Shoup, V. (ed.) CRYPTO 2005. LNCS, vol. 3621, pp. 361–377. Springer, Heidelberg (2005). doi:10.1007/11535218_22

6. Bellare, M., Pointcheval, D., Rogaway, P.: Authenticated key exchange secure against dictionary attacks. In: Preneel, B. (ed.) EUROCRYPT 2000. LNCS, vol. 1807, pp. 139–155. Springer, Heidelberg (2000). doi:10.1007/3-540-45539-6_11

7. Bellare, M., Rogaway, P.: Random oracles are practical: a paradigm for designing efficient protocols. In: Proceedings of the 1st ACM Conference on Computer and Communications Security, pp. 62–73. ACM (1993)

8. Boyd, C., Nieto, J.M.G.: Round-optimal contributory conference key agreement. In: Desmedt, Y.G. (ed.) PKC 2003. LNCS, vol. 2567, pp. 161–174. Springer, Heidelberg (2003). doi:10.1007/3-540-36288-6_12

9. Bresson, E., Chevassut, O., Pointcheval, D.: Group Diffie-Hellman key exchange secure against dictionary attacks. In: Zheng, Y. (ed.) ASIACRYPT 2002. LNCS, vol. 2501, pp. 497–514. Springer, Heidelberg (2002). doi:10.1007/3-540-36178-2_31

10. Bresson, E., Chevassut, O., Pointcheval, D., Quisquater, J.J.: Provably authenticated group Diffie-Hellman key exchange. In: Proceedings of the 8th ACM Conference on Computer and Communications Security, pp. 255–264. ACM (2001)

11. Camenisch, J., Kiayias, A., Yung, M.: On the portability of generalized Schnorr proofs. In: Joux, A. (ed.) EUROCRYPT 2009. LNCS, vol. 5479, pp. 425–442. Springer, Heidelberg (2009). doi:10.1007/978-3-642-01001-9_25

12. Camenisch, J., Lehmann, A., Neven, G.: Optimal Distributed Password Verification. In: Proceedings of the 22nd ACM Conference on Computer and Communications Security, pp. 182–194. ACM (2015)

13. Camenisch, J., Shoup, V.: Practical verifiable encryption and decryption of discrete logarithms. In: Boneh, D. (ed.) CRYPTO 2003. LNCS, vol. 2729, pp. 126–144. Springer, Heidelberg (2003). doi:10.1007/978-3-540-45146-4_8

14. Camenisch, J., Stadler, M.: Efficient group signature schemes for large groups. In: Kaliski, B.S. (ed.) CRYPTO 1997. LNCS, vol. 1294, pp. 410–424. Springer, Heidelberg (1997). doi:10.1007/BFb0052252

15. Canetti, R.: Universally composable security: a new paradigm for cryptographic protocols. In: 42nd IEEE Symposium on Foundations of Computer Science, pp. 136–145. IEEE (2001)

16. Canetti, R., Halevi, S., Katz, J., Lindell, Y., MacKenzie, P.: Universally composable password-based key exchange. In: Cramer, R. (ed.) EUROCRYPT 2005. LNCS, vol. 3494, pp. 404–421. Springer, Heidelberg (2005). doi:10.1007/11426639_24

17. Canetti, R., Lindell, Y., Ostrovsky, R., Sahai, A.: Universally Composable Two-party and Multi-party Secure Computation. In: Proceedings of the Thirty-fourth Annual ACM Symposium on Theory of Computing, pp. 494–503. ACM (2002)

18. Canetti, R., Rabin, T.: Universal composition with joint state. In: Boneh, D. (ed.) CRYPTO 2003. LNCS, vol. 2729, pp. 265–281. Springer, Heidelberg (2003). doi:10.1007/978-3-540-45146-4_16

19. Fiat, A., Shamir, A.: How to prove yourself: practical solutions to identification and signature problems. In: Odlyzko, A.M. (ed.) CRYPTO 1986. LNCS, vol. 263, pp. 186–194. Springer, Heidelberg (1987). doi:10.1007/3-540-47721-7_12

20. Hofheinz, D., Müller-Quade, J.: Universally composable commitments using random oracles. In: Naor, M. (ed.) TCC 2004. LNCS, vol. 2951, pp. 58–76. Springer, Heidelberg (2004). doi:10.1007/978-3-540-24638-1_4

21. Jarecki, S., Kiayias, A., Krawczyk, H., Xu, J.: Highly-efficient and composable password-protected secret sharing (or: how to protect your bitcoin wallet online). In: 2016 IEEE European Symposium on Security and Privacy, pp. 276–291 (2016)

22. Jarecki, S., Kiayias, A., Krawczyk, H.: Round-optimal password-protected secret sharing and T-PAKE in the password-only model. In: Sarkar, P., Iwata, T. (eds.) ASIACRYPT 2014. LNCS, vol. 8874, pp. 233–253. Springer, Heidelberg (2014). doi:10.1007/978-3-662-45608-8_13

23. Katz, J., Shin, J.S.: Modeling insider attacks on group key-exchange protocols. In: Proceedings of the 12th ACM Conference on Computer and Communications Security, pp. 180–189. ACM (2005)

24. Katz, J., Yung, M.: Scalable protocols for authenticated group key exchange. In: Boneh, D. (ed.) CRYPTO 2003. LNCS, vol. 2729, pp. 110–125. Springer, Heidelberg (2003). doi:10.1007/978-3-540-45146-4_7

25. Liskov, M., Rivest, R.L., Wagner, D.: Tweakable block ciphers. In: Yung, M. (ed.) CRYPTO 2002. LNCS, vol. 2442, pp. 31–46. Springer, Heidelberg (2002). doi:10.1007/3-540-45708-9_3

26. Paillier, P.: Public-key cryptosystems based on composite degree residuosity classes. In: Stern, J. (ed.) EUROCRYPT 1999. LNCS, vol. 1592, pp. 223–238. Springer, Heidelberg (1999). doi:10.1007/3-540-48910-X_16

27. Xu, J., Hu, X.-X., Zhang, Z.-F.: Round-optimal password-based group key exchange protocols in the standard model. In: Malkin, T., Kolesnikov, V., Lewko, A.B., Polychronakis, M. (eds.) ACNS 2015. LNCS, vol. 9092, pp. 42–61. Springer, Heidelberg (2015). doi:10.1007/978-3-319-28166-7_3

Side-Channel Attacks

Score-Based vs. Probability-Based Enumeration – A Cautionary Note

Marios O. Choudary[1]([✉]), Romain Poussier[2], and François-Xavier Standaert[2]

[1] University Politehnica of Bucharest, Bucharest, Romania
marios.choudary@cs.pub.ro
[2] ICTEAM - Crypto Group, Université Catholique de Louvain,
Louvain-la-Neuve, Belgium

Abstract. The fair evaluation of leaking devices generally requires to come with the best possible distinguishers to extract and exploit side-channel information. While the need of a sound model for the leakages is a well known issue, the risks of additional errors in the post-processing of the attack results (with key enumeration/key rank estimation) are less investigated. Namely, optimal post-processing is known to be possible with distinguishers outputting probabilities (e.g. template attacks), but the impact of a deviation from this context has not been quantified so far. We therefore provide a consolidating experimental analysis in this direction, based on simulated and actual measurements. Our main conclusions are twofold. We first show that the concrete impact of heuristic scores such as produced with a correlation power analysis can lead to non-negligible post-processing errors. We then show that such errors can be mitigated in practice, with Bayesian extensions or specialized distinguishers (e.g. on-the-fly linear regression).

1   Introduction

Side-channel attacks are powerful tools to recover the secret keys of cryptographic implementations. When an attacker has physical access to the device (e.g. a smart card) running the cryptographic algorithm, he may extract information using side-channels such as the power consumption or the electromagnetic radiation related to the algorithm being executed. In the following, we refer to this information as a trace or a leakage. Examples of attacks exploiting this type of leakages include (but are not limited to) Kocher et al.'s Differential Power Analysis (DPA) [11], Brier et al.'s Correlation Power Analysis (CPA) [3] or Chari et al.'s Template Attack (TA) [4]. Applied on symmetric block ciphers, they usually allow recovering information on independent parts of the full key which we shall further call subkeys. Typically, an attacker obtains a probability (or a score) for each of the possible value a subkey can take. In order to recover the full key, he has to recombine the information from the different subkeys. In this context we can distinguish three cases. Firstly, if the attacker has enough information on each subkey (i.e. the correct subkey is ranked first), he

© Springer International Publishing AG 2016
O. Dunkelman and S.K. Sanadhya (Eds.): INDOCRYPT 2016, LNCS 10095, pp. 137–152, 2016.
DOI: 10.1007/978-3-319-49890-4_8

can recombine them by simply taking the most likely hypothesis for each subkey. Secondly, if at least one subkey does not provide enough information, a search into the key space needs to be performed. For this purpose, a key enumeration algorithm can be used in order to output the full key in decreasing order of likelihood [2,6,12,17–19,21]. The position of the actual key in this ordering will be denoted as the rank. Finally, when the rank is beyond the attacker's computational power (e.g. $>2^{50}$), a complete key enumeration is not possible anymore. In that case, one can only estimate a bound on the rank using a rank estimation algorithm, which requires knowledge of the actual key (thus is only suitable for an evaluator) [1,8,10,17,18,22,23].

In this paper, we deal with the general issue of evaluating the security of a leaking device in the most accurate manner. This is an important concern since an error might lead to an overestimation of the rank and thus a false sense of security. Sources of errors that can affect the quality of a side-channel security evaluation have been recently listed in [18] and are summarized in Fig. 1, where we can see the full process of a divide-and-conquer side-channel attack from the trace acquisition to the full key recovery, along with the errors that might be introduced. A typical (and well understood [9]) example is the case of model errors, which can significantly affect the efficiency of the subkey recoveries. Interestingly, the figure also suggests that some errors may only appear during the enumeration/rank estimation phase of the side-channel security evaluation. Namely, *combination ordering errors*, which are the core this study, arise from the fact that combining the information obtained for different subkeys can be easily done in a sound manner if the subkey distinguisher outputs probabilities (which is typically the case of optimal distinguishers such as TA), while performing this combination with heuristic scores (e.g. correlation values) is not obvious. This is an interesting observation since it goes against the equivalence

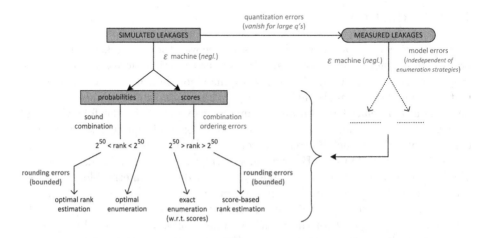

Fig. 1. Errors & bounds in key enumeration and rank estimation.

of TA and CPA (under the assumption that they use identical leakage models) that holds for first-order side-channel attacks [15].

In general, the soundest way to evaluate security is to use a Bayesian distinguisher (i.e. a TA) and to output probabilities. Yet, a possible concrete drawback of such an attack is that it requires a good enough model of the leakages Probability Density Function (PDF), i.e. some profiling. For this reason, non-profiled (and possibly sub-optimal) attacks such as the CPA are also frequently considered in the literature. As previously mentioned, distinguishers that output probabilities allow a straightforward combination of the subkey information. This is because probabilities are known to have a multiplicative relationship. By contrast, the combination of heuristic scores does not provide such an easy relationship, and therefore requires additional assumptions, which may lead to combination ordering errors. This is easily illustrated with the following example. Let us first assume two 1-bit subkeys k_1 and k_2 with probability lists $[p_1^1, p_1^2]$ and $[p_2^1, p_2^2]$ with $p_1^1 >, p_1^2$ and $p_2^1 > p_2^2$. Secondly, let us assume the same subkeys with score lists $[s_1^1, s_1^2]$ and $[s_2^1, s_2^2]$ with $s_1^1 >, s_1^2$ and $s_2^1 > s_2^2$. On the one hand, it is clear that the most (respectively least) probable key is given by $\{p_1^1, p_2^1\}$ or $\{s_1^1, s_2^1\}$ (respectively $\{p_1^2, p_2^2\}$ or $\{s_1^2, s_2^2\}$). On the other hand, only probabilities allow to compare the pair $\{p_1^1, p_2^2\}$ and $\{p_1^2, p_2^1\}$. Since we have no sound way to combine scores, we have no clue how to compare $\{s_1^1, s_2^2\}$ and $\{s_1^2, s_2^1\}$.

In the following, we therefore provide an experimental case study allowing us to discuss these combination ordering errors. For this purpose, we compare the results of side-channel attacks with enumeration in the case of TA, CPA (possibly including a bayesian extension [21]) and Linear Regression (LR) [20] and in particular its non-profiled variant put forward in [7], using both simulations and concrete measurements. Our main conclusions are that (i) combination ordering errors can have a significant impact when the leakage model of the different subkeys in an attack differ, (ii) bayesian extensions of the CPA can sometimes mitigate these drawbacks, yet include additional assumptions that may lead to other types of errors, (iii) "on-the-fly" linear regression is generally a good (and more systematic) way to avoid these errors too, but come at the cost of a model estimation phase, and (iv) only TA are guaranteed to lead to optimal results. The rest of the paper is structured as follows. In Sect. 2, we present the different attacks we analyse, together with a concise description of the key enumeration and rank estimations algorithms. Then, in Sect. 3, we present our experimental results. Conclusions are in Sect. 4.

2   Background

2.1   Attacks

This section describes the different attacks we used along with the tool for full key recovery analysis. The first one we consider is the correlation power analysis [3], along with its bayesian extension to produce probabilities. The second one is the template attack [4]. The last one is the "on-the-fly" stochastic attack [7]. We target a b-bit master key k, where an attacker recovers information on N_s subkeys $k_0, \ldots, k_{N_s-1}$ of length $a = \frac{b}{N_s}$ bits (for simplicity, we assume that a

divides b). Our analysis targets the S-box computation (S) of a cryptographic algorithm. That is, we apply a divide-and-conquer attack where a subkey k is manipulated during a computation $S(x \oplus k)$ for an input x. Given n executions with inputs $\mathbf{x} = (x_i)$, $i \in [1, n]$ (where the bold notation is for vectors), we record (or simulate) the n side-channel traces $\mathbf{l} = (l_i^{x,k})$ (e.g. power consumption), corresponding to the computation of the S-box output value $v_{x,k} = S(x \oplus k)$ with key (subkey) k. We then use these traces in our side-channel attacks.

Correlation Power Analysis (CPA). From a given leakage model $\mathbf{m}^\star = (m_{x,k^\star})$ (corresponding to a key hypothesis $k^\star$) and n traces $\mathbf{l} = (l_i^{x,k})$, we compute Pearson's correlation coefficient $\rho_k^\star = \rho(\mathbf{l}, m_{x,k^\star})$ for all candidates $k^\star$ as shown by (1):

$$\rho_{k^\star} = \frac{\sum_{i=1}^{n}(l_i - \mathsf{E}(\mathbf{l})) \cdot (m_{x_i, k^\star} - \mathsf{E}(\mathbf{m}^\star))}{\mathsf{Std}(\mathbf{l}) \cdot \mathsf{Std}(\mathbf{m}^\star)}, \tag{1}$$

where E and Std denote the sample mean and standard deviation. If the attack is successful, the subkey $k = \arg\max_{k^\star}(\rho_{k^\star})$ is the correct one. This generally holds given a good model and a large enough number of traces. Concretely, our following experiments will always consider CPA with a Hamming weight leakage model, which is a frequent assumption in the literature [14], and was sufficient to obtain successful key recoveries in our case study.

CPA with Bayesian Extension (BCPA). In order to improve the results of key enumeration algorithms with CPA, we may use Fisher's Z-transform (2) on the correlation values $\rho_{k^\star}$ obtained from the CPA attack (see above) [13]:

$$z(\rho_{k^\star}) = \frac{1}{2} \log \frac{1 + \rho_{k^\star}}{1 - \rho_{k^\star}}. \tag{2}$$

Under some additional assumptions, this transformation can help us transform the vector of correlations into a vector of probabilities, which may be more suitable for the key enumeration algorithms, hence possibly leading to improved results.

More precisely, if we let $z_{k^\star} = z(\rho_{k^\star})$ be the z-transform of the correlation for the candidate key $k^\star$, then ideally we can exploit the fact that this value is normally distributed, and compute the probability:

$$\Pr[l|k^\star] = \mathcal{N}_{\mu_z, \sigma_z^2}(z_{k^\star}), \tag{3}$$

where

$$\mu_z = \frac{1}{2} \log \frac{1 + \rho}{1 - \rho}$$

is the z-transform of the real correlation ρ and

$$\sigma_z^2 = \frac{1}{n - 3}$$

is the estimated variance for this distribution (with n the number of attack samples). The main problem, however, is that we do not know the actual value of the correlation ρ (for the correct and wrong key candidates), so we cannot determine μ_z and σ_z^2 without additional assumptions.

A possible solution for the mean μ_z is to assume that the incorrect keys will have a correlation close to zero (so the z-transform for these keys will also be close to zero), while the correct key will have a correlation as far as possible from zero. In this case, we can take the absolute value of the correlation values, $|\rho_{k^*}|$, to obtain $z_{k^*} = z(|\rho_{k^*}|)$. Then, we can use the normal cumulative density function (cdf) $\mathcal{CN}_{0,\sigma_z^2}$ with mean zero to obtain a function that provides larger values for the correct key (i.e. where the z-transform of the absolute value of the correlation has higher values). After normalising this function, we obtain the associated probability of each candidate k^* as:

$$\Pr(k^*|l) = \frac{\mathcal{CN}_{0,\frac{1}{n-3}}(z_{k^*})}{\sum_{k'=0}^{2^a-1} \mathcal{CN}_{0,\frac{1}{n-3}}(z_{k'})}. \tag{4}$$

Alternatively, we can also use the mean of the transformed values $\bar{z} = \mathbf{E}_{k^*}(z_{k^*})$ as an approximation for μ_z (since we expect that $\mu_z > \bar{z} > 0$, so using this value should lead to a better heuristic than just assuming that all incorrect keys have a correlation close to zero).[1]

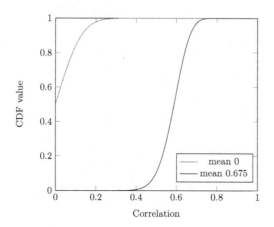

Fig. 2. Assumptions for μ_z in the CPA Bayesian extension.

---

[1] Yet another possible solution, which does not require knowledge of the key either, is to assume that the absolute correlation will be the highest for the correct key candidate, so the z-transform of the absolute correlation should also be the highest. Then, we can simply use the maximum among all the z-transformed values as the mean of the normal distribution for the z-transformed data (i.e. $\mu_z = \max_{k^*}(z_{k^*})$), hence obtaining $\Pr(k^*|l) = \frac{\mathcal{N}_{\mu_z,\frac{1}{n-3}}(z_{k^*})}{\sum_{k'=0}^{2^a-1} \mathcal{N}_{\mu_z,\frac{1}{n-3}}(z_{k'})}$. This did not lead to serious improvements in our experiments, but might provide better results in other contexts.

For illustration, Fig. 2 shows the evolution of the cdf-based probability function for these two choices of correlation mean (for an arbitrary variance of 0.01, corresponding to 103 attack traces taken from the following section). Since in this example, $\bar{z} = 0.6$, many incorrect keys have a correlation higher than 0.2. Hence, in this case, the zero-mean assumption should lead to considerably worse results than the $\bar{z}$ assumption, which we consistently observed in our experiments. In the following, we only report on this second choice.

Next, we also need an assumption for the variance σ_z^2. While the usual (statistical) approach would suggest to use $\sigma_z^2 = 1/(n - 3)$, we observed that in practice, a variance $\sigma_z^2 = 1$ gave better results in our experiments. Since this may be more surprising, we report on both assumptions, and next refer to the attack using $\sigma_z^2 = 1/(n - 3)$ as BCPA1 and to the one using $\sigma_z^2 = 1$ as BCPA2.

For illustration, Fig. 3 shows the evolution of the cdf-based probability function for BCPA1 and BCPA2 with different number of attack samples. The blue curve represents the BCPA2 method, which is invariant with the number of attack samples. The green, red and black curves shows the impact of the number of attack samples on the cdf. As we can see, the slope becomes more and more steep when the number of attack samples increases. This makes the key discrimination harder since at some point the variance becomes too small to explain the errors due to our heuristic assumption in the mean μ_z.

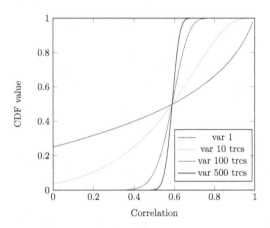

Fig. 3. Normal CDF with mean $\bar{z} = 0.6$ for variance 1 (blue), variance $\frac{1}{7}$ (green, 10 attack samples), variance $\frac{1}{97}$ (red, 100 attack samples) and variance $\frac{1}{497}$ (black, 500 attack samples). (Color figure online)

Summarizing, the formulae for the heuristics BCPA1 and BCPA2 are given by Eqs. (5) and (6):

$$\text{BCPA1} \; : \; \Pr[k^\star | l] = \frac{\mathcal{CN}_{\bar{z}, \frac{1}{n-3}}(z_{k^\star})}{\sum_{k'=0}^{2^a-1} \mathcal{CN}_{\bar{z}, \frac{1}{n-3}}(z_{k'})}. \tag{5}$$

$$\text{BCPA2} \ : \ \Pr[k^\star|l] = \frac{\mathcal{CN}_{\bar{z},1}(z_{k^\star})}{\sum_{k'=0}^{2^a-1} \mathcal{CN}_{\bar{z},1}(z_{k'})}. \tag{6}$$

Gaussian Template Attack (GTA). From a first set of profiling traces $l_p = (l_i^{x,k})$, the attacker estimates the parameters of a normal distribution. He computes a model $\mathbf{m} = (m_{x,k^\star})$, where $m_{x,k^\star}$ contains the mean vector and the covariance matrix associated to the computation of $v_{x,k^\star}$. We consider the approach where a single covariance is computed from all the samples [5]. The probability associated to a subkey candidate $k^\star$ is computed as shown by (7):

$$\Pr[k^\star|l] = \prod_{i=1}^{n} \frac{\mathcal{N}_{m_{x_i,k^\star}}(l_i)}{\sum_{k'=0}^{2^a-1} \mathcal{N}_{m_{x_i,k'}}(l_i)}, \tag{7}$$

where $\mathcal{N}_m$ is the normal probability density function (pdf) with mean and covariance given by the model m.[2] As mentioned in introduction, it is important to note that combining probabilities from different subkeys implies a multiplicative relationship. Since the division in the computation of $\Pr[k^\star|l]$ only multiplies all the probabilities by a constant, this has no impact on the ordering when combining subkeys. We can simply omit this division, thus only multiplying the values of $\Pr[l_i|k^\star] = \mathcal{N}_{m_{x_i,k^\star}}(l_i)$. This may allow us to use an efficient template attack based on the linear discriminant [5], adapted so we can use different plaintexts (see Sect. 2.3 for more details).

Unprofiled Linear Regression (LR). Stochastic attacks [20] aim at approximating the deterministic part of the real leakage function $\theta(v_{x,k})$ with a model $\theta^*(v_{x,k})$ using a basis $g(v_{x,k}) = \{g_0(v_{x,k}), \ldots, g_b(v_{x,k})\}$ chosen by the attacker. For this purpose, he uses linear regression to find the basis coefficients $c = \{c_0, \ldots, c_b\}$ so that the leakage function $\theta(v_{x,k})$ is approximated by $\theta^*(v_{x,k}) = c_0 \cdot g_0(v_{x,k}) + \ldots + c_b \cdot g_b(v_{x,k})$. The unprofiled version of this attack simply estimates on the fly the models $\theta^*(v_{x,k^\star})$ for each key candidate [7]. The probability associated to a subkey candidate $k^\star$ is computed as shown by (8):

$$\Pr[k^\star|l] = \frac{\text{Std}(1 - \theta^*(v_{x,k^\star}))^{-n}}{\sum_{k'=0}^{2^a-1} \text{Std}(1 - \theta^*(v_{x,k'}))^{-n}}, \tag{8}$$

where Std denotes the sample standard deviation. As detailed in [21], a significant advantage of this non-profiled attack is that is produces sound probabilities without additional assumptions on the distribution of the distinguisher. So it does not suffer from combination ordering errors. By contrast, it may be slower to converge (i.e. lead to less efficient subkey recovery) since it has to estimate a model on the fly. In our experiments, we consider two linear basis. The first one uses the hamming weight (HW) of the sensitive value: $g(v_{x,k}) = \{1, \text{HW}(v_{x,k})\}$. The second one uses the bits of the sensitive value $v_{x,k}$, denoted as $b_i(v_{x,k})$. We refer to the first one as LRH and to the second one as LRL.

---

[2] In our experiments with GTA, we only considered one leakage sample per trace, so our distribution is univariate and the covariance matrix becomes a variance.

2.2   Key Enumeration

After the subkey recovery phase of a divide-and-conquer attack, the full key is trivially recovered if all the correct subkeys are ranked first. If this is not the case, a key enumeration algorithm has to be used by the attacker in order to output all keys from the most probable one to the least probable one. In this case, the number of keys having a higher probability than the actual full key (plus one) is called the key rank. An optimal key enumeration algorithm was first described in [21]. This algorithm is limited by its high memory cost and its sequential execution. Recently, new algorithms based on rounded log probabilities have emerged [2,12,17,19]. They overcome these memory and sequential limitations at the cost of non optimality (which is a parameter of these algorithms). Different suboptimal approaches can be found in [6,18].

2.3   Rank Estimation

Rank estimation algorithms are the tools associated to key enumeration from an evaluation point-of-view. Taking advantage of the knowledge of the correct key, they output an estimation of the full key rank which would have been outputted by a key enumeration algorithm with an unbounded computational power. The main difference with key enumeration algorithms is that they are very fast and allow to estimate a rank that would be unreachable with key enumeration (e.g. 2^{100}). These algorithms have attracted some attention recently, as suggested by various publications [1,8,10,17,18,22,23]. For convenience, and since our following experiments are in an evaluation context, we therefore used rank estimation to discuss combination ordering errors. More precisely, we used the simple algorithm of Glowacz et al. [10] (using [1] or [17] would not affect our conclusions since these references have similar efficiencies and accuracies). With this algorithm, all our probabilities are first converted to the log domain (since the algorithm uses an additive relationship between the log probabilities). Then, for each subkey, the algorithm uses a histogram to represent all the log probabilities. The subkey combination is done by histogram convolution. Finally, the estimated key rank is approximated by summing up the number of keys in each bin from the last one to the one containing the actual key. We apply the same method for the correlation attacks, thus assuming a multiplication relationship for these scores too. Admittedly, and while it seems a reasonable assumption for the bayesian extension of CPA, we of course have no guarantee for the standard CPA. Note also that in this case the linear discriminant approach for template attacks in [5] becomes particularly interesting, because the linear discriminant is already in the logarithm domain, so can be used directly with the rank estimation algorithm, while providing comparable results to a key enumeration algorithm using probabilities derived from the normal distribution. Hence, this algorithm also avoids all the numerical problems related to the multivariate normal distribution when using a large number of leakage samples.

3   Experiments

In order to evaluate the results of the different attack approaches, we first ran simulated experiments. We target the output of an AES S-box leading to 16 univariate simulated leakages of the form $HW(S(x_i \oplus k_i)) + \mathcal{N}_i$ for $i \in [0, 15]$. HW represents the hamming weight function and $\mathcal{N}_i$ is a random noise following a zero-mean Gaussian distribution. For a given set of parameters, we study the full key rank using 250 attack repetitions, and compute the ranking entropy as suggested by [16]. By denoting R_i the rank of the full key for a given attack, we compute the expectation of the logarithm of the ranks given by $E_i(\log_2(R_i))$ (and not the logarithm of the average ranks[3] equals to $\log_2(E_i(R_i))$). The ranking entropy is closer to the median than the mean which gives smoother curves. However, we insist that using one or the other metric does not change the conclusions of our experiments.

3.1   Simulations with Identical S-Box Leakages

We first look at the case where the noise is constant for all attacked bytes, meaning $\mathcal{N}_i = \mathcal{N}_j$ (we used a noise variance of 10). We want to investigate if the full key rank is impacted by the way an attacker produces his probabilities or scores. Figure 4 (left) shows the full key rank in function of the number of attack traces for the different methods. One can first notice the poor behavior of BCPA1 (in red). Although it starts by working slightly better than CPA (in blue), it quickly becomes worse.

This is due to the variance that quickly becomes too small and does not allow to discriminate between the different key hypotheses. Secondly, we see that CPA and BCPA2 give very similar results in term of full key success rate. This can be explained since the noise level is constant for the different subkeys in this experiment. Hence, they are equally difficult to recover which limits the combination ordering errors. Thirdly, both LRH (purple) and LRL (purple, dashed) are not impacted by combination ordering errors. However, they suffer from the need to estimate a model. As expected, their slower convergence impacts LRL more, because of its large basis. By contrast, LRH produces the same results as the CPA and BCPA2. This suggest that the estimation errors due to a slower model convergence are more critical than the ordering errors in this experiment. (Note that none of the models suffer from assumption errors in this section, since we simulate Hamming weight leakages). Finally, the GTA (in black) provides the best result overall. Since the GTA is not impacted by combination ordering errors nor by model convergence issues, this curve can be seen as an optimal reference.

---

[3] Using the ranking entropy lowers the impact of an outlier when averaging the result of many experiments. As an example, let's assume that an evaluator does 4 experiments where the key is ranked 1 three times and 2^{24} one time. The ranking entropy would be equal to 6 while the logarithm of the average ranks would be equal to 22, thus being more affected by the presence of an outlier.

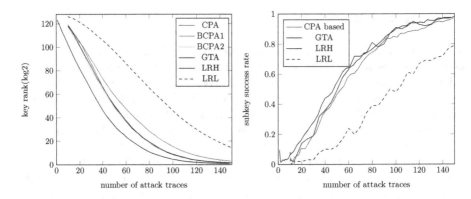

Fig. 4. Key rank (left) and success rate for the byte 0 (right) for constant noise variance in function of the number of attack traces for the different attack methods. CPA (blue), BCPA1 (red), BCPA2 (green), GTA (black), LRH (purple) and LRL (dashed purple). (Color figure online)

For completeness, Fig. 4 (right) shows the success rates of the different methods for the byte 0 (other key bytes lead to similar results in uur experiments were the SNR of different S-boxes was essentially constant). The single byte success rate does not suffer from ordering errors. Thus, this figure is a reference to compare the attack efficiencies before these errors occur. As expected, all the CPA methods (in blue) provide the same single byte success rate. More interestingly, we see that CPA methods, GTA (in black) and LRH (in purple) provide a quite similar success rate. This confirms that the differences in full key recovery for these methods are mainly due to ordering errors in our experiment. By contrast, LRL (in purple, dashed) provides the worst results because of a slower model estimation.

Overall, this constant noise (with correct assumptions) scenario suggests that applying a bayesian extension to CPA is not required as long as the S-boxes leak similarly (and therefore the combination ordering errors are low). In this case, model estimation errors are dominating for all the non-profiled attacks. (Of course, model assumption errors would also play a role if incorrect assumptions were considered).

3.2   Simulations with Different S-Box Leakages

We are now interested in the case where the noise level of each subkey is considerably different[4]. To simulate such a scenario, we chose very different noise levels in order to observe how these differences affect the key enumeration. Namely, we set the noise variances of the subkeys to $[20, 10, 5, 4, 2, 1, 0.67, 0.5, 0.33, 0.25, 0.2, 0.17, 0.14, 0.125, 0.11, 0.1]$. The results of the subkey recoveries for

---

[4] This could happen for example in a hardware implementation of a cryptographic algorithm, where each S-box lookup could involve different transistors.

each subkey will now differ because of the different noise levels. Figure 5 shows the rank of the different methods in function of the number of attack traces for this case. The impact of the combination ordering errors is now higher, as we can see from the gap between the different methods. BCPA1 is still worse than CPA, but this time with a bigger gap. Interestingly, we now clearly see an improvement when using the BCPA2 over the CPA (gain of roughly 2^{10} up to 40 attack traces). Moreover, the gap between BCPA2 and GTA remains the same as it was in the constant noise experiments as soon as the key rank is lower than 2^{80}. This confirms the good behavior of the BCPA2 method, which limits the impact of the ordering errors.

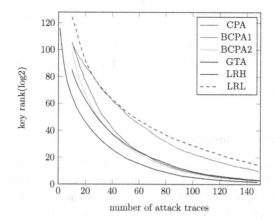

Fig. 5. Key rank (y coordinate) in function of the number of attack traces (x coordinate) with different noise variance for the different attack methods. CPA (blue), BCPA1 (red), BCPA2 (green), GTA (black), LRH (purple) and LRL (dashed purple). (Color figure online)

As for the linear regression, we again witness the advantage of a good starting assumption (with the difference between LRL and LRH). More interestingly, we see that for large key ranks, LRH leads to slightly better results than BCPA2, which suggests that combination ordering errors dominate for these ranks. By contrast, the two methods become again very similar when the rank decreases (which may be because of lower combination ordering errors or model convergence issues). And of course, GTA still works best.

We again provide the single bytes success rates for both byte 0 (left) and byte 8 (right) in Fig. 6. The gap between all the methods tends to be similar as in the constant noise case, where the CPAs, GTA and LRH perform similarly and LRL is less efficient. This confirms that the differences between the constant and different noise scenarios are due to ordering errors.

Overall, this experiment showed that the combination ordering errors can be significant in case the S-boxes leak differently. In this case, Bayesian extensions gain interest. BCPA2 provides a heuristic way to mitigate this drawback. Linear regression is a more systematic alternative (since it does not require assumptions in the distribution of the distinguisher).

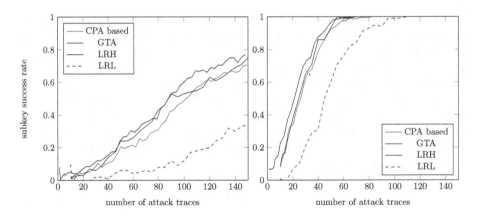

Fig. 6. Success rate (y coordinate) in function of the number of attack traces (x coordinate) for the byte 0 (left) and the byte 8 (right) (different noise variance case) for the different attack methods. CPA-like (blue), GTA (black), LRH (purple) and LRL (dashed purple). (Color figure online)

3.3   Actual Measurements

In order to validate our results, we ran actual attacks on an unprotected software implementation of the AES 128, implemented on a 32-bits ARM microcontroller (Cortex-M4) running at 100 MHz. We performed the trace acquisition using a Lecroy WaveRunner HRO 66 ZI oscilloscope running at 200 megasamples per second. We monitored the voltage variation using a 4.7 Ω resistor set in the supply circuit of the chip. We acquired 50,000 profiling traces using random plaintexts and keys. We also acquired 37,500 attack traces (150 traces per attack with 250 repetitions). For each AES execution, we triggered the measurement at the beginning of the encryption and recorded approximately the execution of one round. The attacks target univariate samples selected using a set of profiling traces as the one giving the maximum correlation with the Hamming weight of the S-box output.

Figure 7 again shows the key rank in function of the number of attack traces for the different attacks against the Cortex-M4 microcontroller. Interestingly, we can still see a small improvement when using the BCPA2 method over the CPA (around 2^5). Looking at the linear regression results suggests that the Hamming weight model is reasonably accurate in these experiments (since LRH is close to GTA). Yet, we noticed that the correlation values for each S-box varied by approximately 0.1 around an average value of 0.45. We conjecture that this 2^5 factor comes from small combination ordering errors. The figure also confirms the good behavior of the BCPA2 and LRH methods. By contrast, LRL is now even slower than BCPA1, which shows that the model estimation errors dominate in this case.

Once more, Fig. 8 shows the individual success rates for bytes 0 and 15. It confirms a that the different distinguishers behave in a similar way as in

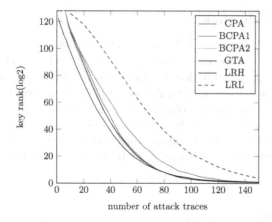

Fig. 7. Key rank (y coordinate) in function of the number of attack traces (x coordinate) on the cortex-m4 microcontroller for the different attack methods. CPA (blue), BCPA1 (red), BCPA2 (green), GTA (black), LRH (purple) and LRL (dashed purple). (Color figure online)

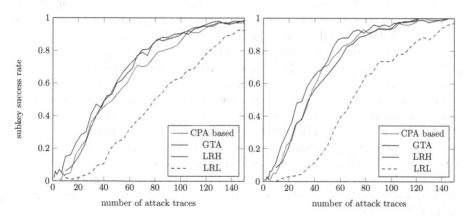

Fig. 8. Success rate (y coordinate) in function of the number of attack traces (x coordinate) for the byte 0 (left) and the byte 15 (right) (cortex-m4 microcontroller case) for the different attack methods. CPA-like (blue), GTA (black), LRH (purple) and LRL (dashed purple). (Color figure online)

simulations and therefore that the differences between the CPAs, GTA and LRH are again dominated by ordering errors. It also confirms that the errors from the LRL method come from model estimation.

Eventually, we stress again that the results would have been different if the real leakages were less close to the Hamming weight leakage function. In that case, model assumption errors would have additionally affected the efficiency of the non-profiled attacks.

3.4   Additional Heuristics

Before concluding, we note that various other types of heuristics could be considered to manipulate the scores produced by a CPA. For example, we assumed a multiplicative relation for those scores, but since we do not know how to combine them in a theoretically sound manner, one could also try to use other heuristics for the subkeys combinations in this case (such as summing the scores, summing the square of the scores, ...). We made some experiments in this direction, without any significant conclusions. These alternative heuristics sometimes came close to the efficiency of BCPA2 or LRH, but never provided better results.

4   Conclusions

Evaluating the security level of a leaking device is a complex task. Various types of errors can limit the confidence in the evaluation results, including the model assumption and estimation errors discussed in [9], and the combination ordering errors discussed in this paper. Overall, our results suggest that using GTA remains the method of choice to evaluate the worst-case security level of a leaking device (both to avoid incorrect a priori assumptions on the model and for optimal enumeration/rank estimation). Yet, good heuristics also exist for non-profiled attacks, in particular BCPA2 and LRH in our experiments.

Concretely, our results lead to the informal conclusion that model estimation and assumption errors usually dominate over combination ordering errors. This conclusion should be amplified for more complicated attacks (e.g. higher order attacks against masked implementations) where the model estimation is generally more challenging. So while being aware of combination ordering errors is important from a theoretical point-of-view, our experiments suggest that they are rarely the bottleneck in a security evaluation (which incidentally confirms previous works in the field, where enumeration and rank estimation were based on heuristic scores).

Acknowledgments. François-Xavier Standaert is a research associate of the Belgian Fund for Scientific Research (FNRS-F.R.S.). This work has been funded in parts by the CHIST-ERA project SECODE and by the ERC project 280141. Marios O. Choudary has been funded in part by University Politehnica of Bucharest, Excellence Research Grants Program, UPB – GEX 2016, contract number 17.

References

1. Bernstein, D.J., Lange, T., van Vredendaal, C.: Tighter, faster, simpler side-channel security evaluations beyond computing power. IACR Cryptology ePrint Archive, 2015:221 (2015)
2. Bogdanov, A., Kizhvatov, I., Manzoor, K., Tischhauser, E., Witteman, M.: Fast and memory-efficient key recovery in side-channel attacks. IACR Cryptology ePrint Archive, 2015:795 (2015)

3. Brier, E., Clavier, C., Olivier, F.: Correlation power analysis with a leakage model. In: Joye, M., Quisquater, J.-J. (eds.) CHES 2004. LNCS, vol. 3156, pp. 16–29. Springer, Heidelberg (2004). doi:10.1007/978-3-540-28632-5_2

4. Chari, S., Rao, J.R., Rohatgi, P.: Template attacks. In: Kaliski, B.S., Koç, K., Paar, C. (eds.) CHES 2002. LNCS, vol. 2523, pp. 13–28. Springer, Heidelberg (2003). doi:10.1007/3-540-36400-5_3

5. Choudary, O., Kuhn, M.G.: Efficient template attacks. In: Francillon, A., Rohatgi, P. (eds.) CARDIS 2013. LNCS, vol. 8419, pp. 253–270. Springer, Heidelberg (2014). doi:10.1007/978-3-319-08302-5_17

6. David, L., Wool, A.: A bounded-space near-optimal key enumeration algorithm for multi-dimensional side-channel attacks. IACR Cryptology ePrint Archive, 2015:1236 (2015)

7. Doget, J., Prouff, E., Rivain, M., Standaert, F.-X.: Univariate side channel attacks and leakage modeling. J. Cryptogr. Eng. 1(2), 123–144 (2011)

8. Duc, A., Faust, S., Standaert, F.-X.: Making masking security proofs concrete. In: Oswald, E., Fischlin, M. (eds.) EUROCRYPT 2015. LNCS, vol. 9056, pp. 401–429. Springer, Heidelberg (2015). doi:10.1007/978-3-662-46800-5_16

9. Durvaux, F., Standaert, F.-X., Veyrat-Charvillon, N.: How to certify the leakage of a chip? In: Nguyen, P.Q., Oswald, E. (eds.) EUROCRYPT 2014. LNCS, vol. 8441, pp. 459–476. Springer, Heidelberg (2014). doi:10.1007/978-3-642-55220-5_26

10. Glowacz, C., Grosso, V., Poussier, R., Schüth, J., Standaert, F.-X.: Simpler and more efficient rank estimation for side-channel security assessment. In: Leander, G. (ed.) FSE 2015. LNCS, vol. 9054, pp. 117–129. Springer, Heidelberg (2015). doi:10. 1007/978-3-662-48116-5_6

11. Kocher, P., Jaffe, J., Jun, B.: Differential power analysis. In: Wiener, M. (ed.) CRYPTO 1999. LNCS, vol. 1666, pp. 388–397. Springer, Heidelberg (1999). doi:10. 1007/3-540-48405-1_25

12. Longo, J., Martin, D.P., Mather, L., Oswald, E., Sach, B., Stam, M.: How low can you go? using side-channel data to enhance brute-force key recovery. Cryptology ePrint Archive, Report 2016/609 (2016). http://eprint.iacr.org/

13. Mangard, S.: Hardware countermeasures against DPA – a statistical analysis of their effectiveness. In: Okamoto, T. (ed.) CT-RSA 2004. LNCS, vol. 2964, pp. 222–235. Springer, Heidelberg (2004). doi:10.1007/978-3-540-24660-2_18

14. Mangard, S., Oswald, E., Popp, T.: Power Analysis Attacks - Revealing the Secrets of Smart Cards. Springer, Heidelberg (2007)

15. Mangard, S., Oswald, E., Standaert, F.-X.: One for all - all for one: unifying standard differential power analysis attacks. IET Inf. Secur. 5(2), 100–110 (2011)

16. Martin, D.P., Mather, L., Oswald, E., Stam, M.: Characterisation and estimation of the key rank distribution in the context of side channel evaluations. IACR Cryptology ePrint Archive, 2016:491 (2016)

17. Martin, D.P., O'Connell, J.F., Oswald, E., Stam, M.: Counting keys in parallel after a side channel attack. In: Iwata, T., Cheon, J.H. (eds.) ASIACRYPT 2015. LNCS, vol. 9453, pp. 313–337. Springer, Heidelberg (2015). doi:10.1007/ 978-3-662-48800-3_13

18. Poussier, R., Grosso, V., Standaert, F.-X.: Comparing approaches to rank estimation for side-channel security evaluations. In: Homma, N., Medwed, M. (eds.) CARDIS 2015. LNCS, vol. 9514, pp. 125–142. Springer, Heidelberg (2016). doi:10. 1007/978-3-319-31271-2_8

19. Poussier, R., Standaert, F.-X., Grosso, V.: Simple key enumeration (and rank estimation) using histograms: an integrated approach. IACR Cryptology ePrint Archive, 2016:571 (2016)

20. Schindler, W., Lemke, K., Paar, C.: A stochastic model for differential side channel cryptanalysis. In: Rao, J.R., Sunar, B. (eds.) CHES 2005. LNCS, vol. 3659, pp. 30–46. Springer, Heidelberg (2005). doi:10.1007/11545262_3

21. Veyrat-Charvillon, N., Gérard, B., Renauld, M., Standaert, F.-X.: An optimal key enumeration algorithm and its application to side-channel attacks. In: Knudsen, L.R., Wu, H. (eds.) SAC 2012. LNCS, vol. 7707, pp. 390–406. Springer, Heidelberg (2013). doi:10.1007/978-3-642-35999-6_25

22. Veyrat-Charvillon, N., Gérard, B., Standaert, F.-X.: Security evaluations beyond computing power. In: Johansson, T., Nguyen, P.Q. (eds.) EUROCRYPT 2013. LNCS, vol. 7881, pp. 126–141. Springer, Heidelberg (2013). doi:10.1007/978-3-642-38348-9_8

23. Ye, X., Eisenbarth, T., Martin, W.: Bounded, yet sufficient? How to determine whether limited side channel information enables key recovery. In: Joye, M., Moradi, A. (eds.) CARDIS 2014. LNCS, vol. 8968, pp. 215–232. Springer, Heidelberg (2015). doi:10.1007/978-3-319-16763-3_13

Analyzing the Shuffling Side-Channel Countermeasure for Lattice-Based Signatures

Peter Pessl[(✉)]

IAIK, Graz University of Technology, Graz, Austria
peter.pessl@iaik.tugraz.at

Abstract. Implementation security for lattice-based cryptography is still a vastly unexplored field. At CHES 2016, the very first side-channel attack on a lattice-based signature scheme was presented. Later, shuffling was proposed as an inexpensive means to protect the Gaussian sampling component against such attacks. However, the concrete effectiveness of this countermeasure has never been evaluated.

We change that by presenting an in-depth analysis of the shuffling countermeasure. Our analysis consists of two main parts. First, we perform a side-channel attack on a Gaussian sampler implementation. We combine templates with a recovery of data-dependent branches, which are inherent to samplers. We show that an adversary can realistically recover some samples with very high confidence.

Second, we present a new attack against the shuffling countermeasure in context of Gaussian sampling and lattice-based signatures. We do not attack the shuffling algorithm as such, but exploit differing distributions of certain variables. We give a broad analysis of our attack by considering multiple modeled SCA adversaries.

We show that a simple version of shuffling is not an effective countermeasure. With our attack, a profiled SCA adversary can recover the key by observing only 7 000 signatures. A second version of this countermeasure, which uses Gaussian convolution in conjunction with shuffling twice, can increase side-channel security and the number of required signatures significantly. Here, roughly 285 000 observations are needed for a successful attack. Yet, this number is still practical.

Keywords: Lattice-based cryptography · BLISS · Side-channel analysis · Countermeasures

1 Introduction

Quantum computers are a serious threat to a majority of currently in-use public-key cryptosystems. Although powerful enough quantum computers might not be available in the near future, their possible advent causes concerns and has already led to official recommendations from government bodies, such as the NSA [16,22]. Furthermore, standardization agencies are starting to investigate post-quantum alternatives [6]. Very recently, also Google began experimenting with post-quantum key-exchange algorithms in their Chrome browser [1,3].

© Springer International Publishing AG 2016
O. Dunkelman and S.K. Sanadhya (Eds.): INDOCRYPT 2016, LNCS 10095, pp. 153–170, 2016.
DOI: 10.1007/978-3-319-49890-4_9

Lattice-based cryptography is a very promising candidate for the post-quantum world. It proved to be very versatile and offers practical realizations of many public-key building blocks. When it comes to digital signatures, the Bimodal Lattice Signature Scheme (BLISS), which was presented by Ducas, Durmus, Lepoint, and Lyubashevsky [7] at CRYPTO 2013, is an attractive option. This is due to its efficiency both in terms of runtime and parameter sizes. Signature and public key sizes are in the range of current RSA moduli, which is a significant improvement over many earlier proposals.

There already exists a large body of work targeting efficient implementation of lattice-based primitives. Even when only considering BLISS, these range from hardware implementations [18] to microcontrollers [17,19] and PCs [23]. However, up until very recently the implementation-security aspect was pretty much neglected. The first side-channel attack on a lattice-based signature scheme, namely BLISS, was presented at CHES 2016 by Groot Bruinderink et al. [10]. They use a cache attack to recover some of the outputs of a Gaussian sampling algorithm. By combining information from multiple signatures and respective identified samples, they are able to recover the key. Note that Gaussian samplers play an integral part in most lattice-based schemes and their implementations. Hence, this type of attack might be applicable to a multitude of settings.

Shuffling was proposed by Saarinen [21] as a countermeasure against such an attack. Instead of securing the sampler itself, which would come at a hefty price, one could simply generate n Gaussian samples using an unprotected implementation and then randomly permute them. Shuffling is easy to implement and has a relatively low runtime overhead. This makes it especially attractive for use in low-resource devices, such as microcontrollers. However, the concrete security gains achieved by shuffling have thus far never been analyzed. As a consequence, convincing security arguments are still sorely lacking.

Our Contribution. In this paper, we tackle the above problem and present an in-depth analysis of shuffling in context of lattice-based signatures. Our analysis consists of two main parts, a side-channel analysis and a new attack on shuffling.

In the first part, we perform a side-channel attack on a Gaussian sampler implementation running on an ARM microcontroller. Our attack combines two methods. First, we recover the control flow of the sampling procedure. As samplers, including the one used by us, require data-dependent branches and are not inherently constant runtime, this already allows to narrow down the possible samples. And second, we use templates to uniquely identify the sampled value. While this attack is not able to identify all samples, it can recover certain values with very high confidence.

In the second part of our shuffling analysis, we present a new attack on the countermeasure. We perform an *un-shuffling*, i.e., reassign some recovered samples to the corresponding part of the signature output. After having collected enough matching pairs over multiple signatures, we can recover the private signing key. We stress that we do not attack the shuffling algorithm as such, we do not even consider its leakage in our analysis. Instead, we exploit the difference in distributions of Gaussian samples (high standard deviation) and a specific key-dependent intermediate (low standard deviation).

As we aim for a broad analysis, we evaluate this attack given several modeled side-channel adversaries. They are largely based on the previous side-channel analysis, but to test the theoretical boundaries of the countermeasure we also include an ideal attacker who is able to recover all samples. We also consider two different versions of the shuffling countermeasure. Our analysis shows that the simpler variant does not provide a noteworthy increase in side-channel security. Our ideal attacker succeeds using only 40 signatures. With 7 000 signatures, the modeled adversary which is closest to our side-channel analysis can also easily recover the key. However, the second shuffling version, which uses Gaussian convolution and shuffles twice, can increase the number of observed signatures required for an attack significantly. Yet, with around 260 000 signatures (for both mentioned adversaries) an attack is still practical and possible[1].

Finally, note that while we focus on BLISS, Gaussian sampling is required for most lattice-based schemes. Thus, the shuffling countermeasure and also our attack could be used for a wide range of implementations.

Outline. In Sect. 2, we recall BLISS, discrete Gaussians and proposed samplers. Then, in Sect. 3 we discuss previous work on SCA and countermeasures on BLISS. We evaluate the side-channel leakage of a concrete Gaussian sampler implementation in Sect. 4. Using the results of this side-channel analysis, we present an attack on the shuffling countermeasure and also discuss its outcome in Sect. 5. Finally, we conclude in Sect. 6.

2   BLISS and Gaussian Samplers

We now give a brief description of BLISS [7]. We then go on and describe the discrete Gaussian distribution and methods to sample from it.

2.1   BLISS - Bimodal Lattice Signatures

The most efficient instantiation of BLISS works with polynomials over the ring $\mathcal{R}_q = \mathbb{Z}_q[x]/\langle x^n + 1 \rangle$. We will later use the fact that the multiplication of two polynomials $\mathbf{a}, \mathbf{b} \in \mathcal{R}_q$ can be written as a matrix-vector product, i.e., $\mathbf{ab} = \mathbf{aB} = \mathbf{bA}$, where the columns of matrices $\mathbf{A}, \mathbf{B}$ are negacyclic rotations of $\mathbf{a}$ and $\mathbf{b}$, respectively.

Key generation and signature verification do not play a role in our later analysis, here we refer to [7]. Signature generation is described in Algorithm 1. It takes as input a message μ, a public key $\mathbf{A}$, and a private key $\mathbf{S} = (\mathbf{s}_1, \mathbf{s}_2)$, with $\mathbf{s}_1$ a polynomial with exactly $\delta_1 n$ coefficients in $\{\pm 1\}$, $\delta_2 n$ coefficients in $\{\pm 2\}$, and all other elements being 0. First, two noise polynomials $\mathbf{y}_1, \mathbf{y}_2$ are sampled from a discrete Gaussian distribution D_σ. The intermediate $\mathbf{u}$ is hashed together with the message, where H outputs a bit vector $\mathbf{c}$ of length n and (small) hamming

---
[1] These numbers assume recoverability of bit b for each signature (cf. Sect. 5.3).

Algorithm 1. BLISS Signature Algorithm

Input: Message μ, public key $\mathbf{A} = (\mathbf{a}_1, q - 2)$, private key $\mathbf{S} = (\mathbf{s}_1, \mathbf{s}_2)$
Output: A signature $(\mathbf{z}_1, \mathbf{z}_2^\dagger, \mathbf{c})$
1: $\mathbf{y}_1 \leftarrow D_\sigma^n, \mathbf{y}_2 \leftarrow D_\sigma^n$
2: $\mathbf{u} = \zeta \cdot \mathbf{a}_1 \mathbf{y}_1 + \mathbf{y}_2 \bmod 2q$
3: $\mathbf{c} = \mathrm{H}(\lfloor \mathbf{u} \rceil_d \bmod p || \mu)$
4: Sample a uniformly random bit b
5: $\mathbf{z}_1 = \mathbf{y}_1 + (-1)^b \mathbf{s}_1 \mathbf{c}$
6: $\mathbf{z}_2 = \mathbf{y}_2 + (-1)^b \mathbf{s}_2 \mathbf{c}$
7: Continue with some probability $f(\mathbf{Sc}, \mathbf{z})$, restart otherwise (see [7])
8: $\mathbf{z}_2^\dagger = (\lfloor \mathbf{u} \rceil_d - \lfloor \mathbf{u} - \mathbf{z}_2 \rceil_d)$
9: **return** $(\mathbf{z}_1, \mathbf{z}_2^\dagger, \mathbf{c})$

---

weight κ. The noise polynomials are then added to $\mathbf{s}_1 \mathbf{c}$ and $\mathbf{s}_2 \mathbf{c}$, respectively. A subsequent rejection-sampling step prevents leakage of the key. Finally, the compressed signature is returned[2]. Throughout this paper, we use the BLISS-I parameter set [7] given in Table 1. It provides a security level of 128 bit.

Table 1. BLISS-I parameter set

n	q	σ	δ_1, δ_2	κ	d
512	12289	215.73	0.3, 0	23	10

2.2 Discrete Gaussians

We denote with D_σ the discrete Gaussian distribution with standard deviation σ and zero mean; we use $y \leftarrow D_\sigma$ for variables sampled from this distribution. The probability-mass function $D_\sigma(x) = \rho_\sigma(x)/\rho_\sigma(\mathbb{Z})$, with $\rho_\sigma(x) = \exp(\frac{-x^2}{2\sigma^2})$ and the normalization constant $\rho_\sigma(\mathbb{Z}) = \sum_{k=-\infty}^{\infty} \rho_\sigma(k)$. With D_σ^n, we denote the n-dimensional extension. Samples from D_σ^n can simply be generated by independently sampling n times from D_σ.

Implementation of Gaussian Samplers. The emergence of lattice-based cryptography and its reliance on discrete Gaussian noise led to a large number of proposed sampler architectures. Apart from generic methods like rejection sampling and inversion sampling, these also include, e.g., the Knuth-Yao random walk [8], the Ziggurat method [4], and arithmetic coding [21].

Compared to lattice-based public-key encryption [14], the standard deviation required for BLISS is relatively high. This makes samplers requiring large precomputed tables less attractive, especially for constrained devices and their usually low storage capacities. For this reason, Pöppelmann et al. [18] proposed an optimized sampler which is based on the inversion method. Since their approach is tailored for low-resource devices and also an ideal candidate for use with

---

[2] The constants ζ, d, p are used for compression purposes. For details, see [7].

the shuffling countermeasure, we use their algorithm in our work and now give a more detailed description.

For inversion sampling, one first precomputes a cumulative distribution table (CDT), i.e., a table $T[y] = \mathrm{P}(x < y | x \leftarrow D_\sigma^+)$ for $y \in [0, \tau\sigma]$. Here, τ denotes the tail-cut factor which is required due to the infinite support of D_σ. Thanks to symmetry of D_σ, sampling can be easily reduced to sampling from the one-sided distribution D_σ^+ with support $[0, \tau\sigma]$ and then sampling a random sign bit. As the statistical distance to a true discrete Gaussian must be kept low, the entries of T need to be stored with a very high precision, e.g., 128 bit.

For actual sampling, one generates a uniformly random $r \in [0, 1)$ and returns the y satisfying $T[y] \leq r < T[y + 1]$ (using a binary search in T). To reduce the table size and speed up sampling, Pöppelmann et al. propose the following optimizations. They save memory by using Gaussian convolution. They set $k = 11$, $\sigma' = \sigma/\sqrt{1 + k^2} \approx 19.53$ and sample two values $y', y'' \leftarrow D_{\sigma'}$. They then combine them to $y \leftarrow D_\sigma$ by setting $y = ky' + y''$. Furthermore, they speed up sampling by using a byte-oriented guide table I. Each entry $I[r_0]$ stores the smallest interval $(\min_{r_0}, \max_{r_0})$ with $T[\min_{r_0}] \leq r_0/256$ and $T[\max_{r_0}] \geq (r_0 + 1)/256$. By using this table, the range for the following binary search can be immediately reduced to the interval $[\min_{r_0}, \max_{r_0})$.

The detailed sampling procedure is given in Algorithm 2. It uses a byte-wise approach, where $T_j[i]$ denotes the j-th byte of $T[i]$. To save memory, the table T is stored in floating-point representation, using a mantissa table M and an exponent table E. For efficiency reasons Pöppelmann et al. actually store $T[y] = \mathrm{P}(x \geq y | x \leftarrow D_\sigma^+)$, i.e., $T[0] = 1$ and $T[y] > T[y + 1]$. This is accounted for in the binary-search part. For further explanations we refer to [18].

Algorithm 2. CDT Sampler using Guide Tables [18]

Input: Guide table I, mantissa table M, exponent table E
Output: A value y' sampled according to $D_{\sigma'}$
1: Sample a uniformly random byte r_0
2: $[\min, \max] = I[r_0]$
3: $i = (\min + \max)/2$, $j = 0$, $k = 0$
4: **while** max-min > 1 **do**
5:     $t = T_j[i]$, with $T_j[i] = M_{j - E[i]}[i]$ or 0
6:     **if** $t > r_j$ **then**
7:         $\min = i$, $i = (i + \max)/2$, $j = 0$
8:     **else if** $t < r_j$ **then**
9:         $\max = i$, $i = (\min + i)/2$, $j = 0$
10:     **else**
11:         $j = j + 1$
12:         **if** $k < j$ **then**
13:             Sample uniformly random byte r_j, $k = j$
14: Sample a uniformly random bit s
15: **if** s **then return** $-i$
16: **else return** i

---

3   Side-Channel Attacks and Countermeasures for Gaussian Sampling

When analyzing the components of BLISS for side-channel weaknesses, the Gaussian sampler appears to be a critical and especially hard to protect part. To the best of our knowledge, none of the samplers given in the previous section inherently feature a constant runtime or a complete absence of data-dependent branches. Thus, it should not come as a huge surprise that the first reported side-channel attack on lattice-based signatures, which we will now discuss, targets samplers. We will then also state possible countermeasures, including shuffling.

3.1   A Cache Attack on BLISS

At CHES 2016, Groot Bruinderink et al. [10] presented the first side-channel attack on BLISS. They perform a cache attack, i.e., observe time differences caused by the CPU cache, to partially recover the Gaussian noise vector $\mathbf{y}_1$. They analyze the susceptibility of two sampler implementations to such attacks in depth[3].

Their attack proceeds as follows. First, they need to observe the creation of multiple signatures $(\mathbf{z}_j, \mathbf{c}_j)$, where $\mathbf{z}_j$ refers to only the first signature polynomial $\mathbf{z}_1$ of the j-th signature. They then focus on line 5 of Algorithm 1, i.e., $\mathbf{z}_1 = \mathbf{y}_1 + (-1)^b \mathbf{s}_1 \mathbf{c}$. For each recovered Gaussian sample, an attacker can create an equation of form:

$$z_{ji} = y_{ji} + (-1)^{b_j} \langle \mathbf{s}_1, \mathbf{c}_{ji} \rangle \tag{1}$$

Here, i denotes the index of the recovered Gaussian sample in the signature. z_{ji} and y_{ji} are the i-th coefficients of $\mathbf{z}_1$ and $\mathbf{y}_1$ in the j-th signature. $\mathbf{c}_{ji}$ denotes the i-th column of $\mathbf{C}_j$, which is the matrix used in the matrix-vector representation of polynomial multiplication.

The cache attack does not reveal the random but secret bit b. Therefore, Groot Bruinderink et al. keep only those equations which satisfy $z_{ji} = y_{ji}$, i.e., where $\langle \mathbf{s}_1, \mathbf{c}_{ji} \rangle = 0$ and thus the value of b is irrelevant. They then build a matrix $\mathbf{L}$ where the columns are the filtered $\mathbf{c}_{ji}$. This matrix satisfies $\mathbf{s}_1 \mathbf{L} = 0$. The key $\mathbf{s}_1$ can then be found in the kernel-space of $\mathbf{L}$. $\mathbf{s}_2$ can be reconstructed by using the relation between public and private key.

Groot Bruinderink et al. also consider a scenario where the information on the samples is not exact, but instead a small error is possible. Here, they formulate a lattice problem and use lattice-reduction techniques for key recovery.

3.2   Countermeasures

Protecting samplers from attacks like the one above seems to be difficult. While there exist methods for constant-runtime and protected sampling, they come at a hefty performance impact (see, e.g., [2]). Also, there do exist alternatives

---

[3] Further sampler architectures are discussed in the full version of [10].

to using (high-precision) Gaussian noise. However, they either do not apply to signature schemes [1] or they are suboptimal in terms of security or signature size [7,11].

Instead of protecting the sampler itself, one could also simply use an unprotected (or somewhat protected) sampler implementation to generate n samples and then randomly permute them. This breaks the connection between time of sampling and index in the signature and thus makes attacks more difficult. This shuffling countermeasure was first proposed by Roy et al. [20], albeit in the context of lattice-based public-key encryption. Recently, Saarinen [21] proposed a variant that uses shuffling twice (in conjunction with Gaussian convolution) for use in BLISS. However, neither Roy nor Saarinen provided an analysis of this countermeasure. Thus, its true effectiveness has still been unknown. Below we describe the simple and the two-stage shuffling approaches, where we use Gaussian convolution for both setups. We will later evaluate these two shuffling variants in terms of security.

Single-Stage Shuffling: $\mathbf{y}', \mathbf{y}'' \leftarrow D_{\sigma'}^n$, $\mathbf{y} = \text{Shuffle}(k\mathbf{y}' + \mathbf{y}'')$
Two-Stage Shuffling: $\mathbf{y}', \mathbf{y}'' \leftarrow D_{\sigma'}^n$, $\mathbf{y} = k \cdot \text{Shuffle}(\mathbf{y}') + \text{Shuffle}(\mathbf{y}'')$

4 A Side-Channel Attack on a Gaussian Sampler

Before evaluating the shuffling countermeasure, it is important to understand how much information on Gaussian samples a side-channel attacker can realistically expect. For this reason, we now present a side-channel analysis of a sampler implementation. Recall that Gaussian sampling is a random process that does not involve any keying material. Also, its output is typically used only once. Hence, we are limited to single-trace SPA-style attacks.

4.1 Implementation and Measurement Setup

For our experiments, we implemented the Gaussian sampling procedure proposed by Pöppelmann et al. [18] in software. The contents of all required lookup-tables are directly taken from their open-sourced BLISS FPGA implementation. Note that our analysis focuses solely on sampling from $D_{\sigma'}$ (Algorithm 2), i.e., we do not use any leakage stemming from the Gaussian convolution step.

As a target platform, we chose a Texas Instruments MSP432 (ARM Cortex-M4F) microcontroller on a MSP432P401R LaunchPad development board[4]. For pseudo-random number generation we used the on-chip hardware AES accelerator in counter mode. While this setup is likely susceptible to DPA attacks [12,15], we do not use any leakage of the AES execution.

In our attack we exploit the EM side channel. As shown in Fig. 1, we placed a Langer RF-B 3-2 near-field probe in proximity to the external core-voltage regulation circuitry. Note that for this setup, no spatial profiling of on-chip EM leakage is required. Also, we expect the results of power measurements to be

---

[4] The design files of this development board are available online [24].

Fig. 1. Measurement setup. The EM probe is placed directly to the left of the external core-voltage regulation circuitry.

somewhat similar. For our evaluation, we use a dedicated trigger that signals the start of a sampling procedure. Real-world attackers do not have this option and need to detect the 1024 calls to Algorithm 2 required for sampling $\mathbf{y}_1$. Such adversaries can use, e.g., trace alignment in combination with the methods described in the next section.

4.2 Reconstructing the Control Flow

When analyzing Algorithm 2, it becomes obvious that the data-dependent branches offer a lot of information on the sampled value. In fact, the return value can be uniquely determined by the first random byte r_0 and the control flow.

We recover the control flow using a trace-matching approach. For each possible conditional jump, we record a reference trace by computing the mean of multiple profiling traces at select points in time (in some cases just a single point) near the first occurrence of this branch. During the attack, we then compare these references to the attack trace by computing the mean of squared differences. Figure 2 illustrates that for some branches, the most information lies within a time shift of subsequent operations. In these cases, we use a single reference and match them at both locations. We then use the case with the lowest score. We repeat this matching process until the algorithm exits. The position of the respective next matching process is determined on basis of the previously taken branches. The final branch detection then also reveals the sign of the sampled value.

With the described method, we can reconstruct the control flow with perfect accuracy. This should not be surprising, when, e.g., observing the huge trace differences illustrated in Fig. 2.

Note that, while we use device profiling for deriving the reference traces, there exist non-profiled alternatives. An attacker could, e.g., build the references on the fly after a visual inspection of a limited number of traces. Alternatively, he could use a clustering approach for determining the branches.

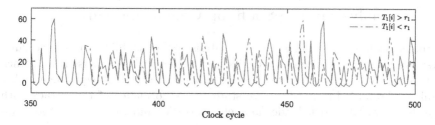

Fig. 2. Demonstration of a timing difference stemming from a branch inside the first loop iteration. After around cycle 420, the trace for $T_1[i] > r_1$ (blue, solid) trails by 8 clock cycles. (Color figure online)

4.3 Determining the Sampled Values via Templates

In order to uniquely determine the sampled value, we recover the value of r_0 using a template attack [5]. For each possible control flow (up to a certain depth), we built templates for each value of r_0 that can potentially result in this flow. The points-of-interest for the attack were determined using a t-test, as proposed by Gierlichs et al. [9]. We limited the maximum number of used points to 8.

The outcome of the template attack is depicted in Fig. 3. There we show a histogram of the maximum classification probabilities. In our implementation, the guide-table lookup already yields the final sample for 206 values of r_0. As seen in Fig. 3a, we cannot determine the correct samples with high confidence in these cases. As our later analysis on the shuffling countermeasure requires such a high confidence, we have to discard these samples. This situation changes in cases that require a single comparison step in the binary-search algorithm, Fig. 3b shows that 6.5% of these samples can be determined with probability close to 1.

If more than a single comparison is required, then the template attack can recover the sampled value almost perfectly. The overall success rate here is 99.5%. If we discard the 1% of samples whose probability is lower than 0.90, then the success rate reaches 99.9%.

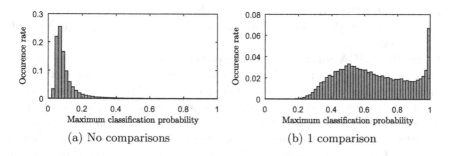

(a) No comparisons

(b) 1 comparison

Fig. 3. Results of the template attacks for no or 1 comparison

5   An Analysis of the Shuffling Countermeasure

In this section, we give an in-depth analysis of the shuffling countermeasure. First, we give a brief discussion on its cost. Afterwards, we present an attack that can circumvent this countermeasure, albeit at the cost of requiring a higher number of recorded signatures. We state the performance of this attack with regards to several modeled side-channel adversaries and variations of the countermeasure.

5.1   Cost

We evaluated the cost of shuffling by implementing the Fisher-Yates shuffling algorithm [13]. When run at 48 MHz, which is the maximum for our MSP432 evaluation platform, shuffling a vector of $n = 512$ entries took 1.5 ms. For comparison, sampling an element from $D_{\sigma'}^n$, which requires 512 calls to Algorithm 2, needs about 2.5 ms. For creating a single signature, 4 elements of $D_{\sigma'}^n$ need to be sampled. The shuffling operation is called either 2 or 4 times, depending on whether single-stage or two-stage shuffling is used. In the latter case, the total runtime of sampling is increased by 57%, which is still relatively little when it comes to SCA countermeasures.

5.2   Considered Attackers

In order to allow a broad analysis of the shuffling countermeasure and to achieve easier reproducibility, we do not directly use the outcome of the attack described in Sect. 4. Instead, we use the results as a basis to model three side-channel adversaries. Each one is based on a different assumption on his capabilities. Note that all following descriptions are in context of sampling from the "small" $D_{\sigma'}$ and thus Algorithm 2, which is called 2048 times during signature generation. We do not use any leakage from the multiplication with k, the addition for Gaussian convolution, and even the shuffling algorithm itself. We do so to keep the analysis as generic and implementation-independent as possible.

A1 - perfect SCA adversary. This attacker is able to recover all generated samples. We use this adversary to evaluate the theoretical limits of the shuffling countermeasure.
A2 - profiled SCA adversary. This attacker is able to profile the device and perform a template attack. We assume that the attacker can correctly determine the entire control flow and is able to correctly classify all samples which required at least 2 comparisons in the binary-search step. For the analysis, we make a further simplification and only use samples with absolute above a certain threshold. This threshold is set so that all samples larger than it require at least 2 comparisons. All samples at and below the threshold are considered to be unknown.

A3 - non-profiled SCA adversary. This attacker is not able to profile and thus cannot perform a template attack. However, he is still able to reconstruct the control flow. All samples which are not uniquely determined by the control flow are considered to be unknown.

Adversary A2 is closest to the side-channel analysis given in the previous section. However, in this model we do not use any potentially classified samples which used only a single comparison (cf. Figure 3b) or the small portion of samples requiring 2 comparisons but being below the threshold. In return, we also ignore the very small error probability and assume that all reconstructed samples are correct.

For our particular BLISS parameter set and sampler implementation, we have the following concrete implications. For A2, the above mentioned threshold is 47, i.e., we say that the adversary can correctly classify all samples with absolute value larger than 47. Approximately 1.5% of the samples from $D_{\sigma'}$ meet this restriction. The adversary A3 can correctly classify all samples with absolute value larger than 54, which amounts to only 0.53% of all samples.

For each modeled adversary, we also evaluate two sub-scenarios with regards to the secret bit b used for computing $(-1)^b \mathbf{s}_1 \mathbf{c}$. First, we consider the case that the adversary can recover this bit with side-channel measurements using, e.g., methods akin to Sect. 4.2. And second, we also evaluate the case that this bit is unknown.

5.3   Attack Without Shuffling

For key recovery, we use the relation also exploited by Groot Bruinderink et al. (cf. Sect. 3.1). We gather equations of the form $z_{ji} = y_{ji} + (-1)^{b_j} \langle \mathbf{s}_1, \mathbf{c}_{ji} \rangle$ and then solve the resulting linear system. We do not consider error correction and require that these equations are correct.

If the entire $\mathbf{y}_1$ is known, which is the case for adversary A1, and no shuffling is used, then key recovery is trivial and requires only a single signature. $\mathbf{s}_1$ (or $-\mathbf{s}_1$) can be computed by solving the linear system given as $\mathbf{z}_1 - \mathbf{y}_1 = (-1)^b \mathbf{s}_1 \mathbf{c}$ for any value of b. Attackers A2 and A3 require multiple signatures in order to recover the key, even in the non-shuffled scenario. As our sampling procedure combines two samples $y', y'' \leftarrow D_{\sigma'}$ to $y = ky' + y''$, we can only recover samples y where the side-channel information reveals both y' and y''. Hence, A2 can recover a portion of $0.015^2 \approx 2.2 \cdot 10^{-4}$ of all samples, whereas for A3 this quantity decreases to $2.2 \cdot 10^{-5}$.

If the b_j are recoverable by using side-channel information, then we can combine n equations $z_{ji} = y_{ji} + (-1)^{b_j} \langle \mathbf{s}_1, \mathbf{c}_{ji} \rangle$ into a linear system which can then simply be solved for the key $\mathbf{s}_1$. The expected number of signatures required to gather $n = 512$ classified samples and corresponding signature values is roughly 4 400 for A2 and 36 000 for A3. Note that in this non-shuffled scenario, the differences between A2 and the SCA from Sect. 4, i.e., not using all classifiable samples, have a significant impact. With that data we would require only around 1 000 signatures to mount this attack.

If the b_j are unknown, then, like also done in [10], we only use those equations where $z_{ji} = y_{ji}$ and thus $\langle s_1, c_{ji} \rangle = 0$. Then, one can search for the key s in the kernel of the matrix composed by the c_{ji}. As seen in Fig. 4a, $\langle s_1, c_{ji} \rangle = 0$ holds for about 15% of all samples, thus the number of required traces needs to be multiplied by 6.6. Hence, 29 000 and 239 000 signatures are required for A2 and A3, respectively. In the remainder of this paper, we give the signature requirements for both cases of b. We state the requirements with a known b first, followed by the unknown case in parentheses.

5.4   An Attack on Shuffling - Basic Concept

If the elements of y_1 are shuffled after sampling, then the above attack is not directly applicable. To still use it, we first need to do an *un-shuffling*, i.e., we need to re-assign recovered Gaussian samples to their respective index in the signature and thus to the correct $z_i \in z_1$.

We do that by exploiting the differing (coefficient-wise) distributions of $s_1 c$ and y_1, they are shown in Fig. 4. The distribution of $s_1 c$, which we denote with $\mathcal{X}_{sc}$, was estimated using a histogram approach, whereas y_1 follows D_σ^n. Observe that the standard deviation of y_1 is much larger than that of $s_1 c$. Thus, we can say that $z_1 \approx y_1$.

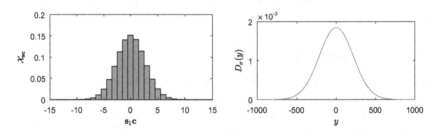

Fig. 4. Comparison of the coefficient-wise distribution of $s_1 c$ ($\mathcal{X}_{sc}$) and y (D_σ^n)

We use this relation as basis of our attack. If we know one particular coefficient y of y_1 but not its position due to shuffling, then we can test all coefficients of the public z_1 for proximity to y. If only a single $z_i \in z_1$ is "close" to y, then we can assign y to the position of z_i and compute $z_i - y$ to retrieve the value of $(-1)^b \langle s_1, c_i \rangle$. As actual metric for closeness, we use $\mathcal{X}_{sc}(z_i - y)$. Observe that this approach is expected to succeed mostly for large absolute values of y and thus z_i, i.e., in the tail of D_σ. Due to the high dimension $n = 512$, there will be many similar values of y and z_i near the center, thus a unique assignment will not be possible in those cases.

5.5   Attack Details

The previous description of our attack is relatively informal, we now give a more in-depth explanation. For now, consider the case of single-stage shuffling, we adapt the approaches to the two-stage variant later on.

Given two values z_i and y, we define $z_i \sim y$ as the event that z_i and y belong to the same index i in the signature. Without considering knowledge of other processed values, we have a likelihood $P(z_i \sim y) = \mathcal{X}_{sc}(z_i - y)$.

When now given the public $\mathbf{z}_1$ and a single sample y of $\mathbf{y}_1$, we can compute, for each $z_i \in \mathbf{z}_1$, $P(z_i \sim y | \mathbf{z}_1)$. We do that by using Bayes' theorem with uniform prior, i.e., $P(z_i \sim y | \mathbf{z}_1) = P(z_i - y) / \sum_{z_j \in \mathbf{z}_1} P(z_j - y)$. Analogously, for a single z_i and a fully reconstructed but shuffled $\mathbf{y}_1$, we can compute $P(z_i \sim y_j | \mathbf{y}_1)$.

We perform this analysis on every possible combination of y and z_i. Thus, we compute a likelihood matrix $\mathbf{L} \in (n \times n)$, with $\mathbf{L}_{i,j} = \mathcal{X}_{sc}(z_i - y_j)$. Afterwards, we apply the Bayesian step to both the columns and the rows of this matrix in order to derive the aforementioned conditional probabilities. Then, we combine both normalized matrices by taking their maximum, i.e., we set $P(z_i \sim y_j) = \max(P(z_i \sim y_j | \mathbf{z}_1), P(z_i \sim y_j | \mathbf{y}_1))$. In other words, we both search for z_i that fit to only one y, and y that fit to only one z_i. Finally, for each $z_i \in \mathbf{z}_1$, we pick the most likely y as $\mathrm{argmax}_{y_j} P(z_i \sim y_j)$. This shuffling analysis is repeated for each recorded signature.

The previously discussed key-recovery algorithm requires errorless information. Thus, we keep only pairs of (z_i, y) that match with very high probability; we set the threshold to 0.99. Even so, matching errors cannot be entirely excluded. However, a small number of errors can be corrected by gathering slightly more than $n = 512$ equations, and then performing the key-recovery procedure multiple times, each time using a random subset of the collected equations. Alternatively, one could use the lattice-based techniques discussed by Groot Bruinderink et al. [10].

Merging Equal y. For key-recovery we compute $z_i - y$ for each recovered pair (z_i, y). Here, the actual *index* of y is irrelevant, only the *value* of y needs to be correct. Consequently, if $\mathbf{y}_1$ contains multiple copies of the same value, then they can be treated as a single entity.

We use this observation as follows. We create a vector $\mathbf{u}$ which contains the unique elements of $\mathbf{y}_1$. We then compute $P(z_i \sim u_j | \mathbf{u})$. For that, we use the number of times each u_j appears in $\mathbf{u}$ as prior probabilities (instead of the uniform distribution). For $\mathbf{y} \leftarrow D_\sigma$, the average number of unique elements in $\mathbf{y}$ is 377. For $\mathbf{y}' \leftarrow D_{\sigma'}$ only 92 elements are unique on average. Especially in the latter case, the merging of equal y' increases the rate of matches and also decreases the computation time of the subsequent analysis. From now on, we will always use this optimization implicitly.

Note that merging equal values of $\mathbf{z}_1$ is not useful. As already hinted by always using subscripts, each z_i is coupled to one specific $\mathbf{c}_i$, i.e., a negacyclic rotation of the signature part $\mathbf{c}$.

Results on Single-Stage Shuffling. We evaluated our described attack against (single-stage) shuffling by running it with 2^{20} signatures. The results for A1 are shown in Fig. 5. 2.5% of all samples match with a probability of at least 0.99. With this number, only 40 (264) signatures are required to gather $n = 512$ equations. As expected, the successfully matched y lie in the tail of D'_σ (Fig. 5b).

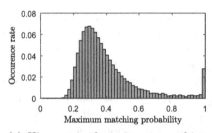
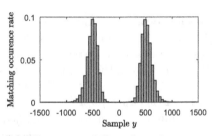

(a) Histogram of maximum matching probabilities.

(b) Histogram of the value of successfully matched y

Fig. 5. Result for the attack on single-stage shuffling, attacker A1

For A2 and A3, we do not know the entire $\mathbf{y}_1$ and so did not compute $P(z_i \sim y_j | \mathbf{y}_1)$. When only using $P(z_i \sim y | \mathbf{z}_1)$, we can match a proportion of $1.4 \cdot 10^{-4}$ (A2) and $2.2 \cdot 10^{-5}$ (A3) of all samples. This translates to requiring 7 000 (46 000) and 46 000 (301 000) signatures, respectively. The number of expected errors in $n = 512$ equations is well below 1 for all considered adversaries.

When compared to the signature requirements without shuffling, one can observe only a marginal increase. All numbers are well low enough to be practical, thus shuffling once is not an effective countermeasure.

5.6   Adaptation to Two-Stage Shuffling

For two-stage shuffling, the $\mathbf{y}', \mathbf{y}''$ are independently permuted. Thus, we cannot compute any elements of $\mathbf{y}_1$ in straight-forward manner which makes the attack from above not directly applicable. A similar one, however, is still possible; we now state the required modifications. As the sampling (and shuffling) process proceeds in two steps, we also adapt a two-stage approach in the attack.

Assume we are given $\mathbf{z}_1$ and the *shuffled* $\mathbf{y}', \mathbf{y}''$, with $\mathbf{y}_1 = k\mathbf{y}' + \mathbf{y}''$ and hence $\mathbf{z}_1 = k\mathbf{y}' + \mathbf{y}'' + (-1)^b \mathbf{s}_1 \mathbf{c}$. We first aim at finding matching pairs for elements of $\mathbf{z}_1$ and $\mathbf{y}'$. Afterwards, for each pair (z_i, y'), we compute $z_i - k y'$ and then match this difference with the elements of the second vector $\mathbf{y}''$. We now explain the details of this process.

First Stage. The first part differs from the previous attack mainly as in we now test the proximity of elements of $\mathbf{z}_1$ to those of $k\mathbf{y}'$. As $\mathbf{z}_1 - k\mathbf{y}' = \mathbf{y}'' + (-1)^b \mathbf{s}_1 \mathbf{c}$, we cannot test proximity with regards to $\mathcal{X}_{sc}$. Instead, we could use the distribution of an $x = x_1 + x_2$, with $x_1 \leftarrow D_{\sigma'}$ and $x_2 \leftarrow \mathcal{X}_{sc}$. We denote it as $\mathcal{X}_{sc+D_{\sigma'}}$. However, as the attacker has (at least partial) knowledge on $\mathbf{y}''$, this would be suboptimal. Hence, we set (with some abuse of notation) $x_2 \leftarrow \mathbf{y}''$, i.e., randomly chosen elements from $\mathbf{y}''$. We call the resulting distribution $\mathcal{X}_{sc+\mathbf{y}''}$ and use it to fill our likelihood matrix $\mathbf{L}$ with $\mathbf{L}_{i,j} = \mathcal{X}_{sc+\mathbf{y}''}(z_i - k y'_j)$. The remainder of the analysis, i.e., the Bayesian steps and picking the maximum, are then the same. Finally, all samples matched with probability greater than 0.99 are fed to the second stage of the recovery.

For A2 and A3, we require additional modifications. First, we cannot compute $\mathcal{X}_{\mathbf{sc}+\mathbf{y}''}$, as $\mathbf{y}''$ is only partially known. Instead, we construct a hybrid distribution that merges $D_{\sigma'}$ (for all unknown samples up to the threshold of 47 and 54, respectively) and the known samples of $\mathbf{y}''$. Then, unlike in the single-stage attack, we would also like to compute (or rather estimate) $P(z_i \sim y_j'|\mathbf{y}')$ despite not having the full $\mathbf{y}'$. We do so by introducing a dummy sample y_d', which represents all (unknown) samples below the model threshold. Thus, we test if z_i matches with any of the known y_j' or with any element below the threshold. We set the likelihood of y_d' as in (2), the remaining steps are then equivalent to those of A1.

$$P(z_i \sim y_d') = \sum_{y=-\text{threshold}}^{\text{threshold}} \mathcal{X}_{\mathbf{sc}+D_{\sigma'}}(z_i - y) \qquad (2)$$

Second Stage. In the second stage, we test each pair (z_i, y_i') found in the previous stage with the elements of $\mathbf{y}''$. We do so by computing $z_i - ky_i'$ and then testing for proximity to the elements of $\mathbf{y}''$ with regards to $\mathcal{X}_{\mathbf{sc}}$.

Even for A1, the expected number of matched pairs per signature in the first stage is relatively small. Thus, we cannot compute $P((z_i - ky_i') \sim y_j''|(\mathbf{z}_1 - k\mathbf{y}'))$ and are left with $P((z_i - ky_i') \sim y_j''|\mathbf{y}'')$. Like in the first stage, A2 and A3 only have partial knowledge of $\mathbf{y}''$. We use the same trick as above and introduce a dummy sample y_d'' representing all elements below the modeled threshold of 47 and 55, respectively. Here we use (3) and then again perform the Bayesian step and a filtering of the most probable matches.

$$P((z_i - ky_i') \sim y_d'') = \sum_{y=-\text{threshold}}^{\text{threshold}} \mathcal{X}_{\mathbf{sc}}(z_i - ky_i' - y) \qquad (3)$$

Results on Two-Stage Shuffling. Like earlier, we evaluated our described attack against two-stage shuffling by running it with 2^{20} signatures. For our ideal adversary A1, we can match 0.26% of samples in the first stage (with probability greater than 0.99). Out of the found pairs, 0.15% can also be matched in the second stage. This results in requiring 260 000 (1 710 000) signatures in order to find $n = 512$ equations.

Interestingly, the losses incurred by the restrictions of A2 are relatively small. We can match 0.25% in the first and 0.15% of samples in the second stage. With 285 000 (1 880 000), the number of required signatures is virtually identical to the previous case. As to be expected, A3 performs worse. The matching rates decrease to 0.18% and 0.10%, respectively. This results in requiring 575 000 (3 800 000) signatures.

Discussion. Apparently, two-stage shuffling can increase the number of required signatures for an attack significantly and thus can be considered an effective countermeasure. For A1, for instance, 260 000 (1 710 000) instead of the previous 40 (264) signatures are required. However, while the given numbers are high, they are still within reach for a dedicated attacker.

This large increase could be explained as follows. For single-stage shuffling, we tested elements from D_σ, with $\sigma \approx 215$, against a distance of $\mathcal{X}_{\mathbf{sc}}$. For the two-stage attack, the ratio of the matched standard deviations is much smaller. For instance, in the second stage we match elements from $D_{\sigma'}$, with $\sigma' \approx 19.5$, against the same $\mathcal{X}_{sc}$. As a result, the matchable samples are even further out in the tail of $D_{\sigma'}$ and so less frequent than was the case for single-stage shuffling. This also explains the compared to A1 maybe surprisingly small losses of A2 and A3. These adversaries can only recover a small number of samples, but the ones they can find are already in tail $D_{\sigma'}$ and thus more likely to be usable. Obviously, the effect of smaller difference of deviations is amplified by requiring two matching steps. So, we can only rewind shuffling for indizes i where *both* y_i' and y_i'' are outliers.

6   Conclusion

Our work shows that shuffling is, at least if done correctly, an effective and cheap countermeasure in the context of lattice-based signatures. However, while it can drastically increase the attack complexity, relying on two-stage shuffling alone might not be enough to protect against attacks on Gaussian samplers. The reported signature requirements for attacks are still practical, at least in the case of a recoverable b. In this regard, recall that we did not use leakage from either multiplication with k and addition of two samples in the Gaussian convolution, the shuffling itself, or from the PRNG. This information can be used to further decrease the number of required signatures. Thus, a mix of countermeasures and reducing the leakage of the sampling algorithm itself is necessary for sufficient protection. For future work, we plan to further investigate and improve the attack technique. Our goal is to eliminate the impact of an unknown b, i.e., to use the same number of signatures as in the known-b case.

Acknowledgements. This work has been supported by the Austrian Research Promotion Agency (FFG) under grant number 845589 (SCALAS). We would also like to thank Leon Groot Bruinderink for his valuable input.

References

1. Alkim, E., Ducas, L., Pöppelmann, T., Schwabe, P.: Post-quantum key exchange - a new hope. In: Holz, T., Savage, S. (eds.) USENIX Security 2016, pp. 327–343. USENIX Association, Berkeley (2016)
2. Bos, J.W., Costello, C., Naehrig, M., Stebila, D.: Post-quantum key exchange for the TLS protocol from the ring learning with errors problem. In: SP 2015, pp. 553–570. IEEE Computer Society (2015)
3. Braithwaite, M.: Experimenting with post-quantum cryptography, July 2016. https://security.googleblog.com/2016/07/experimenting-with-post-quantum.html
4. Buchmann, J., Cabarcas, D., Göpfert, F., Hülsing, A., Weiden, P.: Discrete ziggurat: a time-memory trade-off for sampling from a gaussian distribution over the integers. In: Lange, T., Lauter, K., Lisoněk, P. (eds.) SAC 2013. LNCS, vol. 8282, pp. 402–417. Springer, Heidelberg (2014). doi:10.1007/978-3-662-43414-7_20

5. Chari, S., Rao, J.R., Rohatgi, P.: Template attacks. In: Kaliski, B.S., Koç, K., Paar, C. (eds.) CHES 2002. LNCS, vol. 2523, pp. 13–28. Springer, Heidelberg (2003). doi:10.1007/3-540-36400-5_3
6. Chen, L., Jordan, S., Liu, Y.-K., Moody, D., Peralta, R., Perlner, R., Smith-Tone, D.: NISTIR 8105 DRAFT, Report on Post Quantum Cryptography, February 2016. http://csrc.nist.gov/publications/drafts/nistir-8105/nistir_8105_draft.pdf
7. Ducas, L., Durmus, A., Lepoint, T., Lyubashevsky, V.: Lattice signatures and bimodal gaussians. In: Canetti, R., Garay, J.A. (eds.) CRYPTO 2013. LNCS, vol. 8042, pp. 40–56. Springer, Heidelberg (2013). doi:10.1007/978-3-642-40041-4_3
8. Dwarakanath, N.C., Galbraith, S.D.: Sampling from discrete Gaussians for lattice-based cryptography on a constrained device. Appl. Algebra Eng. Commun. Comput. **25**(3), 159–180 (2014)
9. Gierlichs, B., Lemke-Rust, K., Paar, C.: Templates vs. stochastic methods. In: Goubin, L., Matsui, M. (eds.) CHES 2006. LNCS, vol. 4249, pp. 15–29. Springer, Heidelberg (2006). doi:10.1007/11894063_2
10. Groot Bruinderink, L., Hülsing, A., Lange, T., Yarom, Y.: Flush, gauss, and reload – a cache attack on the BLISS lattice-based signature scheme. In: Gierlichs, B., Poschmann, A.Y. (eds.) CHES 2016. LNCS, vol. 9813, pp. 323–345. Springer, Heidelberg (2016). doi:10.1007/978-3-662-53140-2_16
11. Güneysu, T., Lyubashevsky, V., Pöppelmann, T.: Practical lattice-based cryptography: a signature scheme for embedded systems. In: Prouff, E., Schaumont, P. (eds.) CHES 2012. LNCS, vol. 7428, pp. 530–547. Springer, Heidelberg (2012). doi:10.1007/978-3-642-33027-8_31
12. Jaffe, J.: A first-order DPA attack against AES in counter mode with unknown initial counter. In: Paillier, P., Verbauwhede, I. (eds.) CHES 2007. LNCS, vol. 4727, pp. 1–13. Springer, Heidelberg (2007). doi:10.1007/978-3-540-74735-2_1
13. Knuth, D.E.: The Art of Computer Programming: Seminumerical Algorithms, vol. 2, 3rd edn., pp. 145–146. Addison-Wesley, Salt Lake (1998). Chap. 3
14. Lindner, R., Peikert, C.: Better key sizes (and attacks) for LWE-based encryption. In: Kiayias, A. (ed.) CT-RSA 2011. LNCS, vol. 6558, pp. 319–339. Springer, Heidelberg (2011). doi:10.1007/978-3-642-19074-2_21
15. Moradi, A., Hinterwälder, G.: Side-channel security analysis of ultra-low-power FRAM-based MCUs. In: Mangard, S., Poschmann, A.Y. (eds.) COSADE 2015. LNCS, vol. 9064, pp. 239–254. Springer, Heidelberg (2015). doi:10.1007/978-3-319-21476-4_16
16. NSA/IAD. CNSA Suite and Quantum Computing FAQ, January 2016. https://www.iad.gov/iad/library/ia-guidance/ia-solutions-for-classified/algorithm-guidance/cnsa-suite-and-quantum-computing-faq.cfm
17. Oder, T., Pöppelmann, T., Güneysu, T.: Beyond ECDSA and RSA: lattice-based digital signatures on constrained devices. In: DAC 2014, pp. 110:1–110:6. ACM (2014)
18. Pöppelmann, T., Ducas, L., Güneysu, T.: Enhanced lattice-based signatures on reconfigurable hardware. In: Batina, L., Robshaw, M. (eds.) CHES 2014. LNCS, vol. 8731, pp. 353–370. Springer, Heidelberg (2014). doi:10.1007/978-3-662-44709-3_20
19. Pöppelmann, T., Oder, T., Güneysu, T.: High-performance ideal lattice-based cryptography on 8-Bit ATxmega microcontrollers. In: Lauter, K., Rodríguez-Henríquez, F. (eds.) LATINCRYPT 2015. LNCS, vol. 9230, pp. 346–365. Springer, Heidelberg (2015). doi:10.1007/978-3-319-22174-8_19

20. Roy, S.S., Reparaz, O., Vercauteren, F., Verbauwhede, I.: Compact and side channel secure discrete gaussian sampling. Cryptology ePrint Archive, Report 2014/591 (2014). http://eprint.iacr.org/2014/591
21. Saarinen, M.-J.O.: Arithmetic coding and blinding countermeasures for lattice signatures: engineering a side-channel resistant post-quantum signature scheme with compact signatures. Cryptology ePrint Archive, Report 2016/276 (2016). http://eprint.iacr.org/2016/276
22. Schneier, B.: NSA plans for a post-quantum world, August 2015. https://www.schneier.com/blog/archives/2015/08/nsa_plans_for_a.html
23. strongSwan. strongSwan 5.2.2 Released (2015). https://www.strongswan.org/blog/2015/01/05/strongswan-5.2.2-released.html
24. Texas Instruments. MSP432P401R LaunchPad. http://www.ti.com/tool/msp-exp432p401r

Implementation
of Cryptographic Schemes

Atomic-AES: A Compact Implementation of the AES Encryption/Decryption Core

Subhadeep Banik[1]([✉]), Andrey Bogdanov[2], and Francesco Regazzoni[3]

[1] Temasek Labs, Nanyang Technological University, Singapore, Singapore
bsubhadeep@ntu.edu.sg
[2] DTU Compute, Technical University of Denmark, Lyngby, Denmark
anbog@dtu.dk
[3] ALARI, University of Lugano, Lugano, Switzerland
regazzoni@alari.ch

Abstract. The implementation of the AES encryption core by Moradi et al. at Eurocrypt 2011 is one of the smallest in terms of gate area. The circuit takes around 2400 gates and operates on an 8 bit datapath. However this is an encryption only core and unable to cater to block cipher modes like CBC and ELmD that require access to both the AES encryption and decryption modules. In this paper we look to investigate whether the basic circuit of Moradi et al. can be tweaked to provide dual functionality of encryption and decryption (ENC/DEC) while keeping the hardware overhead as low as possible. As a result, we report an 8-bit serialized AES circuit that provides the functionality of both encryption and decryption and occupies around 2645 GE with a latency of 226 cycles. This is a substantial improvement over the next smallest AES ENC/DEC circuit (Grain of Sand) by Feldhofer et al. which takes around 3400 gates but has a latency of over 1000 cycles for both the encryption and decryption cycles.

Keywords: AES 128 · Serialized implementation

1 Introduction

There has been extensive research into the construction of compact implementations of lightweight block ciphers. This line of research has essentially evolved along two different lines. The first aims to construct proprietary lightweight block ciphers by optimizing one or several parameters in the design spectrum, as has been evidenced by numerous such designs proposed in the past few years: HIGHT [21], KATAN [11], Klein [18], LED [19], Noekeon [13], Present [7], Piccolo [28], Prince [8], Simon/Speck [6] and TWINE [30]. The second aims at attempting to implement standardized ciphers like AES 128 [14] in a lightweight fashion.

There have been several lightweight implementations of AES proposed in literature. Some results like [20] and [10] aim for compact implementations in

© Springer International Publishing AG 2016
O. Dunkelman and S.K. Sanadhya (Eds.): INDOCRYPT 2016, LNCS 10095, pp. 173–190, 2016.
DOI: 10.1007/978-3-319-49890-4_10

ASIC and FPGA platforms respectively (however the work in [20] is for an encryption only core). The works in [23] and [31] aim at lowering critical path and increasing throughput. And the works in [3] and [5] aim to implement circuits with low energy consumption per encryption operation.

For compact implementations of the dual encryption/decryption circuit, the following results are known. In [27], the authors propose a 32-bit serial architecture with optimized tower field implementation of the S-box and a combinatorial optimization of the Mixcolumn circuit. The size of this implementation was around 5400 GE (gate equivalents, i.e. area occupied by an equivalent number of 2-input NAND gates). The "Grain of Sand" implementation [17] by Feldhofer et al. constructs an 8-bit serialized architecture with circuit size of around 3400 GE but a latency of over 1000 cycles for both encryption and decryption. Very recently in [24], the authors report an 8-bit serial implementation that takes 1947/2090 GE for the encryption/decryption circuits respectively. This implementation makes use of intermediate register files that can be synthesized in the ASIC flow using memory compilers.

The implementation by Moradi et al. in [26] with size equal to 2400 GE and encryption latency of 226 cycles is one of the smallest known architectures for AES. The design combines 8-bit and 32-bit serial datapaths in a manner that achieves a surprisingly compact implementation. The design uses scan flip-flops for constructing the registers for the state update and key schedule, a trick that saves 1 GE per flip-flop used. This implementation also uses a 32 bit Mixcolumn circuit instead of the 8-bit serialized structure of [17], because the authors argue that any savings in area achieved by an 8-bit serial circuit is offset by the additional registers required to store its output. Finally since each round function in this circuit is implemented in 21 cycles, the control system is made using a 21 cycle LFSR that generates all timing signals accordingly. However this circuit is an encryption-only core, and therefore can not be used to implement modes like CBC [16], COPA [2], ELmD [15], POET [1] that require access to both AES encryption and decryption functionalities. Therefore area-wise the three smallest known circuits that perform the dual functionalities of both encryption and decryption are

A. Grain of Sand implementation [17] at 3400 GE
B. 8-bit serial implementation in [24] at 4037 GE
C. 32-bit serial implementation in [27] at 5400 GE.

Moreover the Grain of Sand implementation has a latency of over 1000 cycles for both the encryption and decryption operations and so for efficient lightweight implementation of all modes that require access to both AES encryption and decryption it is critical to have an architecture that is both lightweight and incurs minimal latency.

1.1 Contribution and Organization

In this paper we present Atomic-AES, an 8-bit serial architecture that performs the dual functionality of encryption and decryption, and has a circuit size of around 2645 GE and latency of 226 cycles for both encryption and decryption operations. The circuit is closely related to the 8-bit encryption only serial architecture presented in [26], and in fact our architecture has the following additional logic components over the basic circuit proposed by Moradi et al.

1. 2 additional 8-bit multiplexers in the state datapath,
2. 3 additional 8-bit xor gates in the key datapath,
3. 24 additional and gates in the key datapath,
4. 1 additional 8-bit multiplexer, 1 additional 8-bit xor gate, 16 additional and gates during state-key addition,
5. Other additional logic required to implement
   a. S-box and its inverse,
   b. Mixcolumn and its inverse,
   c. Round constants and their inverses.

The paper is organized in the following manner. Section 2 gives some background and description of the architecture presented in [26]. This would be beneficial for the self-sufficiency and better understanding of this paper. Section 3, describes the architecture and functioning of **Atomic-AES** in details, and highlights some issues related to its implementation. Section 4 tabulates all implementation results and compares it with previous architectures present in literature. Section 5 concludes the paper.

2   Background and Preliminaries

In Fig. 1, a pictorial description of the architecture in [26] is given. As can be seen the basic elements of storage are the 16 byte sized registers made of scan flip-flops in the state and key path respectively, used to store the intermediate

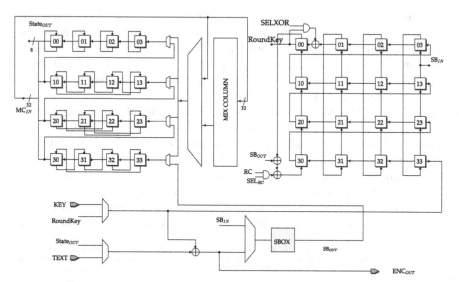

Fig. 1. The 8-bit serial architecture in [26]

states and roundkeys. Each round function is calculated in 21 cycles and so it is important to understand how the data is maneuvered through the registers during this period.[1]

Let us label the 21 cycles per round by the integers 0 to 20. The encryption process starts with the addition of the whitening key and the S-box computation of the first round function. In order to do so the finite sate machine (FSM) generating the round signals is initialized to cycle number 5. So in cycles numbered 5 to 20 (i.e. the very first 16 cycles) the following transformations take place:

Cycles 5 to 20: The 8 bit chunks of plaintext and key are respectively filtered out of the main state and key multiplexers respectively. They are xored, and the resultant signal fed to the S-box. The output of the S-box is fed to the bottom most multiplexer in the state path (marked by SB_{IN}), from where it is shifted serially forward in the next round. Effectively, after the cycle 20 is completed, the state registers would store the value $S(PT \oplus K)$, where $S(\cdot)$ denotes the bytewise application of the AES S-box function. In the same period the 8 bit chunk of the Key is input to key register marked "33", from where it is serially forwarded in the next round, much like in the state register. Therefore, at the end of cycle 20, the Key registers hold the value of the initial whitening key.

After this the cycle counter is automatically reset to 0, and each 21 cycle round function is executed 10 times, thus accounting for a total latency of $16+21*10 = 226$. During this period the order of operations is as follows:

$$\text{Shiftrow} \quad \rightarrow \quad \text{Mixcolumn} \quad \rightarrow \quad \text{Add roundkey} + \text{S-box of next round}$$

To clarify, let us see the cyclewise description of the data movement:

Cycle 0: This cycle is reserved for the Shiftrow operation. Since each 8-bit register in the state and key paths are constructed using scan flip-flops, they have two input data ports which they filter depending on a select signal. As can be seen in Fig. 1, the state registers are connected to facilitate the Shiftrow operation during cycle 0. The key register is "frozen" in this cycle and so no data movement takes place.[2]

Cycles 1 to 4: The Mixcolumn operation is performed during these 4 cycles. The Mixcolumn circuit used in this architecture is a $\{0,1\}^{32} \rightarrow \{0,1\}^{32}$ logic block, and so data from leftmost column (registers marked 00, 10, 20, 30) of the state is fed as input to the Mixcolumn circuit. In the subsequent cycle the Mixcolumn output is driven into the rightmost column (registers marked 03, 13, 23, 33). This operation carried out over 4 cycles computes the Mixcolumn over the entire state. Note that this operation is bypassed in the 10th encryption round as the Mixcolumn function is omitted in the final round.

---

[1] Another important point to note is that this particular architecture interprets the AES input vectors in a row major fashion i.e. the first four bytes are placed in the first row, the second four bytes in the second row so on. Most AES implementations use a column major ordering.

[2] One way to achieve this is to use a gated clock which does not present a leading edge during the shiftrow period.

During this period, the non-linear function of the Keyschedule operation is computed in the Key registers. Recall that the non linear operation in the AES Keyschedule is given as

$$F(K_3) = S(K_3 \lll 8) \oplus RCON_i,$$

where K_3 denotes the third column of the current roundkey, $\lll$ denotes the left rotate operation and $RCON_i$ is the ith round constant (note that the round constant is added to the most significant byte of $S(K_3 \lll 8)$). $(K_3 \lll 8)$ is a 32 bit value and so $S(K_3 \lll 8)$ implies the S-box function applied to each of the 4 bytes of the input. In order to implement the rotation operation, the data is taken from the output of the key register marked "13" and fed to the S-box. Although the architecture uses only one S-box, in cycles 1 to 4, the state path operations do not use the S-box circuit and so the key path S-box operations can be done in this period. The S-box output is xored to the output of the register "00" and the round constant and, in the next cycle is driven into the register marked "30". Note that since there is "vertical" movement of data in the key registers in this period, at the end of cycle 4, the four columns of the key register store the values $K_0 \oplus F(K_3), K_1, K_2, K_3$ respectively, where K_i denotes the ith column of the current roundkey.

Cycles 5 to 20: The bytes of state and roundkey are respectively taken out of the registers marked "00" of both the state and key paths and xored together and fed to the S-box. The output of the S-box is again driven into the bottom most state register "33" and serially shifted forward in the subsequent rounds. This sequence of operations is exactly similar as the ones performed in the very first 16 cycles, with the only exception that an intermediate state and roundkey chunks are xored instead of the raw plaintext and key.

The operations in the Key register are a little more interesting during this period. Note that in order to perform roundkey addition during these cycles, the data emanating from key register "00" be equal to the current roundkey. However we have seen that at the end of cycle 4 the columns of the key registers hold the value $K_0 \oplus F(K_3), K_1, K_2, K_3$. Note that if K_0, K_1, K_2, K_3 and L_0, L_1, L_2, L_3 denote the 4 columns of the current and next roundkey then we have

$$L_0 = K_0 \oplus F(K_3), \quad L_1 = K_1 \oplus L_0, \quad L_2 = K_2 \oplus L_1, \quad L_3 = K_3 \oplus L_2.$$

Thus at the end of cycle 4, only the 0th column holds the correct next round-key L_0. The problem is solved by having an extra xor gate taking inputs from the registers "00" and "01" and output feeding into "00". Since the movement of data is switched to "horizontal", this helps to perform on the fly addition as the key chunks are driven out of the "00" register. The addition is however not executed at cycles 8, 12, 16, 20 by zeroing the SELXOR signal because as previously noted, the 0th column already has the required roundkey. Also after the roundkey addition, each 8-bit key is circularly shifted back into the key registers through register "33" in order to facilitate the operations in the next round function.

The ith round in this architecture computes the Substitution layer for the $(i+1)^{th}$ AES encryption round. This being so, in the tenth and final encryption round the only operations that need be performed are Shiftrows and the final roundkey addition. Thus in the tenth round, the Mixcolumn operation is bypassed in cycles 1–4 and the output ciphertext is available just after the roundkey addition from cycles 5 through 20.

3   Atomic-AES: Architecture and Dataflow

We will now present a full description of the proposed architecture for Atomic-AES which provides dual functionalities for encryption and decryption. A diagram for the proposed architecture is presented in Fig. 2. The architecture builds on the basic circuit in [26], and so the functioning of the circuit during encryption is exactly as described in Sect. 2.

3.1   Issues with the Decryption Circuit

In order to accommodate decryption operation in the basic circuit of [26], there are some principal difficulties. We will list them one by one:

1. **Shiftrows/Inverse Shiftrows**: During the Shiftrow operation the data in the ith row is left-rotated by i bytes ($0 \leq i \leq 3$). Hence the Inverse Shiftrow operation would require the i-byte right-rotation of the ith row data. However in order to accommodate the Inverse Shiftrow and forward Shiftrow simultaneously would potentially require another multiplexer at the input of each 8-bit state register.

2. **Forward/Inverse Keyschedule**: The AES Keyschedule basically has as a non-linear shift register like structure, and it is obvious that the key register structure in [26] was explicitly constructed to accommodate its unique mathematical structure, and at the same time produce the current roundkey in an 8-bit serial fashion. It is not immediately clear how the Inverse Keyschedule could be arranged in such a circuit without increasing the circuit size significantly.

3. **Sequence of Operations During Decryption**: The circuit in [26] requires 21 cycles to complete a round function, with the order of operations being: Shiftrows, Mixcolumn followed by Add roundkey and the S-box layer of the following round. It is however not clear what order of operations would achieve the most efficient circuit for decryption. If one chooses to have roughly the same order of operations i.e. Inverse Shiftrows, Inverse Mixcolumn followed by Add roundkey and Inverse S-box, then as per the specification of the Decryption function, we would require the Inverse Mixcolumn of the roundkey as well (as described in [27]). This would most likely require additional cycles to compute the Inverse Mixcolumn of the roundkey and thus increase the latency.

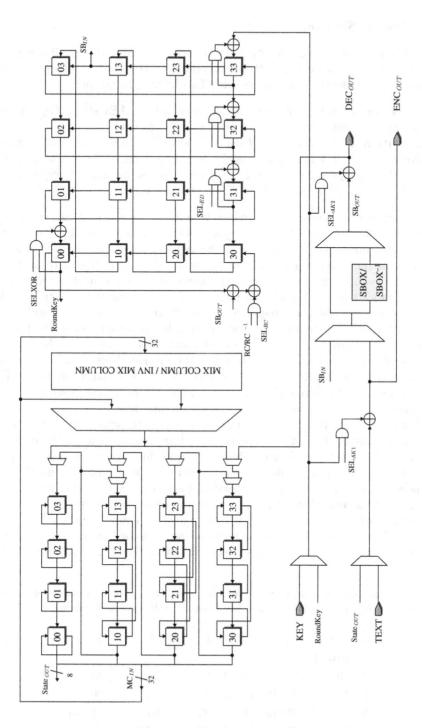

Fig. 2. The AES 8 bit Encryption/Decryption architecture for Atomic-AES

3.2   Inverse Shiftrow

An efficient Encryption/Decryption circuit would need to address all the above issues judiciously. To begin with let us address the issue of Shiftrow/Inverse Shiftrow. We make the following observations before proceeding:

Observation 1: *For the 0th and the 2nd rows of the AES state, Shiftrow and Inverse Shiftrow bring about the same transformation.*

Observation 2: *For the 1st and the 3rd rows of the AES state, Shiftrow and Inverse Shiftrow bring about opposite transformations. Which is to say, that the Shiftrow operation on the 1st row brings about the same transformation as the Inverse Shiftrow on the 3rd row and vice versa.*

A careful examination of the architecture in [26] reveals that each 8-bit register (constructed with scan flip-flops) accepts two inputs (see Fig. 1): one from the register immediately to its right (the rightmost register accepts its input from the leftmost register of the row below it), this connection is to facilitate the serial loading and unloading of the bytes in the state during cycles 5 to 20. The other input facilitates the transfer of data during they Shiftrow cycle. However, for the first three registers of the 1st row (i.e. "10", "11" and "12") the two inputs are actually the same. So in order to accommodate the Inverse Shiftrow, the second input connection of these three registers can be rewired (see Fig. 2) just like in the third row (since the Inverse Shiftrow of the first and Forward Shiftrow of the third row are actually identical transformations). For the last register of this row i.e. "13", an extra multiplexer with input from "10" is required. And that solves the problem for the first row.

For the 3rd row, the situation is even more straightforward. One of the direct results of **Observation 2**, is that the first input connection for the registers "30", "31" and "32" (used primarily for serial loading of data) can be used for the dual purpose of performing Inverse Shiftrow. This being the case there is no need for rewiring the inputs. However just as in the 1st row, for register "33", an extra multiplexer with input from register "30" is required. Also as per **Observation 1**, no change in wiring or logic is required in the 0th and 2nd rows. In Table 1, we summarize the input connections for the first and third row state registers during the various operation stages. For example during serial

Table 1. Input connections to the 1st and 3rd row state registers during various stages of the operation. (SL: Serial Loading, SR: Shiftrow, ISR: Inverse Shiftrow)

#	Register	SL	SR	ISR	#	Register	SL	SR	ISR
	Row 1					**Row 3**			
1	**10**	11	11	13	1	**30**	31	33	31
2	**11**	12	12	10	2	**31**	32	30	32
3	**12**	13	13	11	3	**32**	33	31	33
4	**13**	20	10	12	4	**33**	DEC$_{OUT}$	32	30

loading/unloading, register '13' accepts data coming from register '20', whereas it takes data from '10'/'12' during Shiftrow/Inverse Shiftrow respectively. As seen in Fig. 2, the register '33' takes data from the DEC_{OUT} pin during the serial loading phase (i.e. cycles 5 to 20).

3.3 Inverse Keyschedule

To recall, if K_0, K_1, K_2, K_3 and L_0, L_1, L_2, L_3 denote the 4 columns of the current and next roundkey then we have

$$L_0 = K_0 \oplus F(K_3), \quad L_1 = K_1 \oplus L_0, \quad L_2 = K_2 \oplus L_1, \quad L_3 = K_3 \oplus L_2.$$

During decryption, the roundkeys are generated in reverse order and so in the context of decryption, $\mathbf{L} = L_0, L_1, L_2, L_3$ is essentially the current roundkey and $\mathbf{K} = K_0, K_1, K_2, K_3$ is the key to be generated in the subsequent round. So we rewrite the above relation as

$$K_3 = L_2 \oplus L_3$$
$$K_2 = L_1 \oplus L_2$$
$$K_1 = L_0 \oplus L_1$$
$$K_0 = F(K_3) \oplus L_0 = F(L_2 \oplus L_3) \oplus L_0$$

So in order to have an Encryption/Decryption circuit we need an architecture around the key registers that can both (a) generate $\mathbf{L}$ given $\mathbf{K}$ as input and (b) generate $\mathbf{K}$ given $\mathbf{L}$ as input. The basic architecture in [26] all ready achieves (a) and so we need accommodate (b) i.e. the roundkey generation mechanism during decryption. We offer the following solution. Place three 8-bit xor gates in the 3rd row of Key registers in the following way (refer to Fig. 2).

1. For $1 \leq i \leq 2$, the xor gate takes inputs from the key registers "$3i$" and "$3\overline{i+1}$" and feeds its output into register "$3i$".
2. The third xor gate takes inputs from the registers "33" and the current round-key byte and feeds its output into register "33".
3. For each of these xor gates, the input coming from register "$3i$" is anded with a SEL_{ED} signal. This is done so that serial loading and unloading can be done when required by simply zeroing the SEL_{ED} signal.

To understand how the Inverse Keyschedule works let us look at the flow of data in cycles 5 to 20. For the purpose of simplification let $L_{0i}, L_{1i}, L_{2i}, L_{3i}$ denote the 4 key bytes in the column L_i, and similarly let $K_{0i}, K_{1i}, K_{2i}, K_{3i}$ denote the 4 key bytes in the column K_i. Note that the signal SEL_{ED} is made 1 only during cycles 8, 12, 16, 20 of the decryption phase. The flow of data has been explained in Fig. 3.

It can be seen that at cycle 8, the three rightmost key registers in the bottommost row have the key bytes L_{00}, L_{01}, L_{02}. At this point SEL_{ED} is set to 1. Thus in the next cycle the bottommost key row would contain the bytes

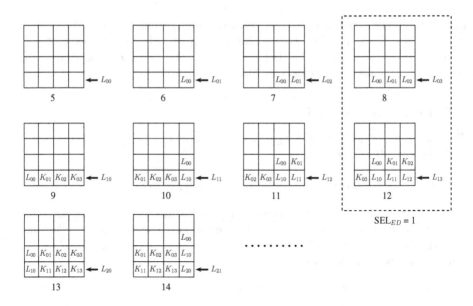

Fig. 3. Data flow in the key registers during Decryption

L_{00}, $K_{01} = L_{00} \oplus L_{01}$, $K_{02} = L_{01} \oplus L_{02}$, $K_{03} = L_{02} \oplus L_{03}$ respectively. Similar additions occur at cycles 12, 16 and 20 and as a result at the beginning of cycle 0 of the next round the four columns of the key register would have the values L_0, K_1, K_2, K_3 respectively. Thereafter in cycles 1 to 4, $F(K_3)$ is computed in the same manner as described in the encryption cycles and added to L_0 in the first column. And as a result at the beginning of cycle 5, the key columns contain $K_0 = L_0 \oplus F(K_3), K_1, K_2, K_3$ which is the complete next roundkey. Since the complete roundkey is already available, the **SELXOR** signal controlling the xor gate in the topmost row is zeroed as the roundkeys are serially driven out for the add roundkey operation. Thus all the functionalities of Inverse Keyschedule are completely accommodated using this architecture. Furthermore the complete decryption roundkey is available from cycles 5 through 20, which is incidentally the period during which we perform the add roundkey operation.

3.4   Sequence of Operations

Unlike ciphers like Midori [4], Prince [8] and Noekeon [13], AES was not designed as an efficiently implementable involutive cipher. As a result, the sequence of operations during the encryption and decryption flow are quite different. The sequence of operation during the encryption flow is as follows:

1. Add whitening key.
2. Rounds 1 to 9
    A. Substitution layer, **B.** Shiftrows, **C.** Mixcolumn, **D.** Add roundkey
3. Round 10
    A. Substitution layer, **B.** Shiftrows, **C.** Add roundkey

As previously mentioned, the 21 cycle encryption phase is arranged as Shiftrow → Mixcolumn → Add roundkey + Substitution layer of next round. The decryption flow of operations must exactly be opposite of encryption. Since the Shiftrows/Inverse Shiftrows can be commuted with S-box/Inverse S-box operation respectively, we can go with the following composition of one decryption round (also used in the architecture in [27]):

Inverse Shiftrow → Inverse Mixcolumn → Add roundkey + Inverse S-box

This sequence is attractive in this particular architecture because it has exactly the same order of operations as in encryption, and so it does not need too many changes in the underlying control system that produces select signals for the various multiplexers in the circuit. However as mentioned in [27], this sequence essentially swaps the order of Add roundkey and Inverse Mixcolumn operations. Since Mixcolumn and hence also Inverse Mixcolumn are linear functions, this requires the Inverse Mixcolumn function to be operated on the current roundkey before using it during the Add roundkey operation (since $MC^{-1}(X + K) = MC^{-1}(X) + MC^{-1}(K)$). There are two ways to achieve this: a) use an additional circuit for Inverse Mixcolumns or b) spend extra cycles to compute the Inverse Mixcolumn of the current roundkey. Option a increases circuit size and option b increases latency.

In this paper we propose an alternate sequence of the decryption cycle that compromises on neither the circuit size nor latency. We propose the following flow:

Inverse Mixcolumn → Inverse Shiftrow → Inverse S-box + Add roundkey

Since this sequence of operations is essentially the mirror inverse of the AES encryption round function, no swapping of Add roundkey and Inverse Mixcolumn is needed, and that obviates the need to calculate the Inverse Mixcolumn of the roundkey. To better explain the operations, let us present a cycle by cycle breakdown of the 21 cycle decryption round function. The decryption starts with the addition of the whitening key. The finite sate machine (FSM) generating the round signals is again initialized to cycle number 5. So in cycles numbered 5 to 20 (i.e. the very first 16 cycles) the following transformations take place:

Cycles 5 to 20: The 8 bit chunks of ciphertext and key are respectively filtered out of the main state and key multiplexers respectively. They are xored, and the resultant signal fed to the state registers. Note that in the corresponding encryption stage, we additionally calculated the S-box of the first round. Hence in order to accommodate both encryption and decryption we need a multiplexer after the S-box circuit as shown in Fig. 2. The Key bytes are input to key register "33", from where it is serially forwarded in the next round. However as mentioned in the previous subsection, the SEL_{ED} signal is set to 1 at rounds 8, 12, 16, 20 due to which at beginning of the next phase, the Key four register columns hold the value L_0, K_1, K_2, K_3 respectively.

After this the cycle counter is automatically reset to 0, and each 21 cycle round function is executed 10 times. Since the data flow in the key registers have already explained in the previous subsection, we concentrate on the state register.

Cycles 0 to 3: These cycles perform the Inverse Mixcolumn operation on the state columns, in exactly the same way forward Mixcolumn is executed in the encryption stage in cycles 1 to 4. However only in the very first round the Inverse Mixcolumn operation is bypassed, as required in AES decryption.

Cycle 4: This cycle is reserved for the Inverse Shiftrow operation.

Cycles 5 to 20: The bytes of state are taken out from register "00" and input into the combined forward and reverse S-box circuit to compute the Inverse S-box operation. The output of the S-box is then xored with the current roundkey byte from the key register "00" and circulated serially back into the state registers via the register marked "33". Note that the order of S-box and Add roundkey in the decryption phase is exactly the opposite as the encryption phase. As a result we employ two 8-bit xor gates, one before and one after the S-box circuit, for key addition in the encryption and decryption stages respectively. The xor gate inputs are controlled by and gates as shown in Fig. 2, in order to bypass the addition operation as required.

In the tenth and final round, the decrypted plaintext is made available from cycles 5 through 20 after the add roundkey operation. The above process is explained pictorially in Fig. 4. We now describe some of the components used in the circuit.

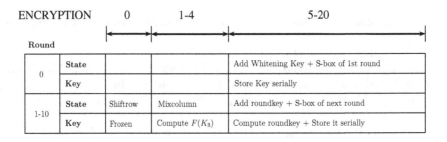

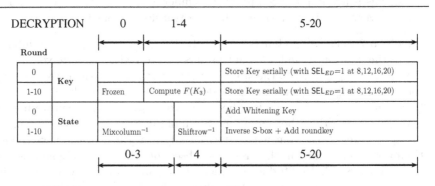

Fig. 4. Operation sequences in the Encryption/Decryption stages

3.5   S-Box

Over the years, there has been substantial research into compact circuit implementations of the AES S-box [9,12,25,27,32]. Almost all of them use the underlying algebraic structure of the AES S-box, that essentially combines an affine transformation with an inverse computation over the AES finite field. However the architecture due to Canright [12] remains one of the smallest in terms of circuit size for the combined Forward and Inverse S-box, and thus this is the architecture we chose for the combined S-box/Inverse S-box circuit.

3.6   Mixcolumn/Inverse Mixcolumn

In [27], the authors use the following decomposition of the Inverse Mixcolumn matrix to achieve an efficient implementation:

$$\begin{pmatrix} 14 & 11 & 13 & 9 \\ 9 & 14 & 11 & 13 \\ 13 & 9 & 14 & 11 \\ 11 & 13 & 9 & 14 \end{pmatrix} = \begin{pmatrix} 2 & 3 & 1 & 1 \\ 1 & 2 & 3 & 1 \\ 1 & 1 & 2 & 3 \\ 3 & 1 & 1 & 2 \end{pmatrix} + \begin{pmatrix} 8 & 8 & 8 & 8 \\ 8 & 8 & 8 & 8 \\ 8 & 8 & 8 & 8 \\ 8 & 8 & 8 & 8 \end{pmatrix} + \begin{pmatrix} 4 & 0 & 4 & 0 \\ 0 & 4 & 0 & 4 \\ 4 & 0 & 4 & 0 \\ 0 & 4 & 0 & 4 \end{pmatrix}$$

The xxtime (i.e. multiplication by 4) operation in AES finite field can be implemented in 5 xor gates as shown ($\boxed{b_6 \oplus b_7}$ is computed just once and the output is reused to construct the 5th LSB)

$$\text{xxtime}(b_7, b_6, \ldots, b_0) \mapsto b_5, b_4, b_3 \oplus b_7, b_2 \oplus \boxed{b_6 \oplus b_7}, b_1 \oplus b_6, b_0 \oplus b_7, \boxed{b_6 \oplus b_7}, b_6$$

Using this implementation of xxtime, the authors proposed a construction of Inverse Mixcolumns using 193 xor gates and a 32 bit multiplexer. However a more efficient implementation is due to Paulo Barreto, which factorizes the Inverse Mixcolumn matrix as:

$$\begin{pmatrix} 14 & 11 & 13 & 9 \\ 9 & 14 & 11 & 13 \\ 13 & 9 & 14 & 11 \\ 11 & 13 & 9 & 14 \end{pmatrix} = \begin{pmatrix} 2 & 3 & 1 & 1 \\ 1 & 2 & 3 & 1 \\ 1 & 1 & 2 & 3 \\ 3 & 1 & 1 & 2 \end{pmatrix} \cdot \begin{pmatrix} 5 & 0 & 4 & 0 \\ 0 & 5 & 0 & 4 \\ 4 & 0 & 5 & 0 \\ 0 & 4 & 0 & 5 \end{pmatrix}$$

To implement the above circuit, we simply premultiply the input column by the Circulant$(5, 0, 4, 0)$ matrix as follows:

$$y_3 = \text{xxtime}(x_3 \oplus x_1) \oplus x_3, \quad y_2 = \text{xxtime}(x_2 \oplus x_0) \oplus x_2$$

$$y_1 = \text{xxtime}(x_3 \oplus x_1) \oplus x_1, \quad y_0 = \text{xxtime}(x_2 \oplus x_0) \oplus x_0$$

where $X = (x_3, x_2, x_1, x_0)$ and $Y = (y_3, y_2, y_1, y_0)$ are the input and output columns of the multiplication block. The multiplication block takes exactly 58 xor gates. Thereafter we choose either X for Mixcolumns or Y for Inverse Mixcolumns, and input the resultant to the AES Mixcolumn circuit, as shown in Fig. 5. Since the Mixcolumn circuit can be efficiently implemented in 108 gates, the combined circuit takes $108 + 58 = 166$ xor gates and a 32 bit multiplexer which is more efficient than the construction in [27].

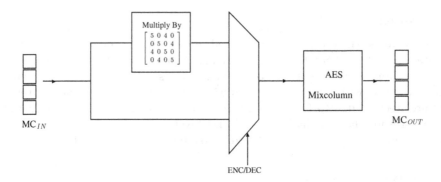

Fig. 5. Mixcolumn/Inverse Mixcolumn circuit

3.7   Round Constants and Control System

We use LUT based round constants. If r is the current round number, then the encryption operation uses $LUT(r)$, while the decryption operation uses $LUT(11 - r)$. The two signals can be input to an 8-bit multiplexer so that one can be chosen over the other as required. To further optimize, one can instead place a multiplexer before the LUT and choose between the 4-bit constants r and $11 - r$, and use the resultant signal as input to the LUT. Since this requires only a 4-bit multiplexer, it saves us additional area equivalent to a 4-bit multiplexer. Furthermore, all control signals are generated using a 21 cycle LFSR as described in [26].

4   Performance Evaluation

In order to perform a fair performance evaluation, we implemented the circuit using VHDL. Thereafter the following design flow was adhered to for all the circuits: a functional verification at the RTL level was first done using Mentor Graphics Modelsim software. The designs were synthesized using the standard cell library of the 90 nm and 65 nm logic process of STM (CORE90GPHVT v 2.1.a and CORE65LPLVT v 5.1) with the Synopsys Design Compiler, with the compiler being specifically instructed to optimize the circuit for area. A timing simulation was done on the synthesized netlist to confirm the correctness of the design, by comparing the output of the timing simulation with known test vectors. The switching activity of each gate of the circuit was collected while running post-synthesis simulation. The average power was obtained using *Synopsys Power Compiler*, using the back annotated switching activity. The results are tabulated in Table 2.

We outline some of the essential lightweight metrics of the known implementations of encryption/decryption architectures of AES and compare it with our own. Energy consumption was listed rather than power as it is a measure of the total electrical work done during one encryption/decryption. Since the circuits

Table 2. Performance comparison of Atomic-AES with previous architectures in literature (Figures separated by '/' indicate corresponding figures for encryption/decryption, E: Encryption only, ED: ENC/DEC)

#	Architecture	Type	Library	Area (GE)	Latency (cycles)	Energy (nJ)	TP_{max} (Mbps)
1	8-bit Serial [26]	E	UMC 180 nm	2400	226	8.4	-
2	Grain of Sand [17]	ED	Philips 350 nm	3400	1032/1165	46.4/52.4	9.9/8.8
3	8-bit Serial [24]	ED	22 nm	4037	336/216	3.9/2.5	432/671
4	32-bit Serial [27]	ED	110 nm	5400	54/54	-	311
5	Atomic-AES	ED	STM 90 nm	**2645**	226/226	3.3	94.4
			STM 65 nm	2976	226/226	2.2	57.8

in Table 2 are implemented using different CMOS logic processes, there are most likely to be wide variations in energy consumption and maximum throughput. For example the throughput of [24] is quite high as it is implemented using the standard cell library of the 22 nm CMOS logic process which is faster than the other logic processes listed in the table. The throughput of [27] is also high as it is a 32-bit serial circuit and thus has considerably lower latency.

In Fig. 6, we present a componentwise breakdown of the circuit size. We use clock gating to generate the clock for the Key registers, since the data movement has to be frozen for one cycle. Apart from the multiplexers included in the implementation of the combined Forward and Inverse S-box, Mixcolumn and Round Constants, a quick glance at Fig. 2, tells us that we need

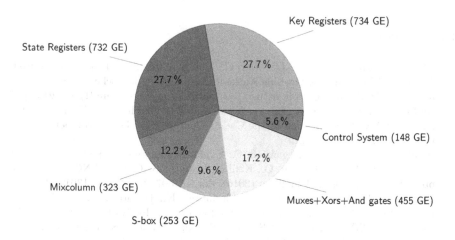

Fig. 6. Area requirements of the individual components

1. Six 8-bit multiplexers around the state register, one 32-bit multiplexer to bypass the Mixcolumn circuit, one 8-bit multiplexer after the S-box, and two 8-bit multiplexers to filter the raw key/plaintext (ciphertext) and the roundkey/state byte respectively.
2. Apart from this six 8-bit xors around the key`registers and two 8-bit xors during state-key addition.
3. One input of five out of the six xor gates is controlled by an and gate.

This adds up to around 455 GE for the multiplexers, xor, and gates in the circuit. The LSFR based control system and the round constants take around 148 GE. Adding up, this leads to 2645 GE for the entire circuit.

5   Conclusion

In this work, we present a compact architecture for AES that performs the dual function of encryption and decryption. Such architectures are useful in lightweight construction of block cipher modes that require access to both the encryption and decryption modules. We build upon the encryption only architecture of [26] and show that certain judicious alterations in logic and wiring can transform the architecture to perform encryption and decryption simultaneously. Our circuit has a size of 2645 GE and has a latency of 226 cycles for both encryption and decryption operations. This is a substantial improvement over the Grain of sand implementation that has an area of 3400 GE but a latency of over 1000 cycles for both encryption and decryption.

Acknowledgement. The authors would like to thank the anonymous reviewers who helped improve the quality and presentation of this paper.

References

1. Abed, F., Fluhrer, S., Foley, J., Forler, C., List, E., Lucks, S., Mcgrew, D., Wenzel, J.: The POET Family of On-Line Authenticated Encryption Schemes. Submission to the CAESAR competition. https://competitions.cr.yp.to/round1/poetv101.pdf
2. Andreeva, E., Bogdanov, A., Luykx, A., Mennink, B., Tischhauser, E., Yasuda, K.: AES-COPA v. 1. Submission to the Caesar Compedition. http://competitions.cr. yp.to/round1/aescopav1.pdf
3. Banik, S., Bogdanov, A., Regazzoni, F.: Exploring energy efficiency of lightweight block ciphers. In: Dunkelman, O., Keliher, L. (eds.) SAC 2015. LNCS, vol. 9566, pp. 178–194. Springer, Heidelberg (2016). doi:10.1007/978-3-319-31301-6_10
4. Banik, S., Bogdanov, A., Isobe, T., Shibutani, K., Hiwatari, H., Akishita, T., Regazzoni, F.: Midori: a block cipher for low energy. In: Iwata, T., Cheon, J.H. (eds.) ASIACRYPT 2015. LNCS, vol. 9453, pp. 411–436. Springer, Heidelberg (2015). doi:10.1007/978-3-662-48800-3_17
5. Banik, S., Bogdanov, A., Regazzoni, F., Isobe, T., Hiwatari, H., Akishita, T.: Round gating for low energy block ciphers. In: IEEE Hardware Oriented Security and Trust (HOST), pp. 55–60 (2016)

6. Beaulieu, R., Shors, D., Smith, J., Treatman-Clark, S., Weeks, B., Wingers, L.: The simon and speck families of lightweight block ciphers. In: IACR eprint archive. https://eprint.iacr.org/2013/404.pdf

7. Bogdanov, A., Knudsen, L.R., Leander, G., Paar, C., Poschmann, A., Robshaw, M.J.B., Seurin, Y., Vikkelsoe, C.: PRESENT: an ultra-lightweight block cipher. In: Paillier, P., Verbauwhede, I. (eds.) CHES 2007. LNCS, vol. 4727, pp. 450–466. Springer, Heidelberg (2007). doi:10.1007/978-3-540-74735-2_31

8. Borghoff, J., Canteaut, A., Güneysu, T., Kavun, E.B., Knezevic, M., Knudsen, L.R., Leander, G., Nikov, V., Paar, C., Rechberger, C., Rombouts, P., Thomsen, S.S., Yalçın, T.: PRINCE – a low-latency block cipher for pervasive computing applications. In: Wang, X., Sako, K. (eds.) ASIACRYPT 2012. LNCS, vol. 7658, pp. 208–225. Springer, Heidelberg (2012). doi:10.1007/978-3-642-34961-4_14

9. Boyar, J., Matthews, P., Peralta, R.: Logic minimization techniques with applications to cryptology. J. Cryptology **26**, 28–312 (2013)

10. Chodowiec, P., Gaj, K.: Very compact FPGA implementation of the AES algorithm. In: Walter, C.D., Koç, Ç.K., Paar, C. (eds.) CHES 2003. LNCS, vol. 2779, pp. 319–333. Springer, Heidelberg (2003). doi:10.1007/978-3-540-45238-6_26

11. Cannière, C., Dunkelman, O., Knežević, M.: KATAN and KTANTAN — a family of small and efficient hardware-oriented block ciphers. In: Clavier, C., Gaj, K. (eds.) CHES 2009. LNCS, vol. 5747, pp. 272–288. Springer, Heidelberg (2009). doi:10.1007/978-3-642-04138-9_20

12. Canright, D.: A very compact S-box for AES. In: Rao, J.R., Sunar, B. (eds.) CHES 2005. LNCS, vol. 3659, pp. 441–455. Springer, Heidelberg (2005). doi:10.1007/11545262_32

13. Daemen, J., Peeters, M., Assche, G.V., Rijmen, V.: Nessie Proposal: NOEKEON. http://gro.noekeon.org/Noekeon-spec.pdf

14. Daemen, J., Rijmen, V.: The Design of Rijndael: AES - The Advanced Encryption Standard. Springer, Heidelberg (2002)

15. Datta, N., Nandi, M.: ELmD v1.0. Submission to the Caesar Compedition. https://competitions.cr.yp.to/round1/elmdv10.pdf

16. Dworkin, M.: Recommendation for Block Cipher Modes of Operation. NIST Special Publication 800–38A. http://csrc.nist.gov/publications/nistpubs/800-38a/spp.800-38a.pdf

17. Feldhofer, M., Wolkerstorfer, J., Rijmen, V.: AES implementation on a grain of sand. IEEE Proc. Inf. Secur. **152**(1), 13–20 (2005)

18. Gong, Z., Nikova, S., Law, Y.W.: KLEIN: a new family of lightweight block ciphers. In: Juels, A., Paar, C. (eds.) RFIDSec 2011. LNCS, vol. 7055, pp. 1–18. Springer, Heidelberg (2012). doi:10.1007/978-3-642-25286-0_1

19. Guo, J., Peyrin, T., Poschmann, A., Robshaw, M.: The LED block cipher. In: Preneel, B., Takagi, T. (eds.) CHES 2011. LNCS, vol. 6917, pp. 326–341. Springer, Heidelberg (2011). doi:10.1007/978-3-642-23951-9_22

20. Hämäläinen, P., Alho, T., Hännikäinen, M., Hämäläinen, T.D.: Design and implementation of low-area and low-power AES encryption hardware core. In: DSD, pp. 577–583 (2006)

21. Hong, D., Sung, J., Hong, S., Lim, J., Lee, S., Koo, B.-S., Lee, C., Chang, D., Lee, J., Jeong, K., Kim, H., Kim, J., Chee, S.: HIGHT: a new block cipher suitable for low-resource device. In: Goubin, L., Matsui, M. (eds.) CHES 2006. LNCS, vol. 4249, pp. 46–59. Springer, Heidelberg (2006). doi:10.1007/11894063_4

22. Kerckhof, S., Durvaux, F., Hocquet, C., Bol, D., Standaert, F.-X.: Towards green cryptography: a comparison of lightweight ciphers from the energy viewpoint. In: Prouff, E., Schaumont, P. (eds.) CHES 2012. LNCS, vol. 7428, pp. 390–407. Springer, Heidelberg (2012). doi:10.1007/978-3-642-33027-8_23

23. Lutz, A.K., Treichler, J., Gürkaynak, F.K., Kaeslin, H., Basler, G., Erni, A., Reichmuth, S., Rommens, P., Oetiker, S., Fichtner, W.: 2Gbit/s hardware realizations of RIJNDAEL and SERPENT: a comparative analysis. In: Kaliski, B.S., Koç, K., Paar, C. (eds.) CHES 2002. LNCS, vol. 2523, pp. 144–158. Springer, Heidelberg (2003). doi:10.1007/3-540-36400-5_12

24. Mathew, S., Satpathy, S., Suresh, V., Anders, M., Kaul, H., Agarwal, A., Hsu, S., Chen, G., Krishnamurthy, R.K.: 340 mV-1.1V, 289 Gbps/W, 2090-gate nanoAES hardware accelerator with area-optimized encrypt/decrypt $GF(2^4)^2$ polynomials in 22 nm tri-gate CMOS. IEEE J. Solid-State Circ. **50**, 1048–1058 (2015)

25. Mentens, N., Batina, L., Preneel, B., Verbauwhede, I.: A systematic evaluation of compact hardware implementations for the rijndael S-box. In: Menezes, A. (ed.) CT-RSA 2005. LNCS, vol. 3376, pp. 323–333. Springer, Heidelberg (2005). doi:10.1007/978-3-540-30574-3_22

26. Moradi, A., Poschmann, A., Ling, S., Paar, C., Wang, H.: Pushing the limits: a very compact and a threshold implementation of AES. In: Paterson, K.G. (ed.) EUROCRYPT 2011. LNCS, vol. 6632, pp. 69–88. Springer, Heidelberg (2011). doi:10.1007/978-3-642-20465-4_6

27. Satoh, A., Morioka, S., Takano, K., Munetoh, S.: A compact rijndael hardware architecture with S-box optimization. In: Boyd, C. (ed.) ASIACRYPT 2001. LNCS, vol. 2248, pp. 239–254. Springer, Heidelberg (2001). doi:10.1007/3-540-45682-1_15

28. Shibutani, K., Isobe, T., Hiwatari, H., Mitsuda, A., Akishita, T., Shirai, T.: *Piccolo*: an ultra-lightweight blockcipher. In: Preneel, B., Takagi, T. (eds.) CHES 2011. LNCS, vol. 6917, pp. 342–357. Springer, Heidelberg (2011). doi:10.1007/978-3-642-23951-9_23

29. Shirai, T., Shibutani, K., Akishita, T., Moriai, S., Iwata, T.: The 128-bit blockcipher CLEFIA (extended abstract). In: Biryukov, A. (ed.) FSE 2007. LNCS, vol. 4593, pp. 181–195. Springer, Heidelberg (2007). doi:10.1007/978-3-540-74619-5_12

30. Suzaki, T., Minematsu, K., Morioka, S., Kobayashi, E.: TWINE: a lightweight block cipher for multiple platforms. In: Knudsen, L.R., Wu, H. (eds.) SAC 2012. LNCS, vol. 7707, pp. 339–354. Springer, Heidelberg (2013). doi:10.1007/978-3-642-35999-6_22

31. Ueno, R., Morioka, S., Homma, N., Aoki, T.: A high throughput/gate AES hardware architecture by compressing encryption and decryption datapaths. In: Gierlichs, B., Poschmann, A.Y. (eds.) CHES 2016. LNCS, vol. 9813, pp. 538–558. Springer, Heidelberg (2016). doi:10.1007/978-3-662-53140-2_26

32. Ueno, R., Homma, N., Sugawara, Y., Nogami, Y., Aoki, T.: Highly efficient $GF(2^8)$ inversion circuit based on redundant GF arithmetic and its application to AES design. In: Güneysu, T., Handschuh, H. (eds.) CHES 2015. LNCS, vol. 9293, pp. 63–80. Springer, Heidelberg (2015). doi:10.1007/978-3-662-48324-4_4

Fast Hardware Architectures for Supersingular Isogeny Diffie-Hellman Key Exchange on FPGA

Brian Koziel[1(✉)], Reza Azarderakhsh[2], and Mehran Mozaffari-Kermani[3]

[1] Texas Instruments, Dallas, USA
kozielbrian@gmail.com
[2] CEECS Department, I-SENSE FAU, Boca Raton, USA
razarderakhsh@fau.edu
[3] EME Department, RIT, Rochester, USA
mmkeme@rit.edu

Abstract. In this paper, we present a constant-time hardware implementation that achieves new speed records for the supersingular isogeny Diffie-Hellman (SIDH), even when compared to highly optimized Haswell computer architectures. We employ inversion-free projective isogeny formulas presented by Costello et al. at CRYPTO 2016 on an FPGA. Modern FPGA's can take advantage of heavily parallelized arithmetic in $\mathbb{F}_{p^2}$, which lies at the foundation of supersingular isogeny arithmetic. Further, by utilizing many arithmetic units, we parallelize isogeny evaluations to accelerate the computations of large-degree isogenies by approximately 57%. On a constant-time implementation of 124-bit quantum security SIDH on a Virtex-7, we generate ephemeral public keys in 10.6 and 11.6 ms and generate the shared secret key in 9.5 and 10.8 ms for Alice and Bob, respectively. This improves upon the previous best time in the literature for 768-bit implementations by a factor of 1.48. Our 83-bit quantum security implementation improves upon the only other implementation in the literature by a speedup of 1.74 featuring fewer resources and constant-time.

Keywords: Post-quantum cryptography · Elliptic curve cryptography · Isogeny-based cryptography · Field programmable gate array

1 Introduction

Post-quantum cryptography (PQC) has been gaining a large amount of interest in the wake of NIST's announcement to standardize post-quantum cryptosystems for use by the US government [1]. Fears of the emergence of a quantum computer that could break today's current cryptosystems and expose a wealth of private information have been increasing the demand for systems to be quantum-safe. Notably, Shor's algorithm [2] could be used in conjunction with a quantum computer to quickly break elliptic curve cryptography (ECC) and RSA. Fortunately, such computers do not currently exist, but it is unclear how long this will last. As such, there is a need to consider viable alternatives to today's

© Springer International Publishing AG 2016
O. Dunkelman and S.K. Sanadhya (Eds.): INDOCRYPT 2016, LNCS 10095, pp. 191–206, 2016.
DOI: 10.1007/978-3-319-49890-4_11

popular cryptosystems before the next major quantum computing breakthrough. Similar to ECC, isogeny-based cryptography also uses points on an elliptic curve to provide security. However, as opposed to security based on the difficulty to factor large point multiplications (which is the case for ECC), isogeny-based cryptography has security based on the difficulty to compute isogenies between supersingular elliptic curves. Currently, this is considered difficult even for quantum computers. An isogeny can be thought of as a unique algebraic map from one elliptic curve to another elliptic curve that satisfies group homomorphism. With the emergence of the supersingular isogeny Diffie-Hellman protocol from Jao and De Feo [3] in 2011, numerous aspects of the protocol have also been studied. Most recently, Costello, Longa, and Naehrig [4] have proposed projective isogeny formulas, which effectively eliminate the numerous inversions in the SIDH protocol and allow for a constant-time implementation. This is naturally immune to most types of simple power analysis and timing analysis. Although the SIDH protocol has been slower than other quantum-resistant schemes, it does feature smaller keys, smaller signatures, and forward secrecy, making it a viable candidate in NIST's PQC standardization workshop. In this paper, we provide the first implementation of the projective isogeny formulas presented in [4] on reconfigurable hardware. This constant-time implementation features 83-bit and 124-bit quantum security. Field programmable gate arrays (FPGA) can take advantage of a large amount of parallelism in basic arithmetic in the extension field $\mathbb{F}_{p^2}$ as well as the computation of large-degree isogenies. Aside from presenting a new speed record for SIDH, the goal of this paper is to show that hardware architectures can take advantage of the large amount of parallelism in SIDH and make it more viable in NIST's PQC workshop. The main contributions of this paper can be summarized as follows: (i) First constant-time SIDH implementation on reconfigurable hardware, 83-bit and 124-bit quantum security levels, utilizing projective isogeny formulas featured in [4], (ii) This SIDH implementation is approximately 50% faster than any other implementation in the literature. (iii) New approach to parallelizing isogeny evaluations to speed-up large-degree isogeny computations by over a factor of 1.5.

2   Preliminaries

Here, we briefly discuss the basis for isogeny-based cryptography. The isogeny-based Diffie-Hellman key exchange was first published by Rostovtsev and Stolbunov in [5]. This was originally defined over ordinary elliptic curves and was thought to feature quantum resistance. However, Childs, Jao, and Stolbunov [6] discovered a quantum algorithm to compute isogenies between ordinary curves in subexponential time. Later, David Jao, Luca De Feo, and Jerome Plut adapted the isogeny-based key exchange to be over supersingular elliptic curves in [3,7], which features no known quantum attack. As we review elliptic curve and isogeny theory, we point the reader to [8] for a much more in-depth explanation of elliptic curve theory.

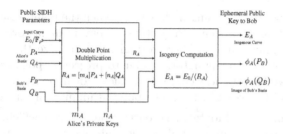

Fig. 1. Alice's first round computations for the SIDH protocol

SIDH Protocol: In the SIDH scheme, Alice and Bob decide on a smooth isogeny prime p of the form $\ell_A^a \ell_B^b \cdot f \pm 1$ where ℓ_A and ℓ_B are small primes, a and b are positive integers, and f is a small cofactor to make the number prime. They further decide on a base supersingular elliptic curve $E_0(\mathbb{F}_q)$ where $q = p^2$. Over this starting supersingular curve E_0, Alice and Bob pick the bases $\{P_A, Q_A\}$ and $\{P_B, Q_B\}$ which generate the torsion groups $E_0[\ell_A^{e_A}]$ and $E_0[\ell_B^{e_B}]$, respectively, such that $\langle P_A, Q_A \rangle = E_0[\ell_A^{e_A}]$ and $\langle P_B, Q_B \rangle = E_0[\ell_B^{e_B}]$. The SIDH protocol proceeds as follows. Alice and Bob each perform a double-point multiplication with two selected private keys that span $\mathbb{Z}/\ell^a\mathbb{Z}$ and $\mathbb{Z}/\ell^b\mathbb{Z}$, respectively. This generates a secret kernel point on each side that is used to efficiently perform a large-degree isogeny. In the first round, Alice calculates $\phi_A : E \to E_A/\langle m_A P_A + n_A P_A \rangle$ and Bob calculates $\phi_B : E \to E_B/\langle m_B P_B + n_B P_B \rangle$, where m and n are the party's secret keys. For the first round, the opposite party's basis points are pushed through the isogeny. At the end of the first round, Alice and Bob each exchange their new supersingular elliptic curve and the basis points of the opposite party on that new curve. With the exchanged information, Alice computes $\phi_{BA} : E_B \to E_{BA}/\langle m_A \phi_B(P_A) + n_A \phi_B(P_A) \rangle$ and Bob computes $\phi_{AB} : E_A \to E_{AB}/\langle m_B \phi_A(P_B) + n_B \phi_A(P_B) \rangle$. The two now share isomorphic curves with a common j-invariant that can be used as a shared secret. We illustrate the computations necessary for the first round from the perspective of Alice in Fig. 1. A round can essentially be broken down into a double point multiplication and a large-degree isogeny computation.

Optimizations to the SIDH Protocol: The supersingular isogeny Diffie-Hellman protocol was first proposed by David Jao and Luca De Feo in [3] in 2011. Since then it has been interesting to see how further papers have improved the protocol. The two main papers that have improved the protocol are [7] by De Feo, Jao, and Plut and [4] by Costello, Longa, and Naehrig. Here, we highlight the main protocol optimizations that we adapt. As introduced in [7], we utilize points on Montgomery curves [9] and optimize arithmetic around them. We define a Montgomery curve, E, as the set of all points (x, y) that satisfy $E_{(A,B)} : By^2 = x^3 + Ax^2 + x$ and a point at infinity. When the value $A_{24} = (A + 2)/4$ is known, these curves feature extremely fast point arithmetic along their Kummer line, $(x, y) \to (X : Z)$, where $x = X/Z$. Isogenies still work for this representation because P and $-P$ generate the same set subgroup of points. This reduces the total number of computations as the y-coordinate does not need to be updated for point arithmetic or when the point is pushed to a new curve by evaluating an isogeny. Projective isogeny formulas over Montgomery curves were introduced in [4]. These formulas projectivize the curve equation with a numerator and denominator, similar to projective point arithmetic. We define a projective Montgomery curve, $\hat{E}$, as the set of all points (x, y) that satisfy $\hat{E}_{(\hat{A},\hat{B},\hat{C})} : \hat{B}y^2 = \hat{C}x^3 + \hat{A}x^2 + \hat{C}x$ and a point at infinity. In this representation, the corresponding affine Montgomery curve would have coefficients $A = \hat{A}/\hat{C}$ and $B = \hat{B}/\hat{C}$. To perform a double point multiplication, we specify that one of Alice and Bob's secret keys is 1, as introduced in [7]. Costello et al. [4] also greatly simplified the starting parameters for SIDH by proposing to use the starting Montgomery curve $E_0/\mathbb{F}_{p^2} : y^2 = x^3 + x$. By specifying points in the base field and trace-zero torsion subgroup, the first round of the SIDH protocol can be performed as a Montgomery [9] ladder followed by a point addition, with all operations in $\mathbb{F}_p$. The second round of the protocol involves a double-point multiplication with elements in $\mathbb{F}_{p^2}$. For this, we utilize the 3-point ladder proposed in [7] that computes $P + mQ$ in $\log_2(m)$ steps. Each step requires 2 point additions and 1 point doubling. We closely follow the projective isogeny formulas presented in [4] for isogenies of degree $\ell_{Alice} = 4$ and $\ell_{Bob} = 3$. For the first round, we push the Kummer coordinates of the other party's basis P, Q, and $Q - P$ through the large-degree isogeny rather than the projective version of P and Q to remove a point subtraction before the 3-point ladder. As proposed by [10], large-degree isogenies can be decomposed into a chain of smaller degree isogeny computations and computed iteratively. From a base curve E_0 and point R of order ℓ^e, we compute a chain of ℓ-degree isogenies: $E_{i+1} = E_i/\langle \ell^{e-i-1}R_i \rangle$, $\phi_i : E_i \to E_{i+1}$, $R_{i+1} = \phi_i(R_i)$. This problem can be visualized as an acylic graph, which is shown in Fig. 3 in Sect. 4.3. In Fig. 4 in Sect. 4.3, we further illustrate a sample strategy to compute each of the ℓ-degree isogenies at the peak of the triangle by saving points at certain nodes to a point queue.

SIDH Protocol Parameters: To make our implementation comparable to the first hardware implementation of affine SIDH in [11] and the first software implementation of projective SIDH in [4], we chose to test our architecture over the primes $p_{503} = 2^{250}3^{159} - 1$ and $p_{751} = 2^{372}3^{239} - 1$. These primes offer 83 and 124 bits of quantum security, respectively.

Similar to the strategy proposed by Costello et al. [4], we begin with a simple Montgomery curve, technically also a short Weierstrass curve: $E_0/\mathbb{F}_{p^2} : y^2 = x^3 + x$. To determine generator points for the torsion subgroups $\ell_A^{e_A}$ and $\ell_B^{e_B}$, we again turn to Costello et al.'s method [4]. For the $\ell_A^{e_A}$-torsion points P_A and Q_A, we find a point $P_A \in E_0(\mathbb{F}_p)[\ell_A^{e_A}]$ as $[f\ell_B^{e_B}](z, \sqrt{z^3 + z})$, where z is the smallest positive integer such that $\sqrt{z^3 + z} \in \mathbb{F}_p$ and P_A has order $\ell_A^{e_A}$. We apply a distortion map over E_0 to P_A to find Q_A such that it is the endomorphism $\tau : E_0(\mathbb{F}_{p^2}) \to E_0(\mathbb{F}_{p^2}), (x+0i, y+0i) \to (-x+0i, 0+iy)$. Thus, $Q_A = \tau(P_A)$. The $\ell_B^{e_B}$-torsion points are found in a similar matter. We find $P_B \in E_0(\mathbb{F}_p)[\ell_B^{e_B}]$ as $[f\ell_A^{e_A}](z, \sqrt{z^3 + z})$, where z is the smallest positive integer such that $\sqrt{z^3 + z} \in \mathbb{F}_p$ and P_B has order $\ell_B^{e_B}$. Lastly, $Q_B = \tau(P_B)$. For the selected primes, our starting parameters are given in Table 1.

Table 1. SIDH public parameters

Curve: $E_0/\mathbb{F}_{p^2}$: $y^2 = x^3 + x$			
Prime	Classical/quantum security (bits)	P_A	P_B
$p_{503} = 2^{250}3^{159} - 1$	125/83	$[3^{159}](14, \sqrt{14^3 + 14})$	$[2^{250}](6, \sqrt{6^3 + 6})$
$p_{751} = 2^{372}3^{239} - 1$	186/124	$[3^{239}](11, \sqrt{11^3 + 11})$	$[2^{372}](6, \sqrt{6^3 + 6})$

3   Proposed Architectures for Isogeny Computations

In this section, we investigate the design of an SIDH core, focusing on optimizing finite-field addition and multiplication. The goal is to design a scalable architecture that features a secure and efficient implementation of SIDH. The proposed projective SIDH formulas presented in [4] make it reasonable to exclude a dedicated inversion module. Further, the simplification of the SIDH parameters allow for a reduction of the number of registers to store the SIDH parameters as well as the ability to perform Montgomery's powering ladder [9] in a base field rather than the 3-point differential Montgomery ladder over a quadratic field first proposed in [3]. In fact, the Montgomery ladder used to perform the first double point multiplication for both Alice and Bob may demonstrate a slight advantage to implementing a more efficient squaring unit. However, this squaring unit would not see much action as it is only used in the ladder of the first round of the key exchange and inversion. A dedicated squaring unit was not implemented for this paper, but should be investigated in the future. The high level design of the isogeny core is depicted in Fig. 2. This core features a single adder unit, multiplier unit with replicated multipliers, dual-port RAM file for registers, and a program ROM file for the controls. The RAM file contained 256 values in $\mathbb{F}_p$, or 256 m-bit entries. For our implementations, $m = 512$ and $m = 752$ for the choices of p_{503} and p_{751}, noted in Sect. 2. The RAM file contains constants for the parameters of the protocol, intermediate values within the protocol, and intermediate values for $\mathbb{F}_{p^2}$ computations. The major constants that are initially put into the RAM file are the constants 0, 1, 2, 4^{-1}, and 6, the base Montgomery

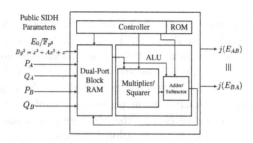

Fig. 2. Proposed High-level architecture of an SIDH core

curve coefficients A, B, and A_{24}, and the basis points P_A, Q_A, $Q_A - P_A$, P_B, Q_B, $Q_B - P_B$. There are more intermediate values necessary for higher key sizes as the graph traversal of the large degree isogeny is more expansive, but 256 values is more than enough, even for 768-bit SIDH, which allows more flexibility and optimization with routines. The program ROM contains the controls for the adder, multiplier, and RAM for every cycle for various SIDH routines (listed in Sect. 4.4). The size of the program ROM unit depends on the number of replicated multipliers as more multipliers will allow for fewer clock cycles. A stall counter was added to the control unit to diminish the impact of stall cycles that fill the program ROM.

3.1 Finite Field Adder

Finite-field addition computes the sum $C = A + B$, where $A, B, C \in \mathbb{F}_p$. If the sum C is greater than p, then there is a reduction by performing the subtraction $C = C - p$ to have $C \in \mathbb{F}_p$. A similar situation occurs for finite-field subtraction, $C = A - B$, where $A, B, C \in \mathbb{F}_p$. An adder can be used as a subtractor if the second operands input bits are flipped. The input operands to our adder/subtractor were selected with two 3:2 multiplexers. Operand 1 could be a value from port A of the RAM, the result from the adder/subtractor, or result from the multiplier. Operand 2 could be a value from port B of the RAM, zero, or the prime. Based on the interface between the RAM unit and the adder/subtractor module, which incurs delays from the register file logic and the 3:2 multiplexer into the adder/subtractor module, we decided to split the addition/subtraction into multiple cycles by cascading multiple, smaller adder/subtractors. We tried to match the critical path delay of the adder with that of the multiplier to ensure that both modules operated efficiently. Our smaller adder/subtractor units were based around 256-bit addition and subtraction. In practice, we utilized 252-bit and 251-bit adder/subtractor units for p_{503} and one 250-bit and two 251-bit adder/subtractor units for p_{751}. Xilinx's default IP was used to create these blocks. Partial sums and operands were pipelined to achieve a high-throughput adder/subtractor. An addition or subtraction was finished in 2 cycles for p_{503} and 3 cycles for p_{751}.

3.2 Field Multiplier

Finite-field multiplication computes the product $C = A \times B$, where $A, B, C \in \mathbb{F}_p$. Since the product is double the size of the inputs, a reduction must be performed so that the product is still within the field. The two known multiplier architectures targeting smooth isogeny primes are in [11,12]. Both utilize Montgomery [13] multiplication and reduction to efficiently perform the large modular multiplications. Montgomery multiplication performs a modular multiplication by transforming integers to m-residues, or the Montgomery domain, and performing multiplications with this representation. Montgomery multiplication converts time-consuming trial divisions to shift operations, which is simple to do in hardware. At the end of computations, the result can be converted out of

---

Algorithm 1. High-Radix Montgomery Multiplication Algorithm [15]

Input: $M = p$, $M' = -M^{-1} \bmod p$, $A = \sum_{i=0}^{m+2} (2^k)^i a_i, a_i \in \{0, 1 \ldots 2^k - 1\}, a_{m+2} = 0$
$B = \sum_{i=0}^{m+1} (2^k)^i b_i, b_i \in \{0, 1 \ldots 2^k - 1\}, \overline{M} = (M' \bmod 2^k) M = \sum_{i=0}^{m+1} (2^k)^i m_i$
$A, B < 2\overline{M}; 4\overline{M} < 2^{km}, R = 2^{\lceil \log_2 p \rceil}$
1. $S_0 = 0$
2. **for** $i = 0$ **to** $m + 2$ **do**
   3. $q_i = (S_i) \bmod 2^k$
   4. $S_{i+1} = (S_i + q_i \overline{M})/2^k + a_i B$
5. **end for**
6. **return** $S_{m+3} = A \times B \times R^{-1} \bmod M$

---

the Montgomery domain with a single Montgomery multiplication. Algorithm 1 demonstrates the Montgomery reduction procedure. In [12], the authors present an efficient method for modular multiplication over smooth isogeny primes of the form $p = 2 \cdot 2^a 3^b - 1$ by using the representation $A = a_1 2^a 3^b + a_2 2^{a/2} 3^{b/2} + a_3$, determining smaller partial products, and then performing an efficient division with some precalculations. The results appear interesting for a software implementation, achieving a 62% speed-up in modular reduction and 43% speed-up in modular multiplication. However, the hardware architecture for the multiplication algorithm appears to suffer. For a 768-bit prime, the Virtex-6 architecture required 11,924 registers and 12,790 lookup-tables, while operating at only 31 MHz and taking 236 cycles per modular multiplication. The other modular multiplier in [11] featured a systolic Montgomery multiplier based on [14]. Using a 2^{16} radix for a 1024-bit modular multiplication, the basic multiplier proposed in [14], operates at a clock frequency of 101.86 MHz, requires 5,709 slices and 131 DSP48's, and performs a modular multiplication in 199 clock cycles, all on a Virtex2 Pro. Further, this multiplier can perform 2 multiplications simultaneously. This already runs rings over the multiplier proposed in [12]. The target of this implementation is a high-throughput and fast multiplier. The implementation in [11] improved this systolic multiplier to allow higher throughput by featuring interleaving multiplications approximately 2/3 of the multiplication latency as well as one fewer stage in the systolic array. Thus, this allows for a 99 cycle multiplication and 68 cycle interleaving for a 512-bit multiplication.

Ultimately, we chose to go with the same interleaved systolic Montgomery multiplier proposed in [11]. This multiplier utilizes the high-radix Montgomery multiplication procedure, which is shown in Algorithm 1. As was originally proposed in [14], we can use a systolic architecture to perform the iterative computations in Algorithm 1. Consider a systolic array of $m + 2$ processing elements that each compute $S_{i+1} = (S_i + q_i \overline{m_j})/2^k + a_i b_j$, where j is the number of the processing element in the array. We can effectively setup a "pump" that pushes a_i and $q_i = (S_i) \bmod 2^k$ from processing element j to processing element $j + 1$. Thus, to perform the high-radix Montgomery multiplication, we start by pushing a 0 through the systolic arrays so that $q_0 = 0$. Following that, we push

a_i through the processing elements, such that it performs $a_i b_j$ and adds that result to $(S_i + q_i \overline{m_j})/2^k$ in each processing element. Essentially, each processing element performs $q_i \overline{m_j}$ and $a_i b_j$ in parallel, and then performs a 4-operand addition with $q_i \overline{m_j}$, $a_i b_j$, S_i, and a carry. After $m + 3$ cycles, the least significant k-bit word of the result is ready. The last word is ready after $3m + 7$ cycles. Interestingly, for a given multiplication, only half of the processing elements are used on a specific cycle. Thus, we can use a single multiplier architecture to handle two multiplications simultaneously, at the cost of multiplexers on the input and output that cycle between an even or odd multiplication. The design in [11] features an interleaved version of [14]. As one multiplication is finishing up, the earlier processing elements are no longer in use. Thus, we can interleave multiplications every $2m + 3$ cycles by gradually filling in these processing elements whose previous task just finished. As is also noted in [11], $\bar{M} = M$ since $M' = 1$ for SIDH primes of the form $2^{e_a} \ell_b^{e_b} f - 1$, which is applicable to both of our test primes. This simplification reduces the total size of the systolic array by one processing element and reduces the latency by 3 cycles. Since a DSP48 block effectively computes up to an 18×18 multiplication, we decided to make our Montgomery multiplier with radix 2^{16}. Using this, we calculated the latency of multiplication and interleaving. For p_{503}, a multiplication required 99 cycles and multiplications could be interleaved 68 cycles into a multiplication. For p_{751}, a multiplication required 144 cycles and multiplications could be interleaved every 98 cycles. We also implemented a larger multiplier unit that featured replicated multiplier units. Multiplications are the main bottleneck in the finite-field operations given by the smooth isogeny primes. As such, we implemented a first-in-first-out circular buffer. Multiplication instructions are issued cyclically starting from multiplier 0 to multiplier $2n - 1$ for n dual multipliers. This comes at the cost of a large multiplexer of size $2n{:}\log_2 2n$ for the output.

4   Parallelizing SIDH

This section details our attempt to maximize the throughput of our architecture throughout the SIDH protocol. Since we used the same even-odd multiplier as [11], we scheduled our instructions with a greedy algorithm that incurs stalls if a multiplication is not on the right even-odd cycle.

4.1   Scheduling

Our program ROM features many different routines such as a small scalar point multiplication or isogeny evaluation of degree 4. Each instruction is 26 bits long and proceeds as follows: bits 0–7 determine the address for port A of the RAM, bits 8–15 determine the address for port B of the RAM, bit 16 signals a write to port A, bits 17–19 indicate the adder operation, bit 20 indicates a read from both RAM ports, bits 21–22 indicate multiplier operation, bits 23 and 24 indicate if operand A and B, respectively, should point to the address of the final point in the isogeny point queue, and bit 25 indicates if the previous bits are a stall counter. We utilized a greedy algorithm to assemble our own assembly code that consists of addition, subtraction, multiplication, and squaring in $\mathbb{F}_p$ or $\mathbb{F}_{p^2}$ to 26-bit aligned instructions. It is assumed that every routine starts on an even cycle. Since a store is the final instruction in a routine, we also reset the multiplier even_odd at the last cycle of a routine so that the next routine starts on an even cycle from the multiplier's perspective. Every instruction was compiled in order, so if an instruction needed the result from a previous instruction, then pipeline stalls were incurred until that value was ready. The greedy algorithm to schedule each operation would check that the RAM, addition, or multiplier unit were available for the particular instruction. For instance, an addition in $\mathbb{F}_p$ could be scheduled if the memory unit at time t, addition unit at time $t +$ mem_latency,addition unit at time $t +$ mem_latency+add_latency, and memory unit at time $t +$ mem_latency $+ 2 *$ add_latency were each available, as the entire operation must go through that exact sequence. Based on the specifications of the dual-port RAM unit, memory load operations require 2 cycles and memory write operations require 1 cycle. The add latency is 2 cycles for p_{503} and 3 cycles for p_{751}. The multiplication and multiplication interleave delays are 99 cycles and 68 cycles for p_{503}, respectively, and 144 cycles and 98 cycles for p_{751}, respectively. If a multiplication occurred on the wrong even_odd cycle, we reschedule the operations by pushing the multiplication a single cycle forward, and pushing any previous instructions that are not a load or multiply by 1 or more cycles, according to the algorithm provided by [11].

4.2 Extension Field Arithmetic

As was previously stated, SIDH operates in the extension field $\mathbb{F}_{p^2}$. For this extension field, we use the irreducible polynomial $x^2 + 1$, applicable to SIDH primes of the form $2^{e_a} \ell_b^{e_b} f - 1$. With this, we propose reduced arithmetic in $\mathbb{F}_{p^2}$ based on fast arithmetic in $\mathbb{F}_p$. These equations were made in a Karatsuba-like fashion to reduce the total number of multiplications and squarings. Let $i = \sqrt{-1}$ be the most significant $\mathbb{F}_p$ in $\mathbb{F}_{p^2}$. Let $A, B \in \mathbb{F}_{p^2}$ and $a_0, b_0, a_1, b_1 \in \mathbb{F}_p$, where $A = a_0 + ia_1$ and $B = b_0 + ib_1$ Then we define the extension field arithmetic $\mathbb{F}_{p^2}$ in terms of $\mathbb{F}_p$ as: $A + B = a_0 + b_0 + i(a_1 + b_1)$, $A - B = a_0 - b_0 + i(a_1 - b_1)$, $A \times B = (a_0 + a_1)(b_0 - b_1) + a_0 b_1 - a_1 b_0 + i(a_0 b_1 + a_1 b_0)$, $A^2 = (a_0 + a_1)(a_0 - a_1) + i2a_0 a_1$, $A^{-1} = (a_0 - ia_1)(a_0^2 + a_1^2)^{-1}$. Based on these representations, parallel calculations could easily be performed for a single operation in $\mathbb{F}_{p^2}$. For instance, three separate multiplications in $\mathbb{F}_p$ could be carried out simultaneously for the calculation of a multiplication in $\mathbb{F}_{p^2}$. With other non-dependent instructions in the scheduling, many multipliers can be used in parallel. Unfortunately, an inversion in $\mathbb{F}_p$ was difficult to parallelize, and suffered as a result. We utilized a k-ary method with $k = 4$ to perform Fermat's little theorem for inversion. We were able to parallelize the generation of the windows $1, 2, 3, \cdots, 2^k - 1$, but after that, the inversion was done serially. k squarings were done in serial followed by a multiplication. The inversion added many lines to the program ROM, and was difficult to parallelize, showing that there may still be some merit to having a dedicated inversion unit.

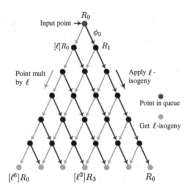

Fig. 3. Acyclic graph structure for performing isogeny computation of ℓ^6.

4.3   Scheduling Isogeny Computations and Evaluations

Large-degree isogeny calculations were performed by traversing a large directed acyclic graph in the shape of a triangle to the leaves, where a smaller degree isogeny was computed. This is illustrated in Fig. 3. From a node in the graph, a point multiplication by ℓ moves to the left and an evaluation of a ℓ-isogeny moves to the right. Based on the cost of an isogeny evaluation and point multiplication, there exists an optimal strategy that traverses the graph to the leave with the minimal computational cost. Notably, an optimal strategy is composed of two optimal sub-strategies. Thus, by recursively optimizing sub-strategies, the overall strategy is determined. We calculated the optimal strategy with the Magma code provided by [4]. In this code, we used the relative ratio of a single point multiplication by ℓ and half of a single ℓ-isogeny evaluation to create an optimal strategy that emphasized point evaluations. In our implementation, we utilize a recursive function to compute the large-degree isogeny with an optimal strategy. We utilized a look-up-table in ROM to hold the optimal strategy and efficiently traverse the acyclic graph. A queue was used to keep track of multiple points on the current curve. As isogenies were computed, these points were pushed through the isogenous mapping to the corresponding point on the new curve. As a method for further parallelization, we noticed that isogeny evaluations have typically been carried out iteratively. Thus, we attempted to parallelize the evaluations by adding additional isogeny evaluation functions for when there were 2 points, 3 points, $\cdots$, up to 9 points in the queue. Specifically, there were no data dependencies between isogeny evaluations of any of the points in the queue. Thus, our assembly code reordered many instructions in a row that had no limiting data dependency, similar to unrolling the loop in a software implementation. We unrolled a max of 6 iterations of the loop at a time to ensure that enough hardware registers were available to hold intermediate values. We found this greatly increase the speed of our isogeny computations. For instance, this method reduced the total time to compute all 4 large-degree isogenies from 7.15 million cycles to 4.54 million cycles for p_{751} and 4 replicated multipliers. We provide an example of isogeny evaluation parallelization in Fig. 3.

Consider computing an ℓ^6-degree isogeny. Following an ℓ-degree isogeny computation, each point in the point queue is pushed through the isogenous mapping. We do this in parallel to utilize our hardware results more effectively. The parallelization is much more evident in larger degree isogeny computations. For instance, there is an average of 4.2 points in Alice's queue after each isogeny computation in our p_{751} implementation. Parallelization of isogeny evaluation could also be applicable to multi-core CPU implementations of SIDH. Our particular hardware implementation was able to parallelize the isogeny evaluations because of the number of multipliers that were readily available. In a software implementation, the multiplication and addition arithmetic might be complex and consume most of the arithmetic units. However, because there is no data dependency, the task to push all of the points through the isogeny could be divided among different cores. For instance, consider pushing 8 points through an isogenous mapping in a quad-core CPU. Each core could evaluate an isogeny for 2 points in the queue to better take advantage of resources. Of course, there would be overhead in distributing the task, but a nice speedup could be achieved when there are several points in the queue.

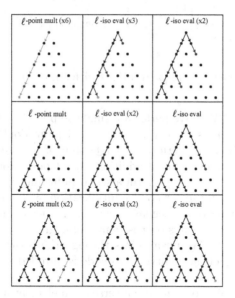

Fig. 4. Performing an isogeny computation of ℓ^6 with a sample strategy and parallel isogeny evaluations.

4.4   Total Cost of Routines

Here, we break up the relative costs of routines within our implementation of the SIDH protocol. Table 2 contains the results of various routines, which closely follows the formulas provided in [4]. $\tilde{A}$, $\tilde{S}$, and $\tilde{M}$ refer to addition, squaring, and

Table 2. Cost of major routines for p_{751}

Routine	Ops in $\mathbb{F}_{p^2}$			#ops in	Latency for n mults (cc)				
	$(\tilde{A})$	$(\tilde{S})$	$(\tilde{M})$	Protocol	2	4	6	8	10
Mont. Ladder Step ($\mathbb{F}_p$)	9	4	5	751	619	495	495	495	495
3-point Ladder Step	14	6	9	751	2181	1329	1120	972	908
Mont Quadruple	11	4	8	1276	1874	1306	1151	1151	1151
Mont Triple	15	5	8	1622	1954	1289	1124	1145	1145
Get 4 Isog	7	5	0	370	586	386	367	363	363
Eval 4 Isog	6	1	9	14	1655	1461	1225	1221	1147
Eval 4 Isog (3 times)	18	3	27	255	4537	2855	2104	1917	1642
Eval 4 Isog (5 times)	30	5	45	98	7427	4212	3036	2489	2215
Eval 4 Isog (7 times)	42	7	63	16	10543	6293	4674	4168	3716
Get 3 Isog	8	3	3	478	833	496	471	434	434
Eval 3 Isog	2	2	6	12	1252	1001	812	810	734
Eval 3 Isog (3 times)	6	6	18	309	3442	2026	1461	1306	1103
Eval 3 Isog (5 times)	10	10	30	112	5638	3123	2229	1776	1535
Eval 3 Isog (7 times)	14	14	42	72	7972	4411	3154	2667	2389
$\mathbb{F}_{p^2}$ Inversion ($\mathbb{F}_p$)	2	757	196	4	142307	142059	142059	141973	141973

multiplication, respectively, in $\mathbb{F}_{p^2}$. Routines with a note of ($\mathbb{F}_p$) count operations in $\mathbb{F}_p$.

- *Mont. Ladder Step* ($\mathbb{F}_p$): We perform a single step of the Montgomery ladder [9] in $\mathbb{F}_p$, which requires 1 point addition and 1 point doubling.
- *3-point Ladder Step*: We perform a single step of the 3-point Montgomery ladder [7], which requires 2 point additions and 1 point doubling.
- *Mont Quadruple/Triple*: We perform a scalar point multiplication by 4 in the case of quadrupling and scalar point multiplication by 3 in the case of tripling.
- *Get ℓ Isog*: We compute an isogeny of degree ℓ. Alice operates over isogenies of degree 4 and Bob operates over isogenies of degree 3.
- *Eval ℓ Isog (x times)*: We push points through the isogenous mapping from their old curve to their new curve. This code is unrolled x times from 1 point to 9 points.
- $\mathbb{F}_{p^2}$ inversion ($\mathbb{F}_p$): We compute the inverse of an element using Fermat's little theorem.

5 FPGA Implementations Results and Discussion

The SIDH core was compiled with Xilinx Vivado 2015.4 to a Xilinx Virtex-7 xc7vx690tffg1157-3 board. All results were obtained after place-and-route. The area and timing results of our SIDH core are shown in Table 3. We focused on 3–5 replicated multipliers in our design to ensure the parallelism in SIDH could

Table 3. Implementation results of SIDH architectures on a Xilinx Virtex-7 FPGA

Type	# Mults	Area					Time			SIDH/s
		# FFs	# LUTs	# Slices	# DSPs	# BRAMs	Freq (MHz)	Latency ($cc \times 10^6$)	Total time (ms)	
p_{503}	6	26,659	19,882	8,918	192	40	181.4	3.80	20.9	47.8
	8	32,541	23,404	11,205	256	37.5	186.8	3.63	19.4	51.5
	10	39,446	28,520	12,962	320	34.5	175.9	3.48	19.8	50.5
p_{751}	6	36,728	25,975	11,801	282	47	177.3	8.21	46.3	21.6
	8	46,857	32,726	15,224	376	45.5	182.1	7.74	42.5	23.5
	10	56,979	40,327	18,094	470	44	172.6	7.41	42.9	23.3

be taken advantage of. The implementation was optimized to reduce the net delay to maximize the clock frequency. These are constant-time results. Our SIDH parameters are discussed in Sect. 2. As these results show, the architectures continue to reduce the total number of clock cycles for SIDH, even at 10 multipliers. This is primarily a result of the parallelism achieved in isogeny evaluation and the 3-point ladder. Furthermore, the architecture appears fairly scalable. Moving from a 503-bit prime to a 751-bit prime did not have much impact on the maximum frequency of the device and added a small proportion of additional resources. For 5 multipliers under the 751-bit prime, approximately 16.71% of the Virtex-7's slices were occupied. Many more resources could be used to attempt more parallelization, but the clock frequency may suffer as a result, which is evident in our implementations of 5 replicated dual-multipliers.

Comparison to Previous Works: The only other hardware implementation is [11], which served as an introductory look into the SIDH protocol on hardware. We provide a rough comparison for 3 replicated multipliers at the 512-bit security level in Table 4. Our architecture performs an entire SIDH key-exchange approximately 1.61 times faster than that of [11]. This is most likely a result of using the new projective isogeny formulas as well as parallelism in the isogeny evaluations. In terms of area, our architecture requires about 15% less flip-flops, look-up-tables, and slices, but requires about 1.5 times as many 36k block RAM modules.

Table 4. Hardware comparison of SIDH architectures on a Virtex-7 with 3 replicated multipliers

Work	Prime (bits)	Area					Time		
		# FFs	# LUTs	# Slices	# DSPs	# BRAMs	Freq (MHz)	Latency ($cc \times 10^6$)	Total time (ms)
Koziel et al. [11]	511	30,031	24,499	10,298	192	27	177	5.967	33.7
This work	503	26,659	19,882	8,918	192	40	181.4	3.80	20.9

Overall, this is to be expected as our architecture does not include an inversion unit. In [11], the $\mathbb{F}_{p^2}$ inversion required about 1886 cycles for each isogeny computation. Our isogeny computations did not require this expensive operation and we were able to parallelize the projective isogeny evaluations that are more complex than their affine isogeny couterparts. The difference in prime sizes does not make much of a difference for area because both are based on a radix 2^{16} multiplier. Most importantly, our implementation is constant-time and the previous one is not, which provides security against simple power analysis and timing attacks. Next, we look at the overall speed of this implementation compared to the state-of-the-art, shown in Tables 5 and 6, which demonstrate the fastest SIDH implementations over approximately 512 and 768-bit keys. These feature approximately 85 and 128-bits of quantum security, respectively. We compare against our implementations with 4 replicated dual-multipliers, which featured the fastest times for our results. These benchmarks have shown that the total time of the SIDH protocol has continued to drop since its inception by Jao and De Feo in [3]. Our 512-bit implementation operated approximately 74% faster than the previous best implementation in hardware in [11]. These results are approximately 48% faster than those of [4], despite the powerful nature of Haswell architectures. Smaller SIDH implementations on ARM also exist [17], but these utilize far fewer resources so it is difficult to make a fair comparison.

Table 5. Comparison to the software implementations of SIDH over 512-bit keys

Work	Platform	Smooth isogeny	Time (ms)				
			Alice	Bob	Alice	Bob	Total
		Prime	Rnd 1	Rnd 1	Rnd 2	Rnd 2	Time
Jao et al. [3]	2.4 GHz Opt	$2^{253}3^{161}7-1$	365	318	363	314	1360
Jao et al. [7]	2.4 GHz Opt	$2^{258}3^{161}186-1$	28.1	28.0	23.3	22.7	102.1
Azarderakhsh et al. [16]	4.0 GHz i7	$2^{258}3^{161}186-1$	-	-	-	-	54.0
Koziel et al. [11]	Virtex-7	$2^{253}3^{161}7-1$	9.35	8.41	8.53	7.41	33.70
This work ($M = 2 \times 4$)	Virtex-7	$2^{250}3^{159}-1$	4.83	5.25	4.41	4.93	19.42

Table 6. Comparison to software implementations of SIDH over 768-bit keys

Work	Platform	Smooth isogeny	Time (ms)				
			Alice	Bob	Alice	Bob	Total
		Prime	Rnd 1	Rnd 1	Rnd 2	Rnd 2	Time
Jao et al. [7]	2.4 GHz Opt	$2^{258}3^{161}186-1$	65.7	54.3	65.6	53.7	239.3
Azarderakhsh et al. [16]	4.0 GHz i7	$2^{386}3^{242}2-1$	-	-	-	-	133.7
Costello et al. [4]	3.4 GHz i7	$2^{372}3^{239}-1$	15.0	17.3	13.8	16.8	62.9
This work ($M = 2 \times 4$)	Virtex-7	$2^{372}3^{239}-1$	10.6	11.6	9.5	10.8	42.5

6   Conclusion

Overall, this paper served as the first constant-time hardware implementation of the supersingular isogeny Diffie-Hellman protocol over projective isogeny formulas. As our results show, our architecture is scalable and is even faster than the previously fastest implementations of the protocol on Haswell PC architectures. Hardware can take advantage of much more parallelism in $\mathbb{F}_{p^2}$ operations and isogeny evaluations over standard software. Our implementation runs at 48% faster than a Haswell architecture running an optimized C version of the same SIDH protocol. By removing the multitude of inversions in the protocol, this new implementation features a faster constant-time performance with less resources than the previous best hardware implementation in the literature. Isogeny-based cryptography represents one possible solution to the impending quantum computing revolution because it features forward-secrecy, small keys, and resembles current protocols based on classical ECC.

Acknowledgment. This material is based upon work supported by the NSF CNS-1464118 and NIST 60NANB16D246 grants awarded to Reza Azarderakhsh.

References

1. Chen, L., Jordan, S.: Report on Post-Quantum Cryptography. NIST IR 8105 (2016)
2. Shor, P.W.: Algorithms for quantum computation: discrete logarithms and factoring. In: 35th Annual Symposium on Foundations of Computer Science (FOCS 1994), pp. 124–134 (1994)
3. Jao, D., Feo, L.: Towards quantum-resistant cryptosystems from supersingular elliptic curve isogenies. In: Yang, B.-Y. (ed.) PQCrypto 2011. LNCS, vol. 7071, pp. 19–34. Springer, Heidelberg (2011). doi:10.1007/978-3-642-25405-5_2
4. Costello, C., Longa, P., Naehrig, M.: Efficient algorithms for supersingular isogeny Diffie-Hellman. In: Robshaw, M., Katz, J. (eds.) CRYPTO 2016. LNCS, vol. 9814, pp. 572–601. Springer, Heidelberg (2016). doi:10.1007/978-3-662-53018-4_21
5. Rostovtsev, A., Stolbunov, A.: Public-Key Cryptosystem Based on Isogenies. IACR Cryptology ePrint Archive 2006, 145 (2006)
6. Childs, A., Jao, D., Soukharev, V.: Constructing Elliptic Curve Isogenies in Quantum Subexponential Time (2010)
7. De Feo, L., Jao, D., Plut, J.: Towards quantum-resistant cryptosystems from supersingular elliptic curve isogenies. J. Math. Crypt. **8**(3), 209–247 (2014)
8. Silverman, J.H.: The Arithmetic of Elliptic Curves. GTM, vol. 106. Springer, New York (1992)
9. Montgomery, P.L.: Speeding the pollard and elliptic curve methods of factorization. Math. Comput. **48**, 243–264 (1987)
10. Couveignes, J.-M.: Hard homogeneous spaces. Cryptology ePrint Archive, Report 2006, 291 (2006)
11. Koziel, B., Azarderakhsh, R., Kermani, M.M., Jao, D.: Post-Quantum Cryptography on FPGA Based on Isogenies on Elliptic Curves. Cryptology ePrint Archive, Report 2016, 672 (2016). http://eprint.iacr.org/2016/672
12. Karmakar, A., Roy, S., Vercauteren, F., Verbauwhede, I.: Efficient finite field multiplication for isogeny based post quantum cryptography. In: International Workshop on the Arithmetic of Finite Fields, WAIFI 2016, to appear

13. Montgomery, P.L.: Modular multiplication without trial division. Math. Comput. **44**(170), 519–521 (1985)
14. McIvor, C., McLoone, M., McCanny, J.V.: High-radix systolic modular multiplication on reconfigurable hardware. In: IEEE International Conference on Field-Programmable Technology, pp. 13–18, December 2005
15. Orup, H.: Simplifying quotient determination in high-radix modular multiplication. In: Proceedings of the 12th Symposium on Computer Arithmetic, ARITH 1995, pp. 193–199. IEEE Computer Society, Washington (1995)
16. Azarderakhsh, R., Jao, D., Kalach, K., Koziel, B., Leonardi, C.: Key compression for isogeny-based cryptosystems. In: Proceedings of the 3rd ACM International Workshop on ASIA Public-Key Cryptography, AsiaPKC 2016, pp. 1–10. ACM, New York (2016)
17. Koziel, B., Jalali, A., Azarderakhsh, R., Jao, D., Mozaffari-Kermani, M.: NEON-SIDH: efficient implementation of supersingular isogeny Diffie-Hellman key exchange protocol on ARM. In: 15th International Conference on Cryptology and Network Security, CANS (2016)

AEZ: Anything-But EaZy in Hardware

Ekawat Homsirikamol and Kris Gaj[✉]

Electrical and Computer Engineering Department,
George Mason University, Fairfax, VA, USA
{ehomsiri,kgaj}@gmu.edu

Abstract. We provide the first hardware implementation of AEZ, a third-round candidate to the CAESAR competition for authenticated encryption. Complex, optimized for software, and impossible to implement in a single pass, AEZ poses significant obstacles for any hardware realization. Still, we find that a hardware implementation of AEZ is quite feasible. On Xilinx Virtex-6 FPGAs, our single-core design has a throughput exceeding 3.4 Gbit/s, and uses about 4600 LUTs and about 1250 CLB slices. In terms of the throughput to area ratio, this performance places it on the 12th position among 28 CAESAR candidate families benchmarked during Round 2 of the competition (assuming the key size of at least 96 bits, and the limit on the message size equal to $2^{11} - 1$ bytes). At the same time, AEZ targets a stronger notion of security against the cipher misuse than all other algorithms implemented and ranked ahead of it in the Round 2 hardware benchmarking study.

Keywords: Authenticated ciphers · AEAD · CAESAR · FPGA · Hardware · MRAE

1 Introduction

Authenticated encryption (AE) has become the preferred approach, in most settings, for achieving symmetric encryption. This paper describes the first hardware implementation of AEZ [9,10], a new AE scheme that targets an unprecedentedly strong security notion. Let us back up and provide a bit of context.

The AE Goal. An AE scheme takes in a key, a nonce, associated data (AD), and a plaintext. For majority of schemes, it returns a ciphertext and a tag. For some schemes, such as AEZ, it returns just a ciphertext (which is then typically longer than the plaintext). Decryption reverses the process, using the same key, nonce, and associated data (AD), as well as the ciphertext and, optionally, the tag returned by encryption, as an input. It returns either a plaintext or an indication of invalidity. There are two aims. *Confidentiality* requires ciphertexts to be computationally indistinguishable from random bits, while *authentication* assures that no one should be able to produce new and valid ciphertexts without knowing the key.

At present, there are just two widely used AE schemes, CCM and GCM. Both are standardized by ISO and NIST, but neither is particularly modern,

© Springer International Publishing AG 2016
O. Dunkelman and S.K. Sanadhya (Eds.): INDOCRYPT 2016, LNCS 10095, pp. 207–224, 2016.
DOI: 10.1007/978-3-319-49890-4_12

efficient, or versatile. To address this, the CAESAR competition for AE schemes began in 2012, attracting some 57 submissions [5].

The AEZ Scheme. AEZ [9, 10] is one of the more unusual CAESAR candidates. Where many submissions tried to excel in hardware efficiency, software efficiency, or both, AEZ focused on a new and unusually strong security notion. That goal, *robust* authenticated encryption (RAE), guarantees all that a conventional AE scheme does and more. First, it must work *as well as possible* even if nonces *do* repeat. That is the goal of *misuse-resistant* authenticated encryption (MRAE) [15]. But an RAE scheme goes further, achieving this as-good-as-possible behavior for any choice of ciphertext expansion (how much longer a ciphertext is than a plaintext), including none at all.

The Cost of RAE. Proponents of RAE and MRAE think that schemes designed to meet these ends will be easier to use and less prone to misuse. But achieving these goals comes at a cost, starting with the fact that they can't be achieved by any one-pass scheme. (A one-pass scheme reads each input and writes each out left-to-right, employing a constant amount of memory.) To encrypt, you must make two passes over the plaintext or employ a buffer as big as the plaintext is long. This is no doubt the reason why, despite the importance of nonce-reuse security, very few CAESAR candidates tried to achieve MRAE. The only Round 2 schemes the authors are aware of are AEZ, Deoxys, HS1-SIV, and Joltik. The comparison among the four of the above schemes in terms of security is beyond the scope of this paper. However, for fairness, it should be mentioned that the security provided by AEZ has a birthday bound of 2^{64} blocks, limited by the state size of the algorithm, which is among the lowest among the Round 2 CAESAR candidates. That means that there are easy distinguishing and forging attacks by the time the adversary queries AEZ with about 2^{64} blocks of message, AD, or nonce. However, the users are protected against these attacks by staying below 2^{48} bytes of data (about 280 TB), by that time, they need to rekey. Increasing this birthday bound was clearly and explicitly a non-goal for the designers of AEZ [10, p. 13]. On the hardware benchmarking side, no VHDL/Verilog implementations of the nonce misuse resistant variants of Deoxys and Joltik, compliant with the CAESAR Hardware API, have been reported to date.

Achieving RAE (which, again, goes beyond MRAE) is an especially tall order, encompassing the ability to encipher arbitrary-length strings. AEZ aims to achieve this with about the efficiency of AES-CTR. The result is the most complicated symmetric encryption scheme we know. AEZ's description spans 1.5 pages of dense pseudocode (excluding the definition of the AES round function and Blake2b) [10].

After explaining that AEZ's name was meant to suggest both *authenticated encryption* (AE) and *easy* (EZ), its authors warn that the alleged easiness refers only to ease of use. "Writing software for AEZ is *not* easy," they write, "while doing a hardware design for AEZ is far worse" [10, p. 2]. After some interaction with us, the AEZ designers added in that "From the hardware designer's perspective, AEZ's name might seem ironic, the name better suggesting *anti-easy*, the *antithesis of easy*, or *anything-but easy*" [10, pp. 2–3]. We note that a prior attempt at implementing AEZ by a Master-level student did not succeed, the designer concluding that AEZ was "hardly suitable for hardware" [3, p. 30].

Contributions. In this paper we overcome these difficulties and develop a fully-functional hardware realization of AEZ. Our realization conforms to the CAE-SAR Hardware API used in the CAESAR competition [11]. We implement everything in the AEZ spec except for the parts that handle arbitrary key lengths, arbitrary ciphertext expansion, and vector-valued AD. Please note that we are not aware of any other Round 2 CAESAR candidate offering arbitrary ciphertext expansion and vector-valued AD.

Our implementation achieves roughly the same throughput as the comparable implementation of AES-GCM, and takes almost the same area as the comparable implementation of OCB. In terms of the throughput to area ratio, our design ranks no. 12 out of 28 benchmarked Round 2 families (assuming the key size greater or equal to 96, and the limit on the message size equal to $2^{11} - 1$ bytes). It trails AES-GCM, only because of the larger area. It outperforms many other AES-based CAESAR candidates, such as CLOC, ELmD, OCB, AES-OTR, SILC, POET, AES-COPA and SHELL.

2   AEZ Overview

AEZ is built on a *generalized block cipher*, Encipher. This object is like a conventional block cipher except that (1) you can feed it any number of bytes (which will get enciphered into the same number of bytes), and (2) you can also provide a *tweak*, which, in this case, is a vector of strings. The tweak is a non-secret value that individualizes the permutation associated to the key.

To create an RAE scheme from its generalized block cipher, AEZ does the following: it takes the input M and it appends to it τ zero bits, where τ is the ciphertext expansion the user wants. Our realization assumes $\tau = 128$. Then you encipher. The result is the final ciphertext. To decrypt with AEZ, reverse the process, deciphering the ciphertext to get an augmented message. If the last τ bits of this augmented message is anything but the all-zero string, the ciphertext is invalid. Otherwise, the rest is the plaintext. For both enciphering and deciphering one uses a tweak that consists of three components (assuming a string-valued AD): an encoding of the ciphertext expansion τ, the nonce N, and the AD A.

Figure 1 describes the generalized block cipher Encipher. The message, M, is already assumed to be extended with τ zeros that we wish to encipher. Initially, attend only to the top-left and top-right portions of the diagram, and assume that $M = M_1 M_1' \cdots M_m M_m' M_u M_v M_x M_y$ has a multiple of 32 bytes (but at least 64 bytes). Each subscripted variable is 16 bytes.

The boxes labeled by pairs (j, i) in the diagram show the application of a *tweakable block cipher* (TBC). The key is always K, the key we wish to encipher

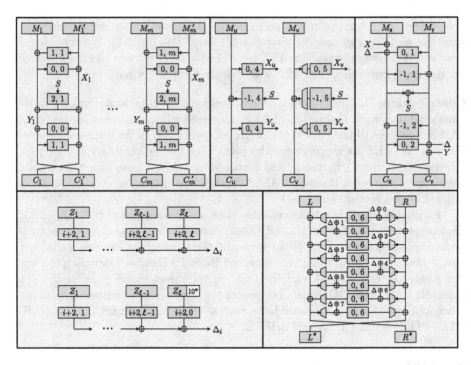

Fig. 1. Illustration of AEZ enciphering, adapted (with permission) from Fig. 5 of the AEZ spec [10]. Rectangles with pairs of numbers are tweakable block ciphers, the pair being that tweak (the key, always K, is not shown). **Top row:** enciphering a message M of (32 or more bytes) with AEZ-core. The i-block (top left) is used for the bulk of the message, but the xy-block (top right) comprises the last 32 bytes, while the uv-block (top middle) comprises the prior 0–31 bytes. (The picture shows a uv-block of 17–31 bytes.) The string X is computed via $X \leftarrow X_1 \oplus \cdots \oplus X_m \oplus X_u \oplus X_v$; if X_u or X_v is undefined then this term is omitted in computing X. The string Y is computed analogously. **Bottom left:** AEZ-hash computes $\Delta = \bigoplus \Delta_i$ from a vector-valued tweak encoding the ciphertext expansion τ, the nonce N, and the AD A. Its i-th component $Z_1 \cdots Z_\ell$ is hashed as shown. **Bottom right:** AEZ-hash, when operating on a string $M = L\,R$ of 16–31 bytes. More rounds are used if M has 1–15 bytes.

under, while the pair in the box is the tweak. Thus the box labeled $(1,1)$ maps K and the 16-byte M_1' to an unnamed 16-byte value that is xor'ed with M_1 to get another 16-byte value, which is fed into the block cipher labeled $(0,0)$, and so on. The figure's caption should make the notation clear.

The middle-top portion of the diagram hints at what happens when the plaintext is not a multiple of 32 bytes. For all *other* AE schemes we know, messages that are not multiples of the block size give rise to a final fragment that includes all the leftover bytes of the message. For AEZ, any leftover bytes form the penultimate chunk instead. This is useful because it ensures that, when 16 or fewer zero bytes are appended to the end of a message, they will always land in a specific block, rather than spanning two. If we want to shortcut the rejection of invalid messages, this feature has a potential to simplify the implementation.

To encipher messages with fewer than 32 bytes, one bypasses the top row of Fig. 1 completely and runs the Feistel network shown at the bottom right instead, splitting the message M into equal-length halves. This algorithm is called AEZ-tiny; the top row is AEZ-core. The two-algorithm approach, one for short messages and another for longer ones, mirrors a large body of cryptographic work in which techniques for "format preserving encryption" (FPE) do not resemble the modes of operation for a "wideblock block cipher."

The TBC used in AEZ is based on AES4 and AES10, which are 4 or 10 round versions of AES. The first is depicted in Fig. 1 as a (j, i)-labeled rectangle for $j \geq 0$; the second is a (j, i)-labeled square for $j = -1$. Neither uses the AES key schedule. In the end, the bulk of the work for enciphering 32 bytes from a long message—one "column" from the top-left of the figure—is the 20 AES rounds associated to the five AES4-based TBC calls. So 10 AES rounds per 16 bytes, the same overhead as AES itself.

One further detail concerns the processing of the empty message $M = \varepsilon$, which AEZ gives special attention to since this is a fairly natural way to realize a message authentication code, the string that is authenticated being the AD.

Among the pleasant characteristics of AEZ is that only the forward direction of the TBC is ever needed, and enciphering and deciphering are virtually identical. Our hardware design benefits from these choices.

3   Hardware Implementation Challenges

AEZ is the most challenging CAESAR candidate to implement in hardware. The reasons for this are summarized below.

Three Algorithms in One. AEZ defines three substantially different algorithms: (a) AEZ-prf to process empty messages, (b) AEZ-tiny to process messages of the size smaller than 32 - authenticator length (in bytes) (= 16 bytes for recommended values of parameters), and (c) AEZ-core to process all remaining message sizes. Although these algorithms share the same major building block, TBC, they have a very different internal structure, and implementation requirements. A hardware designer is faced with the decision to either implement these algorithms separately (without resource sharing), which may substantially increase the circuit area, but simplifies scheduling and control, or base the implementation on a single instance of TBC, which has the opposite implications. In our design, we chose the latter approach in order to address the already quite substantial area requirements of AEZ.

Two-Passes. As shown in Fig. 1, AEZ-core requires two passes. The first pass is used to calculate S, which is a function of the nonce (public message number), all blocks of the associated data, all blocks of the message, the authenticator length τ, and the key. In the second pass, S is used in calculations involving all message blocks. A hardware designer is faced with the decision to either repeat approximately 40 % of computations involving all message blocks, already done in the first pass, or to store intermediate results of the size of the entire message in internal memory. In order to avoid a substantial performance penalty, and keeping in mind that (a) packet sizes in modern communication protocols are relatively small (typically at or below 1500 bytes), and (b) modern FPGAs contains large blocks of memory, which often remain unused by the main cryptographic and data processing tasks, we have decided to follow the latter approach.

Input Re-Blocking. In a typical hardware implementation of an authenticated cipher, input blocks are provided to the cipher module sequentially, one by one. Only one block is processed at a time. All blocks, except the last one have the same length. The last block is often just padded, and then processed similarly (although rarely identically) as other blocks. After each message block is processed, the corresponding ciphertext block leaves the cipher module. As shown in Fig. 1,

- in AEZ-tiny, the blocks L and R have variable length depending on the size of the message, $|M|$,
- in AEZ-core, the blocks M_u and M_v have variable length depending on the size of the message, $|M|$. On top of that (a) neither of these blocks is the last block of the message, and (b) for certain message lengths $|M_u| = 0$. As a result, the implementation of AEZ must internally create and process blocks of data of unconventional sizes, which amounts to input "re-blocking". In hardware, such operation requires variable shifts and rotations, as well as clearing (also known as masking) of variable-size fragments of a block. All these functions have quite substantial area requirements. Additionally, "re-blocking" often requires simultaneous processing of at least two subsequent message blocks, before any of the corresponding ciphertext blocks is released.

Treatment of Incomplete Blocks. The treatment of incomplete blocks is a particularly complex operation in AEZ. As already mentioned in the previous section, these blocks are not the last blocks of the message, and in spite of that still require padding. Additionally, as shown in Fig. 1, they also require substantially different parameters j and i of the Tweakable Block Cipher ($E_K^{j,i}$).

Need for Pre-computations. In order to support the efficient implementation of TBC, the precomputations are highly desirable. The time of these precomputations and the amount of memory required to store the precomputed look-up tables is dependent on the maximum size of the message and the maximum size of associated data. See Sect. 4.2 for details.

Scheduling. As a result of all the aforementioned factors, the complexity of scheduling and the subsequent difficulty of developing a controller for the hardware implementation of AEZ exceeds the difficulty of any other symmetric cryptographic algorithm the authors are aware of, including all other two-pass CAESAR candidates.

4 Design Architecture

4.1 Interface, Protocol, and Design Parameters

Our implementation is based on the CAESAR Hardware API for Authenticated Ciphers, specified in [11], and its Appendix [7]. This API specifies both the interface and the detailed protocol for communication with the core. On top of that, for high-speed implementations, the authors of this API suggest the use of a top-level design, shown in Fig. 2, and provide the corresponding supporting codes implementing the Pre-Processor, Post-Processor, and CMD FIFO. Our implementation takes full advantage of these resources.

Our hardware design is fully optimized for the maximum throughput to area ratio. Its API and performance makes it suitable for use as a part of practical industry-grade systems based on standard bus interfaces such as ARM AXI-4 (Advanced eXtensible Interface 4) [2].

The hardware design presented in this paper aims to be as complete as the software implementation for the Round 2 version of AEZ (v4), developed by the AEZ team [13,14]. One significant difference between the software API and the hardware API is as follows: In the software API [1], the only output from authenticated encryption is the Ciphertext, denoted as c, of the length clen. In our hardware API, the output from authenticated encryption is divided into the Ciphertext and the Tag. In case of AEZ, which does not explicitly specify the tag, the tag is understood as follows. For non-empty messages, the tag is a result of enciphering a sequence of zeros, called an authenticator, of the length of τ bits, using the AEZ Encipher algorithm. For empty messages, the tag is a result of calculating the special AEZ-prf function of nonce, associated data, and the authenticator length τ.

The supported parameters are: key length = 384 bits, nonce length = 96 bits, authenticator length (denoted as ABYTES for the length in bytes and τ for the length in bits) = tag length = 16 bytes = 128 bits, maximum AD = $2^{10} - 1$ bytes, and maximum message/ciphertext size = $2^{11} - 1$ bytes. The maximum sizes of the message, ciphertext, and AD were chosen to support the maximum length of the Ethernet v2 packets [12], equal to 1500 bytes. Additionally our choices

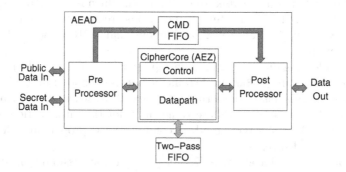

Fig. 2. Top-level design of a two-pass authenticated cipher.

limit the amount of memory required to implement the Two-Pass FIFO. All these choices are fully compliant with the official CAESAR Hardware API for Authenticated Ciphers, approved by the CAESAR Committee [11].

Our design supports both authenticated encryption and authenticated decryption operation, in such a way that only one of these two operations can be executed at a time (half-duplex). This way our design demonstrates the algorithm's ability to share resources between encryption and decryption. Key scheduling, padding and handling of incomplete blocks is implemented fully in hardware. The result of the decrypted message authentication (Success or Failure) is calculated within the core itself. Any unused portions of the last words of outputs are cleared (filled with zeros) before releasing these words outside of the cipher core.

The secret data input ports, used to enter the key, are separated from the public data input ports, used to enter all remaining data. The Public Data Input (PDI) and Data Output (DO) ports have the data port width equal to 64 bits, the Secret Data Input (SDI) port has the width of 32 bits. Our implementation has only one clock and supports only one input stream at a time.

4.2   Tweakable Block Cipher

Design. AEZ is built on top of the Tweakable Block Cipher (TBC) denoted as $E_K^{j,i}$. In Fig. 1, each call to TBC is denoted as a rectangle with parameters (j, i). The parameter j has discrete integer values -1, 0, 1, and 2 for processing message blocks, and values greater or equal to 3 for processing of nonce and associated data. The parameter i has values varying between 0 and m. For processing of messages, the dependence between the message length (in bytes) and m is as follows: $32 \cdot (m+1) \leq message\ length < 32 \cdot (m+2)$. For processing of messages, $m + 1$, is the number of complete 32-byte message block pairs in Message extended with the 16-byte authenticator. For processing of AD, l is the number of complete 16-byte blocks of AD. When processing incomplete AD blocks, as well as when $j = 0$ or -1, i is set to special values shown in Fig. 1.

The block diagram of the TBC module is shown in Fig. 3. Primary ports of the module are shown in bold font: X is the data input, Y is the result, K is the key. The shaded region is used to calculate Δ, which is a variable dependent on the key K and the parameters j and i. The remaining region is used to perform AES calculations on $X \oplus \Delta$, and an optional XOR of the result of these calculations with Δ.

In the shaded region, the x2 module represents the Galois field multiplication by two. I-RAM and J-RAM are two memories used as look-up tables for the precomputed expressions of the form of $2^P I$ and $2^P J$, where $P = 0..15$. The T register is used to store intermediate values used for the initialization of I-RAM and J-RAM. The Δ_{i+1} register is used for computing the proper value of Δ to be used by the unshaded region.

Based on the pseudocode of AEZ [10, p. 7] and our assumption about the size of Nonce (96 bits), Δ can take the following values:

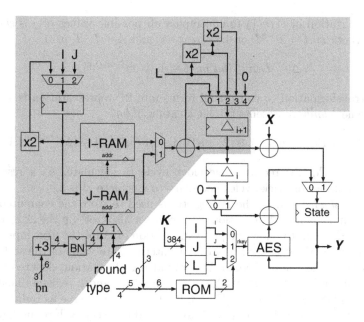

Fig. 3. Block diagram of TBC. Buses have the width of 128 bits unless specified otherwise.

- iJ for $j = -1$, $1 \le i \le 5$
- iI for $j = 0$, $i = 0, 1, 2, 4, 5, 6$
- $(2^{3+\lfloor (i-1)/8 \rfloor} + ((i-1) \bmod 8))I$ for $j = 1, 2$, $1 \le i \le m$
- $2^{j-3}L$ for $j = 4, 5$, $i = 0$
- $2^{j-3}L \oplus (2^{3+\lfloor (i-1)/8 \rfloor} \oplus ((i-1) \bmod 8))J$ for $j = 3, 5$, $1 \le i \le l$.

where,

- $j = 3, 4$, and 5 are used only inside of AEZ-hash(K,T), where $T = ([\tau]_{128}$, N, A).
- $(j = 3, i = 1)$ is used to process the authenticator length, expressed using 128-bits, $[\tau]_{128}$.
- $(j = 4, i = 0)$ is used only to process a 96-bit Nonce, N, i.e., one incomplete block.
- $(j = 5, i \ge 0)$ is used only to process AD, which may include an incomplete block (for which i = 0).

Under the assumption that the maximum AD size is $2^{10} - 1$ bytes and the maximum message size is $2^{11} - 1$ bytes, the maximum value of $bn = i - 1$ is equal to $max(bn) = max(i-1) = max(m-1, l-1) = max(l-1) = \lfloor \frac{2^{10}-1}{2^4} \rfloor - 1 = 2^6 - 1$. Thus, $max(3 + \lfloor \frac{i-1}{8} \rfloor) = 3 + \lfloor \frac{2^6-1}{8} \rfloor = 3 + 7 = 10 \le 15$.

The total number of clock cycles required to pre-compute Δ is based on the number of clock cycles required to calculate the longest possible Δ term, shown in Eq. (1).

$$\Delta \leftarrow 2^{j-3}L \oplus (2^{3+\lfloor (i-1)/8 \rfloor} \oplus ((i-1) \bmod 8))J \qquad (1)$$

The generalization of Eq. (1) to encompass all possible values of j is shown in Eq. (2), where $Init = 2^{j-3}L$ or 0, $bn = i - 1$, and $A = I$, J, or 0.

$$\Delta \leftarrow Init \oplus (bn \bmod 8)A \oplus (2^{3+\lfloor bn/8 \rfloor})A \qquad (2)$$

Further transformation to convert all terms into 2^P representation is shown in Eq. (3), where $bn[b]$ represents the bit location of bn.

$$\Delta \leftarrow Init \oplus (bn[0])A \oplus (2 \cdot bn[1])A \oplus (4 \cdot bn[2])A \oplus (2^{3+bn[6:3]})A \qquad (3)$$

Each term in Eq. (3) requires one clock cycle to calculate. As a result, the maximum number of clock cycles necessary to calculate Δ is 5.

In the *unshaded* region, the Δ_i register is used to store the computed Δ for the final, conditional $\oplus \Delta$ operation. This register also frees up the Δ_{i+1} register in the shaded region to allow the pre-computation of Δ for the next input block.

The State register is used to store an intermediate value of the state, used as an input to the combinational AES round transformation, denoted by AES, or as an output from the entire TBC function. I, J, and L registers hold three separate 128-bit portions of the 384-bit K. These values serve as round keys to the AES round module. The output of ROM is used to select each round key using the 4-bit *round* signal and the 2-bit *type* signal. The *type* is used to select a key set ($\mathbf{k_1}$, $\mathbf{k_2}$, *or* $\mathbf{K}$). The reader should refer to the pseudocode of AEZ, algorithm $\mathrm{E}_K^{j,i}(X)$, for the exact meaning of $\mathbf{k_1}$ and $\mathbf{k_2}$ [10, p. 7]. The total number of clock cycles required to compute the AES-based transformation, $AES10_k$, $AES4_k$, or $AES4_{k_j}$, is equal to the number of AES rounds plus 1. Thus, depending on a particular transformation, this number is equal to either 5 or 11 clock cycles.

Operation. During the one-time pre-calculations, dependent only on the key K, the I, J, and L registers are initialized with the appropriate portions of K. Then, the RAM modules in the shaded region are filled with $2^P \cdot A$, where $A = I$ or J, and $P = 0..15$. The initialization of I-RAM is achieved by loading I to the T register. The T value is then doubled during each of the subsequent 15 clock cycles. All intermediate values of T are stored at the consecutive locations of I-RAM. The counter *round*, incremented from 0 to 15, is used to address I-RAM during these pre-computations. The same procedure is used for the initialization of J-RAM.

Once the look-up tables stored in I-RAM and J-RAM are initialized, the processing of inputs X can start. A typical operation for each 128-bit block X is separated into two stages. The first stage, located in the shaded region of the block diagram, pre-computes the value of Δ, which is dependent on the values of i, j, and K. The second stage, located in the unshaded region, uses the calculated Δ to perform the AES-based computations. The operations of these two stages are categorized into different modes of operation depending on the input parameters j and i, as shown Table 1.

The two stages operate in tandem, with specific actions determined by the *mode*, dependent on the values of j and i, and used by the controller. In case the

Table 1. Modes of operation for TBC. Note: $\alpha = 2^{3+bn[6:3]}A$ where $A = I$ or J. Finalization denotes the final XOR with Δ.

Mode	(j, i)	First stage (pre-computation)			Second stage (main round)		
		Init	I or J	α	Round	Key	Finalization
0	(0, x)	0	I	No	4	k_1	No
1	(1, x)	0	I	Yes	4	k_1	No
2	(2, x)	0	I	Yes	4	k_2	No
3	(3, 1)	L	J	Yes	4	k_1	Yes
4	(4, 0)	2L	J	No	4	k_1	Yes
5	(5, 0)	4L	J	No	4	k_1	Yes
6	(5, x)	4L	J	Yes	4	k_1	Yes
7	(-1, x)	0	J	No	10	K	No

second stage requires a much longer computation time ($mode = 7$), the subsequent operation of the first stage is stalled until the second stage is completed. For each mode of operation, the first stage begins its operation from the initialization of the Δ_{i+1} register with the $Init$ value. If $j > 0$ and $i > 0$, Δ_{i+1} is then XORed with (bn mod 8) A $= 2^{bn[0]}A \oplus 2^{bn[1]}A \oplus 2^{bn[2]}A$ using three clock cycles. In the last clock cycle of the first stage computations, Δ_{i+1} is XORed with α.

The second stage, in the first clock cycle, XORs the pre-computed Δ value with the input X. The remaining clock cycles are spent on computing the AES rounds. Finalization is performed in the last clock cycle, if required.

Both stages operate in parallel, with the second stage performing calculations dependent on the current inputs X, j, and i, and the first stage performing calculations dependent on the next set of inputs j and i.

4.3   CipherCore

The CipherCore Datapath of AEZ is shown in Fig. 4. In order to limit the size of this block diagram and preserve its readability, control signals, serving as inputs to majority of medium-level components, such as TBC, NPAD, MASK and PAD, are not explicitly shown in this diagram.

TBC is the main encryption module. Its internal structure and operation is described in Sect. 4.2. This module serves as a focal point for all processing needs in our design. It processes 128 bits of data at a time (half of a block pair for message/ciphertext and a full block for associated data). The surrounding logic is used to facilitate the transfer of data and storage of intermediate results for the main processor. The following description summarizes the usage of the primary auxiliary units.

The T register holds data that is being operated on by TBC. It is also used as a temporary register to store intermediate values when data shifting is required. The XY register holds the accumulated value of Δ from Fig. 1 or $\Delta \oplus XY$ where $XY = XY_1 \oplus \ldots \oplus XY_m \oplus XY_u \oplus XY_v$ and $XY = X$ for the first pass, and Y for the second pass.

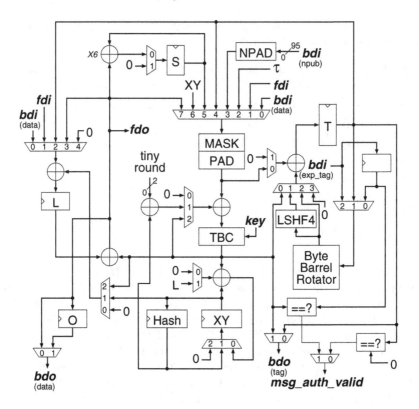

Fig. 4. The CipherCore Datapath of AEZ. Buses have the width of 128 bits unless specified otherwise.

The S register is used to hold the S value calculated at the end of the first pass, during processing of M_x and M_y, as shown in Fig. 1. The O register is used to hold any output that needs to be delayed in order for the output format to be the same as in the software implementation. The NPAD module performs 10* padding for the 96-bit nonce. The MASK and PAD modules are used to perform masking and padding operations required during processing of the last-but-one message block pair with indices u and v, as well as during AEZ-Tiny operations.

The Byte Barrel Rotator module is a variable rotation module. It can rotate by any integer multiple of a full byte. LSHF4 is a 4-bit left shifter used only for the AEZ-Tiny operation. It is required when an input block is of an odd size in bytes, and data needs to be split at a boundary of a nibble.

5   Timing Analysis

5.1   Latency

The design latency is given by Eq. (4). It is a function of T_{Hash}, T_{PRF}, T_{Tiny} and T_{Core}, shown in Eqs. (5), (6), (7), and (8), respectively. T_{Core} is a function of

T_{Full}, T_{UV}, and T_{XY} shown in Eqs. (9), (10), and (11), respectively. In all these equations $|AD|$ and $|M|$ represent the lengths of AD and message, respectively, in bits.

The detailed formulas are important, as they allow the accurate timing analysis for multiple AD and message sizes, and not only for the case of long messages.

$$Latency = T_{keysetup} + T_{Hash} + T_{PRF} + T_{Tiny} + T_{Core}$$
$$= 36 + T_{Hash} + T_{PRF} + T_{Tiny} + T_{Core} \tag{4}$$

$$T_{Hash} = 15 + \left\lceil \frac{|AD|}{128} \right\rceil \cdot 5 \tag{5}$$

$$T_{PRF} = \begin{cases} 0, & \text{if } |M| > 0 \\ 14, & \text{otherwise} \end{cases} \tag{6}$$

$$T_{Tiny} = \begin{cases} 0, & \text{if } |M| \geq 128 \\ 49, & \text{otherwise} \end{cases} \tag{7}$$

$$T_{Core} = \begin{cases} 0, & \text{if } |M| < 128 \\ 12 + T_{XY}, & \text{elif } |M| = 128 \\ 12 + T_{UV} + T_{XY}, & \text{elif } (|M| - 128) < 256 \\ 12 + T_{Full} + T_{XY}, & \text{elif } (|M| - 128) \bmod 256 = 0 \\ 12 + T_{Full} + T_{UV} + T_{XY}, & \text{otherwise} \end{cases} \tag{8}$$

$$T_{Full} = 25 \cdot \left\lfloor \frac{|M| - 128}{256} \right\rfloor + 5 \tag{9}$$

$$T_{UV} = 11 \cdot \left\lceil \frac{(|M| - 128) \bmod 256}{128} \right\rceil + 13 + \begin{cases} 2, & \text{if } (|M| - 128) \bmod 256 = 128 \\ 4, & \text{otherwise} \end{cases} \tag{10}$$

$$T_{XY} = \begin{cases} 38, & \text{if } (|M| - 128) \bmod 256 > 0 \\ 32, & \text{otherwise} \end{cases} \tag{11}$$

In Fig. 5, we illustrate the quite complex dependence of the (a) latency in clock cycles, and (b) number of clock cycles per byte, on the size of the message in bytes, assuming an empty AD. Based on Fig. 5(b), the number of clock cycles per byte reaches the close-to-optimal performance already at message sizes around 50 bytes.

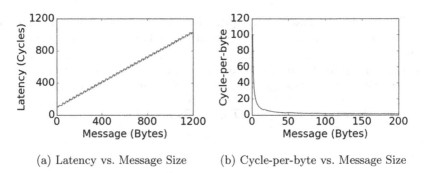

(a) Latency vs. Message Size   (b) Cycle-per-byte vs. Message Size

Fig. 5. The AEZ hardware module latency and the number of cycles per byte as a function of the message size for $|AD| = 0$

5.2 Throughput

Throughput for authenticated encryption and decryption of long messages is given by Eqs. (12) and (13). Equation (12) applies when $|M| = 0$, and $|AD| \gg 0$, where $\gg$ denotes "much bigger". It is based on the time it takes to perform the AEZ Hash operation (bottom left diagram of Fig. 1). Similarly, Eq. (13) applies when $|AD| = 0$, and $|M| \gg 0$. It is based on the time it takes to perform AEZ Core operation on a full block pair (top left diagram of Fig. 1).

$$Throughput_{AD} = \frac{128}{5} \cdot ClkFreq. \tag{12}$$

$$Throughput_M = \frac{256}{25} \cdot ClkFreq. \tag{13}$$

6   Benchmarking in Hardware

6.1   Hardware Results and Comparison with Other CAESAR Candidates

The resource utilization and the maximum clock frequency of the main components of AEZ on Virtex-6 FPGA is shown in Table 2. The TBC module requires about 48 % of the flip-flops and 37 % of the total LUTs as compared to the CipherCore module. The speed of the design is reduced by a factor of 8 % when the unit is integrated with the surrounding logic. The complete unit with the CAESAR Hardware API support (AEAD) requires an additional 15 % of flip-flops and 10 % of LUTs, on top of the resources required by the CipherCore module. The maximum frequency of operation remains exactly the same.

The comparison with all other Round 2 CAESAR candidates (except Tiaoxin), using the same hardware API, is summarized in Table 3. All results have been obtained using exactly the same FPGA device and FPGA tool versions. Benchmarking involved the optimization of tool options using ATHENa [8], with the

Table 2. Components analysis of AEZ unit on Virtex-6 xc6vlx240tff1156-3 FPGA device

	Resource utilization		Frequency
	FFs	LUTs	(MHz)
TBC	927	1527	362
CipherCore	1983	4166	335
AEAD	2347	4597	335

same optimization scheme and effort applied to all candidates. The source of these results is the ATHENa database of results [6], reporting FPGA performance for all implementations of Round 2 candidates submitted for benchmarking in June–August 2016. Each Round 2 CAESAR candidate family (except Tiaoxin) is represented in this study by one or more variants recommended by the submitter teams. For all the candidates and AES-GCM, the throughput is based on either encryption or decryption throughput, whichever is lower. Only the performance of the best variant in terms of the Throughput to Area ratio is reported in [6] and in Table 3, with LUTs used as a primary Area metric.

Since based on the CAESAR Hardware API [11], the implementations of single-pass authenticated ciphers are expected to support all message lengths $\leq 2^{32} - 1$, and implementations of two-pass authenticated ciphers are expected to support all lengths $\leq 2^{11} - 1$, it is natural and fair to compare implementations of both types of ciphers for the maximum message length common for both types of ciphers, which is $2^{11} - 1$.

Additionally, 2 Kbytes is a practical limit for majority of secure networking protocols, such as IPSec – a primary target for high-speed hardware implementations of authenticated encryption. Authenticated encryption without intermediate tags is in general not a good match for applications requiring protection of large volumes of data-at-rest, due to large access times for reading and writing.

The implementers of 7 single-pass authenticated ciphers included in our comparison (AES-GCM, Deoxys, Joltik, OCB, OMD, PAEQ, and SCREAM) specifically supported the two possible maximum AD/message lengths. All corresponding results presented in Table 3 have been generated with the choice of the maximum AD/message equal to $2^{11} - 1$. This choice has appeared to benefit in a noticeable way only the two of them, OCB and OMD, using a precomputed look-up table, with the size dependent on the maximum AD/message length.

For the remaining candidates, we contacted the designers of the implementations listed in Table 3, and asked them explicitly whether they see any way of optimizing their designs (in terms of area and/or maximum clock frequency) in case the maximum AD/message length is smaller or equal to $2^{11} - 1$. None of the designers responded positively to this question. Similarly, our own analysis and preliminary results led to the conclusion that the maximum benefit in terms of the throughput to area ratio, resulting from applying a lower limit on the AD/message length, is not likely to exceed 3 % for any of the remaining one-pass Round 2 CAESAR candidates.

Table 3. Comparison with other CAESAR candidates, with key sizes greater or equal to 96 bits, on Virtex 6 FPGA.

		Frequency (MHz)	Throughput (Mbit/s)	Area		TP/A	
				(LUTs)	(SLICEs)	(Mbit/s/ LUTs)	(Mbit/s/ SLICEs)
1	MORUS	179.7	46002	3898	1216	**11.801**	37.831
2	ACORN	347.7	11127	1194	421	**9.319**	26.430
3	TriviA-ck	300.2	19213	2310	895	**8.317**	21.467
4	ICEPOLE	304.0	44464	5734	1995	**7.754**	22.288
5	AEGIS	203.1	52001	7980	2143	**6.516**	24.266
6	Ketje	229.5	7345	1270	456	**5.783**	16.107
7	NORX	170.5	16368	2968	1022	**5.515**	16.016
8	ASCON	361.0	5134	1620	489	**3.169**	10.499
9	STRIBOB	276.1	11750	4839	1376	**2.428**	8.539
10	Keyak (River)	163.6	7417	6234	1751	**1.190**	4.236
	AES-GCM	*278.3*	*3239*	*3175*	*1053*	*1.020*	*3.076*
11	Deoxys (NR-128-128)	327.3	2793	3142	951	**0.889**	2.937
12	**AEZ**	**335.3**	**3434**	**4597**	**1246**	**0.747**	**2.756**
13	CLOC	254.6	2963	3983	1154	**0.744**	2.568
14	ELmD	247.5	3168	4302	1607	**0.736**	1.971
15	OCB	292.7	3122	4249	1348	**0.735**	2.316
16	PRIMATEs-GIBBON	224.0	1280	1807	653	**0.708**	1.960
17	Joltik (NR-128-64)	439.9	880	1292	524	**0.681**	1.679
18	Minalpher	280.9	1831	2879	1104	**0.636**	1.659
19	PAEQ	258.9	4537	8328	2300	**0.545**	1.973
20	AES-OTR	256.9	2741	5102	1385	**0.537**	1.979
21	SCREAM	170.4	1039	2052	834	**0.506**	1.246
22	Pi-Cipher	170.0	1740	3535	1077	**0.492**	1.616
23	SILC	280.7	1562	3378	989	**0.462**	1.579
24	PRIMATEs-HANUMAN	225.1	693	1769	626	**0.392**	1.107
25	POET	231.2	2959	7695	2444	**0.385**	1.211
26	HS1-SIV	221.7	2769	8392	2219	**0.330**	1.248
27	AES-COPA	214.9	2500	7754	2358	**0.322**	1.060
28	OMD	242.2	940	3562	1243	**0.264**	0.756
29	AES-JAMBU (SIMON)	209.8	186	1376	453	**0.135**	0.411
30	SHELL	16.3	522	81197	22830	**0.006**	0.023

On top of that, both single-pass and two-pass algorithms require *external* memory for the complete functionality, including the temporary storage of decrypted message. In an optimized implementation of the entire system including a two-pass AEAD core, the Two-Pass FIFO and the Output FIFO could be implemented using the same resources. The amount of logic (LUTs) required to multiplex between these two functions of an external memory would be negligible compared to the size of the entire system.

As a result, we believe that the need for an external Two-Pass FIFO, implemented using dedicated FPGA resources, such as Block RAMs, does not put two-pass algorithms in any noticeable disadvantage that could affect the ranking of the candidates (especially to the extent higher than other, more important factors, such as different designer skills and coding styles, different amount of time and effort spent on optimization, etc.)

Based on the results presented in [6], it is fair to say that AEZ outperforms all AES-based CAESAR candidates, other than AEGIS and Deoxys, such as CLOC, ELmD, OCB, AES-OTR, SILC, POET, AES-COPA and SHELL. Our implementation also outperforms the implementation of the only other two-pass Round 2 candidate variant, reported in [6], HS1-SIV. Our implementation of AEZ beats the equivalent implementation of HS1-SIV by a factor of 1.23 in terms of Throughput, 1.83 in terms of Area, and a combined factor of 2.26 in terms of the Throughput/Area ratio. Its Throughput to Area ratio is lower only than that of 11 mostly permutation-based algorithms, none of which fulfills the requirements of robust authenticated encryption (RAE), or even misuse-resistant authenticated encryption (MRAE).

6.2 Comparison with the Optimized Software Implementation

The preliminary results of the software benchmarking using SUPERCOP place AEZ among the top 5 authenticated ciphers on the amd64-architecture platforms [4]. The software benchmark of the optimized software implementation, available at [13], was done on a Skylake-S Intel Core i5-6600 3.3 GHz. The compiler and compilation flags used were: GCC 5.5 with "-march=native -O3". The optimized software implementation was able to achieve the performance of 0.64 cycles-per-byte, equivalent to the throughput of 41.25 Gbit/s for long messages. Comparing to our hardware AEZ core performance on Virtex-6 FPGA, the software is able to achieve approximately 12 times higher throughput, while running at about 10 times higher clock frequency.

Clearly, an optimized software implementation of an AES-based authenticated cipher, running on a modern microprocessor, can easily outperform the corresponding single-core hardware implementation, not just for AEZ, but for majority of other CAESAR candidates. However, one must remember that the hardware resources required by a modern microprocessor, as well as power and energy consumption, are likely much higher than resources required by a single core of AEZ.

On modern FPGAs and All-Programmable Systems on Chip (such as Xilinx Zynq), multiple AEZ cores can be placed and run in parallel to either hard or soft embedded microprocessor core (such as ARM or MicroBlaze). Their availability would free the microprocessor to perform other critical tasks. It would also allow significantly outperforming a single dedicated microprocessor core. For example, the largest Xilinx Virtex-6 FPGA (XC6VLX760) can host up to 95 AEZ Cores, reaching throughput in excess of 326 Gbit/s.

Results of software implementations of AEZ on multiple other platforms, including ARM, can be found in [4].

7   Conclusions

We have developed an efficient implementation of AEZ that outperforms comparable implementations of the majority of other AES-based Round 2 CAESAR candidates. It places 12th in terms of the Throughput to Area ratio, in the ranking of 28 candidates participating in the hardware benchmarking study (assuming the maximum message length of $2^{11} - 1$ bytes), and is outperformed only by single-pass, mostly permutation-based algorithms. Our preliminary analysis strongly suggests that AEZ can outperform majority of the CAESAR candidates and the current standard, AES-GCM, in software, approximately match the performance of AES-GCM in hardware, and at the same time offer a new unprecedented level of resistance against the cipher misuse.

References

1. Caesar call for submissions, final, January 2014. https://competitions.cr.yp.to/caesar-call.html
2. ARM: AMBA Specifications. http://www.arm.com/products/system-ip/amba-specifications.php
3. Arnould, C.: Towards developing ASIC and FPGA architectures of high-throughput CAESAR candidates. Master's thesis, ETH Zurich, March 2015
4. Bernstein, D.J., Lange, T. (eds.): eBACS: ECRYPT Benchmarking of Cryptographic Systems, October 2016. https://bench.cr.yp.to
5. CAESAR: Competition for Authenticated Encryption: Security, Applicability, and Robustness: Cryptographic Competitions, January 2016. http://competitions.cr.yp.to/index.html
6. Cryptographic Engineering Research Group (CERG) at GMU: GMU ATHENa Database of Results, July 2015. https://cryptography.gmu.edu/athenadb/fpga_auth_cipher/rankings_view
7. Cryptographic Engineering Research Group (CERG) at GMU: Addendum to the CAESAR Hardware API v1.0, June 2016. https://cryptography.gmu.edu/athena/index.php?id=CAESAR
8. Gaj, K., Kaps, J.P., Amirineni, V., Rogawski, M., Homsirikamol, E., Brewster, B.Y.: ATHENa - automated tool for hardware evaluation: toward fair and comprehensive benchmarking of cryptographic hardware using FPGAs. In: 20th International Conference on Field Programmable Logic and Applications - FPL 2010, pp. 414–421. IEEE (2010)
9. Hoang, V.T., Krovetz, T., Rogaway, P.: Robust authenticated-encryption AEZ and the problem that it solves. In: Oswald, E., Fischlin, M. (eds.) EUROCRYPT 2015. LNCS, vol. 9056, pp. 15–44. Springer, Heidelberg (2015). doi:10.1007/978-3-662-46800-5_2
10. Hoang, V.T., Krovetz, T., Rogaway, P.: AEZ v4.1: Authenticated Encryption by Enciphering, October 2015. http://web.cs.ucdavis.edu/~rogaway/aez/aez.pdf
11. Homsirikamol, E., Diehl, W., Ferozpuri, A., Farahmand, F., Yalla, P., Kaps, J.P., Gaj, K.: CAESAR Hardware API. Cryptology ePrint Archive, Report 2016/626 (2016). http://eprint.iacr.org/2016/626
12. Hornig, C.: A standard for the transmission of IP datagrams over ethernet networks. STD 41, RFC Editor, April 1984
13. Krovetz, T.: AEZ v4.1 aes-ni version, October 2015. http://www.cs.ucdavis.edu/~rogaway/aez
14. Krovetz, T.: AEZ v4.1 reference code, September 2015. http://www.cs.ucdavis.edu/~rogaway/aez
15. Rogaway, P., Shrimpton, T.: A provable-security treatment of the key-wrap problem. In: Vaudenay, S. (ed.) EUROCRYPT 2006. LNCS, vol. 4004, pp. 373–390. Springer, Heidelberg (2006). doi:10.1007/11761679_23

Functional Encryption

Private Functional Encryption: Indistinguishability-Based Definitions and Constructions from Obfuscation

Afonso Arriaga[1]([✉]), Manuel Barbosa[2], and Pooya Farshim[3]

[1] SnT, University of Luxembourg, Luxembourg City, Luxembourg
afonso.delerue@uni.lu
[2] HASLab - INESC TEC, DCC FC University of Porto, Porto, Portugal
mbb@dcc.fc.up.pt
[3] ENS, CNRS & Inria, PSL Research University, Paris, France
pooya.farshim@gmail.com

Abstract. Private functional encryption guarantees that not only the information in ciphertexts is hidden but also the circuits in decryption tokens are protected. A notable use case of this notion is query privacy in searchable encryption. Prior privacy models in the literature were fine-tuned for specific functionalities (namely, identity-based encryption and inner-product encryption), did not model correlations between ciphertexts and decryption tokens, or fell under strong uninstantiability results. We develop a new indistinguishability-based privacy notion that overcomes these limitations and give constructions supporting different circuit classes and meeting varying degrees of security. Obfuscation is a common building block that these constructions share, albeit the obfuscators necessary for each construction are based on different assumptions. In particular, we develop a composable and distributionally secure hyperplane membership obfuscator and use it to build an inner-product encryption scheme that achieves an unprecedented level of privacy, positively answering a question left open by Boneh, Raghunathan and Segev (ASIACRYPT 2013) concerning the extension and realization of *enhanced security* for schemes supporting this functionality.

Keywords: Function privacy · Functional encryption · Obfuscation · Keyword search · Inner-product encryption

1 Introduction

Standard notions of security for public-key functional encryption [16, 22] do not cover important use cases where, not only encrypted data, but also the circuits (functions) associated with decryption tokens contain sensitive information. The typical example is that of a cloud provider that stores an encrypted data set created by Alice, over which Bob wishes to make advanced data mining queries. Functional encryption provides a solution to this use case: Bob can send a decryption token to the cloud provider that allows it to recover the result of computing

© Springer International Publishing AG 2016
O. Dunkelman and S.K. Sanadhya (Eds.): INDOCRYPT 2016, LNCS 10095, pp. 227–247, 2016.
DOI: 10.1007/978-3-319-49890-4_13

a query over the encrypted data set. Standard security notions guarantee that nothing about the plaintexts beyond query results are revealed to the server. However, they do *not* guarantee that the performed query, which may for example contain a keyword sensitive to Bob, is hidden from the server.

1.1 Function Privacy

Function privacy is an emerging new notion that aims to address this problem. The formal study of this notion begins in the work of Boneh et al. [14], where the authors focused on identity-based encryption (IBE) and presented the first constructions offering various degrees of privacy. From the onset, it became clear that formalizing such a notion is challenging, even for simple functionalities such as IBE, as the holder of a token for circuit C may encrypt arbitrary messages using the public key, and obtain a large number of evaluations of C via the decryption algorithm. Boneh et al. therefore considered privacy for identities with high min-entropy. In general, however, the previous observation implies that (non-trivial) function privacy can only be achieved as long as the token holder is unable to learn C through such an attack, immediately suggesting a strong connection between private functional encryption and obfuscation.

Boneh et al. [14,15] give indistinguishability-based definitions of function privacy for IBE and subspace membership (a generalization of inner-product encryption). Roughly speaking, the IBE model imposes that whenever the token queries of the adversary have high min-entropy (or form a block source), decryption tokens will be indistinguishable from those corresponding to identities sampled from the uniform distribution. For subspace membership, the definition requires the random variables associated with vector components to be a block source.

Tokens for high-entropy identities, however, rarely exist in isolation and are often available in conjunction with ciphertexts encrypted for the very same identities. To address this requirement, the same authors [14] proposed an *enhanced* model for IBE in which the adversary also gets access to ciphertexts encrypted for identities associated with the challenge tokens. This model was subsequently shown in [5] to be infeasible under the formalism of Boneh et al., as correlations with encrypted identities can lead to distinguishing attacks, e.g. via *repetition patterns*. (We will discuss this later in the paper.) Although the model can be salvaged by further restricting the class of admissible distributions, it becomes primitive-specific and formulating a definition for other functionalities is not obvious (and indeed a similar extension was not formalized for subspace membership in [15]). Additionally, this model also falls short of capturing *arbitrary* correlations between encrypted messages and tokens, as it does not allow an adversary to see ciphertexts for identities which, although correlated with those extracted in the challenge tokens, *do not match any of them*.

Very recently, Agrawal et al. [1] have put forth a model for functional encryption that aims to address this problem with a very general UC-style definition (called "wishful security"). The core of the definition is an ideal security notion for functional encryption, which makes it explicit that both data privacy and function privacy should be simultaneously enforced. However, not only is this general simulation-based definition difficult to work with, but also aiming for it would amount to constructing virtual black-box obfuscation, for which strong impossibility results are known [6,20]. Indeed, the positive results of [1] are obtained in idealized models of computation.

1.2 Contributions

The above discussion highlights the need for a general and convenient definition of privacy that incorporates arbitrary correlations between decryption tokens and encrypted messages, and yet can be shown to be feasible without relying on idealized models of computation. The first contribution of our work is an *indistinguishability-based* definition that precisely models arbitrary correlations for general circuits. Our definition builds on a framework for *unpredictable* samplers and unifies within a single definition all previous indistinguishability-based notions.

The second contribution of the paper is four constructions of private functional encryption supporting different classes of circuits and meeting varying degrees of security: (1) a simple and functionality-agnostic construction shown to be secure in the absence of correlated messages, (2) a more evolved and still functionality-agnostic construction (taking advantage of recent techniques from [4]) that achieves function privacy with respect to a general class of samplers that we call *concentrated*; (3) a conceptually simpler construction specific for point functions achieving privacy in the presence of correlated messages beyond all previously proposed indistinguishability-based security definitions; (4) a construction specific for point functions that achieves our strongest notion of privacy (but relies on a more expressive form of obfuscation than the previous construction). We also develop an obfuscator for hyperplane membership that, when plugged into the second construction above gives rise to a private inner-product encryption scheme, answering a question left open by Boneh et al. [15] on how to define and realize *enhanced* security (i.e., privacy in the presence of correlated messages) for schemes supporting this functionality.

THE UNPREDICTABILITY FRAMEWORK. At the core of our definitional work lies a precise definition characterizing which distributions over circuits and what correlated side information can be tolerated by a private FE scheme. We build on ideas from obfuscation [8,10], functional encryption [16,22] and prior work in function privacy [1,5,14,15] to define a game-based notion of *unpredictability* for general functions. Our definition allows a *sampler* S to output a pair of circuit vectors $(\mathbf{C}_0, \mathbf{C}_1)$ and a pair of message vectors $(\mathbf{m}_0, \mathbf{m}_1)$ with arbitrary correlations between them, along with some side information z. Unpredictability then imposes that no predictor $\mathcal{P}$ interacting with oracles computing evaluations on these circuits and messages can find a point x such that $\mathbf{C}_0(x) \neq \mathbf{C}_1(x)$. (We do not impose indistinguishability, which is stronger, results in a smaller class of unpredictable samplers, and hence leads to weaker security.) The predictor $\mathcal{P}$ sees z and the outputs of the sampled circuits on the sampled messages.

It can run in bounded or unbounded time, but it can only make polynomially many oracle queries to obtain additional information about the sampled circuits and messages. To avoid attacks that arise in the presence of computationally unpredictable auxiliary information [9] we adopt *unbounded* prediction later in our analyses.

This formalism fixes the unpredictability notion throughout the paper. We can then capture specific types of samplers by imposing extra structural requirements on them. For instance, we may require the sampler to output a bounded number of circuits and messages, or include specific data in the auxiliary information, or do not include any auxiliary information at all. A sampler outputting circuits *and* messages comes to hand to model the privacy for functional encryption. We emphasize that our definition intentionally does *not* require the messages to be unpredictable. Further discussion on this choice can be found in Sect. 3.

THE PRIV MODEL. Building on unpredictability, we put forth a new indistinguishability-based notion of function privacy. Our notion, which we call PRIV, bears close resemblance to the standard IND-CPA model for functional encryption: it comes with a left-or-right LR oracle, a token-extraction TGEN oracle and the goal of the adversary is to guess a bit. The power of the model lies in that we endow LR with the ability to generate arbitrary messages and circuits via an unpredictable sampler. Trivial attacks are excluded by the joint action of unpredictability and the usual FE legitimacy condition, imposing equality of images on left and right. The enhanced model of Boneh et al. [14] falls in as a special case where the sampler is structurally restricted to be a block source. But our definition goes well beyond this and considers arbitrary and possibly low-entropy correlations. Furthermore, since unpredictability is *not* imposed on messages, PRIV implies IND-CPA security, and consequently it also guarantees *anonymity* for primitives such as IBE and ABE [16]. Correlated circuits may be "low entropy" as long as they are identical on left and right, and since previous definitions adopted a real-or-random definition, they had to exclude this possibility. By giving the sampler the option to omit, manipulate and repeat the messages, our security notion implies previous indistinguishability-based notions in the literature, including those in [1,5,14,15].

2   Preliminaries

2.1   Notation

We denote the security parameter by $\lambda \in \mathbb{N}$ and assume it is implicitly given to all algorithms in unary representation 1^λ. We denote the set of all bit strings of length ℓ by $\{0,1\}^\ell$ and the length of a string x by $|x|$. The bit complement of a string x is denoted by $\overline{x}$. We use the symbol ε to denote the empty string. A vector of strings $\mathbf{x}$ is written in boldface, and $\mathbf{x}[i]$ denotes its ith entry. The number of entries of $\mathbf{x}$ is denoted by $|\mathbf{x}|$. For a finite set X, we denote its cardinality by $|X|$ and the action of sampling a uniformly random element x from X by $x \leftarrow_\$ X$. For a random variable X we denote its support by $[X]$. For a circuit C we denote its size by $|C|$. We call a real-valued function $\mu(\lambda)$ negligible if $\mu(\lambda) \in \mathcal{O}(\lambda^{-\omega(1)})$ and denote the set of all negligible functions by NEGL. Throughput the paper $\perp$ denotes a special failure symbol outside the spaces underlying a cryptographic primitive. We adopt the code-based game-playing framework. As usual "ppt" stands for probabilistic polynomial time.

CIRCUIT FAMILIES. Let $\mathsf{MSp} := \{\mathsf{MSp}_\lambda\}_{\lambda \in \mathbb{N}}$ and $\mathsf{OSp} := \{\mathsf{OSp}_\lambda\}_{\lambda \in \mathbb{N}}$ be two families of finite sets parametrized by a security parameter $\lambda \in \mathbb{N}$. A circuit family $\mathsf{CSp} := \{\mathsf{CSp}_\lambda\}_{\lambda \in \mathbb{N}}$ is a sequence of circuit sets indexed by the security parameter. We assume that for all $\lambda \in \mathbb{N}$, all circuits in CSp_λ share a common input domain MSp_λ and output space OSp_λ. We also assume that membership in sets can be efficiently decided. For a vector of circuits $\mathbf{C} = [C_1, \ldots, C_n]$ and a vector of messages $\mathbf{m} = [m_1, \ldots, m_m]$ we define $\mathbf{C}(\mathbf{m})$ to be an $n \times m$ matrix whose ijth entry is $C_i(m_j)$. When $\mathsf{OSp}_\lambda = \{0, 1\}$ for all values of λ we call the circuit family Boolean.

2.2 Functional Encryption

SYNTAX. A functional encryption scheme FE associated with a circuit family CSp is specified by four ppt algorithms as follows. (1) $\mathsf{FE.Gen}(1^\lambda)$ is the setup algorithm and on input a security parameter 1^λ it outputs a master secret key msk and a master public key mpk; (2) $\mathsf{FE.TGen}(\mathsf{msk}, \mathsf{C})$ is the token-generation algorithm and on input a master secret key msk and a circuit $\mathsf{C} \in \mathsf{CSp}_\lambda$ outputs a token tk for C; (3) $\mathsf{FE.Enc}(\mathsf{mpk}, \mathsf{m})$ is the encryption algorithm and on input a master public key mpk and a message $\mathsf{m} \in \mathsf{MSp}_\lambda$ outputs a ciphertext c; (4) $\mathsf{FE.Eval}(\mathsf{c}, \mathsf{tk})$ is the deterministic evaluation algorithm and on input a ciphertext c and a token tk outputs a value $\mathsf{y} \in \mathsf{OSp}_\lambda$ or failure symbol $\perp$.

We adopt a *computational* notion of correctness for FE schemes and require that no ppt adversary is able to produce a message m and a circuit C that violates the standard correctness property of the FE scheme (that is, $\mathsf{FE.Eval}(\mathsf{FE.Enc}(\mathsf{mpk}, \mathsf{m}), \mathsf{FE.TGen}(\mathsf{msk}, \mathsf{C})) \neq \mathsf{C}(\mathsf{m})$), even with the help of an (unrestricted) token-generation oracle. We also adopt the standard notion of IND-CPA security [16, 22] where an adversary with access to a token-generation oracle cannot distinguish encryptions of messages m_0, m_1 under the standard restriction that it cannot obtain a decryption token for a circuit C for which $\mathsf{C}(m_0) \neq \mathsf{C}(m_1)$.

CORRECTNESS. We will adopt a game-based definition of *computational* correctness for FE schemes which has been widely adopted in the literature [2] and suffices for the overwhelming majority of use cases. Roughly speaking, this property requires that no efficient adversary is able to come up with a message and a circuit which violates the correctness property of the FE scheme, even with the help of an (unrestricted) token-generation oracle. Formally, scheme FE is computationally correct if for all ppt adversaries $\mathcal{A}$

$$\mathbf{Adv}_{\mathsf{FE}, \mathcal{A}}^{\mathrm{cc}}(\lambda) := \Pr\left[\mathrm{CC}_{\mathsf{FE}}^{\mathcal{A}}(1^\lambda)\right] \in \text{NEGL},$$

$CC_{FE}^{\mathcal{A}}(1^\lambda)$:	$IND\text{-}CPA_{FE}^{\mathcal{A}}(1^\lambda)$:	$LR(m_0, m_1)$:
$(msk, mpk) \leftarrow_\$ FE.Gen(1^\lambda)$	$(msk, mpk) \leftarrow_\$ FE.Gen(1^\lambda)$	$c \leftarrow_\$ FE.Enc(mpk, m_b)$
$(m, C) \leftarrow_\$ \mathcal{A}^{TGEN}(mpk)$	$b \leftarrow_\$ \{0, 1\}$	$MList \leftarrow (m_0, m_1) : MList$
$c \leftarrow_\$ FE.Enc(mpk, m)$	$b' \leftarrow_\$ \mathcal{A}^{LR, TGEN}(mpk)$	return c
$tk \leftarrow_\$ FE.TGen(msk, C)$	return $(b = b')$	
$y \leftarrow_\$ FE.Eval(c, tk)$		$TGEN(C)$:
return $(y \neq C(m))$		$tk \leftarrow_\$ FE.TGen(msk, C)$
		$TList \leftarrow C : TList$
$TGEN(C)$:		return tk
$tk \leftarrow_\$ FE.TGen(msk, C)$		
return tk		

Fig. 1. Left: Computational correctness of FE. **Right:** IND-CPA security of FE.

where game $CC_{FE}^{\mathcal{A}}(1^\lambda)$ is shown in Fig. 1 on the left. Perfect correctness corresponds to the setting where the above advantage is required to be zero.

SECURITY. A functional encryption scheme FE is IND-CPA secure [16,22] if for any legitimate ppt adversary $\mathcal{A}$

$$\mathbf{Adv}_{FE,\mathcal{A}}^{ind\text{-}cpa}(\lambda) := 2 \cdot \Pr\left[IND\text{-}CPA_{FE}^{\mathcal{A}}(1^\lambda)\right] - 1 \in \text{NEGL},$$

where game $IND\text{-}CPA_{FE}^{\mathcal{A}}(1^\lambda)$ is defined in Fig. 1 on the right. We say $\mathcal{A}$ is legitimate if for all messages pairs queried to the left-or-right oracle, i.e., for all $(m_0, m_1) \in MList$, and all extracted circuits $C \in TList$ we have that $C(m_0) = C(m_1)$. The IND-CPA notion self-composes in the sense that security against adversaries that place one LR query is equivalent to the setting where an arbitrary number of queries is allowed. It is also well known that IND-CPA security is weaker than generalizations of semantic security for functional encryption [7], but strong impossibility results for the latter have been established.

3   Unpredictable Samplers

The privacy notions that we will be developing in the coming sections rely on multistage adversaries that must adhere to certain high-entropy requirements on the sampled circuits. Rather than speaking about specific distributions for specific circuit classes, we introduce a uniform treatment for *any* circuit class via an unpredictability game. Our framework allows one to introduce restricted classes of samplers by imposing structural restrictions on their internal operation *without changes* to the reference unpredictability game. Our framework extends that of Bellare et al. [8] for obfuscators and also models the challenge-generation phase in private functional encryption in prior works [1,5,14,15].

SYNTAX. A sampler for a circuit family CSp is an algorithm $\mathcal{S}$ that on input the security parameter 1^λ and possibly some state information st outputs a pair of vectors of CSp_λ circuits $(\mathbf{C}_0, \mathbf{C}_1)$ of equal dimension, a pair of vectors of MSp_λ messages $(\mathbf{m}_0, \mathbf{m}_1)$ of equal dimension, and some auxiliary information z. We require the components of the two circuit (resp., message) vectors to be encoded as bit strings of equal length. Input st may encode information about the environment where the sampler is run (e.g., the public parameters of a higher-level protocol) and z models side information available on the sampled circuits or messages.

In the security games we will be considering later on, the goal of adversary will be to distinguish which of two circuit distributions produced by an unpredictable sampler was used to form some cryptographic data (e.g., an obfuscated circuit or an FE token). Our unpredictability definition formalizes the intuition that by examining the input/output behavior of the sampled circuits on messages of choice, the evaluation of legitimate circuits of choice on sampled messages, and the evaluation of sampled circuits on sampled messages, a point leading to differing outputs on some pair of sampled circuits cannot be found. Drawing a parallel to the functional encryption setting, once decryption tokens or encrypted messages become available, the tokens can be used by a legitimate adversary to compute the circuits underneath on arbitrary values, including some special messages that are possibly correlated with the circuits.

UNPREDICTABILITY. A *legitimate* sampler $\mathcal{S}$ is *statistically unpredictable* if for any unbounded *legitimate* predictor $\mathcal{P}$ that places polynomially many queries

$$\mathbf{Adv}^{\mathrm{pred}}_{\mathcal{S},\mathcal{P}}(\lambda) := \Pr\left[\mathrm{Pred}^{\mathcal{P}}_{\mathcal{S}}(1^\lambda)\right] \in \mathrm{NEGL},$$

where game $\mathrm{Pred}^{\mathcal{P}}_{\mathcal{S}}(1^\lambda)$ is shown in Fig. 2. Sampler $\mathcal{S}$ is called legitimate if $\mathbf{C}_0(\mathbf{m}_0) = \mathbf{C}_1(\mathbf{m}_1)$ in game $\mathrm{Pred}^{\mathcal{P}}_{\mathcal{S}}(1^\lambda)$. Predictor $\mathcal{P}$ is legitimate if $\mathsf{C}(\mathbf{m}_0) = \mathsf{C}(\mathbf{m}_1)$ for all queries made to the FUNC oracle.[1]

$$\boxed{\begin{array}{l}
\underline{\mathrm{Pred}^{\mathcal{P}}_{\mathcal{S}}(1^\lambda):} \\[2pt]
(st, st') \leftarrow_\$ \mathcal{P}_1(1^\lambda) \\
(\mathbf{C}_0, \mathbf{C}_1, \mathbf{m}_0, \mathbf{m}_1, z) \leftarrow_\$ \mathcal{S}(st) \\
\mathsf{m} \leftarrow_\$ \mathcal{P}_2^{\mathrm{FUNC}}(1^\lambda, \mathbf{C}_0(\mathbf{m}_0), z, st') \\
\text{return } (\mathbf{C}_0(\mathsf{m}) \neq \mathbf{C}_1(\mathsf{m})) \\[6pt]
\underline{\mathrm{FUNC}(\mathsf{m}, \mathsf{C}):} \\[2pt]
\text{return } (\mathbf{C}_0(\mathsf{m}), \mathsf{C}(\mathbf{m}_0))
\end{array}}$$

Fig. 2. Game defining unpredictability of a sampler $\mathcal{S}$ against $\mathcal{P} = (\mathcal{P}_1, \mathcal{P}_2)$.

We emphasize that the winning condition demands *component-wise* inequality of circuit outputs. In particular the predictor is *not* considered successful if it outputs a message which leads to different outputs on different circuit indices.

---

[1] We *do not* impose that $\mathbf{C}_0(\mathsf{m}) = \mathbf{C}_1(\mathsf{m})$ within the FUNC oracle as this is exactly the event that $\mathcal{P}$ is aiming to invoke to win the game. The restriction we do impose allows for a sampler to be unpredictable while possibility outputting low-entropy messages that might even differ on left and right.

A number of technical choices have been made in devising this definition. By the legitimacy of the sampler $\mathbf{C}_0(\mathbf{m}_0) = \mathbf{C}_1(\mathbf{m}_1)$ and hence only one of these values is provided to the predictor. Furthermore, since the goal of the predictor is to find a differing input, modifying the experiment so that FUNC returns $\mathbf{C}_1(\mathbf{m})$ (or both values) would result in an equivalent definition. Our definition intentionally does *not* consider unpredictability of messages. Instead, one could ask the predictor to output either a message that results in differing evaluations on challenge circuits or a circuit that evaluates differently on challenge messages. This would, however, lead to an excessively restrictive unpredictability notion and excludes many circuit samplers of practical relevance.

A number of special classes of samplers can be defined by imposing structural restrictions on their internal operation. In particular, definitions of high-entropy and block source samplers for keywords [14], block sources for inner products [15], and circuit sampler distributions used in various obfuscation definitions can be seen as particular cases within this framework.

4   Obfuscators

An obfuscator for a circuit family CSp is a uniform ppt algorithm Obf that on input the security parameter 1^λ and the description of a circuit $C \in CSp_\lambda$ outputs the description of another circuit $\overline{C}$. We require any obfuscator to satisfy the following two requirements.

FUNCTIONALITY PRESERVATION: For any $\lambda \in \mathbb{N}$, any $C \in CSp_\lambda$ and any $m \in MSp_\lambda$, with overwhelming probability over the choice of $\overline{C} \leftarrow_{\!s} Obf(1^\lambda, C)$ we have that $C(m) = \overline{C}(m)$.

POLYNOMIAL SLOWDOWN: There is a polynomial poly such that for any $\lambda \in \mathbb{N}$, any $C \in CSp_\lambda$ and any $\overline{C} \leftarrow_{\!s} Obf(1^\lambda, C)$ we have that $|\overline{C}| \leq poly(|C|)$.

Security definitions for obfuscators can be divided into the indistinguishability-based and simulation-based notions. Perhaps the most natural notion is the virtual black-box (VBB) property [6], which requires that whatever can be computed from an obfuscated circuit can be also simulated using oracle access to the circuit. Here, we consider a weakening of this notion, known as *virtual grey-box* (VGB) security [10,11] that follows the VBB approach, but allows simulators to run in *unbounded* time, as long as they make polynomially many queries to their oracles; we call such simulators semi-bounded. Below we present a self-composable strengthening of this notion where the VGB property is required to hold in the presence of multiple obfuscated circuits.

In the context of security definitions for obfuscators, we consider samplers that do not output any messages. Furthermore, we call a sampler *one-sided* if its sampled circuits are identical on left and right with probability 1.

COMPOSABLE VGB. An obfuscator Obf is composable VGB (CVGB) secure if for every ppt adversary $\mathcal{A}$ there exists a semi-bounded simulator Sim such that for every ppt one-sided circuit sampler $\mathcal{S}$ the advantage

$$\mathbf{Adv}^{\mathrm{cvgb}}_{\mathsf{Obf},\mathcal{S},\mathcal{A},\mathsf{Sim}}(\lambda) :=$$
$$\left| \Pr\left[\mathrm{CVGB\text{-}Real}^{\mathcal{S},\mathcal{A}}_{\mathsf{Obf}}(1^\lambda) \right] - \Pr\left[\mathrm{CVGB\text{-}Ideal}^{\mathcal{S},\mathsf{Sim}}_{\mathsf{Obf}}(1^\lambda) \right] \right| \in \mathrm{NEGL},$$

where games $\mathrm{CVGB\text{-}Real}^{\mathcal{S},\mathcal{A}}_{\mathsf{Obf}}(\lambda)$ and $\mathrm{CVGB\text{-}Ideal}^{\mathcal{S},\mathsf{Sim}}_{\mathsf{Obf}}(\lambda)$ are shown in Fig. 3.

$\mathrm{CVGB\text{-}Real}^{\mathcal{S},\mathcal{A}}_{\mathsf{Obf}}(1^\lambda)$:	$\mathrm{CVGB\text{-}Ideal}^{\mathcal{S},\mathsf{Sim}}_{\mathsf{Obf}}(1^\lambda)$:	$\mathrm{DI}^{\mathcal{S},\mathcal{A}}_{\mathsf{Obf}}(1^\lambda)$:
$(\mathbf{C}, z) \leftarrow\!\!{\scriptstyle\$}\ \mathcal{S}(1^\lambda, \varepsilon)$	$(\mathbf{C}, z) \leftarrow\!\!{\scriptstyle\$}\ \mathcal{S}(1^\lambda, \varepsilon)$	$b \leftarrow\!\!{\scriptstyle\$}\ \{0,1\}$
$\overline{\mathbf{C}} \leftarrow\!\!{\scriptstyle\$}\ \mathsf{Obf}(1^\lambda, \mathbf{C})$	$b \leftarrow\!\!{\scriptstyle\$}\ \mathsf{Sim}^{\mathrm{FUNC}}(1^\lambda, 1^{|\mathbf{C}|}, z)$	$b' \leftarrow\!\!{\scriptstyle\$}\ \mathcal{A}^{\mathrm{SAM}}(1^\lambda)$
$b \leftarrow\!\!{\scriptstyle\$}\ \mathcal{A}(1^\lambda, \overline{\mathbf{C}}, z)$	return b	return $(b = b')$
return b		
	$\underline{\mathrm{FUNC}(m):}$	$\underline{\mathrm{SAM}(st):}$
	return $\mathbf{C}(m)$	$(\mathbf{C}_0, \mathbf{C}_1, z) \leftarrow\!\!{\scriptstyle\$}\ \mathcal{S}(1^\lambda, st)$
		$\overline{\mathbf{C}} \leftarrow\!\!{\scriptstyle\$}\ \mathsf{Obf}(1^\lambda, \mathbf{C}_b)$
		return $(\overline{\mathbf{C}}, z)$

Fig. 3. Games defining the CVGB and DI security of an obfuscator.

By considering samplers that only output a single circuit we recover the standard (worst-case) VGB property. The VBB property corresponds to the case where the simulator is required to run in polynomial time. Average-case notions of obfuscation correspond to definitions where the circuit samplers are fixed. A result of Bitansky and Canetti [10, Proposition A.3] on the equivalence of VGB *with* and *without* auxiliary information can be easily shown to also hold in the presence of multiple circuits, from which one can conclude that CVGB *with* auxiliary information is the same as CVGB *without* auxiliary information.

We also introduce the following adaptation of an indistinguishability-based notion of obfuscation introduced in [10] for point functions.

DISTRIBUTIONAL INDISTINGUISHABILITY. An obfuscator Obf is DI secure if, for every unpredictable ppt sampler $\mathcal{S}$ and every ppt adversary $\mathcal{A}$,

$$\mathbf{Adv}^{\mathrm{di}}_{\mathsf{Obf},\mathcal{S},\mathcal{A}}(\lambda) := 2 \cdot \Pr\left[\mathrm{DI}^{\mathcal{S},\mathcal{A}}_{\mathsf{Obf}}(1^\lambda) \right] - 1 \in \mathrm{NEGL},$$

where game $\mathrm{DI}^{\mathcal{S},\mathcal{A}}_{\mathsf{Obf}}(1^\lambda)$ is defined in Fig. 3.

The above definition strengthens the one in [10] and gives the sampler the possibility to leak auxiliary information to the adversary. In particular, we can consider the case where images of an (internally generated) vector of messages that are correlated with the circuits are provided to $\mathcal{A}$. (Our constructions will rely on this property for point obfuscators.) Throughout the paper we consider DI adversaries that place a single query to the SAM oracle. Security with respect to *all* ppt and statistically unpredictable samplers can be shown to be equivalent to a variant definition where the adversary is run *after* the sampler and st is set to the empty string ε.

We recover the definition of indistinguishability obfuscation (iO) [19] when samplers are required to output a single circuit on left and right and include these two circuits explicitly in z. Differing-inputs obfuscation (diO) [3] is obtained if the predictor is also limited to run in polynomial time.

It has been shown that, for *point functions*, the notions of CVGB and DI (without auxiliary information) are equivalent [10, Theorem 5.1]. Following a similar argument to the first part of the proof in [10, Theorem 5.1], we can show that CVGB for *any* circuit family implies distributional indistinguishability even *with auxiliary information* for the same circuit family. Hence, our notion of DI obfuscation is potentially *weaker* than CVGB. This proof crucially relies on the restriction that samplers are required to be unpredictable in the presence of unbounded predictors. The proof of the converse direction in [10, Theorem 5.1] uses techniques specific to point functions and we leave a generalization to wider classes of circuits for future work.

Proposition 1 (CVGB $\implies$ DI). *Any* CVGB *obfuscator for a class of circuits* CSp *is also* DI *secure with respect to all statistically unpredictable samplers for the same class* CSp.

HYPERPLANE MEMBERSHIP. Let $\mathsf{CSp} := \{\mathsf{CSp}_p^d\}$ be a set circuit family of hyperplane membership testing functions that is defined for each value of the security parameter λ such that there is a λ-bit prime p and a positive integer d. Every circuit $\mathsf{C} \in \mathsf{CSp}_p^d$ is canonically represented by a vector $\boldsymbol{a} \in \mathbb{Z}_p^d$ and returns 1 if and only if the input vector $\boldsymbol{x} \in \mathbb{Z}_p^d$ is *orthogonal* to $\boldsymbol{a}$, i.e., $\mathsf{C}[\boldsymbol{a}](\boldsymbol{x}) := 1$ iff $\langle \boldsymbol{x}, \boldsymbol{a} \rangle = 0$.

Canetti et al. [17] presented a virtual black-box obfuscator for the hyperplane membership functionality, which works as follows. Let G be a group of prime order p for which the SVDDH assumption [17] holds. To obfuscate the hyperplane membership circuit represented by a vector $\boldsymbol{a}$, sample a generator g uniformly at random from G, compute $g_i \leftarrow g^{\boldsymbol{a}[i]}$ for $1 \le i \le d$, and construct the circuit that, given a vector $\boldsymbol{x}$, returns 1 if and only if $\prod_{i=1}^d g_i^{\boldsymbol{x}[i]}$ is equal to G's identity element. (Note that $\prod_{i=1}^d g_i^{\boldsymbol{x}[i]} = g^{\langle \boldsymbol{a}, \boldsymbol{x} \rangle}$, so this is the case if $\langle \boldsymbol{a}, \boldsymbol{x} \rangle = 0$.) We assume that the resulting obfuscated circuit is canonically represented by $(g_1, \ldots, g_d)$, generated as described above. This same construction satisfies distributional indistinguishability under a generalization of the SVDDH assumption, a DDH-style assumption we present in Fig. 4. In order to avoid attacks similar to the one described in [9] that puts a one element instance of SVDDH with *arbitrary* auxiliary information (or AI-DHI assumption, as referred to by [9]) in contention with the existence of VGB obfuscators supporting specific classes of circuits, we assume that our generalized SVDDH assumption holds only in the presence of *random* auxiliary information. This immediately translates to an obfuscator that tolerates the same type of leakage, which is enough to serve as a candidate to instantiate our functionality-agnostic constructions and obtain private inner-product encryption schemes, from which it is known how to derive expressive predicates that include equality tests, conjunctions, disjunctions and evaluation of CNF and DNF formulas (among others) [21].

$$\boxed{\begin{array}{l}
\mathsf{Assumption}_{\mathcal{S},\mathcal{A},G,t,d,\mathsf{poly}}(1^\lambda): \\[4pt]
b \leftarrow_\$ \{0,1\}; \; z \leftarrow_\$ \{0,1\}^{\mathsf{poly}(\lambda)} \\
(\mathbf{w}_0, \mathbf{w}_1) \leftarrow_\$ \mathcal{S}(1^\lambda, z, \epsilon) \\
\boldsymbol{g} \leftarrow_\$ G^t \\[4pt]
M_b \leftarrow \begin{bmatrix}
\boldsymbol{g}[1]^{\mathbf{w}_b[1][1]} & \cdots & \boldsymbol{g}[1]^{\mathbf{w}_b[1][d]} \\
\cdots & \cdots & \cdots \\
\boldsymbol{g}[t]^{\mathbf{w}_b[t][1]} & \cdots & \boldsymbol{g}[t]^{\mathbf{w}_b[t][d]}
\end{bmatrix} \\[4pt]
b' \leftarrow_\$ \mathcal{A}(M_b, z) \\
\text{return } (b = b')
\end{array}}$$

Fig. 4. Game defining a DDH-style computational assumption.

5   Function Privacy: A Unified Approach

We now define what function privacy for general functional encryption schemes means and derive the model specific to keyword search schemes by restriction to point circuit families. Our definition follows the indistinguishability-based approach to defining FE security and comes with an analogous legitimacy condition that prevents the adversary from learning the challenge bit simply by extracting a token for a circuit that has differing outputs for the left and right challenge messages. The model extends the IND-CPA game via a left-or-right (LR) oracle that returns ciphertexts *and* tokens for possibly correlated messages and circuits. Since the adversary in this game has access to tokens that depend on the challenge bit, we use the unpredictability framework of Sect. 3 to rule out trivial guess attacks.

The game follows a left-or-right rather than a real-or-random formulation of the challenge oracle [1,5,14,15] as this choice frees the definition from restrictions that must be imposed to render samplers compatible with uniform distribution over circuits. In particular, it allows the sampler to output *low-entropy* circuits as long as they are functionally-equivalent on left and right. It also allows analyzing security under repetitions of functionally-equivalent circuits in the presence of correlated messages, which until now were properties captured separately by *unlinkability* [5] and *enhanced security* [14], and never considered together, not even for the simple case of point functions.

The sampler allows us to model, within a single game, (a) token-only adversarial strategies via samplers that output no message, as the *non-enhanced* security model in [14] and those in [5,15]; (b) adversarial strategies that admit simple correlations between encrypted messages and extracted circuits, as the *enhanced* security model in [14] for point circuits that allows the adversary to obtain ciphertexts that *match* the tokens; (c) adversarial strategies that admit *arbitrary* correlations between extracted circuits and encrypted messages (i.e., not only exact matches).

$\text{PRIV}_{\text{FE}}^{\mathcal{A},\mathcal{S}}(1^\lambda)$:	$LR(st)$:	$\text{TGEN}(C)$:
$(\text{msk}, \text{mpk}) \leftarrow\!\!\text{\$ } \text{FE.Gen}(1^\lambda)$	$(\mathbf{C}_0, \mathbf{C}_1, \mathbf{m}_0, \mathbf{m}_1, z) \leftarrow\!\!\text{\$ } \mathcal{S}(st)$	$\text{TList} \leftarrow \text{TList} : (C, C)$
$b \leftarrow\!\!\text{\$ } \{0,1\}$	$\text{TList} \leftarrow \text{TList} : (\mathbf{C}_0, \mathbf{C}_1)$	$\text{tk} \leftarrow\!\!\text{\$ } \text{FE.TGen}(\text{msk}, C)$
$b' \leftarrow\!\!\text{\$ } \mathcal{A}^{\text{LR},\text{TGEN}}(\text{mpk})$	$\text{MList} \leftarrow \text{MList} : (\mathbf{m}_0, \mathbf{m}_1)$	$\text{return } \text{tk}$
$\text{return } (b = b')$	$\mathbf{tk} \leftarrow\!\!\text{\$ } \text{FE.TGen}(\text{msk}, \mathbf{C}_b)$	
	$\mathbf{c} \leftarrow\!\!\text{\$ } \text{FE.Enc}(\text{mpk}, \mathbf{m}_b)$	
	$\text{return } (\mathbf{tk}, \mathbf{c}, z)$	

Fig. 5. Game defining enhanced privacy of a functional encryption scheme FE.

Our model is functionality-agnostic and unifies all previous indistinguishability-based models in this area. When restricted to point circuits or inner-products families, it gives rise to a new privacy notion that offers significant improvements over those in prior works [5,14,15].

PRIV SECURITY. A functional encryption scheme FE is PRIV secure if, for every unpredictable ppt sampler[2] $\mathcal{S}$ and every ppt adversary $\mathcal{A}$

$$\mathbf{Adv}_{\text{FE},\mathcal{A},\mathcal{S}}^{\text{priv}}(\lambda) := 2 \cdot \Pr\left[\text{PRIV}_{\text{FE}}^{\mathcal{A},\mathcal{S}}(1^\lambda)\right] - 1 \in \text{NEGL},$$

where game $\text{PRIV}_{\text{FE}}^{\mathcal{A},\mathcal{S}}(1^\lambda)$ is defined in Fig. 5. We exclude adversaries $(\mathcal{A}, \mathcal{S})$ that attempt to trivially win the PRIV game via decryption tokens, by either extracting them explicitly via the token-generation oracle, or implicitly via the left-or-right oracle. Formally, the pair $(\mathcal{A}, \mathcal{S})$ is legitimate if, with overwhelming probability $\forall(\mathbf{C}_0, \mathbf{C}_1) \in \text{TList}, \forall(\mathbf{m}_0, \mathbf{m}_1) \in \text{MList} : \mathbf{C}_0(\mathbf{m}_0) = \mathbf{C}_1(\mathbf{m}_1)$.

Note also that for two sampler classes $\mathbb{S}_1$ and $\mathbb{S}_2$ with $\mathbb{S}_1 \subset \mathbb{S}_2$ security with respect to samplers in $\mathbb{S}_2$ is a stronger security guarantee that one for those only in $\mathbb{S}_1$. In particular a stronger restriction on sampler classes results in a weaker definition. Since the definition self-composes for internally stateless samplers, we assume that the adversary places a single query to the LR oracle in the remainder of the paper.

RESTRICTED PRIV AND PRIV-TO. We call an adversary *token-only* if $\mathcal{S}$ does not output any messages, and call the resulting security notion PRIV-TO. Note that, for token-only adversaries, the additional legitimacy constraint above is redundant. We call an adversary *restricted* if for every second-phase TGEN query C_2 there is a first-phase TGEN query C_1 such that $C_2(\mathbf{m}_b) = C_1(\mathbf{m}_b)$ for $b \in \{0,1\}$. Intuitively, this amounts to imposing that images exposed via second-stage queries (i.e., those placed *after* receiving the challenge) can reveal no more than the images obtained in the first stage (i.e., from queries placed *before* receiving the challenge). We call the resulting security notion Res-PRIV.

---

[2] We limit samplers to ppt because in proving the security of our constructions, samplers are used to construct computational adversaries against other schemes. In general, one could consider unbounded samplers.

We emphasize that the Res-PRIV model inherits many of the strengths of the full PRIV model such as arbitrary correlations and a wide range of adaptive token queries.[3]

ON REVEALING IMAGES. The outputs of challenge circuits on challenge messages can be always computed by the adversary, and by imposing equality of images we ensure that they do not lead to trivial distinguishing attacks. (This is similar to the legitimacy condition in FE security models.) It is however less clear why these image values should be explicitly provide to the predictor in the unpredictability game, even when they are equal for left and right circuits-messages pairs. To see this, consider the sampler that for a random word w outputs

$$w_0 = w, \quad w_1 = \overline{w}, \quad m_{0,i} := \begin{cases} w & \text{if } w[i] = 1; \\ \overline{w} & \text{otherwise,} \end{cases} \quad \text{and} \quad m_{1,i} := \begin{cases} \overline{w} & \text{if } w[i] = 1; \\ w & \text{otherwise.} \end{cases}$$

Note that $C[w_0](m_{0,i}) = C[w_1](m_{1,i}) = w[i]$ and hence the images are equal on left and right. Word w_0 can be recovered bit by bit from the image values $C[w_b](m_{0,i})$ and computing $1 - C[w_b](w_0)$ would then reveal the challenge bit b. Finally, without access to the images $C[w_0](m_{0,i})$ the sampler can be shown to be unpredictable as w is chosen randomly. On the other hand, in the presence of images, the sampler is trivially predicable. This counterexample is similar to that briefly discussed in [5] and can be modified to show that the *enhanced* model of Boneh et al. [14] for the so-called $(k_1, \ldots, k_T)$-distributions is not achievable.

RELATIONS AMONG NOTIONS. Clearly PRIV implies its weaker variant Res-PRIV, which in turn implies PRIV-TO. It is not too difficult to see that PRIV also implies IND-CPA.[4] A noteworthy consequence of this is that for *all-or-nothing* functionalities (such as PEKS, IBE or ABE) any PRIV-secure construction is also *index hiding* (aka. anonymous), whereby ciphertexts do not leak any information about their intended recipients (i.e., about tokens that may permit recovering the payload). Res-PRIV would imply a restricted analogue of IND-CPA (where images in the second phase should match one in the first phase), which for point functions is equivalent to the standard IND-CPA model. IND-CPA security does not imply PRIV-TO: consider an IND-CPA-secure scheme that is modified to append circuits in the clear to their tokens. PRIV-TO does not imply IND-CPA either: consider a PRIV-TO-secure scheme that is modified to return messages in the clear with ciphertexts. (Note that these separations hold even for point functions.) Fig. 6 summarizes relations among notions of security.

---

[3] When the restriction here is imposed on the IND-CPA model for point function, the resulting model remains as strong as the full IND-CPA model.

[4] Consider a sampler which does not output any circuits and simply returns (possibly low-entropy) messages included in the state st passed to it. This sampler is trivially unpredictable. Furthermore, the legitimacy conditions in the two games exactly match.

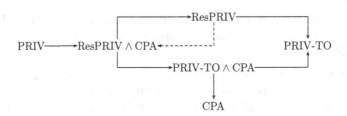

Fig. 6. Relations among security notions for private functional encryption. The dotted implication only holds for keyword search schemes as weak (aka. restricted) and standard IND-CPA security models are equivalent for point circuits.

6   Constructions

6.1   The Obfuscate-Extract (OX) Transform

Our first construction formalizes the intuition that obfuscating circuits before computing a token for them will provide some form of token privacy.

THE OX TRANSFORM. Let Obf be an obfuscator supporting a circuit family CSp and let FE be a functional encryption scheme supporting all polynomial-size circuits. We construct a functional encryption scheme OX[FE, Obf] via the OX transform as follows. Setup, encryption and evaluation algorithms are identical to those of the base functional encryption scheme. The token-generation algorithm creates a token for the circuit that results from obfuscating the extracted circuit, i.e., $OX[FE, Obf].TGen(msk, C) := FE.TGen(msk, Obf(1^\lambda, C))$. Correctness of this construction follows from those of its underlying components. We now show that this construction yields function privacy against PRIV-TO adversaries. Since PRIV-TO does not imply IND-CPA security—see the discussion in Sect. 5—we establish IND-CPA security independently. The proof of the following theorem is straightforward and results from direct reductions to the base FE and Obf schemes used in the construction.

Theorem 1 (OX is PRIV-TO ∧ IND-CPA). *If obfuscator* Obf *is DI secure, then scheme* OX[FE, Obf] *is PRIV-TO secure. Furthermore, if* FE *is IND-CPA secure* OX[FE, Obf] *is IND-CPA secure.*

We note that this proof holds for arbitrary classes of circuits and arbitrary (circuits-only) samplers. Using the composable VGB point-function obfuscator of Bitansky and Canetti [10] and any secure functional encryption scheme that is powerful enough to support one exponentiation and one equality test (e.g., supports $\mathbf{NC}^1$ circuits) we obtain a private keyword search scheme in the presence of tokens for arbitrarily correlated keywords. If the underlying functional encryption scheme supports the more powerful functionality that permits attaching a payload to the point, one obtains a PRIV-TO anonymous identity-based encryption scheme where arbitrary correlations are tolerated. In this case, on input (ID, m), the functionality supported by the underlying FE scheme would return m if $\overline{C}(\mathsf{ID}) = 1$, where $\overline{C}$ was sampled from $\mathsf{Obf}(C[\mathsf{ID}^\star])$ during token generation; it would return $\bot$ otherwise.

The above theorem shows that DI is sufficient to build a PRIV-TO scheme. It is however easy to see that the existence of a single-circuit DI obfuscator is also necessary. Indeed, given any PRIV-TO scheme FE we can DI-obfuscate a single circuit C by generating a fresh FE key pair, and outputting $\mathsf{FE.Eval}(\cdot, \mathsf{tk})$ where tk is a token for C.

Proposition 2 (PRIV-TO vs. DI). *A PRIV-TO-secure functional encryption for a circuits family* CSp *exists if a DI obfuscator for* CSp *exists. Conversely, a single-circuit DI obfuscator for* CSp *exists if a PRIV-TO-secure functional encryption for* CSp *exists.*

A similar line of reasoning shows that the extractor-based constructions of private FE by Boneh et al. [14] and Arriaga et al. [5] give rise to single-circuit DI obfuscators for point functions for the specific classes of samplers considered in those works.

Agrawal et al. [1] have proposed a simulation-based definition of privacy that strikes a different balance between practical relevance and feasibility. However, the definition in [1] implies VBB obfuscation, which is known to be feasible only for restricted classes of circuits [12], in idealized models of computation [13,18] or with restricted forms of auxiliary information. The above proposition shows that our model is closer to the weaker form of DI obfuscation, which as shown in Proposition 1 is implied by VGB (and hence VBB) obfuscation, and is therefore more amenable to instantiations in the standard model.

6.2   The Trojan-Obfuscate-Extract (TOX) Transform

We now present a generic construction that achieves Res-PRIV security for a class of samplers that we call *concentrated*. To this end, we build on the ideas from [4] on converting selective to adaptive security and achieving simulation-based security from IND-CPA security for FE schemes.

THE TOX TRANSFORM. Given a symmetric encryption scheme SE, a general-purpose obfuscator Obf and a functional encryption FE for all circuits, our *Trojan-Obfuscate-Extract* (TOX) transform operates as follows. The master public key of the scheme is the same as that of the base FE scheme. Its master secret key includes a symmetric key k and the master secret key for the base FE scheme. To encrypt a message m we call the base FE encryption routine on $(0, 0^\lambda, m)$. To generate a token for a circuit C, we first generate an obfuscation $\overline{C} \leftarrow_\$ \mathsf{Obf}(C)$, a ciphertext $c \leftarrow_\$ \mathsf{SE.Enc}(k, 0^n)$ and construct the following circuit.

$$\mathsf{Troj}[\overline{C}, c](b, k, m) := \begin{cases} \overline{C}(m) & \text{if } b = 0; \\ C^*(m) & \text{if } b = 1, \text{ where } C^* = \mathsf{SE.Dec}(k, c). \end{cases}$$

Finally, we extract a token for $\mathsf{Troj}[\overline{C}, c]$. Evaluation simply invokes the corresponding operation in the underlying FE.

The correctness and IND-CPA security of this construction follow easily from the correctness and IND-CPA security of the underlying functional encryption scheme via straightforward reductions. Intuitively, during the normal operation of the scheme, the tokens in the construction will simply evaluate an obfuscation of the extracted circuit. In the proof of privacy, however, we will take advantage of the fact that a totally independent circuit can be hidden inside the token within the symmetric encryption ciphertext, and unlocked by a message containing the correct symmetric decryption key. For the proof to go through, the hidden circuit must be carefully selected so that the legitimacy condition is observed throughout. In order to meet this latter restriction, we consider the following constrained class of samplers.

CONCENTRATED SAMPLERS. We say a sampler $\mathcal{S}$ is $\mathcal{S}^*$-concentrated if for all st, all CSp_λ-vectors $\mathbf{C}$ we have that

$$\Pr\left[\mathbf{C}(\mathbf{m}_0) = \mathbf{C}(\mathbf{m}_1) \neq \mathbf{C}(\mathbf{m}^*)\right] \in \text{NEGL} \text{ and } \Pr\left[\mathbf{C}_0(\mathbf{m}_0) \neq \mathbf{C}^*(\mathbf{m}^*)\right] \in \text{NEGL},$$

where the probability space of these is defined by operations $(\mathbf{C}^*, \mathbf{m}^*) \leftarrow_\$ \mathcal{S}^*(z, \mathbf{C})$ and $(\mathbf{C}_0, \mathbf{C}_1, \mathbf{m}_0, \mathbf{m}_1, z) \leftarrow_\$ \mathcal{S}(st)$.

Concentration is a property independent of unpredictability and we will be relying on both in our construction. Unpredictability is used in the reduction to the DI assumption. Concentration guarantees the existence of a sampler $\mathcal{S}^*$ that generates circuits $\mathbf{C}^*$ and messages $\mathbf{m}^*$ which permit decoupling circuits and messages in the security proof. Intuitively, quantification over all $\mathbf{C}$ means that adversarially generated circuits will lead to image matrices that collide with those leaked by the sampler with overwhelming probability. The additional restriction on $\mathbf{C}^*(\mathbf{m}^*)$ guarantees that one can switch from the honest branch of challenge tokens to one corresponding to the trojan branch. Both of these properties are important to guarantee legitimacy when making a reduction to the security of the FE scheme. We however need to impose that legitimacy also holds for second-phase TGEN queries as well, and this is where we need to assume Res-PRIV security: the extra legitimacy condition allows us to ensure that by moving to $\mathbf{m}^*$ the legitimacy condition is not affected in the second phase either. Finally, an important observation is that, because we are dealing with concentrated samplers, our security proof goes through assuming obfuscators that need only tolerate random auxiliary information.

Theorem 2 (TOX is Res-PRIV). *If obfuscator* Obf *is DI secure,* SE *is IND-CPA secure and* FE *is IND-CPA secure, then scheme* TOX[FE, Obf, SE] *is Res-PRIV secure with respect to concentrated samplers.*

Proof. The proof proceeds via a sequence of three games as follows.

Game$_0$: This game is identical to Res-PRIV: challenge vector $\mathbf{C}_b$ is extracted and $\mathbf{m}_b$ is encrypted for a random bit b and for all TGEN queries, string 0^n is encrypted using SE in the trojan branch.

Game$_1$: In this game, instead of 0^n we encrypt the circuits queried to the (first or second-phase) TGEN oracle under a symmetric key k^* in the trojan branch.

In the challenge phase, we sample $(\mathbf{C}^*, \mathbf{m}^*) \leftarrow_\$ \mathcal{S}^*(z, \mathbf{C})$, where $\mathbf{C}$ are all first-phase TGEN queries, and encrypt $\mathbf{C}^*$ under k^* for the challenge circuits in the trojan branch. This transition is negligible down to IND-CPA security of SE.

Game$_2$: In this game, instead of encrypting $(0, 0, \mathbf{m}_b)$ we encrypt $(1, \mathsf{k}^*, \mathbf{m}^*)$ in the challenge phase where the latter is generated using $\mathcal{S}^*(z, \mathbf{C})$. We reduce this hop to the IND-CPA security of FE. We generate a key k^*, answer first-stage TGEN queries using the provided TGEN oracle and encrypt circuits under k^* in the trojan branch to get st. We run $\mathcal{S}(st)$ and get $(\mathbf{C}_0, \mathbf{m}_0, \mathbf{C}_1, \mathbf{m}_1, z)$. We then run $\mathcal{S}^*(z, \mathbf{C})$, where $\mathbf{C}$ are all first-phase TGEN queries, to get $(\mathbf{C}^*, \mathbf{m}^*)$. We prepare challenges tokens by encrypting $\mathbf{C}^*$ under k^* in the trojan branch and using the provided TGEN oracle we generate the challenge tokens. We query the provided LR on $(0, 0, \mathbf{m}_b)$ and $(1, \mathsf{k}^*, \mathbf{m}^*)$ and receive the corresponding vector of ciphertexts. Second-stage TGEN queries are handled using provided TGEN oracle and k^*. Finally, we return the same bit that the distinguisher returns. Legitimacy of first-stage TGEN queries follows from the first condition on concentration that with high probability $\mathbf{C}(\mathbf{m}_b) = \mathbf{C}(\mathbf{m}^*)$. For the challenge tokens, this follows from the second concentration requirement that $\mathbf{C}_b(\mathbf{m}_b) = \mathbf{C}^*(\mathbf{m}^*)$. For the second-stage queries we rely on the restriction on the adversary. Recall that in the Res-PRIV model, any second-stage queries must have an image vector which matches one for a first-stage query. Since the first-stage images match those on $\mathbf{m}^*$ (and hence are legitimate), the second-stage ones will be also legitimate. We output $(b' = b)$ where the distinguisher outputs b'. As a result of this game, the challenge messages no longer depend on b. It is easy to see that according to the IND-CPA challenge bit this reduction interpolates between games Game$_1$ and Game$_2$.

Game$_3$: In this game we use $\mathbf{C}_1$ in challenge token generation even if $b = 0$. We show this hop in unnoticeable down to the security of the obfuscator. We sample an FE key pair and a symmetric key and simulating the first-stage TGEN queries for the adversary as before. We define a DI sampler that outputs the circuits that the Res-PRIV sampler outputs, but extends the circuit list to include another copy of $\mathbf{C}_1$ on both sides. This sampler also outputs as auxiliary information z' the original auxiliary information output by the PRIV sampler, extended with the random coins used to generate the FE key, the symmetric key, and to run the first stage of the adversary (this will allow the second stage DI adversary to reconstruct the keys and first stage TGEN queries). It follows that this sampler is unpredictable as long as the Res-PRIV sampler is. When we receive the obfuscations and z', we generate $(\mathbf{C}^*, \mathbf{m}^*) \leftarrow_\$ \mathcal{S}^*(z, \mathbf{C})$, where $\mathbf{C}$ are all first-phase TGEN queries. We form the challenge tokens using the received obfuscations and $\mathbf{C}^*$, taking the $\mathbf{C}_1$ obfuscations from the duplicated part of the challenge, and the $\mathbf{C}_0$ obfuscations from the original part (these can now be either $\mathbf{C}_0$ or $\mathbf{C}_1$ depending on the external challenge bit). Challenge ciphertexts are generated by encrypting $\mathbf{m}^*$ (rules of Game$_2$). We answer the second-stage TGEN queries using the FE key and the symmetric key. We return whatever the distinguisher returns. It is easy to see that according to the DI challenge bit this reduction interpolates between games Game_2 and Game_3.

In Game_3 both the challenge tokens and challenge ciphertexts are independent of the bit b and hence the advantage of any adversary is 0.   □

EXAMPLES. Consider keyword samplers which output high-entropy keywords and messages with arbitrary image matrices. All such samplers are concentrated around a sampler S^* that outputs uniformly random keywords and messages subject to the same image pattern. The second concentration condition is immediate and the first follows from the fact that all messages and circuits have high entropy and $\mathbf{C}$ is selectively chosen.

As another example, consider hyperplane membership circuits $\mathsf{C}[\mathbf{v}](\mathbf{w})$ that return 1 iff $\langle \mathbf{v}, \mathbf{w} \rangle = 0 \pmod{p}$ for a prime p. Samplers which output n vectors $\mathbf{v}_i \in \mathbb{Z}_p^d$ and m messages $\mathbf{w}_i \in \mathbb{Z}_p^d$ where all vector entries have high entropy can be easily shown to be unpredictable. Given the corresponding $n \times m$ image matrix, whenever $d(n+m) > nm$, a high-entropy pre-image to the image matrix can be sampled as the system will be underdetermined. Under this condition, the second requirement needed for concentration is met, and the first condition follows as this pre-image is high entropy and $\mathbf{C}$ is selectively chosen. This observation implies that a DI obfuscator for the hyperplane membership problem, as that of Canetti et al. [17] shown in Sect. 4, will immediately yield a private functional encryption scheme for the same functionality under arbitrary correlations via the TOX construction, a problem that was left open in [15].

6.3   The Disjunctively-Obfuscate-Extract (DOX) Transform

Our third construction is specific to point functions, and besides being simpler and more efficient, can tolerate arbitrary correlations between challenge keywords and encrypted messages. Put differently this construction removes the concentration restriction on samplers. For this construction we require a functional encryption scheme that supports the OR composition of two DI-secure point obfuscations. The composable VGB point obfuscator of Bitansky and Canetti [10] implies that the required DI point obfuscator exists. Furthermore, we also rely on a standard functional encryption scheme that supports the evaluations of four group operations in a DDH group (corresponding to the disjunction of two point function obfuscations), which is a relatively modest computation. We are, however, unable to lift the mild second-stage restriction.

THE DOX TRANSFORM. Let Obf be an obfuscator supporting a point circuit family CSp over message space MSp. Let FE be a functional encryption scheme supporting general circuits, and let PRP be a pseudorandom permutation. We construct a keyword search scheme KS for keyword space WSp $=$ MSp via the *Disjunctively-Obfuscate-Extract* (DOX) transform as follows. The key-generation algorithm samples a PRP key $\mathsf{k} \leftarrow_{\!\$} \mathsf{K}(1^\lambda)$ and an FE key pair $(\mathsf{msk}, \mathsf{mpk}) \leftarrow_{\!\$} \mathsf{FE.Gen}(1^\lambda)$. It returns $((\mathsf{k}, \mathsf{msk}), \mathsf{mpk})$. The encryption operation is identical to that of the FE scheme. The test algorithm is identical to the evaluation algorithm of FE. The token-generation algorithm computes $\mathsf{FE.TGen}(\mathsf{msk}, \mathsf{Obf}(1^\lambda, C[w]) \vee \mathsf{Obf}(1^\lambda, C[E(k, w)]))$. The FE-extracted circuits are two-point circuits implemented as the disjunction of two obfuscated point functions. One of the points will correspond to the searched query, whereas the other point will be pseudorandom and will be only used for proofs of security.

The proof of Res-PRIV security of this construction involves an intricate game hopping argument, in order to deal with all possible correlations allowed by the Res-PRIV model (which are the same as those allowed by full PRIV). We refer to the full version of this paper for a detailed description and security analysis of this construction.

6.4   The Verifiably-Obfuscate-Encrypt-Extract (VOEX) Transform

Our last construction lifts the second-stage restriction at the cost of relying on more expressive forms of obfuscators. The novelty in this construction resides in the observation that, in order to offer the keyword search functionality, it suffices to encrypt information that enables equality checks between words and messages to be carried out. In our fourth construction we encode a message m as an *obfuscation* of the point function $C[m]$. Concretely, we obfuscate words before extraction *and* messages before encryption. Equality with w can be checked using a circuit $D[w]$ that on input an obfuscated point function $\mathsf{Obf}(C[m])$ returns $\mathsf{Obf}(C[m])(w)$. We emphasize that $D[w]$ is *not* a point function. We also need to ensure that an attacker cannot exploit the $D[w]$ circuits by, say, encrypting obfuscations of malicious circuits of its choice. We do this using NIZK proofs to ensure the outputs of the point obfuscator are *verifiable*: one can publicly verify that an obfuscation indeed corresponds to some point function. To summarize, our construction relies on a DI obfuscator supporting point functions $C[m](w) := (m = w)$ and circuits $D[w](\overline{C}) := \overline{C}(w)$ and a general-purpose FE. The circuits $C[m]$ and $D[w]$ were used negatively by Barak et al. [6] to launch generic attacks against VBB and VGB obfuscators. Here, the restrictions imposed on legitimate PRIV samplers ensure that these attacks cannot be carried out in our setting, and obfuscators supporting them can be used positively to build private FE schemes. The detailed description of the scheme and its security analysis can be found in the full version of this paper.

Acknowledgements. Afonso Arriaga was supported by the National Research Fund, Luxembourg (AFR Grant No. 5107187). Manuel Barbosa was funded by project "NanoSTIMA: Macro-to-Nano Human Sensing: Towards Integrated Multimodal Health Monitoring and Analytics/NORTE-01-0145-FEDER-000016", which is financed by the North Portugal Regional Operational Programme (NORTE 2020), under the PORTU-GAL 2020 Partnership Agreement, and through the European Regional Development Fund (ERDF). Pooya Farshim was supported in part by grant ANR-14-CE28-0003 (Project EnBid).

References

1. Agrawal, S., Agrawal, S., Badrinarayanan, S., Kumarasubramanian, A., Prabhakaran, M., Sahai, A.: On the practical security of inner product functional encryption. In: Katz, J. (ed.) PKC 2015. LNCS, vol. 9020, pp. 777–798. Springer, Heidelberg (2015). doi:10.1007/978-3-662-46447-2_35

2. Abdalla, M., Bellare, M., Catalano, D., Kiltz, E., Kohno, T., Lange, T., Malone-Lee, J., Neven, G., Paillier, P., Shi, H.: Searchable encryption revisited: consistency properties, relation to anonymous IBE, and extensions. J. Cryptol. **21**(3), 350–391 (2008)

3. Ananth, P., Boneh, D., Garg, S., Sahai, A., Zhandry, M.: Differing-inputs obfuscation and applications. IACR Cryptology ePrint Archive, Report 2013/689 (2013)

4. Ananth, P., Brakerski, Z., Segev, G., Vaikuntanathan, V.: From selective to adaptive security in functional encryption. In: Gennaro, R., Robshaw, M. (eds.) CRYPTO 2015. LNCS, vol. 9216, pp. 657–677. Springer, Heidelberg (2015). doi:10.1007/978-3-662-48000-7_32

5. Arriaga, A., Tang, Q., Ryan, P.: Trapdoor privacy in asymmetric searchable encryption schemes. In: Pointcheval, D., Vergnaud, D. (eds.) AFRICACRYPT 2014. LNCS, vol. 8469, pp. 31–50. Springer, Heidelberg (2014). doi:10.1007/978-3-319-06734-6_3

6. Barak, B., Goldreich, O., Impagliazzo, R., Rudich, S., Sahai, A., Vadhan, S., Yang, K.: On the (im)possibility of obfuscating programs. In: Kilian, J. (ed.) CRYPTO 2001. LNCS, vol. 2139, pp. 1–18. Springer, Heidelberg (2001). doi:10.1007/3-540-44647-8_1

7. Barbosa, M., Farshim, P.: On the semantic security of functional encryption schemes. In: Kurosawa, K., Hanaoka, G. (eds.) PKC 2013. LNCS, vol. 7778, pp. 143–161. Springer, Heidelberg (2013). doi:10.1007/978-3-642-36362-7_10

8. Bellare, M., Stepanovs, I., Tessaro, S.: Poly-many hardcore bits for any one-way function and a framework for differing-inputs obfuscation. In: Sarkar, P., Iwata, T. (eds.) ASIACRYPT 2014. LNCS, vol. 8874, pp. 102–121. Springer, Heidelberg (2014). doi:10.1007/978-3-662-45608-8_6

9. Bellare, M., Stepanovs, I., Tessaro, S.: Contention in cryptoland: obfuscation, leakage and UCE. In: Kushilevitz, E., Malkin, T. (eds.) TCC 2016. LNCS, vol. 9563, pp. 542–564. Springer, Heidelberg (2016). doi:10.1007/978-3-662-49099-0_20

10. Bitansky, N., Canetti, R.: On strong simulation and composable point obfuscation. J. Cryptol. **27**(2), 317–357 (2014)

11. Bitansky, N., Canetti, R., Kalai, Y.T., Paneth, O.: On virtual grey box obfuscation for general circuits. In: Garay, J.A., Gennaro, R. (eds.) CRYPTO 2014. LNCS, vol. 8617, pp. 108–125. Springer, Heidelberg (2014). doi:10.1007/978-3-662-44381-1_7

12. Brakerski, Z., Rothblum, G.N.: Black-box obfuscation for d-CNFs. In: ITCS 2014, pp. 235–250. ACM (2014)

13. Brakerski, Z., Rothblum, G.N.: Virtual black-box obfuscation for all circuits via generic graded encoding. In: Lindell, Y. (ed.) TCC 2014. LNCS, vol. 8349, pp. 1–25. Springer, Heidelberg (2014). doi:10.1007/978-3-642-54242-8_1

14. Boneh, D., Raghunathan, A., Segev, G.: Function-private identity-based encryption: hiding the function in functional encryption. In: Canetti, R., Garay, J.A. (eds.) CRYPTO 2013. LNCS, vol. 8043, pp. 461–478. Springer, Heidelberg (2013). doi:10.1007/978-3-642-40084-1_26

15. Boneh, D., Raghunathan, A., Segev, G.: Function-private subspace-membership encryption and its applications. In: Sako, K., Sarkar, P. (eds.) ASIACRYPT 2013. LNCS, vol. 8269, pp. 255–275. Springer, Heidelberg (2013). doi:10.1007/978-3-642-42033-7_14

16. Boneh, D., Sahai, A., Waters, B.: Functional encryption: definitions and challenges. In: Ishai, Y. (ed.) TCC 2011. LNCS, vol. 6597, pp. 253–273. Springer, Heidelberg (2011). doi:10.1007/978-3-642-19571-6_16

17. Canetti, R., Rothblum, G.N., Varia, M.: Obfuscation of hyperplane membership. In: Micciancio, D. (ed.) TCC 2010. LNCS, vol. 5978, pp. 72–89. Springer, Heidelberg (2010). doi:10.1007/978-3-642-11799-2_5

18. Canetti, R., Vaikuntanathan, V.: Obfuscating branching programs using black-box pseudo-free groups. IACR Cryptology ePrint Archive, Report 2013/500 (2013)

19. Garg, S., Gentry, C., Halevi, S., Raykova, M., Sahai, A., Waters, B.: Candidate indistinguishability obfuscation and functional encryption for all circuits. In: FOCS 2013, pp. 40–49. IEEE Computer Society (2013)

20. Goldwasser, S., Kalai, Y.T.: On the impossibility of obfuscation with auxiliary input. In: FOCS 2005, pp. 553–562. IEEE Computer Society (2005)

21. Katz, J., Sahai, A., Waters, B.: Predicate encryption supporting disjunctions, polynomial equations, and inner products. J. Cryptol. **26**(2), 191–224 (2013)

22. O'Neill, A.: Definitional issues in functional encryption. IACR Cryptology ePrint Archive, Report 2010/556 (2010)

Revocable Decentralized Multi-Authority Functional Encryption

Hikaru Tsuchida[1]($^{\boxtimes}$), Takashi Nishide[2], Eiji Okamoto[2], and Kwangjo Kim[3]

[1] NEC Corporation, 1753, Shimonumabe, Nakahara-Ku,
Kawasaki, Kanagawa 211-8666, Japan
h-tsuchida@bk.jp.nec.com
[2] Faculty of Engineering, Information and Systems, University of Tsukuba,
1-1-1 Tennodai, Tsukuba, Ibaraki 305-8573, Japan
{nishide,okamoto}@risk.tsukuba.ac.jp
[3] Computer Science Department, KAIST, 291 Daehak-ro, Yuseong-gu,
Daejeon 305-701, Korea
kkj@kaist.ac.kr

Abstract. Attribute-Based Encryption (ABE) is regarded as one of the most desirable cryptosystems realizing data security in the cloud storage systems. Functional Encryption (FE) which includes ABE and the ABE system with multiple authorities are studied actively today. However, ABE has the attribute revocation problem. In this paper, we propose a new revocation scheme using update information, i.e., revocation patch (not update key), in which an encryptor does not need to care about the revocation list. We propose an FE scheme with multiple authorities and no central authority supporting revocation by using revocation patch. Our proposal realizes the revocation on the attribute level. More precisely, we introduce the new concept, i.e., the revocation on the category level that is a generalization of attribute level. We prove that our construction is adaptively secure against chosen plaintext attacks and static corruption of authorities based on the decisional linear (DLIN) assumption.

Keywords: Functional encryption · Access control · Multiple authorities · Revocation · Attribute-level

1 Introduction

1.1 Background

In recent years, outsourcing data storage to cloud service providers has been increasing. Due to this change, there are frequent leaks of confidential data in cloud storage system. Therefore, data security in the cloud server is required. Attribute-Based Encryption (ABE) [3,6,9,13,14,19,22] is regarded as one of the

---

This work was completed while the corresponding author was a graduate student at University of Tsukuba.

O. Dunkelman and S.K. Sanadhya (Eds.): INDOCRYPT 2016, LNCS 10095, pp. 248–265, 2016.
DOI: 10.1007/978-3-319-49890-4_14

most desirable cryptosystems realizing data security in the cloud storage systems. ABE systems can provide data security and access control without a trusted server by using access policies and associated attributes among ciphertexts and private keys. For example, if the data owner encrypts data with an access policy like ("the sales department" OR ("the development department" AND "chief")), only a staff member in the sales department and a chief of the development department can decrypt the data in a Ciphertext-Policy ABE (CP-ABE) system. Furthermore, Functional Encryption (FE) [5,16,18] which includes ABE and the ABE system with multiple authorities [6,13,14,18] are proposed.

However, CP-ABE has the attribute revocation problem. For example, if a staff member in the sales department got fired and still has a decryption key related to the attribute of "the sales department" illegally, he/she may be able to decrypt encrypted data associated with an access policy related to "the sales department". Accordingly, a CP-ABE system needs a user (attribute) revocation scheme. In previous research, there are two types of revocation schemes: indirect revocation [4,11,21] and direct revocation [1]. The former scheme requires an update key for revocation issued by an authority, but an encryptor does not have to care about a revocation list in the indirect revocation system. The latter scheme can revoke users without using an update key because an encryptor specifies revoked users for ciphertexts by using the revocation list which may be specified and given by the authority or may be specified freely by the encryptor. In other words, in the direct revocation system, an encryptor has to care about a revocation list. That is, in indirect (direct) revocation, users are revoked by an authority (encryptor resp.). The direct revocation with multiple authorities was already proposed in [10], but the indirect one is not proposed.

To achieve attribute-level revocation, encryptors will expect that individual authorities such as universities and government maintain revocation lists rather than specifying the revocation list for ciphertexts by themselves. Furthermore, it is also desirable that a new cryptosystem has expressiveness of access policies (e.g. an FE system) and practical attribute management (e.g. multi-authority ABE system). For this reason, we propose FE with multiple authorities supporting indirect revocation, i.e., an encryptor does not have to care about revocation list.

1.2 Our Results

We propose a new revocation scheme using what we call revocation patch, *patch revocation scheme*, by combining indirect revocation and Decentralized Multi-Authority Functional Encryption (DMA-FE) [18] (which realizes non-monotone access structures using inner-product relations with multiple authorities and no central authority), i.e. the first DMA-FE scheme supporting indirect revocation. Our proposed scheme realizes the revocation on the category level that is a generalization of indirect revocation on the attribute level. (For more information about revocation on the attribute level, see the full version of this paper [23].)

Here, we give the intuitive explanation of how to specify revocation with a toy example of revocation on the category level in our scheme. Suppose that, the attribute category t_1 includes the attribute values $A_{t_1}, A'_{t_1}, A''_{t_1}$ and the attribute category t_2 includes the attribute values $B_{t_2}, B'_{t_2}, B''_{t_2}$. We would like to specify $(\neg A_{t_1} \wedge \neg B_{t_2})$ as a ciphertext policy as described below. Then, we specify revocation information in the ciphertext on the category level, as, for example, $((\neg A_{t_1} \wedge t_1[v_{t_1}]) \wedge (\neg B_{t_2} \wedge t_2[v_{t_2}]))$. Here, $(\neg A_{t_1} \wedge t_1[v_{t_1}])$ means that the decryptor needs to have attribute of t_1 except A_{t_1}, i.e., needs to have A'_{t_1} or A''_{t_1}. Moreover, the decryptor's attribute of t_1 needs to be not revoked before issuing the revocation patch of version v_{t_1} by the authority that manages attribute category t_1. $(\neg B_{t_2} \wedge t_2[v_{t_2}])$ means the similar condition about t_2 as for t_1. If we would like to specify a non-monotone access structure for a ciphertext, the revocation information is required to be on the category level, not attribute level. The revocation on the attribute level considers that the user's attribute which is associated with an access policy is valid or revoked, but does not consider the other attributes of the category to which the user's attribute belongs. For this reason, our scheme supporting non-monotone access control realizes the revocation on the category level, rather than attribute level. We also prove that our construction is adaptively secure against chosen plaintext attacks and static corruption of authorities based on the DLIN assumption. (We note that DMA-FE [18] of Okamoto and Takashima does not achieve the security against static corruption of authorities.)

We show a comparison with previous works in Tables 1 and 2. In Table 1, LSSS means Linear Secret Sharing Scheme. In Table 2, SD method means Subset Difference method in [15]. Std. model, GBGM and ROM mean standard

Table 1. Comparison with previous works

Schemes	Authority (central authority)	Policy	Access structure
AI09 [2]	Single	Key-Policy	Monotone (LSSS)
DDM15 [7]	Single	Key-Policy	Non-monotone (LSSS)
H15 [10]	Multiple (Y)	Ciphertext-Policy	Monotone (LSSS)
L12 [13]	Multiple (N)	Ciphertext-Policy	Monotone (LSSS)
OT13 [18]	Multiple (N)	Ciphertext-Policy	Non-monotone (LSSS & Inner-Product)
This work	Multiple (N)	Ciphertext-Policy	Non-monotone (LSSS & Inner-Product)

Table 2. Comparison with previous works (cont.)

Schemes	Revocation	Revocation level	Security model	Assumption
AI09 [2]	Direct/Indirect (CS method)	User level	selective (Std. model)	DBDH
DDM15 [7]	Direct (SD method)	User level	full (Std. model)	DLIN
H15 [10]	Direct	User level	full$^+$ (GBGM & ROM)	-
L12 [13]	-	-	full$^+$ (ROM)	DLIN
OT13 [18]	-	-	full (ROM)	DLIN
This work	Patch (CS method)	Category level	full$^+$ (ROM)	DLIN

model, generic bilinear group model and random oracle model, respectively. The "full" and "full$^+$" mean "adaptively payload-hiding against chosen plaintext attacks" and "adaptively payload-hiding against chosen plaintext attacks and static corruption of authorities". DBDH means the Decisional Bilinear Diffie-Hellman assumption. Tables 1 and 2 show that our proposal has expressiveness of access policy and practical attribute management. It also shows that our proposed scheme realizes that each of the authorities is able to revoke user's attribute by themselves (not an encryptor).

1.3 Key Techniques

Overview. Our scheme is based on DMA-FE [18]. In DMA-FE [18], there are roughly two types of ciphertexts: the encrypted message and the headers for access control. Only the user who has attribute keys associated with the access policy can restore the secret (or session key) from the headers and decrypt the encrypted message by using it. We add keys and headers for attribute revocation to DMA-FE [18] by introducing the basis of dual pairing vector spaces (DPVS) for attribute revocation. Due to this, only the user who has attribute keys associated with the access policy and keys for attribute revocation which are not revoked can get the message. A key for attribute revocation is like an attribute key in DMA-FE [18], so an attribute vector is embedded in a key for attribute revocation. Each key for attribute revocation is tied to every attribute key. Therefore, if the key for attribute revocation is revoked, the attribute key to which it is tied also becomes the invalid key.

How to Revoke. We realize a mechanism that we call patch that provides the same functionality as indirect revocation by using DPVS and devising the encoding of an attribute vector like a full binary tree. In the patch revocation scheme, an attribute authority prepares a full binary tree of users for every attribute category and issues the latest revocation patch associated with a covering node of the full binary tree (by running PUpdate algorithm) when the event of user's attribute revocation occurred. The revocation patch is the update information and equivalent to update keys in indirect revocation. Issuing the latest revocation patches of each attribute by each attribute authority realizes the revocation on the category level. When an encryptor generates the ciphertext with the access policy, he/she obtains the latest revocation patches of each attribute associated with the access policy and applies it to the ciphertext, i.e., makes the headers for attribute revocation by using the latest revocation patches. If the product of attribute vector (which represents the user's label) in the key for attribute revocation and the header's attribute vector (which represents the path of the covering node) is not zero, the key for attribute revocation is revoked.

Comparison Between Patch Revocation and Indirect Revocation. If there are many decryptors, the patch revocation scheme is superior to the indirect revocation scheme because the patch revocation scheme can reduce the communication cost and process of decryptors. For details of comparison between patch revocation and indirect revocation, see the full version of this paper [23].

How to Prove Security. We employ the Dual System Encryption (DSE) methodology in [18] to prove the adaptive security. However, we cannot apply the DSE directly because the attacker of our proposal can request user's attribute keys and keys for attribute revocation that satisfy the challenge access structure (but some user's private key is revoked). That is, we cannot use the key query restriction in the security proof straightforwardly. To solve this problem, we use the secret sharing and the proof methodology in [7]. Furthermore, to prove the security against the static corruption of authorities, we use the technique in [13].

1.4   Related Works

ABE (with Single Authority): Sahai and Waters introduced Fuzzy Identity-Based Encryption (FIBE) [22] that is a special type of ABE. The only access structure supported in FIBE is "threshold". In FIBE, ciphertexts and user's private key are associated with a set of attributes ω and both a threshold parameter d and another set of attributes ω', respectively. Then, if $|\omega \cap \omega'| \geq d$ holds, the user can decrypt ciphertexts and get the plaintext. Some ABE is studied and developed actively after [22] is introduced. ABE can provide data security and access control without a trusted server by using access policies and associated attributes among ciphertexts and user's private keys. Key-Policy ABE (KP-ABE) [9] introduced by Goyal et al. is the scheme that supports an access structure in user's private key. CP-ABE [3] introduced by Bethencourt et al. is the scheme that supports an access structure in ciphertexts. Ostrovsky et al. [19] proposed a scheme that supports non-monotone access structure where negated attributes are available. In recent years, FE [5,16] including ABE is proposed.

Multi-Authority ABE: Chase proposed the first multi-authority ABE [6] that extends FIBE. After that, Müller et al. proposed the multi-authority ABE [14] that extends CP-ABE. In recent years, Lewko proposed an adaptively secure multi-authority CP-ABE against static corruption of authorities [12,13] and Okamoto and Takashima proposed multi-authority functional encryption without a central authority (DMA-FE) [18].

Revocation: Boldyreva et al. introduced the IBE supporting revocation by update key [4]. After that, Sahai et al. proposed the ABE supporting revocation by update keys and updating ciphertext [21]. Recently, Lee et al. introduced a new cryptographic primitive realizing a time-evolution mechanism [11], in other words, Lee et al. proposed a new revocation scheme with modularity. Meanwhile, Attrapadung et al. proposed the ABE supporting revocation without update keys which specifies revoked users for ciphertexts directly [1]. Attrapadung et al. also proposed the ABE supporting (user-level) direct/indirect revocation [2]. Qian et al. proposed the KP-ABE supporting direct revocation and achieving adaptive security in composite order bilinear groups [20]. González-Nieto et al. proposed the full-hiding revocable predicate encryption supporting direct revocation where the revocation list is hidden specified for ciphertexts [8][1]. Datta et al. proposed the (unbounded) KP-ABE supporting direct revocation by using a subset difference method in prime order bilinear groups [7]. Horváth proposed multi-authority ABE (with a central authority) which specifies revoked users for ciphertexts directly [10].

1.5 Notations

We follow the notations in [11,18].

When A is a random variable or distribution, $y \xleftarrow{\mathsf{R}} A$ denotes that y is randomly selected from A according to its distribution. When A is a set, $y \xleftarrow{\mathsf{U}} A$ denotes that y is uniformly selected from A. We denote the finite field of order q by $\mathbb{F}_q$, and $\mathbb{F}_q \setminus \{0\}$ by $\mathbb{F}_q^\times$. A vector symbol denotes a vector representation over $\mathbb{F}_q$, e.g., $\vec{x}$ denotes $(x_1, \ldots, x_n) \in \mathbb{F}_q^n$. For two vectors $\vec{x} = (x_1, \ldots, x_n)$ and $\vec{v} = (v_1, \ldots, v_n)$, $\vec{x} \cdot \vec{v}$ denotes the inner-product $\sum_{i=1}^n x_i v_i$. The vector $\vec{0}$ is abused as the zero vector in $\mathbb{F}_q^n$ for any n. X^T denotes the transpose of matrix X. A bold face letter denotes an element of vector space $\mathbb{V}$, e.g., $\boldsymbol{x} \in \mathbb{V}$. When $\boldsymbol{b}_i \in \mathbb{V}(i = 1, \ldots, n)$, $\mathsf{span}\langle \boldsymbol{b}_1, \ldots, \boldsymbol{b}_n \rangle \subseteq \mathbb{V}$ (resp. $\mathsf{span}\langle \vec{x}_1, \ldots, \vec{x}_n \rangle$) denotes the subspace generated by $\boldsymbol{b}_1, \ldots, \boldsymbol{b}_n$ (resp. $\vec{x}_1, \ldots, \vec{x}_n$). For bases $\mathbb{B} := (\boldsymbol{b}_1, \ldots, \boldsymbol{b}_N)$ and $\mathbb{B}^* := (\boldsymbol{b}_1^*, \ldots, \boldsymbol{b}_N^*)$, $(x_1, \ldots, x_N)_\mathbb{B} := \sum_{i=1}^N x_i \boldsymbol{b}_i$ and $(y_1, \ldots y_N)_{\mathbb{B}^*} := \sum_{i=1}^N y_i \boldsymbol{b}_i^*$. For a format of attribute vectors $\vec{n} := (d; n_{A,1}, \ldots, n_{A,d}, n_{R,1}, \ldots, n_{R,d})$ that indicates dimensions of vector spaces, $\vec{e}_{\mathsf{f},t,j}$ denotes the canonical basis vector

$$(\overbrace{0, \ldots, 0}^{j-1}, 1, \overbrace{0, \ldots, 0}^{n_{\mathsf{f},t}-j}) \in \mathbb{F}_q^{n_{\mathsf{f},t}} \text{ for } \mathsf{f} = A, R; t = 1, \ldots, d; j = 1, \ldots, n_{\mathsf{f},t},$$

where f and t represent the functionality (A represents the access control and R represents the revocation) and the attribute authority. $GL(n, \mathbb{F}_q)$ denotes the general linear group of degree n over $\mathbb{F}_q$.

For a string $L \in \{0,1\}^n$, let $L[i]$ be the ith bit of, L, and $L|_i$ be the prefix of L with i-bit length. For example, if $L = 010$, then $L[0] = *, L[1] = 0, L[2] = 1, L[3] = 0$, and $L|_0 = *, L|_1 = 0, L|_2 = 01, L|_3 = 010$.

1.6 Preliminaries

We use DPVS introduced by Okamoto and Takashima [16] and general predicates (non-monotone access structures using inner-product relations). We also use the subset-cover revocation framework introduced by Naor et al. [15]. For these preliminaries, see the full version of this paper [23].

---

[1] The scheme of [8] can hide the revocation list (i.e., identities of revoked users) specified for ciphertexts in a provably secure way, but an encryptor needs to care about revocation lists. We note that an encryptor does not have to care about the revocation list in the schemes supporting indirect revocation [4,11,21] and our scheme. However, we note that the aim of the indirect revocation [4,11,21] and our scheme is not to hide the revocation list specified for ciphertexts in a provably secure way.

2 Revocable Decentralized Multi-Authority Functional Encryption (R-DMA-FE)

2.1 Definitions of R-DMA-FE

Definition 1 (Revocable Decentralized Multi-Authority Functional Encryption). *A revocable decentralized multi-authority functional encryption (R-DMA-FE) scheme consists of the following algorithms. These are randomized algorithms except for* Dec.

1. GSetup(1^λ)

   The GSetup *algorithm takes as input a security parameter* λ *and outputs a global parameter* gparam.

2. ASetup$(\text{gparam}, t, n_{A,t}, N_{max,t}, \varphi_t)$

   The ASetup *algorithm takes as input a global parameter* gparam, *an attribute authority (or category)* t $(1 \le t \le d)$, *a dimension of attribute vector space* $n_{A,t}$, *the maximum number* $N_{max,t}$ *of users for the attribute in the category* t *and the upper bound* φ_t *for the number of subsets in the cover. It outputs an attribute-authority public key* apk_t, *an attribute-authority secret key* ask_t, *an revocation public key* rpk_t *and an revocation secret key* rsk_t.

3. PUpdate$(t, \text{rpk}_t, \text{rsk}_t, r\ell_{v_t}{}^2, v_t)$

   The PUpdate *takes as input an attribute authority (or category)* t, *a revocation public key* rpk_t, *a revocation secret key* rsk_t, *the latest revocation list* $r\ell_{v_t}{}^2$ *and the latest version number for the revocation patch* v_t. *It outputs the latest revocation patch* CP_{v_t}.

4. KeyGen$(\text{gparam}, t, \text{ask}_t, \text{rsk}_t, \text{gid}, \vec{x}_{A,t})$

   The KeyGen *takes as input a global parameter* gparam, *an attribute authority (or category)* t, *a revocation secret key* rsk_t, *the user* gid *and an attribute vector* $\vec{x}_{A,t}$. *It outputs the user secret key* $\text{usk}_{\text{gid},(t,\vec{x}_{A,t}),\text{rt}}$ *where* rt *represents the number of return (after* gid*'s* $(t,\vec{x}_{A,t})$ *revocation[3]).*

5. Enc$(\{\text{apk}_t, \text{rpk}_t, \text{CP}_{v_t}\}, m, \mathbb{S})$

   The Enc *takes as inputs a set of public keys from relevant authorities* $\{\text{apk}_t, \text{rpk}_t\}$, *a set of the latest revocation patches from relevant authorities* $\{\text{CP}_{v_t}\}$, *a message* $m \in \mathbb{G}_T$, *and an access structure* $\mathbb{S}$. *It outputs a ciphertext* $\text{ct}_{\mathbb{S},\{v_t\}}$.

6. Dec$(\text{gparam}, \{\text{apk}_t, \text{rpk}_t, \text{usk}_{\text{gid},(t,\vec{x}_{A,t}),\text{rt}}\}, \text{ct}_{\mathbb{S},\{v_t\}})$

   The Dec *takes as inputs a set of public keys from relevant authorities* $\{\text{apk}_t, \text{rpk}_t\}$ *and secret keys* $\{\text{usk}_{\text{gid},(t,\vec{x}_{A,t}),\text{rt}}\}$ *corresponding to user* gid *and pair of attributes and number of return after revocation* $\{((t,\vec{x}_{A,t}),\text{rt})\}$ *and a ciphertext* $\text{ct}_{\mathbb{S},\{v_t\}}$. *It outputs a message* m *or a special symbol* $\perp$.

An R-DMA-FE scheme should have the following correctness property: for all security parameter λ, *all attribute sets* $\Gamma := \{(t,\vec{x}_{A,t})\}$, *all* gid, *all the number of return (after* gid*'s* $(t,\vec{x}_{t,A})$ *revocation)* rt, *all messages* m, *all*

---

[2] We define a user's attribute revocation list with its version v_t: $r\ell_{v_t} \subseteq \{1, \ldots, N_{max,t}\}$.

[3] We assume that a revoked user can become unrevoked again (possibly several times) after the user was revoked.

access structures $\mathbb{S}$ *and all the latest revocation lists* $r\ell_{v_t}$, *it holds that* $m = \mathsf{Dec}(\mathsf{gparam}, \{\mathsf{apk}_t, \mathsf{rpk}_t, \mathsf{usk}_{\mathsf{gid},(t,\vec{x}_{A,t})\in\Gamma}, \mathsf{rt}\}, \mathsf{ct}_{\mathbb{S},\{v_t\}})$ *with overwhelming probability, if* $\mathbb{S}$ *accepts* Γ *and* $\forall\delta$ *related with* Γ, *i.e.,* $\vec{1} \in \mathsf{span}\langle M_\delta\rangle$ *s.t.* $M_\delta := (M_j)_{\gamma(j)=1}$, *there exists no* j *s.t.* $\mathsf{FindNode}(\mathsf{gid}_i, (t,\vec{x}_{A,t}), \mathsf{rt}) \in r\ell_{v_t} \in \{r\ell_{v_t}\}_t$ [4] *and* $\rho(j) = (t,\vec{x}_{A,t})$ *or* $\neg(t,\vec{x}_{A,t})$, *where*

$$\mathsf{gparam} \xleftarrow{\mathsf{R}} \mathsf{GSetup}(1^\lambda),$$

$$(\mathsf{apk}_t, \mathsf{ask}_t, \mathsf{rpk}_t, \mathsf{rsk}_t) \xleftarrow{\mathsf{R}} \mathsf{ASetup}(\mathsf{gparam}, t, n_{A,t}, N_{max,t}, \varphi_t),$$

$$\mathsf{CP}_{v_t} \xleftarrow{\mathsf{R}} \mathsf{PUpdate}(t, \mathsf{rpk}_t, \mathsf{rsk}_t, r\ell_{v_t}, v_t),$$

$$\mathsf{usk}_{\mathsf{gid},(t,\vec{x}_{A,t}),\mathsf{rt}} \xleftarrow{\mathsf{R}} \mathsf{KeyGen}(\mathsf{gparam}, t, \mathsf{ask}_t, \mathsf{rsk}_t, \mathsf{gid}, \vec{x}_{A,t}),$$

$$\mathsf{ct}_{\mathbb{S},\{v_t\}} \xleftarrow{\mathsf{R}} \mathsf{Enc}(\{\mathsf{apk}_t, \mathsf{rpk}_t, \mathsf{CP}_{v_t}\}, m, \mathbb{S}),$$

We let S *be the set of authorities. We assume each attribute is assigned to one authority and an attribute is considered to be of the form of* $(t, \vec{x}_t)$. *For simplicity, we also assume that each authority manages only one attribute category.* [5]

Definition 2 (Security of R-DMA-FE). *For an adversary, we define* $\mathsf{Adv}_{\mathcal{A}}^{\mathsf{R-DMA-FE,PHCA}}(\lambda)$ *to be the advantage in the following experiment for any security parameter* λ. *An R-DMA-FE scheme is adaptively payload-hiding secure against chosen plaintext attacks and static corruption of authorities if the advantage of any polynomial-time adversary is negligible:*

Setup

Given 1^λ, *the challenger gives* $\mathsf{gparam} \xleftarrow{\mathsf{R}} \mathsf{GSetup}(1^\lambda)$ *to adversary* $\mathcal{A}$. $\mathcal{A}$ *specifies a set* $\mathcal{S}' \subset \mathcal{S}$ *of corrupt authorities, where* $\mathcal{S}(:= \{1, \dots, d\})$ *is the set of all the authorities in the system. For good authority* $t \in \mathcal{S} \setminus \mathcal{S}'$, *the challenger runs* $(\mathsf{apk}_t, \mathsf{ask}_t, \mathsf{rpk}_t, \mathsf{rsk}_t) \xleftarrow{\mathsf{R}} \mathsf{ASetup}(\mathsf{gparam}, t, n_{A,t}, N_{max,t}, \varphi_t)$ *and gives* $\{\mathsf{apk}_t, \mathsf{rpk}_t\}_{t\in\mathcal{S}\setminus\mathcal{S}'}$ *to* $\mathcal{A}$.

Phase 1

The adversary is allowed to issue a polynomial number of queries, $(\mathsf{gid}, (t, \vec{x}_{A,t}))$, *to the challenger or oracle* $\mathsf{KeyGen}(\mathsf{gparam}, t, \mathsf{ask}_t, \mathsf{rsk}_t, \cdot, \cdot)$ *for private keys, attribute secret key* $\mathsf{usk}_{\mathsf{gid},(t,\vec{x}_{A,t}),\mathsf{rt}}$, *where* t *is an attribute category belonging to a good authority,* gid *is an global identifier and* rt *is the number of return after* gid's $(t, \vec{x}_{A,t})$ *revocation.* [6] *The adversary is also allowed*

---

[4] Here, we define $\mathsf{FindNode} : \{0,1\}^* \times \{(t, \vec{x}_{A,t})\} \times \mathbb{N} \cup \{0\} \to \{1, \dots, N_{max,t}\}$. The $\mathsf{FindNode}$ is not a priori function. An attribute authority assigns $(\mathsf{gid}, (t, \vec{x}_{A,t}), \mathsf{rt})$ to the $\mathsf{FindNode}(\mathsf{gid}, (t, \vec{x}_{A,t}), \mathsf{rt})$-th leaf node newly and uniquely every time the user key is issued. We remark that an attribute authority can decide how to choose a leaf by itself as long as the assignment is unique. Then, let "user u" in the subset-cover revocation framework equal $\mathsf{FindNode}(\mathsf{gid}, (t, \vec{x}_{A,t}), \mathsf{rt})$. That is, $\mathsf{FindNode}(\mathsf{gid}, (t, \vec{x}_{A,t}), \mathsf{rt}) = u \in \{1, \dots, N_{max,t}\}$.

[5] We note that actually each authority can manage several attribute categories.

[6] For example, a user is initially unrevoked, and the user may be revoked. If the user becomes unrevoked again, then rt is 1.

to issue a polynomial number of queries, $(\{(\mathsf{gid}, (t, \vec{x}_{A,t}), \mathsf{rt})\}_{t \in \mathcal{S} \backslash \mathcal{S}'}, \mathsf{v}_t)$, to the challenger or oracle $\mathsf{PUpdate}(t, \mathsf{rpk}_t, \mathsf{rsk}_t, \{\mathsf{FindNode}(\cdot, \cdot, \cdot)\}, \cdot)$ for revocation patch $\mathsf{CP}_{\mathsf{v}_t}$. Note that the adversary is allowed to query only one revocation patch for each t and v_t.

Challenge

Let $\Gamma_{\mathsf{gid}_i} := \{(t, \vec{x}_{A,t})\}(i = 1, \dots, \nu)$ be the queries set to the KeyGen oracle with gid_i. The adversary submits two messages $m^{(0)}, m^{(1)}$, an access structure $\mathbb{S} := (M, \rho)$ and the pair of revocation lists for relevant good authorities and the number of version $\{(RL_t, \mathsf{v}_t) \mid RL_t := \{(\mathsf{gid}, (t, \vec{x}_{A,t}), \mathsf{rt})\}\}_{t \in \mathcal{S} \backslash \mathcal{S}'}$. We note that for a valid matrix (i.e., matrix used to specify an access structure by using a linear secret sharing scheme) in the security game, the rows associated with corrupt authorities cannot include the target vector $\vec{1}$ in their span. The access structure and revocation history must satisfy at least one of the following restrictions for each i:

Restriction I
$\Gamma_{\mathsf{gid}_i} \cup \Gamma'$ must fail to satisfy $\mathbb{S}$, where $\Gamma' := \{(t', \vec{x}_{A,t'}) \mid t' \in \mathcal{S}'\}$.

Restriction II
$\forall \delta$ related with $\Gamma_{\mathsf{gid}_i} \cup \Gamma'$, when $\mathbb{S}$ accepts δ, i.e., $\vec{1} \in \mathrm{span}\langle M_\delta \rangle$ s.t. $M_\delta := (M_j)_{\gamma(j)=1}$, there exists j s.t. $\mathsf{FindNode}(\mathsf{gid}_i, (t, \vec{x}_{A,t}), \mathsf{rt}) \in r\ell_{\mathsf{v}_t} = \{\mathsf{FindNode}(\mathsf{rl}_t) \mid \mathsf{rl}_t \in RL_t\}$ for any rt and $\mathsf{usk}_{\mathsf{gid}_i, (t, \vec{x}_{A,t}), \mathsf{rt}}$ which is given to $\mathcal{A}$, $t \in \mathcal{S} \backslash \mathcal{S}'$, $\rho(j) = (t, \vec{x}_{A,t})$ or $\neg(t, \vec{x}_{A,t})$.

The adversary must also give the challenger the public keys and the revocation patches for any corrupt authorities whose attributes appear in the access structure. Given it, the challenger flips a random coin $b \xleftarrow{\mathsf{U}} \{0, 1\}$, and sends the adversary $\mathsf{ct}^{(b)}_{\mathbb{S}, \{\mathsf{v}_t\}}$ (obtained by running $\mathsf{PUpdate}$ and Enc).

Phase 2

The adversary is allowed to issue a polynomial number of queries, $(\mathsf{gid}, (t, \vec{x}_{A,t}))$, to the challenger or oracle $\mathsf{KeyGen}(\mathsf{gparam}, t, \mathsf{ask}_t, \mathsf{rsk}_t, \cdot, \cdot)$ for private keys, user secret key $\mathsf{usk}_{\mathsf{gid}_i, (t, \vec{x}_{A,t}), \mathsf{rt}}$ subject to the same restriction as before. The adversary is also allowed to issue a polynomial number of queries, $(\{(\mathsf{gid}, (t, \vec{x}_{A,t}), \mathsf{rt})\}_{t \in \mathcal{S} \backslash \mathcal{S}'}, \mathsf{v}_t)$, to the challenger or oracle $\mathsf{PUpdate}(t, \mathsf{rpk}_t, \mathsf{rsk}_t, \{\mathsf{FindNode}(\cdot, \cdot, \cdot)\}, \cdot)$ for revocation patch $\mathsf{CP}_{\mathsf{v}_t}$ subject to the same restriction as before.

Guess

The adversary outputs a guess b' of b.

The advantage of an adversary $\mathcal{A}$ in the above game is defined as $\mathsf{Adv}^{\mathsf{R-DMA-FE, PHCA}}_{\mathcal{A}}(\lambda) := |\Pr[b' = b] - 1/2|$ for any security parameter λ. An R-DMA-FE scheme is adaptively payload-hiding secure against chosen plaintext attacks and static corruption of authorities if all polynomial time adversaries have at most a negligible advantage in the above game.

<u>Remark:</u> We show toy examples of adversary's key queries. Suppose that, there are three attribute authorities t_1, t_2 and t_3. t_1, t_2 and t_3 manage each attribute A_{t_1}, B_{t_2}, and C_{t_3} respectively. We would like to specify $(A_{t_1} \wedge B_{t_2}) \vee (A_{t_1} \wedge C_{t_3})$ as the challenge access structure.

Case 1. If the adversary gets the valid attribute key of A_{t_1}, the security game can be continued. The adversary follows the restriction I.

Case 2. If the adversary gets the valid attribute key of A_{t_1} and the revoked key of B_{t_2}, the security game can be continued. The adversary follows not the restriction I but the restriction II.

Case 3. If the adversary gets the valid attribute keys of A_{t_1} and C_{t_3} and the revoked key of B_{t_2}, the security game is aborted. The adversary's keys satisfy not $(A_{t_1} \wedge B_{t_2})$ but $(A_{t_1} \wedge C_{t_3})$. That is, the adversary does not follow the restriction I or II. The restriction II means that the adversary must have at least one revoked attribute key for each combination of attribute keys which satisfies the challenge access structure. If the adversary has the valid attribute key of A_{t_1} and the revoked keys of B_{t_2} and C_{t_3}, the security game can be continued. The adversary follows not the restriction I but the restriction II.

2.2 Construction

Our proposal is based on DMA-FE [18], so we follow the notations in [18].

We define function $\tilde{\rho} : \{1, \dots, \ell\} \to \{1, \dots, d\}$ by $\tilde{\rho}(i) := t$ if $\rho(i) = (t, \vec{v})$ or $\rho(i) := \neg(t, \vec{v})$, where ρ is given in access structure $\mathbb{S} := (M, \rho)$. In the proposed scheme, we assume that $\tilde{\rho}$ is injective for $\mathbb{S} := (M, \rho)$ with ciphertext $\mathsf{ct}_{\mathbb{S}, \{v_t\}}$. In the description of the scheme, we assume that input vector $\vec{x}_{\mathsf{f},t} := (x_{\mathsf{f},t,1}, \dots, x_{\mathsf{f},t,n_{\mathsf{f},t}})$ is normalized such that $x_{\mathsf{f},t,1} := 1$. (If $\vec{x}_{\mathsf{f},t}$ is not normalized, we can change it to a normalized one by $(1/x_{\mathsf{f},t,1}) \cdot \vec{x}_{\mathsf{f},t}$ assuming that $x_{\mathsf{f},t,1}$ is non-zero). In addition, we assume that input vector $\vec{v}_{R,\tilde{\rho}(i),v_{\tilde{\rho}(i)},j} := (v_{R,\tilde{\rho}(i),v_{\tilde{\rho}(i)},j,1}, \dots, v_{R,\tilde{\rho}(i),v_{\tilde{\rho}(i)},j,2h_{\tilde{\rho}(i)}+4})$ satisfies that $v_{R,\tilde{\rho}(i),v_{\tilde{\rho}(i)},j,2h_{\tilde{\rho}(i)}+4} \neq 0$. We use the notations in [23] (which is the full version of this paper) for DPVS, e.g., $(x_1, \dots, x_N)_{\mathbb{B}}$, $(y_1, \dots, y_N)_{\mathbb{B}^*}$ for $x_i, y_i \in \mathbb{F}_q$, and $\vec{e}_{\mathsf{f},t,j}$. For matrix, $X := (\chi_{i,j})_{i,j=1,\dots,N} \in \mathbb{F}_q^{N \times N}$ and element $\boldsymbol{v}$ in N−dimensional $\mathbb{V}$, $X(\boldsymbol{v})$ denotes $\sum_{i=1,j=1}^{N,N} \chi_{i,j} \phi_{i,j}(\boldsymbol{v})$ using canonical maps $\{\phi_{i,j}\}$. Similarly, for matrix $(\vartheta_{i,j}) := (X^{-1})^T$, $(X^{-1})^T(\boldsymbol{v}) := \sum_{i=1,j=1}^{N,N} \vartheta_{i,j} \phi_{i,j}(\boldsymbol{v})$. It holds that $e(X(\boldsymbol{x}), (X^{-1})^T(\boldsymbol{y})) = e(\boldsymbol{x}, \boldsymbol{y})$ for any $\boldsymbol{x}, \boldsymbol{y} \in \mathbb{V}$. In this paper, $\mathsf{f}(\in \{A, R\})$ is the subscript related to each functionality, that is, $\mathsf{f} = A$ is the subscript related to access control functionality and $\mathsf{f} = R$ is related to revocation functionality. The mapping $\psi_t : \{0,1\}^* \times \{(t, \vec{x}_{A,t})\} \times (\mathbb{N} \cup \{0\}) \to \{0,1\}^{h_t}$ takes a user's global identifier gid, user's attribute $(t, \vec{x}_{A,t})$ and the number of return (after gid's $(t, \vec{x}_{A,t})$ revocation) rt, and outputs a user's label $L_{\mathsf{gid},(t,\vec{x}_{A,t}),\mathsf{rt}}$ assigned to the leaf node of a user binary tree managed by an attribute authority t, where h_t represents the height of the user binary tree. The one-to-one mapping $\Phi_t : \{*, 0, 1\} \times \{0, \dots, h_t\} \to \{3, \dots, 2h_t + 3\}$ takes 0, 1 (assigned to the edges of the user binary tree managed by an attribute authority t) or $*$ (assigned to the root node of it) and the depth of it, then outputs the positions of non-zero elements of the vector $\vec{v}_{R,\tilde{p}(i),v_{\tilde{p}(i)},j}$. For example, $\varPhi_t(*,0) := 3$, and, in general, $\varPhi_t(a,b) := 2(b+1) + a$ where $a = 0,1$ and $b = 1,\ldots,\mathsf{h}_t$. We also defines L-list as the history of issuing user's key.

$\mathsf{GSetup}(1^\lambda)$:

    $\mathsf{param}_{\mathbb{G}} := (q, \mathbb{G}, \mathbb{G}_T, G, e) \xleftarrow{\mathsf{R}} \mathcal{G}_{\mathsf{bpg}}(1^\lambda),\ H : \{0,1\}^* \to \mathbb{G};$

  return $\mathsf{gparam} := (\mathsf{param}_{\mathbb{G}}, H)$.

    Remark: Given gparam, the following values can be computed by anyone and

          shared by all parties: $G_0 := H(0^\lambda),\ G_1 := H(0^{\lambda-1} \| 1),\ g_T := e(G_0, G_1)$.

$\mathsf{ASetup}(\mathsf{gparam}, t, n_{A,t}, N_{max,t}(= 2^{\mathsf{h}_t}), \varphi_t)$:

    $\mathsf{param}_{\mathbb{V}_{A,t}} := (q, \mathbb{V}_{A,t}, \mathbb{G}_T, \mathbb{A}_{A,t}, e) := \mathcal{G}_{\mathsf{dpvs}}(1^\lambda, 6n_{A,t} + 1, \mathsf{param}_{\mathbb{G}}),$

    $X_{A,t} \xleftarrow{\mathsf{U}} GL(6n_{A,t} + 1, \mathbb{F}_q),\ \boldsymbol{b}_{A,t,i} := X_{A,t}((0^{i-1}, G_0, 0^{6n_{A,t}+1-i}))$ for $i = 1,\ldots,6n_{A,t} + 1,$

    Set $\mathbb{B}_{A,t} := (\boldsymbol{b}_{A,t,i})_{i=1,\ldots,6n_{A,t}+1},$

    $\hat{\mathbb{B}}_{A,t} := (\boldsymbol{b}_{A,t,1}, \ldots, \boldsymbol{b}_{A,t,2n_{A,t}}, \boldsymbol{b}_{A,t,5n_{A,t}+1}, \ldots, \boldsymbol{b}_{A,t,6n_{A,t}+1}),$

    $\mathsf{ask}_t := X_{A,t},\ \mathsf{apk}_t := (\mathsf{param}_{\mathbb{V}_{A,t}}, \hat{\mathbb{B}}_{A,t}),$

    Run $\mathbf{CS.Setup}(N_{max,t})$ which takes the maximum number of users and outputs

    the user's binary tree. Then, assign a random value $\varsigma_{\nu_i} \in \mathbb{F}_q^\times$ to each leaf node ν_i in $\mathcal{BT}_t$.[7]

    $n_{R,t} := 4 + 2\log_2 N_{max,t} + \varphi_t,$

    $\mathsf{param}_{\mathbb{V}_{R,t}} := (q, \mathbb{V}_{R,t}, \mathbb{G}_T, \mathbb{A}_{R,t}, e) := \mathcal{G}_{\mathsf{dpvs}}(1^\lambda, 6n_{R,t} + 1, \mathsf{param}_{\mathbb{G}}),$

    $X_{R,t} \xleftarrow{\mathsf{U}} GL(6n_{R,t} + 1, \mathbb{F}_q),\ \boldsymbol{b}_{R,t,i} := X_{R,t}((0^{i-1}, G_0, 0^{6n_{R,t}+1-i}))$ for $i = 1,\ldots,6n_{R,t} + 1,$

    Set $\mathbb{B}_{R,t} := (\boldsymbol{b}_{R,t,i})_{i=1,\ldots,6n_{R,t}+1},$

    $\hat{\mathbb{B}}_{R,t} := (\boldsymbol{b}_{R,t,1}, \boldsymbol{b}_{R,t,2\mathsf{h}_t+5}, \ldots, \boldsymbol{b}_{R,t,n_{R,t}}, \boldsymbol{b}_{R,t,n_{R,t}+1}, \boldsymbol{b}_{R,t,n_{R,t}+2\mathsf{h}_t+5}, \ldots, \boldsymbol{b}_{R,t,2n_{R,t}},$

        $\boldsymbol{b}_{R,t,5n_{R,t}+1}, \boldsymbol{b}_{R,t,5n_{R,t}+2\mathsf{h}_t+5}, \ldots, \boldsymbol{b}_{R,t,6n_{R,t}+1}),$

    $\mathsf{rsk}_t := (X_{R,t}, \mathcal{BT}_t, \varPhi_t, \psi_t),\ \mathsf{rpk}_t := (\mathsf{param}_{\mathbb{V}_{R,t}}, \hat{\mathbb{B}}_{R,t}, \varphi_t),$

  return $(\mathsf{ask}_t, \mathsf{apk}_t, \mathsf{rsk}_t, \mathsf{rpk}_t)$.

$\mathsf{PUpdate}(t, \mathsf{rpk}_t, \mathsf{rsk}_t, r\ell_{v_t} := \{\mathsf{FindNode}(\mathsf{gid}, (t, \vec{x}_{A,t}), \mathsf{rt})\}, \mathsf{v}_t)$:

    Run $\mathbf{CS.Cover}(\mathcal{BT}_t, r\ell_{v_t})$ (which takes the user's binary tree and the revocation list)

    and outputs the covering set $CV_{r\ell_{v_t}} = \{S_{i'_1}, \ldots, S_{i'_{m'_{v_t}}}\},$

    for $j = 1,\ldots,m'_{v_t}(\le \varphi_t),$

    $\eta^{[1]}_{t,v_t,j}, \eta^{[2]}_{t,v_t,j}, \eta^{[3]}_{t,v_t,j} \xleftarrow{\mathsf{U}} \mathbb{F}_q,$

    $d_j, d_{j,a}, r_{t,v_t,j} \xleftarrow{\mathsf{U}} \mathbb{F}_q^\times$ s.t. $d_{j,0} + \ldots + d_{j,|ID(i'_j)|} = d_j$, for $a = 0,\ldots,|ID(i'_j)|,$

    for $1 \le z \le 2\mathsf{h}_t + 4,$

$$v_{R,t,v_t,j,z} = \begin{cases} d_j & (z = 2) \\ -d_{j,a} & (z = \varPhi_t(ID(i'_j)[a], a); 0 \le a \le |ID(i'_j)|) \\ r_{t,v_t,j} & (z = 2\mathsf{h}_t + 4) \\ 0 & (else) \end{cases}$$

---

[7] $N_{max,t}$ is smaller than q for assigning ς_{ν_i} to each leaf node uniquely.

$$\boldsymbol{p}_{t,\mathsf{v}_t,j}^{[1]} = (\overbrace{\vec{v}_{R,t,\mathsf{v}_t,j}}^{n_{R,t}}, \overbrace{0^{\varphi t}}^{n_{R,t}}, \overbrace{0^{n_{R,t}}}^{3n_{R,t}}, \overbrace{0^{3n_{R,t}}}^{n_{R,t}}, \overbrace{0^{n_{R,t}}}^{1}, \eta_{t,\mathsf{v}_t,j}^{[1]})_{\mathbb{B}_{R,t}},$$

$$\boldsymbol{p}_{t,\mathsf{v}_t,j}^{[2]} = (\overbrace{0^{n_{R,t}}}^{n_{R,t}}, \overbrace{\vec{v}_{R,t,\mathsf{v}_t,j}}^{n_{R,t}}, \overbrace{0^{\varphi t}}^{3n_{R,t}}, \overbrace{0^{3n_{R,t}}}^{n_{R,t}}, \overbrace{0^{n_{R,t}}}^{1}, \eta_{t,\mathsf{v}_t,j}^{[2]})_{\mathbb{B}_{R,t}},$$

$$\boldsymbol{p}_{t,\mathsf{v}_t,j}^{[3]} = (\overbrace{0^{n_{R,t}}}^{n_{R,t}}, \overbrace{0^{n_{R,t}}}^{n_{R,t}}, \overbrace{0^{3n_{R,t}}}^{3n_{R,t}}, \overbrace{\vec{v}_{R,t,\mathsf{v}_t,j}}^{n_{R,t}}, \overbrace{0^{\varphi t}}^{1}, \eta_{t,\mathsf{v}_t,j}^{[3]})_{\mathbb{B}_{R,t}},$$

return $\mathsf{CP}_{\mathsf{v}_t} = (\mathsf{v}_t, \{\boldsymbol{p}_{t,\mathsf{v}_t,j}^{[1]}, \boldsymbol{p}_{t,\mathsf{v}_t,j}^{[2]}, \boldsymbol{p}_{t,\mathsf{v}_t,j}^{[3]}\}_{j=1}^{m'_{\mathsf{v}_t}})$.

$\mathsf{KeyGen}(\mathsf{gparam}, t, \mathsf{ask}_t, \mathsf{rsk}_t, \mathsf{gid}, \vec{x}_{A,t} := (x_{A,t,1}, \ldots, x_{A,t,n_{A,t}}) \in \mathbb{F}_q^{n_{A,t}} \setminus \{\vec{0}\}$ s.t. $x_{A,t,1} := 1)$:

$G_{\mathsf{gid}}(= \delta G_1) := H(\mathsf{gid}) \in \mathbb{G}, \vec{\varphi}_{f,t} := (\varphi_{f,t,1}, \ldots, \varphi_{f,t,n_{f,t}}) \xleftarrow{\mathsf{U}} \mathbb{F}_q^{n_{f,t}}$ for $f = A, R$,

$\boldsymbol{b}_{f,t,i}^* := (X_{f,t}^{-1})^T((0^{i-1}, G_1, 0^{6n_{f,t}+1-i}))$ for $f = A, R$,

Set $\mathbb{B}_{f,t}^* = (\boldsymbol{b}_{f,t,i}^*)_{i=1,\ldots,6n_{f,t}+1}$ for $f = A, R$,

If $(\mathsf{gid}, (t, \vec{x}_{A,t}), \mathsf{rt}')$ exists in L-list, then set $\mathsf{rt} = \mathsf{rt}' + 1$ and change $(\mathsf{gid}, (t, \vec{x}_{A,t}), \mathsf{rt}')$
to $(\mathsf{gid}, (t, \vec{x}_{A,t}), \mathsf{rt})$ in L-list,

If $(\mathsf{gid}, (t, \vec{x}_{A,t}), \mathsf{rt}')$ does not exist in L-list, then set $\mathsf{rt} = 0$ and add $(\mathsf{gid}, (t, \vec{x}_{A,t}), \mathsf{rt})$ to L-list,

Assign $(\mathsf{gid}, (t, \vec{x}_{A,t}), \mathsf{rt})$ to $\mathsf{FindNode}(\mathsf{gid}, (t, \vec{x}_{A,t}), \mathsf{rt})$ $-$ th leaf node of $\mathcal{BT}_t$,

$L_{\mathsf{gid},(t,\vec{x}_{A,t}),\mathsf{rt}} = \psi_t(\mathsf{gid}, (t, \vec{x}_{A,t}), \mathsf{rt})$,

Retrieve $\varsigma_{ID^{-1}(L_{\mathsf{gid},(t,\vec{x}_{A,t}),\mathsf{rt}})}$ from $\mathcal{BT}_t$ and we define it as $\varsigma_{\mathsf{gid},(t,\vec{x}_{A,t}),\mathsf{rt}}$,

$$\boldsymbol{k}_{A,t}^* := (\overbrace{\vec{x}_{A,t}}^{n_{A,t}}, \overbrace{\delta\vec{x}_{A,t}}^{n_{A,t}}, \overbrace{0^{2n_{A,t}}}^{2n_{A,t}}, \overbrace{\vec{\varphi}_{A,t}}^{n_{A,t}}, \overbrace{\varsigma_{\mathsf{gid},(t,\vec{x}_{A,t}),\mathsf{rt}}\vec{x}_{A,t}}^{n_{A,t}}, \overbrace{0}^{1})_{\mathbb{B}_{A,t}^*},$$

$\mathsf{uak}_{\mathsf{gid},(t,\vec{x}_{A,t})} := (\mathsf{gid}, (t, \vec{x}_{A,t}), \boldsymbol{k}_{A,t}^*)$,

$\vec{x}_{R,t} := (x_{R,t,1}, \ldots, x_{R,t,2\mathsf{h}_t+4})$,

for $z = 1, \ldots, 2\mathsf{h}_t + 4(= 4 + 2\log_2 N_{max,t})$,

$$x_{R,t,z} = \begin{cases} 1 & (z = 1) \\ \gamma & (z = 2, \Phi_t(L_{\mathsf{gid},(t,\vec{x}_{A,t}),\mathsf{rt}}[a], a); 0 \le a \le |L_{\mathsf{gid},(t,\vec{x}_{A,t}),\mathsf{rt}}|(= \mathsf{h}_t) \\ 0 & (else) \end{cases}$$

$$\boldsymbol{k}_{R,t}^* := (\overbrace{\vec{x}_{R,t}, 0^{\varphi t}}^{n_{R,t}}, \overbrace{\delta\vec{x}_{R,t}, 0^{\varphi t}}^{n_{R,t}}, \overbrace{0^{2n_{R,t}}}^{2n_{R,t}}, \overbrace{\vec{\varphi}_{R,t}}^{n_{R,t}}, \overbrace{\varsigma_{\mathsf{gid},(t,\vec{x}_{A,t}),\mathsf{rt}}\vec{x}_{R,t}, 0^{\varphi t}}^{n_{R,t}}, \overbrace{0}^{1})_{\mathbb{B}_{R,t}^*},$$

$\mathsf{uik}_{L_{\mathsf{gid},(t,\vec{x}_{A,t}),\mathsf{rt}}} := (\mathsf{gid}, (t, \vec{x}_{A,t}), \mathsf{rt}, \boldsymbol{k}_{R,t}^*)$,

return $\mathsf{usk}_{L_{\mathsf{gid},(t,\vec{x}_{A,t}),\mathsf{rt}}} := (\mathsf{rt}, \mathsf{uak}_{\mathsf{gid},(t,\vec{x}_{A,t})}, \mathsf{uik}_{L_{\mathsf{gid},(t,\vec{x}_{A,t}),\mathsf{rt}}})$.

$\mathsf{Enc}(\{\mathsf{apk}_t, \mathsf{rpk}_t, \mathsf{CP}_{\mathsf{v}_t}\}, m, \mathbb{S})$:

$w_i \xleftarrow{\mathsf{U}} \mathbb{F}_q^\times$ for $i = 1, \ldots, \ell, \; s'_0 \xleftarrow{\mathsf{U}} \mathbb{F}_q$,

$$\vec{f}_A \xleftarrow{\mathsf{U}} \mathbb{F}_q^r, \ \vec{s}_A^T := (s_{A,1}, \dots, s_{A,\ell})^T := M \cdot \vec{f}_A^T, \ s_{A,0} := \vec{1} \cdot \vec{f}_A^T,$$

$$\vec{f}'_A \xleftarrow{\mathsf{R}} \mathbb{F}_q^r \text{ s.t. } \vec{1} \cdot \vec{f}'_A = s'_0, \ \vec{s}'_A^T := (s'_{A,1}, \dots, s'_{A,\ell})^T := M \cdot \vec{f}'_A^T,$$

$$\vec{f}_R \xleftarrow{\mathsf{U}} \mathbb{F}_q^r, \ \vec{s}_R^T := (s_{R,1}, \dots, s_{R,\ell})^T := M \cdot \vec{f}_R^T, \ s_{R,0} := \vec{1} \cdot \vec{f}_{R,0}^T,$$

$$\vec{f}'_R \xleftarrow{\mathsf{R}} \mathbb{F}_q^r \text{ s.t. } \vec{1} \cdot \vec{f}'_R = -s'_0, \ \vec{s}'_R^T := (s'_{R,1}, \dots, s'_{R,\ell})^T := M \cdot \vec{f}'_R^T,$$

for $i = 1, \dots, \ell$,

$$\eta_{A,i}, \ \theta_{A,i}, \ \theta'_{A,i}, \ \theta''_{A,i} \xleftarrow{\mathsf{U}} \mathbb{F}_q,$$

if $\rho(i) = (t, \vec{v}_{A,i} := (v_{A,i,1}, \dots, v_{A,i,n_{A,t}}) \in \mathbb{F}_q^{n_{A,t}} \setminus \{\vec{0}\}$ s.t. $v_{A,i,n_{A,t}} \neq 0)$,

$$\boldsymbol{c}_{A,i} := (\overbrace{s_{A,i}\vec{e}_{A,t,1} + \theta_{A,i}\vec{v}_{A,i}}^{n_{A,t}}, \overbrace{s'_{A,i}\vec{e}_{A,t,1} + \theta'_{A,i}\vec{v}_{A,i}}^{n_{A,t}}, \overbrace{0^{2n_{A,t}}}^{2n_{A,t}}, \overbrace{0^{n_{A,t}}}^{n_{A,t}},$$
$$\overbrace{w_i\vec{e}_{A,t,1} + \theta''_{A,i}\vec{v}_{A,i}}^{n_{A,t}}, \overbrace{\eta_{A,i}}^{1})_{\mathbb{B}_{A,t}},$$

if $\rho(i) = \neg(t, \vec{v}_{A,i})$,

$$\boldsymbol{c}_{A,i} := (\overbrace{s_{A,i}\vec{v}_{A,i}}^{n_{A,t}}, \overbrace{s'_{A,i}\vec{v}_{A,i}}^{n_{A,t}}, \overbrace{0^{2n_{A,t}}}^{2n_{A,t}}, \overbrace{0^{n_{A,t}}}^{n_{A,t}}, \overbrace{w_i\vec{v}_{A,i}}^{n_{A,t}}, \overbrace{\eta_{A,i}}^{1})_{\mathbb{B}_{A,t}},$$

for $j = 1, \dots, m'_{\mathsf{v}_{\tilde{\rho}(i)}}$ (where $t = \tilde{\rho}(i)$),

$$\eta_{R,i,j}, \ \theta_{R,i,j}, \ \theta'_{R,i,j}, \ \theta''_{R,i,j}, \ \tau_{i,j}, \ \tau'_{i,j}, \ \tau''_{i,j} \xleftarrow{\mathsf{U}} \mathbb{F}_q,$$

$$\boldsymbol{c}_{R,i,j} = s_{R,i}\boldsymbol{b}_{R,t,1} + \theta_{R,i,j}\boldsymbol{p}_t^{[1]}, \mathsf{v}_{t,j} + \tau_{i,j}\boldsymbol{b}_{R,t,j+2h_t+4}$$
$$+ s'_{R,i}\boldsymbol{b}_{R,t,n_{R,t}+1} + \theta'_{R,i,j}\boldsymbol{p}_{t,\mathsf{v}_t,j}^{[2]} + \tau'_{i,j}\boldsymbol{b}_{R,t,j+n_{R,t}+2h_t+4}$$
$$+ (-w_i)\boldsymbol{b}_{R,5n_{R,t}+1} + \theta''_{R,i,j}\boldsymbol{p}_{t,\mathsf{v}_t,j}^{[3]} + \tau''_{i,j}\boldsymbol{b}_{R,t,j+5n_{R,t}+2h_t+4}$$
$$+ \eta_{R,i,j}\boldsymbol{b}_{R,6n_{R,t}+1}$$

(Let $\eta'_{R,i,j} = \theta_{R,i,j}\eta_{t,\mathsf{v}_t,j}^{[1]} + \theta'_{R,i,j}\eta_{t,\mathsf{v}_t,j}^{[2]} + \theta''_{R,i,j}\eta_{t,\mathsf{v}_t,j}^{[3]} + \eta_{R,i,j}$)

(We already defined $\vec{v}_{R,t,\mathsf{v}_t,j}$ in PUpdate,)

$$= (\overbrace{s_{R,i}\vec{e}_{R,t,1} + \theta_{R,i,j}\vec{v}_{R,t,\mathsf{v}_t,j}}^{n_{R,t}}, 0^{j-1}, \tau_{i,j}, 0^{\varphi_t-j},$$
$$\overbrace{s'_{R,i}\vec{e}_{R,t,1} + \theta'_{R,i,j}\vec{v}_{R,t,\mathsf{v}_t,j}}^{n_{R,t}}, 0^{j-1}, \tau'_{i,j}, 0^{\varphi_t-j}, \overbrace{0^{2n_{R,t}}}^{2n_{R,t}}, \overbrace{0^{n_{R,t}}}^{n_{R,t}},$$
$$\overbrace{-w_i\vec{e}_{R,t,1} + \theta''_{R,i,j}\vec{v}_{R,t,\mathsf{v}_t,j}}^{n_{R,t}}, 0^{j-1}, \tau''_{i,j}, 0^{\varphi_t-j}, \overbrace{\eta'_{R,i,j}}^{1})_{\mathbb{B}_{R,t}},$$

$c_{d+1} := g_T^{s_{A,0}+s_{R,0}}m, \ \mathsf{ct}_{\mathbb{S},\{\mathsf{v}_t\}} := (\mathbb{S}, \{\boldsymbol{c}_{A,i}, \{\boldsymbol{c}_{R,i,j}, \boldsymbol{p}_{t,\mathsf{v}_t,j}^{[1]}\}_{j=1}^{m'_{\mathsf{v}_t}}\}_{i=1}^{\ell}, c_{d+1})$ [8]

return $\mathsf{ct}_{\mathbb{S},\{\mathsf{v}_t\}}$.

---

[8] If an attribute authority t continues to make the past revocation patch available, $\mathsf{ct}_{\mathbb{S},\{\mathsf{v}_t\}}$ does not have to include $\{\boldsymbol{p}_{t,\mathsf{v}_t,j}^{[1]}\}_{j=1}^{m'_{\mathsf{v}_t}}$. If $\mathsf{ct}_{\mathbb{S},\{\mathsf{v}_t\}}$ includes it, an attribute authority t has only to publish the latest revocation patch.

$\mathsf{Dec}(\mathsf{gparam}, \{\mathsf{apk}_t, \mathsf{rpk}_t, \mathsf{usk}_{\mathsf{gid},(t,\vec{x}_{A,t}),\mathsf{rt}}\}, \mathsf{ct}_{\mathbb{S},\{v_t\}})$:

If $\mathbb{S} := (M, \rho)$ accepts $\Gamma := \{(t, \vec{x}_{A,t}) \in \mathsf{usk}_{\mathsf{gid},(t,\vec{x}_{A,t}),\mathsf{rt}}\}$,

then compute I and $\{\alpha_i\}_{i \in I}$ s.t. $\vec{1} = \sum_{i \in I} \alpha_i M_i$, where M_i is the i-th row of M, and

$$I \subseteq \{i \in \{1, \ldots, \ell\} \mid [\rho(i) = (t, \vec{v}_{A,i}) \wedge (t, \vec{x}_{A,t}) \in \Gamma \wedge \vec{v}_{A,i} \cdot \vec{x}_{A,t} = 0]$$
$$\vee [\rho(i) = \neg(t, \vec{v}_{A,i}) \wedge (t, \vec{x}_{A,t}) \in \Gamma \wedge \vec{v}_{A,i} \cdot \vec{x}_{A,t} \neq 0]\},$$

for each $i \in I$ (where $t = \tilde{\rho}(i)$),

pick $\{\boldsymbol{p}_{t,v_t,j}^{[1]}\}_{j=1}^{m'_{v_t}}$ from $\mathsf{ct}_{\mathbb{S},\{v_t\}}$,

$K_i := e(\boldsymbol{c}_{R,i,j}, \boldsymbol{k}_{R,t}^*)$ for j s.t. $e(\boldsymbol{p}_{t,v_t,j}^{[1]}, \boldsymbol{k}_{R,t}^*) = 1$,

$$K := \prod_{i \in I \wedge \rho(i)=(t,\vec{v}_{A,i})} (e(\boldsymbol{c}_{A,i}, \boldsymbol{k}_{A,t}^*) \cdot K_i)^{\alpha_i} \cdot \prod_{i \in I \wedge \rho(i)=\neg(t,\vec{v}_{A,i})} (e(\boldsymbol{c}_{A,i}, \boldsymbol{k}_{A,t}^*)^{1/(\vec{v}_{A,i} \cdot \vec{x}_{A,t})} \cdot K_i)^{\alpha_i}$$

return $m' := c_{d+1}/K$.

[Correctness]

Here, the value g_T^s is written as $\mathsf{g_T}(s)$ in the way that the function e^x is written as $\exp(x)$.

$$K := \mathsf{g_T}(\sum_{i \in I \wedge \rho(i)=(t,\vec{v}_{A,i})} \alpha_i(s_{A,i} + \delta s'_{A,i} + w_i \varsigma_{\mathsf{gid},(t,\vec{x}_{A,t}),\mathsf{rt}}))$$

$$\mathsf{g_T}(\sum_{i \in I \wedge \rho(i)=\neg(t,\vec{v}_{A,i})} \alpha_i(\vec{v}_{A,i} \cdot \vec{x}_{A,t})^{-1}(s_{A,i} + \delta s'_{A,i} + w_i \varsigma_{\mathsf{gid},(t,\vec{x}_{A,t}),\mathsf{rt}})(\vec{v}_{A,i} \cdot \vec{x}_{A,t}))$$

$$\mathsf{g_T}(\sum_{i \in I} \alpha_i(s_{R,i} + \delta s'_{R,i} - w_i \varsigma_{\mathsf{gid},(t,\vec{x}_{A,t}),\mathsf{rt}}))$$

$$= \mathsf{g_T}(\sum_{i \in I} (\alpha_i(s_{A,i} + s_{R,i}) + \delta \alpha_i(s'_{A,i} + s'_{R,i}))) = \mathsf{g_T}(s_{A,0} + s_{R,0}) = g_T^{s_{A,0}+s_{R,0}}$$

since $\sum_{i \in I} \alpha_i s_{\mathsf{f},i} = s_{\mathsf{f},0}$ for $\mathsf{f} = A, R$, $\sum_{i \in I} \alpha_i s'_{A,i} = s'_0$, $\sum_{i \in I} \alpha_i s'_{R,i} = -s'_0$.

Encoding of Attribute Vector for Revocation and Toy Example. The user is assigned a unique leaf node of the attribute binary tree used to manage users having the attribute. Then, the label is given to the user according to the path from the root node to the leaf node. At that time, the attribute vector is constructed according to the user's label and the covering node in the complete subtree method. The mapping Φ_t works to associate the basis in DPVS and the root node and the edges. The encoding of the attribute vector is as follows:

$$\vec{x}_{R,t} := (\overbrace{1}^{1}, \overbrace{\text{random value } \gamma}^{1}, \overbrace{\text{root node and each edge}(\gamma \text{ or } 0)}^{2h_t+1}, \overbrace{0}^{1}),$$

$$\vec{v}_{R,t,v_t,j} := (\overbrace{0}^{1}, \overbrace{\text{random value } d}^{1}, \overbrace{\text{root node and each edge}(-(d - \text{shared value})d_a \text{ or } 0)}^{2h_t+1},$$
$$\overbrace{\text{random value } r}^{1}),$$

We show a toy example in Fig. 1.

$h_t := 2, v_t = 1$

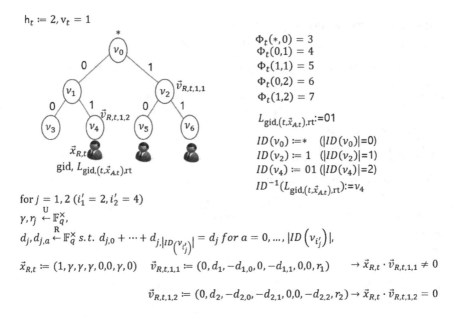

$\Phi_t(*, 0) = 3$
$\Phi_t(0,1) = 4$
$\Phi_t(1,1) = 5$
$\Phi_t(0,2) = 6$
$\Phi_t(1,2) = 7$

$L_{\mathrm{gid},(t,\vec{x}_{A,t}),\mathrm{rt}} := 01$

$ID(v_0) := *$ $(|ID(v_0)| = 0)$
$ID(v_2) := 1$ $(|ID(v_2)| = 1)$
$ID(v_4) := 01$ $(|ID(v_4)| = 2)$
$ID^{-1}(L_{\mathrm{gid},(t,\vec{x}_{A,t}),\mathrm{rt}}) := v_4$

for $j = 1, 2$ $(i_1' = 2, i_2' = 4)$
$\gamma, r_j \xleftarrow{U} \mathbb{F}_q^\times$,
$d_j, d_{j,a} \xleftarrow{R} \mathbb{F}_q^\times$ s.t. $d_{j,0} + \cdots + d_{j,|ID(v_{i'_j})|} = d_j$ for $a = 0, \ldots, |ID(v_{i'_j})|$,

$\vec{x}_{R,t} := (1, \gamma, \gamma, \gamma, 0, 0, \gamma, 0)$ $\vec{v}_{R,t,1,1} := (0, d_1, -d_{1,0}, 0, -d_{1,1}, 0, 0, r_1)$ $\rightarrow \vec{x}_{R,t} \cdot \vec{v}_{R,t,1,1} \neq 0$

$\vec{v}_{R,t,1,2} := (0, d_2, -d_{2,0}, -d_{2,1}, 0, 0, -d_{2,2}, r_2) \rightarrow \vec{x}_{R,t} \cdot \vec{v}_{R,t,1,2} = 0$

Fig. 1. Encoding of vector for revocation as a toy example

Comparison with the DMA-FE Scheme [18]. We show comparison with the DMA-FE scheme [18] in the full version of this paper [23].

2.3   Performance

We show a comparison of parameter size with previous works in Tables 3, 4 and 5. The construction of revocable-storage ABE supporting indirect revocation [11,21] is built in composite order bilinear groups. (The construction of [2,7,10, 13,18] and our scheme is built in prime order bilinear groups.) Therefore, those are outside the scope of comparison. In Tables 3, 4 and 5, SK, PK, MSK, CT and UI represent the bit length of (user's) Secret Key, Public Key, Master Secret Key, Ciphertext and Update Information, respectively. UI is update key (in [2]) or revocation patch (in our scheme). $|\mathbb{G}|, |\mathbb{G}_T|, |\mathbb{Z}_q|$ and $|\mathbb{F}_q|$ represent the bit length of element in $\mathbb{G}, \mathbb{G}_T, \mathbb{Z}_q$ and $\mathbb{F}_q$, respectively. CA and AA represent Central Authority and Attribute Authority. Γ_{max} represents the maximum number of attributes in the system. Γ represents the number of attribute in user's secret keys or ciphertexts. ℓ is the size of rows in the LSSS matrix. h is the height of user's (binary) tree in the system. h_t is the height of user's (binary) tree managed by an attribute authority t. φ is the upper bound for the number of subsets in the cover in the system. φ_t is the upper bound for the number of subsets in the cover for an attribute category t. R means the number of revoked user in the system. R_t means the number of revoked user's attribute managed by an attribute authority t. n_t is the dimension of attribute vector in the category t. We define function

$\tilde{\rho} : \{1, \ldots, \ell\} \to \{1, \ldots, d\}$ by $\tilde{\rho}(i) := t$ if $\rho(i) = (t, \vec{v})$ or $\rho(i) := \neg(t, \vec{v})$, where ρ is given in access structure $\mathbb{S} := (M, \rho)$. We assume $\tilde{\rho}$ is injective for access structure with ciphertext. We also define $\hat{\rho} : \{1, \ldots, \Gamma\} \to \{1, \ldots, d\}$. Tables 1, 2, 3, 4 and 5 show that our proposal has more advantageous funtionalities in exchange for increasing the parameter size.

Table 3. Comparison of parameter size with previous works

Schemes	PK	MSK
AI09 [2]	$|\mathbb{G}_T| + (\Gamma_{max} + \varphi + 1)|\mathbb{G}|$	$2^{h+1}|\mathbb{Z}_q|$
DDM15 [7]	$|\mathbb{G}_T| + 111|\mathbb{G}|$	$111|\mathbb{G}|$
H15 [10]	$2|\mathbb{G}|$ (CA) $|\mathbb{G}_T| + |\mathbb{G}|$ (AA)	$2|\mathbb{Z}_q|$ (CA) $2|\mathbb{Z}_q|$ (AA)
L12 [13]	$2|\mathbb{G}_T| + 48|\mathbb{G}|$ (AA)	$24|\mathbb{G}| + 48|\mathbb{Z}_q|$ (AA)
OT13 [18]	$(10n_t^2 + 7n_t + 1)|\mathbb{G}|$ (AA)	$(25n_t^2 + 10n_t + 1)|\mathbb{F}_q|$ (AA)
This work	$(18n_t^2 + 9n_t + 18\varphi_t^2 + 99\varphi_t + 36h_t\varphi_t + 48h_t + 101)|\mathbb{G}|$ (AA)	$(18n_t^2 + 9n_t + 14h_t^2 + \varphi_t^2 + 6h_t\varphi_t + 16h_t + 8\varphi_t + 2^{h_t} + 17)|\mathbb{F}_q|$ (AA)

Table 4. Comparison of parameter size with previous works (cont.)

Schemes	SK	CT
AI09 [2]	$2(\ell + 1) \log(2^h)|\mathbb{G}|$	$|\mathbb{G}_T| + (\Gamma + 1 + \mathsf{R} \log(2^h/\mathsf{R}))|\mathbb{G}|$ (Direct) $|\mathbb{G}_T| + (\Gamma + 2)|\mathbb{G}|$ (Indirect)
DDM15 [7]	$(5 + 16\ell + 32 \log^2(2^h))|\mathbb{G}|$	$|\mathbb{G}_T| + (16\Gamma + 64\mathsf{R} - 27)|\mathbb{G}|$
H15 [10]	$(1 + \Gamma)|\mathbb{G}|$	$(1 + \ell)|\mathbb{G}_T| + 2(\ell + \mathsf{R})|\mathbb{G}|$
L12 [13]	$12\Gamma|\mathbb{G}|$	$(\ell + 1)|\mathbb{G}_T| + 12\ell|\mathbb{G}|$
OT13 [18]	$\sum_{i'=1}^{\Gamma} n_{\hat{\rho}(i')}|\mathbb{F}_q| + \sum_{i'=1}^{\Gamma} (5n_{\hat{\rho}(i')} + 1)|\mathbb{G}|$	$|\mathbb{G}_T| + \sum_{i=1}^{\ell} (5n_{\tilde{\rho}(i)} + 1)|\mathbb{G}|$
This work	$\sum_{i'=1}^{\Gamma} n_{\hat{\rho}(i')}|\mathbb{F}_q| + \sum_{i'=1}^{\Gamma} (6n_{\hat{\rho}(i')} + 12h_t + 6\varphi_t + 26)|\mathbb{G}|$	$|\mathbb{G}_T| + \sum_{i=1}^{\ell} (6n_{\tilde{\rho}(i)} + 1 + \mathsf{R}_{\tilde{\rho}(i)} \log(2^{h_{\tilde{\rho}(i)}} / \mathsf{R}_{\tilde{\rho}(i)})(24h_{\tilde{\rho}(i)} + 12\varphi_{\tilde{\rho}(i)} + 50))|\mathbb{G}|$

Table 5. Comparison of parameter size with previous works (cont.)

Schemes	UI
AI09 [2]	$2(\mathsf{R} \log(2^h/\mathsf{R}))|\mathbb{G}|$
DDM15 [7]	-
H15 [10]	-
L12 [13]	-
OT13 [18]	-
This work	$(36h_t\mathsf{R}_t \log(2^{h_t}/\mathsf{R}_t) + 18\varphi_t\mathsf{R}_t \log(2^{h_t}/\mathsf{R}_t) + 75\mathsf{R}_t \log(2^{\mathsf{R}_t}/\mathsf{R}_t))|\mathbb{G}|$ (AA)

2.4 Security of the Proposed R-DMA-FE

The DLIN assumption is given in the full version of this paper [23].

Theorem 1. *The proposed R-DMA-FE scheme is adaptively payload-hiding against chosen plaintext attacks and static corruption of authorities under the DLIN assumption in the random oracle model.*

The proof of Theorem 1 is given in the full version of this paper [23].

3 Conclusion

In this paper, we proposed the first DMA-FE scheme supporting patch revocation in which an encryptor does not have to care about the revocation list. Our proposed scheme realizes the revocation on the category level that is a generalization of attribute level for the first time. We proved that our construction is adaptively secure against chosen plaintext attacks and static corruption of authorities based on the DLIN assumption. (We note that DMA-FE [18] of Okamoto and Takashima does not achieve the security against static corruption of authorities.) In the future, we will try to apply new techniques called indexing and consistent randomness amplification [17] to reduce the size of public parameters.

Acknowledgements. This work was supported in part by JSPS KAKENHI Grant Number 26330151 and JSPS and DST under the Japan - India Science Cooperative Program. The authors would like to thank anonymous reviewers for their useful comments.

References

1. Attrapadung, N., Imai, H.: Conjunctive broadcast and attribute-based encryption. In: Shacham, H., Waters, B. (eds.) Pairing 2009. LNCS, vol. 5671, pp. 248–265. Springer, Heidelberg (2009). doi:10.1007/978-3-642-03298-1_16
2. Attrapadung, N., Imai, H.: Attribute-based encryption supporting direct/indirect revocation modes. In: Parker, M.G. (ed.) IMACC 2009. LNCS, vol. 5921, pp. 278–300. Springer, Heidelberg (2009). doi:10.1007/978-3-642-10868-6_17
3. Bethencourt, J., Sahai, A., Waters, B.: Ciphertext-policy attribute-based encryption. In: 2007 IEEE Symposium on Security and Privacy, pp. 321–334 (2007)
4. Boldyreva, A., Goyal, V., Kumar, V.: Identity-based encryption with efficient revocation. In: ACM CCS 2008, pp. 417–426 (2008)
5. Boneh, D., Sahai, A., Waters, B.: Functional encryption: definitions and challenges. In: Ishai, Y. (ed.) TCC 2011. LNCS, vol. 6597, pp. 253–273. Springer, Heidelberg (2011). doi:10.1007/978-3-642-19571-6_16
6. Chase, M.: Multi-authority attribute based encryption. In: Vadhan, S.P. (ed.) TCC 2007. LNCS, vol. 4392, pp. 515–534. Springer, Heidelberg (2007). doi:10.1007/978-3-540-70936-7_28
7. Datta, P., Dutta, R., Mukhopadhyay, S.: Adaptively secure unrestricted attribute-based encryption with subset difference revocation in bilinear groups of prime order. In: Pointcheval, D., Nitaj, A., Rachidi, T. (eds.) AFRICACRYPT 2016. LNCS, vol. 9646, pp. 325–345. Springer, Heidelberg (2016). doi:10.1007/978-3-319-31517-1_17

8. González-Nieto, J.M., Manulis, M., Sun, D.: Fully private revocable predicate encryption. In: Susilo, W., Mu, Y., Seberry, J. (eds.) ACISP 2012. LNCS, vol. 7372, pp. 350–363. Springer, Heidelberg (2012). doi:10.1007/978-3-642-31448-3_26
9. Goyal, V., Pandey, O., Sahai, A., Waters, B.: Attribute-based encryption for fine-grained access control of encrypted data. In: ACM CCS 2006, pp. 89–98 (2006)
10. Horváth, M.: Attribute-based encryption optimized for cloud computing. In: Italiano, G.F., Margaria-Steffen, T., Pokorný, J., Quisquater, J.-J., Wattenhofer, R. (eds.) SOFSEM 2015. LNCS, vol. 8939, pp. 566–577. Springer, Heidelberg (2015). doi:10.1007/978-3-662-46078-8_47
11. Lee, K., Choi, S.G., Lee, D.H., Park, J.H., Yung, M.: Self-updatable encryption: time constrained access control with hidden attributes and better efficiency. In: Sako, K., Sarkar, P. (eds.) ASIACRYPT 2013. LNCS, vol. 8269, pp. 235–254. Springer, Heidelberg (2013). doi:10.1007/978-3-642-42033-7_13
12. Lewko, A., Waters, B.: Decentralizing attribute-based encryption. In: Paterson, K.G. (ed.) EUROCRYPT 2011. LNCS, vol. 6632, pp. 568–588. Springer, Heidelberg (2011). doi:10.1007/978-3-642-20465-4_31
13. Lewko, A.B.: Functional encryption: new proof techniques and advancing capabilities. Ph.D. thesis, The University of Texas (2012)
14. Müller, S., Katzenbeisser, S., Eckert, C.: Distributed attribute-based encryption. In: Lee, P.J., Cheon, J.H. (eds.) ICISC 2008. LNCS, vol. 5461, pp. 20–36. Springer, Heidelberg (2009). doi:10.1007/978-3-642-00730-9_2
15. Naor, D., Naor, M., Lotspiech, J.: Revocation and tracing schemes for stateless receivers. In: Kilian, J. (ed.) CRYPTO 2001. LNCS, vol. 2139, pp. 41–62. Springer, Heidelberg (2001). doi:10.1007/3-540-44647-8_3
16. Okamoto, T., Takashima, K.: Fully secure functional encryption with general relations from the decisional linear assumption. In: Rabin, T. (ed.) CRYPTO 2010. LNCS, vol. 6223, pp. 191–208. Springer, Heidelberg (2010). doi:10.1007/978-3-642-14623-7_11
17. Okamoto, T., Takashima, K.: Fully secure unbounded inner-product and attribute-based encryption. In: Wang, X., Sako, K. (eds.) ASIACRYPT 2012. LNCS, vol. 7658, pp. 349–366. Springer, Heidelberg (2012). doi:10.1007/978-3-642-34961-4_22
18. Okamoto, T., Takashima, K.: Decentralized attribute-based signatures. In: Kurosawa, K., Hanaoka, G. (eds.) PKC 2013. LNCS, vol. 7778, pp. 125–142. Springer, Heidelberg (2013). doi:10.1007/978-3-642-36362-7_9
19. Ostrovsky, R., Sahai, A., Waters, B.: Attribute-based encryption with non-monotonic access structures. In: ACM CCS 2007, pp. 195–203 (2007)
20. Qian, J., Dong, X.: Fully secure revocable attribute-based encryption. J. Shanghai Jiaotong Univ. (Sci.) 16(4), 490–496 (2011)
21. Sahai, A., Seyalioglu, H., Waters, B.: Dynamic credentials and ciphertext delegation for attribute-based encryption. In: Safavi-Naini, R., Canetti, R. (eds.) CRYPTO 2012. LNCS, vol. 7417, pp. 199–217. Springer, Heidelberg (2012). doi:10.1007/978-3-642-32009-5_13
22. Sahai, A., Waters, B.: Fuzzy identity-based encryption. In: Cramer, R. (ed.) EUROCRYPT 2005. LNCS, vol. 3494, pp. 457–473. Springer, Heidelberg (2005). doi:10.1007/11426639_27
23. The full version of this paper. It will appear in the IACR Cryptology ePrint Archive. https://eprint.iacr.org/

Symmetric-Key Cryptanalysis

On Linear Hulls and Trails

Tomer Ashur$^{(\boxtimes)}$ and Vincent Rijmen

Department of Electrical Engineering, ESAT/COSIC,
KU Leuven, and iMinds, Leuven, Belgium
{tomer.ashur,vincent.rijmen}@esat.kuleuven.be

Abstract. This paper improves the understanding of linear cryptanalysis by highlighting some previously overlooked aspects. It shows that linear hulls are sometimes formed already in a single round, and that overlooking such hulls may lead to a wrong estimation of the linear correlation, and thus of the data complexity. It shows how correlation matrices can be used to avoid this, and provides a tutorial on how to use them properly. By separating the input and output masks from the key mask it refines the formulas for computing the expected correlation and the expected linear potential. Finally, it shows that when the correlation of a hull is not properly estimated (e.g., by using the correlation of a single trail as the correlation of the hull), the success probability of Matsui's Algorithm 1 drops, sometimes drastically. It also shows that when the trails composing the hull are properly accounted for, more than a single key bit can be recovered using Algorithm 1. All the ideas presented in this paper are followed by examples comparing previous methods to the corrected ones, and verified experimentally with reduced-round versions of Simon32/64.

Keywords: Linear cryptanalysis · Linear hulls · Simon

1 Introduction

Linear cryptanalysis is introduced by Matsui and applied to DES in [11]. The formalism of linear cryptanalysis is extended in [4,6,13]. These works emphasize the similarity with the formalism for differential cryptanalysis that existed before. The *linear hull* is introduced as the counterpart of a differential. It is often used to prove the security of block ciphers against cryptanalysis, e.g. in [10]. A critical study of the *linear hull effect* is presented in [12]. A different framework for linear cryptanalysis, called *correlation matrices*, is introduced in [7].

In this paper, we revisit [7] and apply it to the block cipher Simon reduced to 3 rounds. Firstly, Simon's simple structure allows to construct simple and illustrative examples to highlight the similarities and differences between the two formalisms for linear cryptanalysis in practice. Secondly, the structure of Simon is sufficiently different from other mainstream ciphers to highlight the impact of some theoretical observations.

© Springer International Publishing AG 2016
O. Dunkelman and S.K. Sanadhya (Eds.): INDOCRYPT 2016, LNCS 10095, pp. 269–286, 2016.
DOI: 10.1007/978-3-319-49890-4_15

In Sect. 3 we follow the 'classical' formalism and show that the round function of Simon exhibits one-round hulls. In Sect. 4 we repeat the analysis using correlation matrices and illustrate that these matrices can facilitate the automatic analysis of ciphers, even when one-round hulls exist.

In Sect. 5 we present our first theoretical observation. We use the theoretical contributions of [7] to discuss the validity of a popular method to compute the *potential* of a linear hull.

In Sect. 6 we present our second theoretical observation. When several trails with correlation contributions of comparable magnitude and different signs exist, the performance of Matsui's Algorithm 1 strongly depends on the values of some roundkey bits. When this dependency is taken into account, the algorithm can be extended to recover multiple roundkey bits [14]. When this fact is neglected, the average success rate of Algorithm 1 drops, sometimes dramatically. Furthermore, we show a case where increasing the number of known plaintexts beyond a certain value, leads to a *decrease* in the success probability of the attack.

2   Notation and Terminology

In this section, we recall some definitions and terminology of linear cryptanalysis [4,6,7,10,11,13].

2.1   Boolean Functions

We denote the field with two elements by $GF(2)$ and the vector space of dimension n over this field by $GF(2)^n$. We use $+$ to denote addition in some field. The field in which the addition is made is always clear from the context.

A boolean function $y = f(x)$ is a function $f : GF(2)^n \rightarrow GF(2)$ mapping a vector of size n with binary components into a single bit. A boolean vector function $y = F(x)$ is a function $F : GF(2)^n \rightarrow GF(2)^m$ that maps a binary vector of size n into a binary vector of size m. A permutation is an invertible boolean vector function. A boolean vector function $y = F(x)$ with output size m can be viewed as the parallel execution of m boolean functions such that $y_i = F_i(x)$ where $0 \leq i \leq m - 1$ denotes the bit position.

A keyed boolean vector function $y = F(x, k) = F_k(x)$ is a family of boolean vector functions, indexed by a key k. An iterative block cipher with r rounds is a composition of r permutations $F_{k_{r-1}} \circ F_{k_{r-2}} \circ \ldots \circ F_{k_0}(x)$. Observe that many r-round ciphers contain in fact a reduced extra round, consisting only of an extra key addition. We will ignore this. In this paper we will assume that the roundkeys k_i are independent. Hence the key of a block cipher, denoted by k, is defined as the vector consisting of the concatenation of the r roundkeys k_i.

2.2   Masks and Approximations

Let a, b be two vectors of size n. Then $a^t x = \sum_{i=0}^{n-1} a_i \cdot x_i$. We will call a the mask of x. In practical examples, the masks will often contain many zero bits.

In order to emphasize which bits are nonzero, we will sometimes use the following set notation:

$$a = \{i_1, i_2, \ldots, i_u\} \Leftrightarrow \begin{cases} a_j = 1, \forall j \in \{i_1, i_2, \ldots, i_u\} \\ a_j = 0, \forall j \notin \{i_1, i_2, \ldots, i_u\} \end{cases}$$

Using this notation, the addition (XOR) of two masks corresponds to the symmetric difference operation on the sets.

A linear approximation for a keyed boolean permutation is a tuple (a, b, c) such that a, b and c are masks for the input, the output and the key, respectively. Let p be the fraction of inputs x for which the equation $a^t x + b^t F_k(x) + c^t k = 0$ holds. The correlation of the linear approximation (a, b, c) is defined as $\text{cor}(a, b, c) = 2 \cdot (p - \frac{1}{2}) = 2p - 1$. In general, both p and $\text{cor}(a, b, c)$ will depend on the value of k. When $c = 0$, we abbreviate the notation $(a, b, 0)$ and $\text{cor}(a, b, 0)$ to (a, b) and $\text{cor}(a, b)$.

2.3   Linear Hulls and Trails

A (linear) trail Ω covering r rounds of an iterative block cipher is a concatenation of linear approximations each covering a single round such that the output mask of round i equals the input mask of round $i + 1$. Hence we can identify the trail with a vector of $r + 1$ masks $\omega_i, 0 \le i \le r$ $\Omega = (\omega_0, \omega_1, \ldots, \omega_r)$. Round i has input mask ω_i and output mask ω_{i+1}. The correlation contribution of a trail Ω is the product of the correlations of the individual rounds: $\text{cor}_p(\Omega) = \prod_{i=0}^{r-1} \text{cor}_{\text{round } i}(\omega_i, \omega_{i+1})$. In a key-alternating cipher the round consists of a fixed part g followed by an addition with the round key. We can write:

$$\text{cor}_{\text{round } i}(\omega_i, \omega_{i+1}) = (-1)^{\omega_i^t k_i} \text{cor}_g(\omega_i, \omega_{i+1}). \tag{1}$$

Note, however, that this notation implicitly assumes that to each bit of the round input a different bit of the roundkey is added. We will say more on this in Sect. 5.3. We obtain:

$$\text{cor}_p(\Omega) = \prod_i (-1)^{\omega_i^t k_i} \text{cor}_g(\omega_i, \omega_{i+1}) = |\text{cor}_p(\Omega)| \cdot (-1)^{d_\Omega + \sum_i \omega_i^t k_i}, \tag{2}$$

with $d_\Omega = 1$ if $\prod_i \text{cor}_g(\omega_i, \omega_{i+1})$ is negative; otherwise $d_\Omega = 0$.

A linear hull covering r rounds of a block cipher is a tuple (α, β). The hull is composed of a set of linear trails all having the same input mask and output mask but that can differ in the intermediate masks. The correlation of a linear hull is

$$\text{cor}(\alpha, \beta) = \sum_{\substack{\Omega \\ \omega_0 = \alpha, \omega_r = \beta}} \text{cor}_p(\Omega) \tag{3}$$

3   One-Round Hulls in Simon

In this section, we briefly recall the definition of Simon's round function. We prove the existence of one-round hulls, which impact the computation of correlations of multi-round hulls.

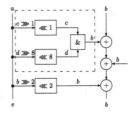

Fig. 1. Trail through one round of Simon (without the final swap operation). The dashed box indicates the part of the round that we discuss in Sect. 4.

3.1 Simon

Simon is a family of lightweight block ciphers designed by the US National Security Agency and published in 2013 [3]. The Simon$2n/mn$ family of lightweight block ciphers has 10 members differing in the block and key sizes. All members of the family have a Feistel structure with round function R employing a non-linear function f. In each round i, R receives two n-bit input words X^i and Y^i, and outputs two n-bit words X^{i+1} and Y^{i+1}. The round function uses three operations: addition in $GF(2)^n$, bitwise AND, and a left circular shift by j positions, which we denote by $+$, $\&$, and $\lll j$, respectively. The internal non-linear function f is defined as:

$$f(X^i) = [(X^i \lll 1)\&(X^i \lll 8)] + (X^i \lll 2).$$

The output of the round function R on input block (X^i, Y^i) is: $R^i(X^i, Y^i) = (Y^i + f(X^i) + k^i, X^i)$, where i is the round number. The entire cipher is $R^{r-1} \circ R^{r-2} \circ \ldots \circ R^0(X^0, Y^0)$. The structure of the round function of Simon is depicted in Fig. 1.

3.2 Linear Hulls and Trails Through One Round of Simon

We use the notation (a, b, c, d, e) to describe a linear trail through one round of Simon. Here a and b denote the left and right input masks; c and d denote the masks at the outputs of the two topmost rotations; e and b denote the left and right output masks (before the swap operation), cf. Fig. 1.

We now study the behavior of linear trails over one round of Simon using the rules of propagation of linear trails introduced in [4,6]. The rule for trail propagation over the branch operation implies the following constraint on a, b, c, d, e:

$$a + e = (b \ggg 2) + (c \ggg 1) + (d \ggg 8) \tag{4}$$

Note that the rule for trail propagation over the addition operation is already implicit in the way we propagate the b mask through Fig. 1. The output bit z of a bitwise AND operation $z = x$ AND y is correlated to the 4 linear functions of the two input bits:

$$\text{cor}(z, 0) = \text{cor}(z, x) = \text{cor}(z, y) = 1/2, \ \text{cor}(z, x + y) = -1/2.$$

It follows that the AND operation in Simon leads to the following constraints on b, c, d: if a bit in c or d is set, then the bit in b at the corresponding position needs to be set. This translates to:

$$\bar{c} \text{ OR } b = 1 \tag{5}$$

$$\bar{d} \text{ OR } b = 1 \tag{6}$$

The following lemma expresses that some one-round trails come in groups.

Lemma 1. *Let (a, b, c, d, e) be a one-round trail over Simon. If there exists an index i such that $b_i = b_{i+7} = 1$, then the trail (a, b, c, d, e) satisfies the constraints (4)–(6) if and only if the trail $(a, b, c + (1 \lll i), d + (1 \lll (i + 7)), e)$ satisfies the constraints (4)–(6).*

Proof. The proof can be found in the full version of this paper [2].

Since the trails in Lemma 1 have the same input masks (a, b) and the same output masks (e, b), they are in the same one-round linear hull. Figures 2, 3 and 4 each show two trails derived from one another by means of Lemma 1. Notice that in each set both trails select exactly the same bits of the roundkeys.

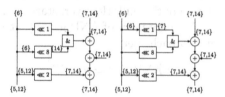

Fig. 2. Two trails of a one-round hull.

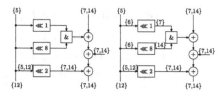

Fig. 3. Two trails of a second one-round hull.

3.3 Correlations and Correlation Contributions

We now want to compute the correlation contributions of the trails of Figs. 2, 3 and 4. The usual rule is to assume that all nonlinear functions act independently and to multiply all the correlations. This results in the following values for the correlation contributions of the six trails:

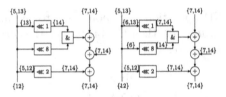

Fig. 4. Two trails of a third one-round hull. The trails have nonzero contributions of the same magnitude and opposite sign. The hull has correlation zero.

	c	d	cor
Fig. 2	$\varnothing$	$\{14\}$	2^{-2}
	$\{7\}$	$\varnothing$	2^{-2}
Fig. 3	$\varnothing$	$\varnothing$	2^{-2}
	$\{7\}$	$\{14\}$	2^{-2}
Fig. 4	$\{14\}$	$\varnothing$	2^{-2}
	$\{7,14\}$	$\{14\}$	-2^{-2}

In each case by adding the correlation contributions of the two trails we obtain the correct correlation of the hull. However, starting from the observation that when $b_i = b_{i+7} = 1$, there are pairs of AND gates that share one input bit, we can follow a different approach. Let

$$s = x \text{ AND } y, \quad t = y \text{ AND } z$$

Then we have

$$s + t = y \text{ AND } (x + z),$$

which implies the following:

$$\text{cor}(s + t, x + z) = \text{cor}(s + t, 0) = \text{cor}(s + t, y) = 1/2$$
$$\text{cor}(s + t, x + y + z) = -1/2$$
$$\text{cor}(s + t, x + y) = \text{cor}(s + t, y + z) = 0$$
$$\text{cor}(s + t, x) = \text{cor}(s + t, z) = 0$$

These values can be used to derive immediately the exact correlations of the linear hulls of Figs. 2, 3 and 4. Observe that the linear hull of Fig. 4 has correlation zero, while both trails have a nonzero correlation contribution. Hence, mounting an attack and using the correlation contribution of a trail as an estimate for the correlation of this linear hull will likely lead to wrong results.

4   Correlation Matrices

In this section, we follow the alternative approach of [7] to compute correlations and correlation contributions.

4.1   Correlation Matrix for Simon

In order not to repeat too much from the previous approach, we concentrate on the most interesting part of the round function: the AND function combined with the preceding expanding linear function $\text{lin}(x) = (x \lll 1, x \lll 8)$. This part is indicated by a dashed box in Fig. 1. The correlation matrix of a map f is defined as follows:

Definition 1 (Correlation matrix [7]).

$$\mathbf{C}_{uw}^f := \text{cor}(u^t f(x), w^t x)$$

For a linear map $y = \mathbf{M}x$, we have: $\mathbf{C}_{uw} = \delta(\mathbf{M}^t u + w)$, where δ is the Kronecker-delta function (which is defined by $\delta(0) = 1$ and $\delta(x) = 0, \forall x \neq 0$). This gives for $\text{lin}(x)$:

$$\mathbf{C}_{uv,w}^{\text{lin}} = \delta(w + (u \ggg 1) + (v \ggg 8)),$$

where we denote the row index of $\mathbf{C}^{\text{lin}}$ by uv in order to make it more clear from the notation this is an expansion function, and hence, the row index of the matrix (i.e., the output) is twice as long as the column index.

The correlation matrix for a 1-bit AND operation $z = x$ AND y is given by:

$$\mathbf{C}^A = \begin{bmatrix} \text{cor}(0,0) & \text{cor}(0,x) & \text{cor}(0,y) & \text{cor}(0,x+y) \\ \text{cor}(z,0) & \text{cor}(z,x) & \text{cor}(z,y) & \text{cor}(z,x+y) \end{bmatrix} = \begin{bmatrix} 1 & 0 & 0 & 0 \\ \frac{1}{2} & \frac{1}{2} & \frac{1}{2} & -\frac{1}{2} \end{bmatrix}$$

We can express the matrix elements by means of the following formula:

$$\mathbf{C}_{a,bc}^A = (1-a)(1-b)(1-c) + \frac{1}{2}a(-1)^{bc}$$

The 16-bit parallel AND operation is a special case of the boxed map discussed in [7]. Hence, we obtain:

$$\mathbf{C}_{a,bc}^{\text{AND}} = \prod_i \mathbf{C}_{a_i,b_i c_i}^A = \prod_i \left((1-a_i)(1-b_i)(1-c_i) + \frac{1}{2}a_i(-1)^{b_i c_i} \right)$$

In order to compute the correlation matrix for a combined map, we only have to multiply the correlation matrices of its components [7]:

$$\mathbf{C}^{f_2 \circ f_1} = \mathbf{C}^{f_2} \times \mathbf{C}^{f_1}$$

For the combination of $\text{lin}(x)$ and AND, we obtain:

$$\mathbf{C}_{u,w} = \sum_{xy} \mathbf{C}_{u,xy}^{\text{AND}} \mathbf{C}_{xy,w}^{\text{lin}}$$

$$= \sum_{xy} \prod_i \left((1-u_i)(1-x_i)(1-y_i) + \frac{1}{2}u_i(-1)^{x_i y_i} \right) \delta(w + (x \ggg 1) + (y \ggg 8))$$

The δ-function is nonzero only when $y = (w \lll 8) + (x \lll 7)$. Hence, we obtain:

$$\mathbf{C}_{u,w} = \sum_x \prod_i \left((1-u_i)(1-x_i)(1-(w_{i-8} + x_{i-7})) + \frac{1}{2}u_i(-1)^{x_i(w_{i-8}+x_{i-7})} \right)$$

$$\tag{7}$$

4.2 Examples

We now apply (7) to compute the correlations and correlation contributions of the linear hulls, respectively trails, shown in Fig. 2, 3 and 4. Remember that we consider only the combination of the linear map lin and the AND operation. We denote the input mask for this combined map by w and the output mask by u. They are related as follows to the masks (a, b, c, d, e) defining a trail over one round, cf. Fig. 1:

$$w = a + e + (b \ggg 2)$$
$$u = b$$

The hull of Fig. 2 has input $w = \{6\} = 0040_x$ and output $u = \{7, 14\} = 4080_x$. Filling out these values in (7), we obtain

$$C_{4080,0040} = \sum_x \prod_{i=7,14} \left(\frac{1}{2}(-1)^{x_i(w_{i-8}+x_{i-7})} \right) \prod_{i \neq 7,14} (1 - x_i)(1 - (w_{i-8} + x_{i-7})) \ .$$

From the first factor of the product on the right, we see that in order to obtain a nonzero contribution, x_i must equal 0 for all i except $i = 7, 14$. Combined with the second factor we obtain that x_7 is free and all other $x_i = 0$. Hence we obtain:

$$C_{4080,0040} = \sum_{x_7} \prod_{i=7,14} \frac{1}{2}(-1)^{x_i(w_{i-8}+x_{i-7})}$$

$$= \underbrace{\frac{1}{4}(-1)^0(-1)^0}_{x_7=0,\text{trail of Fig. 2, left}} + \underbrace{\frac{1}{4}(-1)^0(-1)^0}_{x_7=1,\text{trail of Fig. 2, right}} = \frac{1}{2}$$

The two terms in the sum are the correlation contributions of the two trails that are shown in Fig. 2 and that together form the one-round hull.

The hull of Fig. 3 has $w = \varnothing = 0000_x$ and $u = \{7, 14\} = 4080_x$. We obtain:

$$C_{4080,0000} = \sum_x \prod_{i=7,14} \left(\frac{1}{2}(-1)^{x_i x_{i-7}} \right) \prod_{i \neq 7,14} (1 - x_i)(1 - x_{i-7})$$

From the product on the right, we obtain that x_7 is free and all other $x_i = 0$. Hence we obtain

$$C_{4080,0000} = \sum_{x_7} \prod_{i=7,14} \frac{1}{2}(-1)^{x_i x_{i-7}}$$

$$= \underbrace{\frac{1}{4}(-1)^0(-1)^0}_{x_7=0,\text{trail of Fig. 3, left}} + \underbrace{\frac{1}{4}(-1)^0(-1)^0}_{x_7=1,\text{trail of Fig. 3, right}} = \frac{1}{2}$$

The hull of Fig. 4 has $w = \{13\} = 2000_x$ and $u = \{7, 14\} = 4080_x$. We obtain:

$$C_{4080,2000} = \sum_x \prod_{i=7,14} \left(\frac{1}{2}(-1)^{x_i(w_{i-8}+x_{i-7})} \right) \prod_{i \neq 7,14} (1 - x_i)(1 - (w_{i-8} + x_{i-7}))$$

From the product on the right, we obtain that x_7 is free, $x_{14} = 1$ and all other $x_i = 0$.

$$C_{4080,2000} = \sum_{x_7} \prod_{i=7,14} \frac{1}{2}(-1)^{x_i(w_{i-8}+x_{i-7})}$$

$$= \frac{1}{4}(-1)^0(-1)^0 + \frac{1}{4}(-1)^0(-1)^1 = 0$$

We see that the two trails of this hull have opposite contributions, resulting in a correlation zero for the hull.

4.3 Conclusion

As expected, this method gives the same results as the method of Sect. 3. However, observe that by using correlation matrices, the dependence between the inputs of the AND operation is taken care of automatically. Observe also that while the end result of this method is the correlation of the linear hull, we also obtain the correlation contributions of all the individual trails as the nonzero terms in the final sum.

5 Expected Correlation and Potential

Several recent works provide bounds for the security of ARX ciphers and other ciphers defined at bit-level against linear cryptanalysis by bounding the potential of linear hulls [15–17]. The bounds on the hulls are computed by summing the squares of the expected values of the correlation contributions of the linear trails, which are constructed automatically using mixed-integer linear programming (MILP) techniques.

Several of these works mention the problem that may arise in the computation of the correlation contribution of a linear trail when non-linear functions share inputs. We showed in Sect. 4 that correlation matrices don't have this problem.

In this section we address a second problem with the computation of the potential. Note that this problem doesn't occur for differential characteristics and differentials. It is one reason why we do not agree that differential and linear trails can be treated in exactly the same way, as is claimed e.g. in [17].

5.1 Expected Correlation

For a key-alternating cipher, the expected value (over all roundkeys) of the correlation contribution of a linear trail equals

$$\mathrm{E}[\mathrm{cor_p}(\Omega)] = 0 \tag{8}$$

This follows directly by taking the expectation of (2). Intuitively, (8) might look contradictory to [11], in particular to Algorithm 1. The apparent contradiction can be solved as follows. Although [7] writes:

The multiple-round linear expressions described in [11] *correspond with what we call linear trails.*

there is in fact a difference. The approximations of [11] are linear expressions in terms of plaintext bits, ciphertext bits and roundkey bits. In the trails of [7], the roundkey bits are left out of the expression. It follows that the expected value of the correlation contribution becomes zero. By (3) we obtain that the expected value over all roundkeys of the correlation of a linear hull is

$$E[cor(a, b)] = 0 .$$

5.2   Potential

Since the data complexity of a linear attack is inversely proportional to the square of the correlation, it is of importance to know or to bound the value $E[(cor(a, b))^2]$. In [13], Nyberg calls this quantity the *potential* of the linear hull, and gives the following formula to compute it:

$$E[(cor(a, b))^2] = \sum_{\substack{\Omega \\ \omega_0=a, \omega_r=b}} (cor_p(\Omega))^2 \tag{9}$$

The potential is also called the Expected Linear Probability (ELP). We briefly recall here the proof for (9), using our own notation. By definition of expectation, we have:

$$E[(cor(a, b))^2] = \frac{1}{K} \sum_k \left(\sum_{\substack{\Omega \\ \omega_0=a, \omega_r=b}} cor_p(\Omega) \right) \left(\sum_{\substack{\Omega' \\ \omega'_0=a, \omega'_r=b}} cor_p(\Omega') \right)$$

Using (2):

$$= \frac{1}{K} \sum_\Omega \sum_{\Omega'} \left(\sum_k (-1)^{d_\Omega + d_{\Omega'} + \sum_i (\omega_i + \omega'_i)^t k_i} |cor_p(\Omega)| |cor_p(\Omega')| \right)$$

Since

$$\sum_k (-1)^{\sum_i (\omega_i + \omega'_i)^t k_i} = \begin{cases} K & \text{if } \omega_i = \omega'_i, \forall i, \\ 0 & \text{else,} \end{cases} \tag{10}$$

we have:

$$E[(cor(a, b))^2] = \sum_\Omega (cor_p(\Omega))^2 . \tag{11}$$

$\square$

5.3 Additions/Corrections

We will now show that if a cipher exhibits one-round hulls as described in Sect. 3, Formula (11) is no longer correct. The existence of one-round hulls implies that we can have more than one trail corresponding to the same linear mask of the roundkey. For example, Figs. 2, 3 and 4 each show two different trails corresponding to the same linear mask of the roundkey.

In order to explain the consequences, (1) has to be slightly rewritten, using a different notation. In fact, we need to distinguish between *trails* and *masks for the roundkey*. From now on, we use κ_i to denote the mask for the roundkey of round i, and $\mathcal{K}$ to denote the vector of roundkey masks. We use W to denote the vector of the data masks required to uniquely define the trail: $W = (w_0, w_1, \ldots, w_r)$. Note that the domain of the w_i may be larger than the domain of the κ_i. For example, in Fig. 1, the data mask w_i contains a, b, c and d, while the roundkey mask κ_i needs to contain only b.

We denote by l, respectively L, the functions mapping w_i to the corresponding κ_i, respectively W to the corresponding $\mathcal{K}$. These functions are specific to the cipher. With this notation, (1) becomes:

$$\text{cor}_{\text{round } i}(w_i, w_{i+1}) = (-1)^{\kappa_i^t k_i} \text{cor}_g(w_i, w_{i+1}), \text{ with } \kappa_i = l(w_i).$$

When L is one-to-one, formula (11) applies without modifications. However, if L is a non-injective map, then the sum of (10) become nonzero as soon as $\mathcal{K} = \mathcal{K}'$, which still allows $W \neq W'$. Hence (11) becomes:

$$E[(\text{cor}(a,b))^2] = \sum_{\mathcal{K}} \sum_{\substack{W, W' \\ L(W) = L(W') = \mathcal{K}}} (\text{cor}_p(W))(\text{cor}_p(W')) \ .$$

Converting back, we obtain:

$$E[(\text{cor}(a,b))^2] = \sum_{\mathcal{K}} \left(\sum_{\substack{W \\ L(W) = \mathcal{K}}} \text{cor}_p(W) \right)^2 \tag{12}$$

Comparing (9) to (12), we see that the difference between the two values can take positive as well as negative values. In particular when there are several trails with correlation contributions of comparable magnitude, the difference can be significant. Applied to the one-round hulls of Figs. 2, 3 and 4, we get the following results:

(a, b)	$E[(\text{cor}(a,b))^2]$ with (9)	$E[(\text{cor}(a,b))^2]$ with (12)
$(\{6;7,14\},\{5,12;7,14\})$	2^{-3}	2^{-2}
$(\{5;7,14\},\{12;7,14\})$	2^{-3}	2^{-2}
$(\{5,13;7,14\},\{12;7,14\})$	2^{-3}	0

We performed practical experiments and confirmed the values in the right-most column.

5.4   Conclusion

Finally, we would like to discuss when (12) has to be used instead of (11), or in other words: "For which ciphers is the map L from data-input masks to roundkey masks not one-to-one?" We already demonstrated that Simon is such a cipher. Also Speck and ciphers using Substitution-Permutation-Substitution (SPS) round functions like Camellia [1] are in this category.

Perhaps we should conclude that the difference between (11) and (12) points to a problem with the methodology being used to construct linear trails. Indeed, it would be possible to define a linear trail by its roundkey mask, and then adapt the method to compute its correlation contribution to make sure that all terms are included.

6   On Matsui's Algorithm 1

In this section, we investigate how the success rate of Matsui's Algorithm 1 is influenced by all the trails in the same linear hull. As described already in [14], this phenomenon can be used to extend Matsui's Algorithm 1 and to extract multiple key-bits. We illustrate this for reduced Simon in Sect. 6.5.

In Sects. 6.3 and 6.4 we study another consequence of this phenomenon: sometimes, the success rate of Matsui's Algorithm 1 will be worse than the estimate based on the study of a single trail. Somewhat counter-intuitively, the success rate of an attack may even decrease when the number of known plaintexts is increased! As far as we know, this is the first time that an explanation for such an effect is provided.

First, we describe the background for this special phenomenon: the 4 trails that constitute a hull over three rounds of Simon (Sect. 6.1) and the key-dependence of their correlations (Sect. 6.2).

6.1   Four Trails Through Three Rounds of Simon

Figure 5 shows two trails through three rounds of Simon-32. Both trails start from the plaintext bits $\{8, 10, 16, 28\}$ and end in the ciphertext bit $\{16\}$. Hence they belong to the same 3-round linear hull. Two more trails belonging to this 3-round linear hull are presented in the full version of the paper [2]. It can be shown that this 3-round hull doesn't have any other trails with nonzero correlation contribution. These 4 linear trails are linearly dependent: denoting the vector of roundkey masks of Trail i by Ω_i, we have

$$\Omega_1 + \Omega_2 + \Omega_3 + \Omega_4 = 0$$

All trails involve bits $\{0, 12\}$ from the first roundkey, bit $\{14\}$ from the second, and bit $\{0\}$ from the third roundkey. Additionally, each of these trails have the following bits involved:

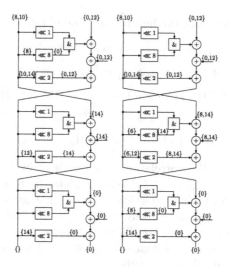

Fig. 5. Two trails in a 3-round linear hull. Trail 1 is shown on the left, Trail 2 on the right.

$$\text{Trail 1: } \emptyset$$
$$\text{Trail 2: bit } 8$$
$$\text{Trail 3: bit } 15$$
$$\text{Trail 4: bits } 8, 15$$

We denote by Z the sum of the roundkey bits involved in all trails. The sum of the roundkey bits involved in Trail 2, 3 and 4, we denote respectively by $Z + z_0$, $Z + z_1$ and $Z + z_0 + z_1$.

6.2   Correlation Contributions of the Trails

Straightforward computations similar to the computations in Sect. 3 and Sect. 4 show that the trails have the following correlation contributions:

$$\text{Trail 1: } \mathrm{cor}_{\mathrm{p}}^{(1)} = (-1)^Z \cdot 2^{-4}$$
$$\text{Trail 2: } \mathrm{cor}_{\mathrm{p}}^{(2)} = (-1)^{Z+z_0} \cdot 2^{-5}$$
$$\text{Trail 3: } \mathrm{cor}_{\mathrm{p}}^{(3)} = (-1)^{Z+z_1+1} \cdot 2^{-5}$$
$$\text{Trail 4: } \mathrm{cor}_{\mathrm{p}}^{(4)} = (-1)^{Z+z_0+z_1} \cdot 2^{-5}$$

Note that these correlation contributions exist only as intermediate mathematical results. An attacker who can observe only plaintext and ciphertext bits, can measure only the sum of the four correlation contributions, i.e. the correlation of the hull. We suspect that this fact forms the basis of Murphy's argument against *the probability statements made in the usual definition of a linear hull* [12].

We denote the correlation of the hull by cor_h and obtain:

$$\text{cor}_h = (-1)^Z \cdot 2^{-4} + (-1)^{Z+z_0} 2^{-5} + (-1)^{Z+z_1+1} 2^{-5} + (-1)^{Z+z_0+z_1} 2^{-5} \quad (13)$$

$$= (-1)^Z \cdot 2^{-5} \left(2 + (-1)^{z_0} + (-1)^{z_1+1} + (-1)^{z_0+z_1} \right) \quad (14)$$

$$= (-1)^{Z+z_0} \cdot 2^{-5} \left((-1)^{z_0} \cdot 2 + 1 + (-1)^{z_1+z_0+1} + (-1)^{z_1} \right) \quad (15)$$

$$= (-1)^{Z+z_1} \cdot 2^{-5} \left((-1)^{z_1} \cdot 2 + (-1)^{z_0+z_1} - 1 + (-1)^{z_0} \right) \quad (16)$$

$$= (-1)^{Z+z_0+z_1} \cdot 2^{-5} \left((-1)^{z_0+z_1} \cdot 2 + (-1)^{z_1} + (-1)^{z_0+1} + 1 \right) \quad (17)$$

From (13) we see that the correlation is determined by the values of $Z, Z + z_0, Z + z_1 + 1$, and $Z + z_0 + z_1$. Table 1 considers the 8 possible assignments for these variables and their correlations. We see that for a fixed Z, the value $(-1)^Z \cdot 3 \cdot 2^{-5}$ is three times more likely to occur than the value $(-1)^{Z+1} \cdot 2^{-5}$. In the following, we will investigate how likely each value is, and show how different values affect the success rate of Matsui's Algorithm 1 when different trails are considered as if they are the only trails.

Table 1. The possible values for cor_h obtained from (13).

Z	z_0	z_1	cor_h	Z	z_0	z_1	cor_h
0	0	0	$3 \cdot 2^{-5}$	1	0	0	$-3 \cdot 2^{-5}$
0	0	1	$3 \cdot 2^{-5}$	1	0	1	$-3 \cdot 2^{-5}$
0	1	0	-2^{-5}	1	1	0	2^{-5}
0	1	1	$3 \cdot 2^{-5}$	1	1	1	$-3 \cdot 2^{-5}$

6.3 Knowing Trail 1 only

We adopt the figures of [11, Table 2] to express the relation between correlation of a hull, the number of known plaintext and the success rate. Concretely, we derive from the table that if the hull has correlation c, then using c^{-2}, $4c^{-2}$ and $8c^{-2}$ known plaintexts, the algorithm achieves success rates of respectively 84%, 98% and 100%.

In order to apply Algorithm 1 using Trail 1, the adversary first computes the correlation contribution of Trail 1:

$$\text{cor}_p^{(1)} = (-1)^Z \cdot 2^{-4} \quad (18)$$

Using the assumption that the correlation of the hull is approximately equal to the correlation contribution of Trail 1, the adversary concludes that a sample of $N = 2^{10}$ known plaintexts should be sufficient to estimate Z with a success rate of 98%.

Subsequently, the adversary collects a sample of N known plaintexts and uses them to compute the experimental correlation $\hat{c}$. Depending on the value of $\hat{c}$, the adversary "guesses" a value for the sum (XOR) of the roundkey bits associated with the trail. Using (18) the adversary is led to believe that the actual bias can only take the values 2^{-4} and -2^{-4} and so the obvious decision rule is to guess for the XOR of the roundkey bits the value 1 if $\hat{c} < 0$, and the value 0 if $\hat{c} > 0$. From (14), however, we obtain:

$$z_0 = 0, z_1 = 0 \rightarrow \text{cor}_h = (-1)^Z \cdot 3 \cdot 2^{-5}$$
$$z_0 = 0, z_1 = 1 \rightarrow \text{cor}_h = (-1)^Z \cdot 3 \cdot 2^{-5}$$
$$z_0 = 1, z_1 = 0 \rightarrow \text{cor}_h = (-1)^Z \cdot (-1) \cdot 2^{-5}$$
$$z_0 = 1, z_1 = 1 \rightarrow \text{cor}_h = (-1)^Z \cdot 3 \cdot 2^{-5}$$

In the first, the second and the last case, the actual correlation is $(-1)^Z \cdot 3 \cdot 2^{-5}$, which is 50% larger than the value that was obtained using Trail 1 only. Using 2^{10} known plaintexts, the success rate of Algorithm 1 increases from the predicted 98% to 100%.

In the third case, however, not only the magnitude of the correlation has decreased, but also the sign has changed. This means that Algorithm 1's estimate for Z will be *usually wrong*! The success rate drops from the predicted 98% to $100 - 84 = 16\%$. We conclude that the average success rate of Matsui's Algorithm 1 drops from the predicted 98% to

$$0.75 \cdot 100\% + 0.25 \cdot 16\% = 79\%.$$

When the data complexity is increased, the estimate of the actual correlation through the sample correlation is improved. This means that the first term in the sum increases, while the second one decreases. The success probability in the general case is given by:

$$1 - 0.75 \cdot \phi \left(\frac{-\left(\frac{N}{2} + 3 \cdot N \cdot 2^{-6} - \frac{N}{2}\right)}{\sqrt{\frac{N}{4} - 9 \cdot N \cdot 2^{-12}}} \right) + 0.25 \cdot \phi \left(\frac{-\left(\frac{N}{2} - N \cdot 2^{-6} - \frac{N}{2}\right)}{\sqrt{\frac{N}{4} + \cdot N \cdot 2^{-12}}} \right)$$

Differentiating with respect to N shows that the function is maximized with a success rate of 80% when $N = 2^{9.12}$, and tends to 75% as N tends to 2^{32}. So we get the following observation.

Observation: In an attack based on (the original, non-extended version of) Matsui's Algorithm 1 the optimal number of plaintexts can be smaller than the full codebook. Increasing the number of plaintexts beyond this optimal number may *decrease* the success rate of the attack.

6.4 Knowing only One of the Trails 2–4

Similar to the case of Trail 1, we can use the individual correlations presented in Sect. 6.2. Hence, for Trail 2, the adversary will compute

$$\text{cor}_p^{(2)} = (-1)^{Z+z_0} 2^{-5}$$

and conclude that 2^{12} known plaintexts should be sufficient to estimate $Z + z_0$ with a success rate of 98%. Since the predicted correlation differs only in sign, the decision rule for the guessed sum of the roundkey bits is as before. Repeating the success rate analysis and using the numbers from Table 1, we learn that the success rate with 2^{12} known plaintexts drops from the predicted 98% to

$$0.5 \cdot 100\% + 0.25 \cdot 98\% + 0.25 \cdot 0\% = 74.5\%$$

The success rate is maximized and saturates with 75% when N grows beyond $2^{12.1}$ as the middle term tends to 100% and the others stay steady. Similar computations give for Trail 4 the same result as for Trail 2. For Trail 3, setting $N = 2^{12}$ gives a reduced success rate of:

$$0.25 \cdot 100\% + 0.5 \cdot 0\% + 0.25 \cdot 2\% = 25.5\%.$$

A success rate below 50% means that Algorithm 1 will more often produce the wrong answer.

6.5 Knowing All Trails

When all trails are taken into account, Matsui's Algorithm 1 can be extended and recover more than a single bit, cf. also [14]. The approach can be summarized as follows. The adversary knows now that the correlation of the hull can take 4 values, distanced $2 \cdot 2^{-5}$ apart, cf. (13) and Table 1. The adversary divides the space of possible $\hat{c}$ outcomes into four regions instead of just two. After collecting N plaintexts, the adversary computes $\hat{c}$ and guesses for the key bits the values that produce the correlation the closest to $\hat{c}$. We can compute the success rate as follows.

If $Z = 0$ and $z_0 z_1 \in \{00, 01, 11\}$, then $\mathrm{cor}_h = 3 \cdot 2^{-5}$. The attack will be successful if $\hat{c} > 2^{-4}$. When $N = 2^{12}$, this happens with probability 0.98. The adversary obtains $1 + 3(-1/3 \log_2(1/3)) = 1 + \log_2(3) \approx 2.6$ bits of information.

If $Z = 0$ and $z_0 z_1 = 10$, then $\mathrm{cor}_h = -1 \cdot 2^{-5}$. The attack will be successful if $-2^{-4} < \hat{c} < 0$. When $N = 2^{12}$, this happens with probability 0.95. The adversary obtains 3 bits of information.

If $Z = 1$ and $z_0 z_1 = 10$, then $\mathrm{cor}_h = 2^{-5}$. The attack will be successful if $0 < \hat{c} < 2^{-4}$. When $N = 2^{12}$, this happens with probability 0.95. The adversary obtains 3 bits of information.

If $Z = 1$ and $z_0 z_1 \in \{00, 01, 11\}$, then $\mathrm{cor}_h = -3 \cdot 2^{-5}$. The attack will be successful if $\hat{c} < -2^{-4}$. When $N = 2^{12}$, this happens with probability 0.98. The adversary obtains 2.6 bits of information.

6.6 Conclusion

It has been observed before that the accuracy of linear attacks can be improved if multiple trails are taken into account [5,8,9]. The example that we treated in this section illustrates this for the specific case where we use Matsui's Algorithm 1 and all trails are in the same linear hull.

When we take the dependencies on the roundkey bits into account, we can use Algorithm 1 to recover more than 1 key bit as in [14]. However, when we do not take into account these dependencies, there are cases where Algorithm 1 systematically provides the wrong outcome, no matter how much we increase the number of known plaintexts. In fact, there are cases where increasing the number of known plaintexts beyond a certain value will result in a decrease of the attack's success rate. Future work should revisit attacks that were using the correlation of a single trail as an estimate for the correlation of the hull, as well as attacks using Matsui's Algorithm 1, to see whether the data complexity needs to be modified, and whether more key bits can be recovered. In previous sections we showed how this can be done using correlation matrices, taking into account conflicting effects that were previously overlooked.

It remains to be investigated how we can extend this analysis to hulls over more rounds, when it becomes infeasible to enumerate all the trails. Secondly, it would be interesting to investigate the consequences for Matsui's Algorithm 2. Algorithm 2 tries to find the last-round keys that minimize the distance between the correlation over $R - x$ rounds that is predicted by the adversary and the experimental correlation computed from ciphertexts and known plaintexts. If the actual correlation is very far from the predicted correlation, as we observed here, there could be many wrong keys ranked above the correct key.

Acknowledgments. The authors would like to thank Kaisa Nyberg and the anonymous reviewers for their comments. This work was partially supported by the Research Council KU Leuven, OT/13/071.

References

1. Aoki, K., Ichikawa, T., Kanda, M., Matsui, M., Moriai, S., Nakajima, J., Tokita, T.: Camellia: a 128-bit block cipher suitable for multiple platforms — design and analysis. In: Stinson, D.R., Tavares, S. (eds.) SAC 2000. LNCS, vol. 2012, pp. 39–56. Springer, Heidelberg (2001). doi:10.1007/3-540-44983-3_4
2. Ashur, T., Rijmen, V.: On linear hulls and trails in simon. IACR Cryptology ePrint Archive 2016, 88 (2016). http://eprint.iacr.org/2016/088
3. Beaulieu, R., Shors, D., Smith, J., Treatman-Clark, S., Weeks, B., Wingers, L.: The SIMON and SPECK families of lightweight block ciphers. Cryptology ePrint Archive, Report 2013/404 (2013). http://eprint.iacr.org/
4. Biham, E.: On Matsui's linear cryptanalysis. In: Santis, A. (ed.) EUROCRYPT 1994. LNCS, vol. 950, pp. 341–355. Springer, Heidelberg (1995). doi:10.1007/BFb0053449
5. Biryukov, A., Cannière, C., Quisquater, M.: On multiple linear approximations. In: Franklin, M. (ed.) CRYPTO 2004. LNCS, vol. 3152, pp. 1–22. Springer, Heidelberg (2004). doi:10.1007/978-3-540-28628-8_1
6. Chabaud, F., Vaudenay, S.: Links between differential and linear cryptanalysis. In: Santis, A. (ed.) EUROCRYPT 1994. LNCS, vol. 950, pp. 356–365. Springer, Heidelberg (1995). doi:10.1007/BFb0053450

7. Daemen, J., Govaerts, R., Vandewalle, J.: Correlation matrices. In: Preneel, B. (ed.) FSE 1994. LNCS, vol. 1008, pp. 275–285. Springer, Heidelberg (1995). doi:10.1007/3-540-60590-8_21

8. Hermelin, M., Cho, J.Y., Nyberg, K.: Multidimensional linear cryptanalysis of reduced round serpent. In: Mu, Y., Susilo, W., Seberry, J. (eds.) ACISP 2008. LNCS, vol. 5107, pp. 203–215. Springer, Heidelberg (2008). doi:10.1007/978-3-540-70500-0_15

9. Kaliski, B.S., Robshaw, M.J.B.: Linear cryptanalysis using multiple approximations. In: Desmedt, Y.G. (ed.) CRYPTO 1994. LNCS, vol. 839, pp. 26–39. Springer, Heidelberg (1994). doi:10.1007/3-540-48658-5_4

10. Keliher, L., Meijer, H., Tavares, S.: New method for upper bounding the maximum average linear hull probability for SPNs. In: Pfitzmann, B. (ed.) EUROCRYPT 2001. LNCS, vol. 2045, pp. 420–436. Springer, Heidelberg (2001). doi:10.1007/3-540-44987-6_26

11. Matsui, M.: Linear cryptanalysis method for DES cipher. In: Helleseth, T. (ed.) EUROCRYPT 1993. LNCS, vol. 765, pp. 386–397. Springer, Heidelberg (1994). doi:10.1007/3-540-48285-7_33

12. Murphy, S.: The effectiveness of the linear hull effect. J. Math. Cryptol. **6**(2), 137–147 (2012). http://dx.doi.org/10.1515/jmc-2011-0025

13. Nyberg, K.: Linear approximation of block ciphers. In: Santis, A. (ed.) EUROCRYPT 1994. LNCS, vol. 950, pp. 439–444. Springer, Heidelberg (1995). doi:10.1007/BFb0053460

14. Röck, A., Nyberg, K.: Generalization of Matsui's algorithm 1 to linear hull for key-alternating block ciphers. Des. Codes Cryptograph. **66**(1–3), 175–193 (2013). http://dx.doi.org/10.1007/s10623-012-9679-1

15. Shi, D., Hu, L., Sun, S., Song, L.: Linear (hull) cryptanalysis of round-reduced versions of KATAN. Cryptology ePrint Archive, Report 2015/964 (2015). http://eprint.iacr.org/

16. Shi, D., Hu, L., Sun, S., Song, L., Qiao, K., Ma, X.: Improved linear (hull) cryptanalysis of round-reduced versions of SIMON. Cryptology ePrint Archive, Report 2014/973 (2014). http://eprint.iacr.org/

17. Sun, S., Hu, L., Wang, M., Wang, P., Qiao, K., Ma, X., Shi, D., Song, L., Fu, K.: Towards finding the best characteristics of some bit-oriented block ciphers and automatic enumeration of (related-key) differential and linear characteristics with predefined properties. Cryptology ePrint Archive, Report 2014/747 (2014). http://eprint.iacr.org/

Related-Key Cryptanalysis of Midori

David Gérault[✉] and Pascal Lafourcade

University Clermont Auvergne, Clermont-Ferrand, France
{david.gerault,pascal.lafourcade}@udamail.fr

Abstract. Midori64 and Midori128 [2] are lightweight block ciphers, which respectively cipher 64-bit and 128-bit blocks. While several attack models are discussed by the authors of Midori, the authors made no claims concerning the security of Midori against related-key differential attacks. In this attack model, the attacker uses *related-key differential characteristics*, *i.e.*, tuples $(\delta_P, \delta_K, \delta_C)$ such that a difference (generally computed as a XOR) of δ_P in the plaintext coupled with a difference δ_K in the key yields a difference δ_C after r rounds with a good probability. In this paper, we propose a constraint programming model to automate the search for optimal (in terms of probability) related-key differential characteristics on Midori. Using it, we build related-key distinguishers on the full-round Midori64 and Midori128, and mount key recovery attacks on both versions of the cipher with practical time complexity, respectively $2^{35.8}$ and $2^{43.7}$.

Keywords: Midori · Related-key attack · Constraint programming

1 Introduction

The increasing usage of embedded devices led to a lot of research on how to adapt existing cryptographic primitives for the low power and energy constraints associated with the internet of things. Lightweight block ciphers follow this trend, and aim at providing energy efficient ways to ensure confidentiality for fixed size block messages. In 2015, the authors of [2] consider the challenging task of minimizing the energy cost for a lightweight block cipher. They proposed a lightweight symmetric block cipher scheme called *Midori*, composed of two versions Midori64 and Midori128, which respectively cipher 64- and 128-bit message blocks.

In this paper, we challenge the related-key security of both versions of Midori. In the related-key model, introduced independently by Biham [3] and Knudsen [12], the attacker is allowed to require the encryption of messages of his choice under the secret key, but also under other keys which have a relation to the original one. For instance, if K is the secret key, the attacker can require the

This research was conducted with the support of the FEDER program of 2014–2020, the region council of Auvergne, and the Digital Trust Chair of the University of Auvergne.

© Springer International Publishing AG 2016
O. Dunkelman and S.K. Sanadhya (Eds.): INDOCRYPT 2016, LNCS 10095, pp. 287–304, 2016.
DOI: 10.1007/978-3-319-49890-4_16

encryption of a message m under K, but also under another key K^*, computed as $K \oplus \delta_K$, where $\oplus$ is the XOR operation and δ_K is a bit string chosen by the attacker. In an ideal block cipher, the distribution of the resulting ciphertext difference should be uniform and independent from the input difference. However, in real ciphers, there exist *related-key differential characteristics*, *i.e.*, difference propagation patterns, which happen with higher probabilities. In a related-key differential attack, the attacker requires the encryption of message pairs satisfying a difference δ_P, under keys satisfying a difference δ_K, expecting an output difference δ_C.

One of the main applications of related-key cryptanalysis is finding collisions on hash functions built from block ciphers (*e.g.*, with the Davies-Meyer construction). For instance, the hash function used in Microsoft's Xbox was broken due to the related-key vulnerability of the underlying block cipher TEA [19], leading to a hack of the system [20]. Related-key attacks were not taken into account in the design of Midori, and the authors made no claim on its security in this model. As Midori is designed for embedded devices, it could however be used to build a hash function, which motivates scrutinizing its security in the related key setting.

The search for related-key differential characteristics is however difficult. Following the idea of [9], we use constraint programming (CP) to tackle this problem. In this programming paradigm, instead of providing an imperative algorithm, the programmer describes the problem to be solved as a set of variables linked together by *constraints* (for instance, $x + y = 10$), and the exploration of the search space is left to the solver. While an overwhelming part of the cryptanalysis literature relies on custom algorithms, we believe that a more generic approach is very promising. In particular, as shown in [9], constraint programming seems less error prone than custom code.

Contributions

- We provide constraint programming models to find optimal related key differential characteristics on both versions of Midori.
- Using our models, we give the optimal $R - 1$ rounds related-key differential characteristics of both versions of Midori, with probability 2^{-14} for Midori64, and 2^{-38} for Midori128.
- We then mount practical time key recovery attacks requiring $2^{35.8}$ operations with 20 related keys for Midori64, and $2^{43.7}$ encryptions with 16 related keys for Midori128.
- We also provide a related-key distinguisher of probability 2^{-16} for Midori64 and 2^{-40} for Midori128.

Related Work: Most results in the literature using constraint programming for the cryptanalysis of block ciphers use Mixed Integer Linear Programming (MILP). In [14], the authors use MILP to mount a linear cryptanalysis on a stream cipher and on the block cipher AES, both in the single key setting. In [17,18], the authors use MILP to find the best related-key differential characteristics on several bit oriented block ciphers, but they do not treat Midori. As opposed to MILP, CP supports *table constraints* defining tuples of authorized

Table 1. Summary of the attacks against Midori.

Type	Rounds	Data	Time	Reference
Midori64				
Impossible differential	10	$2^{62.4}$	$2^{80.81}$	[6]
Meet-in-the-middle	12	$2^{55.5}$	$2^{125.5}$	[13]
Invariant subspace*	full(16)	2	2^{16}	[11]
Related-key differential	14	2^{59}	2^{116}	[7]
Related-key differential	**full(16)**	$2^{23.75}$	$2^{35.8}$	Sect. 5
Midori128				
Related-key differential	**full(20)**	$2^{43.7}$	$2^{43.7}$	Sect. 5

*Note that this attack only works if a key from the weak class is used

values, which provides a rather efficient way to model the non linear SBs. To the best of our knowledge, only [9] uses classical CP instead of MILP. The authors present a model for finding optimal related-key differential characteristics against AES, using a method similar to the one presented in this paper.

The existing attacks against Midori are summed up in Table 1.

In [6], the authors propose an impossible differential attack on 10 rounds of Midori64. In [13], Li Lin and Wenling Wu describe a meet-in-the-middle attack on 12-round Midori64. In [11], the authors exhibit a class of 2^{32} weak keys which can be distinguished with a single query. Assuming a key from this class is used, then it can be recovered with as little as 2^{16} operations, and a data complexity of 2^1. Finally, a related-key cryptanalysis of Midori64 is performed in [7]. It covers 14 rounds and has a complexity of 2^{116}, as opposed to $2^{35.8}$ for ours. This difference is due to their differential characteristics being far from optimal.

As for Midori128, to the best of our knowledge, no cryptanalysis on it has been published yet. We fill this gap by mounting a key recovery attack on the whole cipher, requiring $2^{43.7}$ encryptions.

Outline: In Sect. 2, we give a brief description of Midori. We then remind the basics of related-key cryptanalysis and introduce our notations in Sect. 3. We present our CP models in Sect. 4. Finally, we detail our results in Sect. 5, before concluding in the last section.

2   Description of Midori Encryption Scheme

Both versions of Midori, Midori64 and Midori128, use 128-bit keys. In both versions, the blocks are treated as 4×4 matrices of words of m bits, with $m = 4$ for Midori64 and $m = 8$ for Midori128. The encryption process consists in applying a round function that updates an internal state S, represented as shown on Fig. 1 (where the s_i are 4-bit words for Midori64 and

$$S = \begin{pmatrix} s_0 & s_4 & s_8 & s_{12} \\ s_1 & s_5 & s_9 & s_{13} \\ s_2 & s_6 & s_{10} & s_{14} \\ s_3 & s_7 & s_{11} & s_{15} \end{pmatrix}$$

Fig. 1. Representation of the state in Midori.

8-bit words (bytes) for Midori128), for a given number of rounds R. For Midori64, R is equal to 16, whereas for Midori128 R is 20^1.

The round function is composed of the following consecutive operations:

SubCell (SB) substitutes every cell of the state, using a non linear *Substitution Box*, denoted Sbox. The Sbox of Midori64 is given as example in Fig. 2(a). For Midori128, 4 different Sboxes are used (one for each line of the state)[2].

ShuffleCell (SC) operates a permutation of the cells of the state. On input $(s_0, \ldots, s_{15})$, it applies the following permutation:

$$(s_0, s_{10}, s_5, s_{15}, s_{14}, s_4, s_{11}, s_1, s_9, s_3, s_{12}, s_6, s_7, s_{13}, s_2, s_8).$$

MixColumns (MC) multiplies the state by the symmetric matrix given in Fig. 2(b), thus applying a linear transformation on each column independently. It has the quasi-MDS property if $MC(0,0,0,0) = (0,0,0,0)$ or $|X| + |MC(X)| = 0$ or $|X| + |MC(X)| \geq 4$, where $|X|$ denotes the number of non-zero words in a column X of the state.

KeyAdd (KA) is a XOR between S and a round key derived from the initial key.

The Midori encryption process works as follows: an initial KeyAdd, using the whitening key WK, is applied. Then, the round function is executed $R-1$ times. Finally, a final SubCell is applied to the resulting state, and a new KeyAdd is performed, again using WK. The round key derivation is very straightforward: the key for each round i is obtained by XORing the initial key with a predefined 4×4 constant matrix α_i. For Midori64, the 128-bit key is considered as two 4×4 matrices of 4-bit words K_0 and K_1, and WK is computed as $K_0 \oplus K_1$. The round key for round i is computed as $K_{i \mod 2} \oplus \alpha_i$. For Midori128, K is a single 4×4 bytes matrix, and $WK = K$. The round key for round i is then simply computed as for Midori64: $K \oplus \alpha_i$.

3 Related-Key Cryptanalysis

Differential cryptanalysis studies the propagation of the differences, generally computed as a XOR, between two plaintexts ciphered with the same key.

x	0	1	2	3	4	5	6	7	8	9	a	b	c	d	e	f
$SB(x)$	c	a	d	3	e	b	f	7	8	9	1	5	0	2	4	6

$$\begin{pmatrix} 0 & 1 & 1 & 1 \\ 1 & 0 & 1 & 1 \\ 1 & 1 & 0 & 1 \\ 1 & 1 & 1 & 0 \end{pmatrix}$$

(a) The Sbox of Midori64.

(b) The MixColumns matrix of Midori.

Fig. 2. Midori description.

---

[1] The full specification is presented in [2].
[2] The Sboxes of Midori 128 are given in [11].

Related-key cryptanalysis, which was independently introduced by Biham [3] and Knudsen [12], additionally considers the case where the two plaintexts are ciphered with different keys. A tuple $(\delta_{in}, \delta_K, \delta_{out})$ is an *n-rounds related-key differential* for a keyed round function f_K, for which f_K^i denotes the output after round i (starting from 0), if for some plaintext P and key K it holds that $f_K^{n-1}(P) \oplus f_{K \oplus \delta_K}^{n-1}(P \oplus \delta_{in}) = \delta_{out}$. Similarly, if $X_i^{P,K}$ denotes the *internal state* of the round function with inputs P and K at round i, a tuple $(\delta_{in}, \delta_K, \delta_{X_0} \ldots \delta_{X_{n-1}}, \delta_{out})$ is an *n-rounds related-key differential characteristics* if $(\delta_{in}, \delta_{out}, \delta_K)$ is an n-rounds related-key differential and, for all i from 0 to $n-1$, it holds that $X_i^{P,K} \oplus X_i^{P \oplus \delta_{in}, K \oplus \delta_K} = \delta_{X_i}$.

The differences δ_X are composed of differential words, defined as $\delta_X[i][j] = X[i][j] \oplus X'[i][j]$, where $X[i][j]$ (resp. $X'[i][j]$) denotes a word at position i, j (where $(i, j) \in [0; 3]^2$) of a matrix X (resp. X').

The probability $p = Pr[(\delta_{in}, \delta_K) \to \delta_{out}]$ denotes the probability that a related-key differential $(\delta_{in}, \delta_k, \delta_{out})$ holds, *i.e.*, for P and K drawn uniformly at random, $f_K(P) \oplus f_{K \oplus \delta_K}(P \oplus \delta_{in}) = \delta_{out}$.

Note that, by definition, for the linear parts L of the cipher, we have $L(P) \oplus L(P \oplus \delta) = L(\delta)$, for any P and δ. On the other hand, for the non linear parts NL, $NL(P) \oplus NL(P \oplus \delta)$ is generally different from $NL(\delta)$. Hence, to handle the non linear parts of block ciphers (namely the Sboxes), related-key differential cryptanalysis usually uses a Differential Distribution Table (DDT) to derive the probability $Pr[\delta_{in} \to \delta_{out}]$ that for a random word w, $SB(w) \oplus SB(w \oplus \delta_{in}) = \delta_{out}$. For any differential words δ_{in} and δ_{out}, $DDT[\delta_{in}][\delta_{out}]$ gives the number of words w satisfying this relation, and the probability is computed as $\frac{DDT[i][j]}{2^{|W|}}$, where $|W|$ denotes the bit length of the words. When the Sboxes are bijective[3], they do not introduce nor remove differences. More formally, it holds that for any word w, $SB(w) \oplus SB(w \oplus \delta) \neq 0$ if $\delta \neq 0$, and $SB(w) \oplus SB(w \oplus \delta) = 0$ if $\delta = 0$. Said otherwise, for a given Sbox, $Pr[0 \to 0] = 1$. Hence, the probability of a related-key differential characteristic is only affected by *active Sboxes*, *i.e.*, Sboxes which have a non-zero difference at their input. Thus, the probability p for a related-key differential characteristic to hold for random P and K is computed as the product of the probabilities associated with each the active Sboxes it contains. We have $p = \prod_{i=1}^x p_i$, where p_i denotes the probability that the transition $\delta_{in i} \to \delta_{out i}$, defined by the related-key differential characteristic, holds for the ith active Sbox (among x). Since the complexity of a related-key differential key recovery attack is directly related to the probability of the related-key differential that is used, characteristics with the least possible active Sboxes are generally the most interesting. The crucial point of this type of cryptanalysis is then to determine high probability related-key differential characteristics.

However, exhaustive search on all possible input differences is not practical for Midori because the size of the input is 128 bits for the key and 64 or 128 bits for the plaintext. Hence, a common method is to solve the problem in two steps (*e.g.*, [5,8,9]). The first step does not consider the value of the differential

---

[3] It is the case for most block ciphers, including Midori.

bytes, but only the positions of non-zero differences. During the first step, the differential words are abstracted to a compact representation. In the *compact representation*, the differential words are abstracted to differential bits. The differential bit Δ representing a differential word δ in the compact representation is defined by $\delta = 0 \Rightarrow \Delta = 0$ and $\delta \neq 0 \Rightarrow \Delta = 1$. We denote a differential word by δ, and a differential bit by Δ.

An n-round *compact related-key differential characteristic* Δ abstracting a related-key differential characteristic $\delta = (\delta_{in}, \delta_k, \delta_{X_0}, \ldots, \delta_{X_{n-1}}, \delta_{out})$ is the tuple $(\Delta_{in}, \Delta_k, \Delta_{X_0}, \ldots, \Delta_{X_{n-1}}, \Delta_{out})$, where $\Delta_{in} = 0$ if $\delta_{in} = 0$ and $\Delta_{in} = 1$ if $\delta_{in} \neq 0$ (and similarly for $\Delta_{X_0}, \ldots, \Delta_{X_{n-1}}$ and Δ_{out}).

The idea of working with two steps is that related-key differential characteristics with the best probabilities are generally the ones with the least active Sboxes. Hence, a lot of filtering can be done by simply starting by working on differential bits and minimizing the number of active Sboxes.

Once compact related-key differential characteristics minimizing the number of active Sboxes are obtained, a second step is run to build full related-key differential characteristics built with differential words. Note that not all compact related-key differential characteristics can be instantiated with differential words. The main reason is that a given input difference to an Sbox can only yield a limited number of output differences. For instance, in Midori64, if $\delta_X = 0x1$ (where $0x$ denotes hexadecimal representation), there exists no word X such that $SB(X) \oplus SB(X \oplus 0x1) = 0x9$, according to the Sboxes of Midori64 given in Fig. 2(a). Moreover, the coefficients of the MixColumns matrix cannot be directly taken into account with differential bits, nor can the equalities of the corresponding differential words. This yields transitions that are correct when working on a bit related-key differential characteristic, but not with differential words. Such solutions are said to be *inconsistent*, otherwise, they are *consistent*.

4   Constraint Programming Model

We describe our constraint programming models to find related-key differential characteristics with optimal probability on Midori. This process is decomposed in two steps: the first one aims at lower bounding the number of active Sboxes. It only considers compact related-key differential characteristics. The second step transforms the solutions to Step 1 into word related-key differential characteristics when it is possible. In other words, during Step 1, we simply find the positions of the differences in the related-key differential characteristic, and in Step 2, we assign actual values to these differences.

4.1   Step 1

Variables and Objective Function: In Step 1, we consider the propagation of differences through the cipher by working on compact related-key differential characteristics. Let n denote the number of times the full round function is applied, *i.e.*, we neglect the initial KeyAdd and the final KeyAdd and SB. For

Midori64, $n = 15$, and for Midori128, $n = 19$. When no information about the format is provided, the variables are $n \times 4 \times 4$ binary arrays, *i.e.*, one 4×4 matrix per round. Each of the following variables represent differential bits:

Δ_K represents the differential bits of the key. For Midori64, it is modeled as a $2 \times 4 \times 4$ binary array (as the initial key is composed of two 4×4 matrices). For Midori128, it is represented as a 4×4 binary matrix.

Δ_{SB} represents the state after the SB operation. Note that since this operation does not introduce differences, this variable is somehow redundant. We however use it for readability.

Δ_{KA} represents the state after the KeyAdd operation.

Δ_{MC} represents the state after the MixColumns operation.

Δ_{SC} represents the state after the ShuffleCell operation.

The relations between these variables for a given round r is:

$$\Delta_{SB}[r] \xrightarrow{SC} \Delta_{SC}[r] \xrightarrow{MC} \Delta_{MC}[r] \xrightarrow{KA} \Delta_{KA}[r] \xrightarrow{SB} \Delta_{SB}[r+1]$$

Our aim is to minimize the number of active Sboxes, *i.e.*, Sboxes with non zero differences. Hence, we ask the solver to minimize the sum of all $\Delta_{SB}[r]$, which constitutes our *objective function*:

$$Minimize \left(\sum_{r=0}^{n-1} \sum_{i=0}^{3} \sum_{j=0}^{3} \Delta_{SB}[r][i][j] \right)$$

Constraints: Since we work with differential bits representing the presence or absence of difference, we cannot use the regular XOR operation between such values for KeyAdd nor MixColumns. Let Δ_0 and Δ_1 denote two differential bits. We remind that Δ_0 (resp. Δ_1) is 1 if $\delta_0 \neq 0$ (resp. $\delta_1 \neq 0$). The compact representation contains no information about the actual values of δ_0 and δ_1 when they are non-zero. This abstraction leads us to define the following constraint that describes the xor between several differential bits $x_1, \ldots, x_{q-1}$ where x_q is the result:

$$XOR(x_1, \ldots, x_q) \Leftrightarrow \{x_1 + \ldots + x_q \neq 1\}$$

where $+$ denotes the integer addition and $x_1, \ldots, x_k \in \{0, 1\}$. Intuitively, it states that the xor of the $q - 1$ corresponding words is known to be 0 when all the differential bits are zero, or only one is non zero, but can be either 0 or 1 otherwise.

For ShuffleCell, we simply apply the permutation given in Sect. 2 to build $\Delta_{SC}[r]$ from $\Delta_{SB}[r]$.

$\Delta_{SC}[r][0][0] = \Delta_{SB}[r][0][0], \Delta_{SC}[r][1][0] = \Delta_{SB}[r][2][2], \Delta_{SC}[r][2][0] = \Delta_{SB}[r][1][1],$
$\Delta_{SC}[r][3][0] = \Delta_{SB}[r][3][3], \Delta_{SC}[r][0][1] = \Delta_{SB}[r][2][3], \Delta_{SC}[r][1][1] = \Delta_{SB}[r][0][1],$
$\Delta_{SC}[r][2][1] = \Delta_{SB}[r][3][2], \Delta_{SC}[r][3][1] = \Delta_{SB}[r][1][0], \Delta_{SC}[r][0][2] = \Delta_{SB}[r][1][2],$
$\Delta_{SC}[r][1][2] = \Delta_{SB}[r][3][0], \Delta_{SC}[r][2][2] = \Delta_{SB}[r][0][3], \Delta_{SC}[r][3][2] = \Delta_{SB}[r][2][1],$
$\Delta_{SC}[r][0][3] = \Delta_{SB}[r][3][1], \Delta_{SC}[r][1][3] = \Delta_{SB}[r][1][3], \Delta_{SC}[r][2][3] = \Delta_{SB}[r][2][0],$
$\Delta_{SC}[r][3][3] = \Delta_{SB}[r][0][2].$

The constraint for MC contains two parts, where r varies from 0 to $n-1$ and j varies from 0 to 3.

Firstly the quasi-MDS property directly gives the following constraint:

$$\left(\sum_{i=0}^{3} \Delta_{SC}[r][i][j] + \Delta_{MC}[r][i][j]\right) \in \{0, 4, 5, 6, 7, 8\}$$

Then, we model the fact that $MC(0,0,0,0) = (0,0,0,0)$ as follows:

$$\left(\sum_{i=0}^{3} \Delta_{SC}[r][i][j] = 0\right) \Leftrightarrow \left(\sum_{i=0}^{3} \Delta_{MC}[r][i][j] = 0\right).$$

The second part directly implements the product of the vector Δ_{SC} with the matrix given in Midori to get Δ_{MC}. It is modeled as follows:

$$\mathsf{XOR}(\Delta_{SC}[r][1][j], \Delta_{SC}[r][2][j], \Delta_{SC}[r][3][j], \Delta_{MC}[r][0][j])$$
$$\mathsf{XOR}(\Delta_{SC}[r][0][j], \Delta_{SC}[r][2][j], \Delta_{SC}[r][3][j], \Delta_{MC}[r][1][j])$$
$$\mathsf{XOR}(\Delta_{SC}[r][0][j], \Delta_{SC}[r][1][j], \Delta_{SC}[r][3][j], \Delta_{MC}[r][2][j])$$
$$\mathsf{XOR}(\Delta_{SC}[r][0][j], \Delta_{SC}[r][1][j], \Delta_{SC}[r][2][j], \Delta_{MC}[r][3][j])$$

For KA, following the rules of Midori and the XOR constraint described earlier[4], we have, for r from 0 to $n-1$, and i and j from 0 to 3: For Midori64:

$$\mathsf{XOR}(\Delta_{MC}[r][i][j], \Delta_{K}[r \mod 2][i][j], \Delta_{KA}[r][i][j])$$

and for Midori128:

$$\mathsf{XOR}(\Delta_{MC}[r][i][j], \Delta_{K}[i][j], \Delta_{KA}[r][i][j])$$

4.2   Step 2

Variables: In addition to the variables from Step 1, new ones are introduced to represent the differential words in the whitening key, the plaintext, the result of the initial KeyAdd, and the probabilities for each Sbox. When no information about the format is provided, the following variables are $n \times 4 \times 4$ word arrays, i.e., one 4×4 matrix per round.

δ_K represents the differential words in the key. It is modeled as a $2 \times 4 \times 4$ array of 4-bit words for Midori64, as a 4×4 byte matrix for Midori128.
δ_{SB} represents the state after the SB operation.
δ_{KA} represents the state after the KeyAdd operation.
δ_{MC} represents the state after the MixColumns operation.
δ_{SC} represents the state after the ShuffleCell operation.

---

[4] Note that the XOR operations between the key and constants at each rounds are not taken into account when working at a differential level. This is because the constants are canceled, i.e., for two different keys K^0 and K^1, and a constant c, $(K^0 \oplus c) \oplus (K^1 \oplus c) = K^0 \oplus K^1$.

δ_{WK} represents the whitening key, which is $\delta_K[0] \oplus \delta_K[1]$ for Midori64, and δ_K for Midori128.

δ_P represents the plaintext and $\delta_{P'}$ the state after the initial KeyAdd.

P is a $n \times 4 \times 4$ matrix used to compute the final probability, where $P[r][i][j] = 0$ if $\delta_{SB}[r][i][j] = 0$, and $log_2(DDT[\delta_{KA}[r][i][j]][\delta_{SB}[r+1][i][j]])$ otherwise. For Midori64, the domain of this variable is $\{0, 1, 2\}$, whereas for Midori128 it is $\{0, 1, 2, 3, 4, 5, 6\}$.

To find the optimal related-key differential characteristics, we need to maximize the sum of the P variables, hence our objective function is

$$Maximize \left(\sum_{r=0}^{n-1} \sum_{i=0}^{3} \sum_{j=0}^{3} P[r][i][j] \right)$$

Constraints: The first constraints aim at linking each variable of the input (from Step 1) to the variables of Step 2: for instance, for δ_{KA}, we have $\forall r \in [0..n-1], \forall i \in [0..3], \forall j \in [0..3]$:

$$\text{if } \Delta_{KA}[r][i][j] == 0 \text{ then } \delta_{KA}[r][i][j] = 0, \text{ else } \delta_{KA}[r][i][j] > 0$$

Similar constraints are defined for the other input variables.

The constraint for SC is exactly the same as in Step 1, except that it is on the δ variables (differential words) instead of the Δ variables (differential bits).

For the other operations, we make use of *table* constraints to model the Sboxes and the XOR operations[5]. Intuitively, table constraints tell the solver which tuples of values are allowed.

We denote tupleXOR the set of all tuples (X, Y, Z) that satisfying $Z = X \oplus Y$. Similarly we denote tupleSB$_S$ the tuples modeling the DDT for the Sbox S, *i.e.*, for every couple of words $\delta_{in}, \delta_{out}$ there is a tuple $(\delta_{in}, \delta_{out}, log_2(DDT_S[\delta_{in}][\delta_{out}]))$, where DDT_S is the DDT of the Sbox S.

We also denote by TABLE$((x, y, z), SET)$, the constraint that tells the solver that the values x, y and z must for a valid tuple with regards to a given SET.

We define XORbyte$(x, y, z) := $ TABLE$((x, y, z),$ tupleXOR$)$ and extend it to XORbyte$_3(x, y, z, w) := $ XORbyte(t, z, w), where t is defined by XORbyte(x, y, t).

The MixColumns operation can then be expressed, according the specification of the cipher, by: $\forall r \in [0..n-1], \forall j \in [0..3]$:

$$XORbyte_3(\delta_{SC}[r][1][j], \delta_{SC}[r][2][j], \delta_{SC}[r][3][j], \delta_{MC}[r][0][j])$$
$$XORbyte_3(\delta_{SC}[r][0][j], \delta_{SC}[r][2][j], \delta_{SC}[r][3][j], \delta_{MC}[r][1][j])$$
$$XORbyte_3(\delta_{SC}[r][0][j], \delta_{SC}[r][1][j], \delta_{SC}[r][3][j], \delta_{MC}[r][2][j])$$
$$XORbyte_3(\delta_{SC}[r][0][j], \delta_{SC}[r][1][j], \delta_{SC}[r][2][j], \delta_{MC}[r][3][j])$$

We now define δ_{WK}. For Midori64, δ_{WK} is defined by: $\forall i \in [0..3], \forall j \in [0..3]$:

$$XORbyte(\delta_K[0][i][j], \delta_K[1][i][j], \delta_{WK}[i][j])$$

---

[5] As the operation XOR is not by default implemented in the solver.

For Midori128, we simply have $\delta_{WK} = \delta_K$.

The initial KeyAdd operation then is modeled as: $\forall i \in [0..3], \forall j \in [0..3]$:

$$\mathsf{XORbyte}(\delta_P[r][i][j], \delta_{WK}[i][j], \delta_{P'}[i][j])$$

For the other KeyAdd operations, we have $\forall r \in [0..n-1], \forall i \in [0..3], \forall j \in [0..3]$:

$$\mathsf{XORbyte}(\delta_{MC}[r][i][j], \delta_K[r \mod 2][i][j], \delta_{KA}[r][i][j])$$

for Midori64, and

$$\mathsf{XORbyte}(\delta_{MC}[r][i][j], \delta_K[i][j], \delta_{KA}[r][i][j])$$

for Midori128. We finally model the SB operations. In the model for Midori128, where 4 different Sboxes are used, we use Sbox_i, where i is the number of the line. For readability, the number of the Sbox is omitted in what follows.

The initial SB is modeled as follows: $\forall i \in [0..3], \forall j \in [0..3]$:

$$\mathsf{TABLE}((\delta_{P'}[i][j], \delta_{SB}[0][i][j], P[0][i][j]), \mathsf{tupleSB})$$

Then, for the other rounds, we have $\forall r \in [1..n-1], \forall i \in [0..3], \forall j \in [0..3]$:

$$\mathsf{TABLE}((\delta_{KA}[r-1][i][j], \delta_{SB}[r][i][j], P[r][i][j]), \mathsf{tupleSB})$$

5   Results

Step 1 was implemented in the Minizinc[6] language, and solved using the solver Chuffed[7], which minimizes the objective function in around 3 h for Midori64 and in around 10 h for Midori128[8]. Step 2 was solved using Choco3 [15], which finds the best related-key differential characteristic (when it exists) for each input from Step 1 within 10 s.

The results are given in Table 2.

Note that since all Sboxes in our related-key differential characteristics have the best possible probability (2^{-2}), any related-key differential characteristic with more Sboxes has a lower probability.

5.1   Key Recovery Attacks

Our goal is to recover the secret key K. We use an *encryption oracle* $\mathsf{Enc}_K(x, m)$ that encrypts a message m with the key $K \oplus x$. Our attacks are different for Midori64 and Midori128.

---

[6] http://www.minizinc.org/.

[7] https://github.com/geoffchu/chuffed.

[8] We run our experiments on an Intel i7-4790, 3.6 Ghz with 16 GB RAM.

Table 2. The results obtained by the solvers, both for full-round and $n-1$ rounds of both versions of Midori. The number of Sboxes is the result of Step 1, and the probability is the result of Step 2.

Version	Number of rounds	Number of Sboxes	Probability
Midori64	15	7	2^{-14}
	16	8	2^{-16}
Midori128	19	19	2^{-38}
	20	20	2^{-40}

Midori64: For this attack, we first recover WK, one word at a time, using 16 15-rounds related-key differential characteristics with $16 \cdot 2^{19.32} = 2^{23.32}$ operations. Then we use 4 14-rounds related-key differential characteristics to recover $K[0]$ in $2^{35.8}$ operations. By combining them, we obtain $K[1] = K[0] \oplus WK$ and deduce K (composed of $K[0]$ and $K[1]$), for a total complexity of $2^{23.32} + 2^{35.8} \approx 2^{35.8}$.

Recovery of WK: The solvers give 16 different 15-rounds related-key differential characteristics, the corresponding related-key differentials are given in Appendix A. Each of them contains only one non-zero difference at the end of the 15th round (corresponding to $\delta_{KA}[14]$ in Fig. 3), all at different positions.

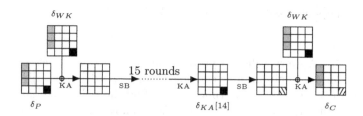

Fig. 3. An example of a related-key differential characteristic provided by the solver.

To be complete we need to give all the details our related-key differential characteristics. Minizinc finds optimal 2-rounds patterns with 1 active Sbox in all the odd rounds and none in the even rounds, as described in Fig. 4. Then the missing steps in Fig. 3 are 7.5 times the characteristic given in Fig. 4.

In order to recover one word of WK, we use the corresponding values of δ_K, δ_P and $\delta_{KA}[14]$ given by the related-key differential characteristics[9]. First we randomly choose some plaintext P, and query the oracle for $C = \mathsf{Enc}_K(0, P)$, and $C^* = \mathsf{Enc}_K(\delta_K, P\delta_P)$. We compute $\delta_C = C \oplus C^*$. We say that δ_C is *valid* iff $\forall i, j \in [0,3], \delta_{KA}[14][i][j] = 0 \Rightarrow \delta_C[i][j] = \delta_{WK}[i][j]$. If δ_C is valid then we compute $\delta_{SB}[i][j] = \delta_C[i][j] \oplus \delta_{WK}[i][j]$ (where (i,j) is the positions of the non-null difference). We now use the fact that for Midori64 the maximum value in

---

[9] From δ_K which is composed of $\delta_{K[0]}$ and $\delta_{K[1]}$, we can compute $\delta_{WK} = \delta_{K[0]} \oplus \delta_{K[1]}$.

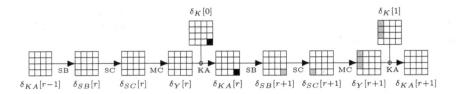

Fig. 4. An optimal 2-rounds related-key differential characteristic from Set 1 for Midori64, where r is an even round. Non-zero differences are represented as black and gray squares. It has 1 active Sbox, for instance with ▦ $= 0x1$ and ■ $= 0x2$.

Input: $\delta_K, \delta_P, \forall k \in [0,14]\delta_{KA}[k], i, j$
$P \overset{\$}{\leftarrow} \{0, 2^{64} - 1\}$;
$C = Enc_K(0, P); \; C^* = Enc_K(\delta_K, P \oplus \delta_P)$;
$\delta_C = C^* \oplus C$;
if δ_C *is valid* **then**
   $\delta_{SB}[i][j] = \delta_C[i][j] \oplus \delta_{WK}[i][j]$;
   for $\forall x \in DDTS(\delta_{KA}[14][i][j], \delta_{SB}[i][j])$ **do**
      $WK[i][j] = SB[i][j] \oplus C[i][j]$;
      $CPT[WK[i][j]]++$;
   end
end

Algorithm 1: How to recover WK in Midori64, where $DDTS(a,b) = \{x : SB(x) \oplus SB(x \oplus a) = b\}$, and CPT a table that is initialized to zero and stores the occurrences of possible candidates for WK.

the DDT is 4, *i.e.* every valid δ_C yields at most 4 possible values for $SB[i][j]$ (x in Algorithm 1), to obtain four candidates for $WK[i][j] = SB[i][j] \oplus C[i][j]$. By repeating this process several times we can find the right candidate: it is the word that has the most occurrences[10]. This is formally described in Algorithm 1. This is done 16 times, one for each word of WK.

Complexity Analysis of Algorithm 1: Our aim is to determine how many times we need to repeat our attack in order to have the true key. To determine precisely this value denoted T, we follow the approach given in [16]. It uses the signal to noise ratio S/N introduced by Biham in [4]. It is defined as $S/N = \frac{2^k \cdot p}{\alpha \cdot \beta}$, where k is the number of key bits that we want to recover (in our case, $k = 4$ since we aim to recover a word of 4 bits of the key), p is the probability of the related-key differential characteristic (for us $p = 2^{-14}$), α is the number of key candidates suggested for each good pair (using the DDT, we have $\alpha = 4$), and β is the ratio of the pairs that are not discarded. For β we have $2^{-14} + 2^{-60}$ since 2^{-14} is the probability given by the solvers and 2^{-60} corresponds to the false positives, *i.e.*, pairs having the same difference pattern, with 4 bits of undetermined difference. Then we obtain $S/N = \frac{2^k \cdot p}{\alpha \cdot \beta} = \frac{2^4 \cdot 2^{-14}}{4 \cdot (2^{-14} + 2^{-60})} = \frac{2^{-10}}{2^{-12} + 2^{-58}} \approx 4$. We denote by

---

[10] Indexes of the cell having the maximum values in the tables CPT.

P_S the probability to obtain the true key. We use the Eq. (19) of [16], where Φ denote the density probability function of the standard normal distribution, and Φ^{-1} its inverse: $P_S = \Phi\left(\dfrac{\sqrt{T \cdot S/N} - \Phi^{-1}(1-2^{-k})}{\sqrt{S/N+1}}\right)$ (19). Then we can obtain P_S for given values of T, S/N and α. Note that since we repeat the analysis 16 times (one for each word of WK), we need to have P_S^{16} sufficiently large as well. By numerical approximation we obtain $T = 20 \approx 2^{4.32}$, which gives $P_S > 0.99$, and $P_S^{16} > 0.99$. Hence, using $T \cdot p^{-1}$ plaintext pairs, we recover a key word with a probability greater than 0.99. The corresponding data complexity is then $16 \cdot 2 \cdot 20 \cdot 2^{14} \approx 2^{23.32}$ chosen plaintexts, as well as 1 related key, for each related-key differential characteristic used.

Recovery of $K[0]$: Using WK previously computed thanks to the 15-rounds related-key differential characteristics, we decrypt the last round of Midori and obtain the state of the 14th round $\delta_{KA}[14]$. Now we use other four 14-rounds related-key differential characteristics outputted by the solvers, one for each column of $K[0]$, the corresponding related-key differentials are given in Appendix B. They have only one active Sbox in the last round and there is a characteristic for each position of the active word. Hence we obtain the value of $\delta_{KA}[13]$. Similarly as in the case of Midori64, we can use the DDT to obtain 4 possibilities for a word of $SB[13]$. In the encryption function of Midori64, we have to apply the ShuffleCell (which does not influence our attack), then MixColumns which propagates the position of the 4 possibles values into different position. Then we

---

Input: $\delta_K, \delta_P, \forall k \in [0,14]\delta_{KA}[k], i, j, WK$
$P \xleftarrow{\$} \{0, 2^{64} - 1\}$;
$C = Enc_K(0, P)$; $C^* = Enc_K(\delta_K, P \oplus \delta_P)$;
$\delta_C = C^* \oplus C$;
if $\delta_C = \delta_{WK}$ **then**
 $SB = C \oplus WK$; $SB^* = C^* \oplus WK^* \oplus \delta_{WK}$;
 $X = Inv_{SB}(SB)$; $X^* = Inv_{SB}(SB^*)$;
 $\delta_X = X^* \oplus X$;
 $\delta_{SB}[14] = Inv_{SC}(Inv_{MC}(\delta_X))$;
 for $\forall x \in DDTS(\delta_{KA}[13], \delta_{SB}[14])$ **do**
  for $\forall u, v, w \in \{0, 2^4\}^3$ **do**
   $K[0][.][j'] = MC[14][.][j'] \oplus KA[.][j']$;
   $CPT[K[0][.][j']] + +$;
  end
 end
end

Algorithm 2: How to recover $K[0]$ in Midori64, where Inv_{SB} (resp. Inv_{MC} and Inv_{SC}) is the inverse of the Sbox function (resp. MC and SC), $\delta_{WK} = \delta_{K[0]} \oplus \delta_{K[1]}$, CPT a table that is initialized to 0 and stores the occurrences of possible candidates for $K[0]$, and j' is the index of the column where the 4 possibles values appeared after the MixColumn. The value of j' is deterministic and only depends of the positions i, j.

need to guess all the remaining values for these 3 words of 4 bits in the column where the value has been shifted after the ShuffleCell. This leads us to a total of $2^4 \cdot 2^4 \cdot 2^4 \cdot 4 = 2^{14}$ possibilities. Each of these possibilities gives a candidate for 4 words of the key $K[0]$ by xoring the obtained column of $MC[13]$ and the corresponding column of $KA[14]$, as described in Algorithm 2.

Complexity Analysis of Algorithm 2: This time $k = 4 \times 4 = 16$, $p = 2^{-14}$, $\alpha = 2^{14}$, $\beta = 2^{-14} + 2^{-60}$, then $S/N = \frac{2^k \cdot p}{\alpha \cdot \beta} = \frac{2^{16} \cdot 2^{-14}}{2^{14} \cdot (2^{-14} + 2^{-60})} = \frac{2^{-12}}{2^{-14} + 2^{-58}} \approx 4$. Then by numerical approximation we find $T = 28 \approx 2^{4.8}$, which gives $\mathsf{P_S} > 0.99$, and $\mathsf{P_S}^4 > 0.99$. It gives us the time complexity of $2 \cdot 4 \cdot 28 \cdot 2^{14} \cdot 2^{14} = 2^{35.8}$ and a data complexity of $2 \cdot 4 \cdot 28 \cdot 2^{14} = 2^{21.8}$.

Midori128: The constraints programming solvers find 16 patterns similar to the one of Fig. 5, each of them having a different position for the active Sbox. The corresponding related-key differentials are given in Appendix C. Hence, we can build 16 different 19-rounds related-key differential characteristics with 19 active Sboxes each, and each happening with probability 2^{-38}, to recover one word of WK per characteristic. We use a similar technique as for Midori64 for WK, since in Midori128 the key is K exactly WK.

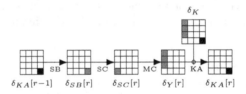

Fig. 5. The optimal 1-round related-key differential characteristic for Midori128. There is 1 active Sbox that can be repeated to cover 19 rounds of Midori128, for instance with ▨ = $0x9$ and ■ = $0x1$.

Complexity Evaluation: Here, we only need to use 16 related-key differential characteristics, so we want $\mathsf{P_S}^{16} \geq 0.99$. To compute S/N, we use $k = 8$ as we recover a 8-bit word of the key for each related-key differential characteristic, $p = 2^{-38}$, $\alpha = 64$ (according to the DDT), and $\beta = 2^{-38} + 2^{-120}$. With these values, we have $S/N = 4$, and need $T = 25 \approx 2^{4.7}$ to have $\mathsf{P_S}^{16} > 0.99$. Thus, the data complexity of the attack is $2 \cdot 25 \cdot 2^{38} \approx 2^{43.7}$ plaintexts and 16 related keys, and we need $2^{43.7}$ encryptions.

5.2   Related-Key Distinguishers

A related-key distinguisher aims at distinguishing a cipher scheme from a Pseudo-Random Function (PRF) that represents an ideal cipher. We construct two distinguishers for Midori64 and for Midori128.

Input: δ_K, δ_P, δ_C
for $i = 1$ to $2^{18.5}$ **do**

  $P \xleftarrow{\$} \{0, 2^{64} - 1\}$;
  $C = Enc_K(0, P)$; $C^* = Enc_K(\delta_K, P \oplus \delta_P)$;
  if $C \oplus C^* = \delta_C$ **then** return 1;

end
return 0;

   Algorithm 3: Algorithm for a distinguisher for Midori64.

Midori64: Midori64 has related-key differential characteristics with 8 active Sboxes, all of which can be crossed with the maximal probability 2^{-2}. Following the bound given in [1], we have that only $\frac{\sqrt{2}}{p} = \frac{\sqrt{2}}{2^{-18}} = 2^{18.5}$ are needed to distinguish Midori64 from a PRF, using the distinguisher given in Fig. 3. The equivalent complexity to find for a PRF is 2^{26} operations (following the formula given in [10]). Thus, we are able to distinguish Midori64 from a PRF[11].

Midori128: As for Midori64, there exist patterns that can be repeated to cover the whole cipher. The optimal ones only contain one active Sbox per round, *e.g.*, Fig. 5, hence leading to 20-rounds distinguishers with 20 active Sboxes. Since these Sboxes can be crossed with maximal probability (2^{-2}), the probability of the distinguisher is 2^{-40}. Hence, distinguishing Midori128 from a PRF, using Algorithm 3, can be done with $\frac{\sqrt{2}}{p} = 2^{39.5}$ encryptions of plaintext pairs whereas the equivalent complexity to find such a structure for a PRF is 2^{52}, again with the formula given in [10].

6 Conclusion

In this paper, we give a practical related-key attack on Midori64, improving the existing key recovery attack from 2^{116} for 14 rounds to $2^{35.8}$ for the full 16 rounds cipher. We also are able to provide the first related-key attack on Midori128 with a complexity of $2^{43.7}$. In order to construct such impressive practical attacks, we model Midori with constraint programming. The constraint programming solvers help us determine the minimal number of active Sboxes in a few hours, and then to derive optimal related-key differential characteristics in a few seconds.

 Finally we propose two efficient distinguishers for Midori64 and Midori128. In the future, we aim at exploring how CP can be used to perform some related-key cryptanalysis on other symmetric encryption schemes.

Acknowledgement. We would like to thank Marine Minier for her valuable advice.

---

[11] In Appendix D, we provide an example of values that satisfy the distinguisher built using the pattern given in Fig. 4 with ▨ = $0xa$ and ■ = $0xa$.

A   16 Related-Key Differentials for WK for Midori64

In the following table we give the 16 differential found by the solver that we used for recovery WK for Midori64.

n^o	δ_P	δ_K	$\delta_{KA}[14]$
1	1110000000000002	0000000000000002111000000000000	0000000000000002
2	0000110100020000	0000000000020000000110100000000	0000000000020000
3	0000000200000111	0000000200000000000000000000111	0000000200000000
4	0002000010110000	0002000000000000000000010110000	0002000000000000
5	0000022200000010	0000000000000010000022200000000	0000000000000010
6	1011000000200000	0000000000200000101100000000000	0000000000200000
7	0000002011100000	0000002000000000000000011100000	0000002000000000
8	0010000000002202	0010000000000000000000000002202	0010000000000000
9	0000000000001211	0000000000000020000000000001011	0000000000000200
10	0000000003110000	0000000002000000000000001110000	0000000002000000
11	1101020000000000	0000020000000001101000000000000	0000020000000000
12	0100222000000000	0100000000000000000222000000000	0100000000000000
13	0000000011012000	0000000000020000000000011010000	0000000000002000
14	0000000020001110	0000000020000000000000000001110	0000000020000000
15	0000301100000000	0000200000000000000101100000000	0000200000000000
16	2111000000000000	2000000000000000111000000000000	2000000000000000

B   4 Related-Key Differentials for $K[0]$ for Midori64

In the following table we give the 4 related-key differential characteristics found by the solver, that we used for recovery $K[0]$ for Midori64.

n^o	δ_P	δ_K	$\delta_{KA}[13]$
1	0111000000000000	0111000000000001000000000000000	1000000000000000
2	0000101100000000	0000101100000000000100000000000	0000100000000000
3	0000000001110000	0000000001110000000000001000000	0000000001000000
4	0000000000001011	0000000000001011000000000000100	0000000000000100

C   16 Related-Key Differentials for Midori128

In the following table, we give the 16 related-key differential characteristics found by the solver, that we used for recovery K for Midori128. The hexadecimal values of corresponding bytes are separated by a coma for more clarity.

n^o	δ_P	δ_K	δ_{KA}[18]
1	33, 1, 1, 1, 0, 0, 0, 0, 0, 0, 0, 0, 0, 0, 0, 0	32, 1, 1, 1, 0, 0, 0, 0, 0, 0, 0, 0, 0, 0, 0, 0	32, 0, 0, 0, 0, 0, 0, 0, 0, 0, 0, 0, 0, 0, 0, 0
2	0, 3, 0, 0, 2, 2, 2, 0, 0, 0, 0, 0, 0, 0, 0, 0	0, 1, 0, 0, 2, 2, 2, 0, 0, 0, 0, 0, 0, 0, 0, 0	0, 1, 0, 0, 0, 0, 0, 0, 0, 0, 0, 0, 0, 0, 0, 0
3	0, 0, 6, 0, 0, 0, 0, 0, 0, 0, 0, 1, 1, 0, 1	0, 0, 7, 0, 0, 0, 0, 0, 0, 0, 0, 1, 1, 0, 1	0, 0, 7, 0, 0, 0, 0, 0, 0, 0, 0, 0, 0, 0, 0, 0
4	0, 0, 0, 8, 0, 0, 0, 0, 1, 0, 1, 1, 0, 0, 0	0, 0, 0, 9, 0, 0, 0, 0, 1, 0, 1, 1, 0, 0, 0	0, 0, 0, 9, 0, 0, 0, 0, 0, 0, 0, 0, 0, 0, 0, 0
5	0, 0, 0, 0, 32, 0, 1, 1, 0, 0, 0, 0, 0, 0, 0	0, 0, 0, 0, 33, 0, 1, 1, 0, 0, 0, 0, 0, 0, 0	0, 0, 0, 0, 32, 0, 0, 0, 0, 0, 0, 0, 0, 0, 0
6	1, 1, 0, 1, 0, 3, 0, 0, 0, 0, 0, 0, 0, 0, 0	1, 1, 0, 1, 0, 2, 0, 0, 0, 0, 0, 0, 0, 0, 0	0, 0, 0, 0, 0, 2, 0, 0, 0, 0, 0, 0, 0, 0, 0
7	0, 0, 0, 0, 0, 0, 6, 0, 1, 1, 1, 0, 0, 0, 0	0, 0, 0, 0, 0, 0, 7, 0, 1, 1, 1, 0, 0, 0, 0	0, 0, 0, 0, 0, 0, 7, 0, 0, 0, 0, 0, 0, 0, 0
8	0, 0, 0, 0, 0, 0, 0, 8, 0, 0, 0, 0, 0, 9, 9, 9	0, 0, 0, 0, 0, 0, 0, 1, 0, 0, 0, 0, 0, 9, 9, 9	0, 0, 0, 0, 0, 0, 0, 1, 0, 0, 0, 0, 0, 0, 0, 0
9	0, 0, 0, 0, 0, 0, 0, 0, 1, 2, 2, 0, 0, 0, 0	0, 0, 0, 0, 0, 0, 0, 0, 3, 2, 2, 0, 0, 0, 0	0, 0, 0, 0, 0, 0, 0, 0, 1, 0, 0, 0, 0, 0, 0
10	0, 0, 0, 0, 0, 7, 7, 7, 0, 0, 0, 0, 0, 0, 6, 0	0, 0, 0, 0, 0, 7, 7, 7, 0, 0, 0, 0, 0, 0, 1, 0	0, 0, 0, 0, 0, 0, 0, 0, 0, 0, 0, 0, 0, 0, 1, 0
11	0, 0, 0, 0, 0, 0, 0, 0, 0, 0, 0, 1, 3, 1, 1	0, 0, 0, 0, 0, 0, 0, 0, 0, 0, 0, 1, 2, 1, 1	0, 0, 0, 0, 0, 0, 0, 0, 0, 0, 0, 0, 2, 0, 0
12	0, 0, 0, 0, 0, 0, 0, 0, 32, 32, 0, 32, 33, 0, 0, 0	0, 0, 0, 0, 0, 0, 0, 0, 32, 32, 0, 32, 1, 0, 0, 0	0, 0, 0, 0, 0, 0, 0, 0, 0, 0, 0, 0, 0, 1, 0, 0, 0
13	0, 0, 0, 0, 0, 0, 0, 0, 33, 0, 0, 1, 1, 1, 0	0, 0, 0, 0, 0, 0, 0, 0, 32, 0, 0, 1, 1, 1, 0	0, 0, 0, 0, 0, 0, 0, 0, 32, 0, 0, 0, 0, 0, 0
14	0, 0, 0, 0, 1, 1, 0, 1, 0, 0, 0, 8, 0, 0, 0, 0	0, 0, 0, 0, 1, 1, 0, 1, 0, 0, 0, 9, 0, 0, 0, 0	0, 0, 0, 0, 0, 0, 0, 0, 0, 0, 0, 9, 0, 0, 0, 0
15	1, 1, 1, 0, 0, 0, 0, 0, 0, 0, 0, 0, 0, 0, 8	1, 1, 1, 0, 0, 0, 0, 0, 0, 0, 0, 0, 0, 0, 9	0, 0, 0, 0, 0, 0, 0, 0, 0, 0, 0, 0, 0, 0, 9
16	1, 0, 1, 1, 0, 0, 0, 0, 0, 6, 0, 0, 0, 0, 0	1, 0, 1, 1, 0, 0, 0, 0, 0, 7, 0, 0, 0, 0, 0	0, 0, 0, 0, 0, 0, 0, 0, 0, 7, 0, 0, 0, 0, 0

D   Example of Related-Key Distinguisher for Midori64

In Table 3, we give an example of values that satisfy the related-key distinguisher built using the pattern given in Fig. 4 with ▨ $= 0xa$ and ■ $= 0xa$.

Table 3. A pair of plaintext/key couples following the related key distinguisher on Midori64, where Y can be P, K or C.

	Plaintext (P)	Key (K)	Ciphertext (C)
Y	cdd0776b6777667e	fffefefeefeeffe7 5448eff3eeffffff	99471218a32ea67c
Y^*	6770776b67776674	fffefefeefeeffed fee8eff3eeffffff	33e71218a32ea67c
$\delta_Y = Y \oplus Y^*$	aaa000000000000a	00000000000a aaa000000000	aaa0000000000000

References

1. Baignères, T., Sepehrdad, P., Vaudenay, S.: Distinguishing distributions using chernoff information. In: Heng, S.-H., Kurosawa, K. (eds.) ProvSec 2010. LNCS, vol. 6402, pp. 144–165. Springer, Heidelberg (2010). doi:10.1007/978-3-642-16280-0_10

2. Banik, S., Bogdanov, A., Isobe, T., Shibutani, K., Hiwatari, H., Akishita, T., Regazzoni, F.: Midori: a block cipher for low energy. In: Iwata, T., Cheon, J.H. (eds.) ASIACRYPT 2015. LNCS, vol. 9453, pp. 411–436. Springer, Heidelberg (2015). doi:10.1007/978-3-662-48800-3_17

3. Biham, E.: New types of cryptanalytic attacks using related keys. In: Helleseth, T. (ed.) EUROCRYPT 1993. LNCS, vol. 765, pp. 398–409. Springer, Heidelberg (1994). doi:10.1007/3-540-48285-7_34

4. Biham, E., Shamir, A.: Differential Cryptanalysis of the Data Encryption Standard. Springer, London (1993)

5. Biryukov, A., Nikolić, I.: Automatic search for related-key differential characteristics in byte-oriented block ciphers: application to AES, Camellia, Khazad and others. In: Gilbert, H. (ed.) EUROCRYPT 2010. LNCS, vol. 6110, pp. 322–344. Springer, Heidelberg (2010). doi:10.1007/978-3-642-13190-5_17

6. Chen, Z., Wang, X.: Impossible differential cryptanalysis of midori. IACR Cryptology ePrint Archive 2016, 535 (2016)
7. Dong, X.: Cryptanalysis of reduced-round midori64 block cipher. Cryptology ePrint Archive, Report 2016, 676 (2016). http://eprint.iacr.org/2016/676
8. Fouque, P.-A., Jean, J., Peyrin, T.: Structural evaluation of AES, and chosen-key distinguisher of 9-round AES-128. In: Canetti, R., Garay, J.A. (eds.) CRYPTO 2013. LNCS, vol. 8042, pp. 183–203. Springer, Heidelberg (2013). doi:10.1007/978-3-642-40041-4_11
9. Gerault, D., Minier, M., Solnon, C.: Constraint programming models for chosen key differential cryptanalysis. In: The 22nd International Conference on Principles and Practice of Constraint Programming, Toulouse, France (2016)
10. Gilbert, H., Peyrin, T.: Super-sbox cryptanalysis: improved attacks for AES-like permutations. In: Hong, S., Iwata, T. (eds.) FSE 2010. LNCS, vol. 6147, pp. 365–383. Springer, Heidelberg (2010). doi:10.1007/978-3-642-13858-4_21
11. Guo, J., Jean, J., Nikolić, I., Qiao, K., Sasaki, Y., Sim, S.M.: Invariant subspace attack against full midori64. Cryptology ePrint Archive, Report 2015, 1189 (2015). http://eprint.iacr.org/
12. Knudsen, L.R.: Cryptanalysis of LOKI 91. In: Seberry, J., Zheng, Y. (eds.) AUSCRYPT 1992. LNCS, vol. 718, pp. 196–208. Springer, Heidelberg (1993). doi:10.1007/3-540-57220-1_62
13. Lin, L., Wu, W.: Meet-in-the-middle attacks on reduced-round midori-64. Cryptology ePrint Archive, Report 2015, 1165 (2015). http://eprint.iacr.org/
14. Mouha, N., Wang, Q., Gu, D., Preneel, B.: Differential and linear cryptanalysis using mixed-integer linear programming. In: Wu, C.-K., Yung, M., Lin, D. (eds.) Inscrypt 2011. LNCS, vol. 7537, pp. 57–76. Springer, Heidelberg (2012). doi:10.1007/978-3-642-34704-7_5
15. Prud'homme, C., Fages, J.-G., Lorca, X.: Choco Documentation. TASC, INRIA Rennes, LINA CNRS UMR 6241, COSLING S.A.S. (2016)
16. Selçuk, A.A.: On probability of success in linear and differential cryptanalysis. J. Crypt. **21**(1), 131–147 (2008)
17. Sun, S., Hu, L., Wang, M., Yang, Q., Qiao, K., Ma, X., Song, L., Shan, J.: Extending the applicability of the mixed-integer programming technique in automatic differential cryptanalysis. In: Lopez, J., Mitchell, C.J. (eds.) ISC 2015. LNCS, vol. 9290, pp. 141–157. Springer, Heidelberg (2015). doi:10.1007/978-3-319-23318-5_8
18. Sun, S., Hu, L., Wang, P., Qiao, K., Ma, X., Song, L.: Automatic security evaluation and (Related-key) differential characteristic search: application to SIMON, PRESENT, LBlock, DES(L) and other bit-oriented block ciphers. In: Sarkar, P., Iwata, T. (eds.) ASIACRYPT 2014. LNCS, vol. 8873, pp. 158–178. Springer, Heidelberg (2014). doi:10.1007/978-3-662-45611-8_9
19. Wheeler, D.J., Needham, R.M.: TEA, a tiny encryption algorithm. In: Preneel, B. (ed.) FSE 1994. LNCS, vol. 1008, pp. 363–366. Springer, Heidelberg (1995). doi:10.1007/3-540-60590-8_29
20. ZDNet: New xbox security cracked by linux fans. http://www.zdnet.com/article/new-xbox-security-cracked-by-linux-fans

Some Proofs of Joint Distributions of Keystream Biases in RC4

Sonu Jha[1], Subhadeep Banik[2(✉)], Takanori Isobe[3], and Toshihiro Ohigashi[4]

[1] Grocme Ltd, Kolkata, India
jhasonu1987@yahoo.com
[2] Temasek Labs, Nanyang Technological University, Singapore, Singapore
bsubhadeep@ntu.edu.sg
[3] Kobe University, Kobe, Japan
takanori.isobe@jp.sony.com
[4] Tokai University, Tokyo, Japan
ohigashi@tsc.u-tokai.ac.jp

Abstract. In Usenix Security symposium 2015, Vanhoef and Piessens published a number of results regarding weaknesses of the RC4 stream cipher when used in the TLS protocol. The authors unearthed a number of new biases in the keystream bytes that helped to reliably recover the plaintext using a limited number of TLS sessions. Most of these biases were based on the joint distribution successive/non-successive keystream bytes. Moreover, the biases were reported after experimental observations and no theoretical explanations were proffered. In this paper, we provide detailed proofs of most of these biases, and provide certain generalizations of the results reported in the above paper. We also unearth new biases based on the joint distributions of three consecutive bytes.

Keywords: RC4 · TLS · Distinguisher

1 Introduction

RC4, designed by Rivest in 1987, was not very long ago one of the most widely used stream ciphers in the world. It was adopted in many software applications and standard protocols such as SSL/TLS, WEP, Microsoft Lotus and Oracle secure SQL. After the disclosure of its algorithm in 1994, RC4 attracted intense cryptanalytic efforts by the cryptographic community. Since then, there have been numerous cryptanalytic attempts to discover weaknesses in the ultra simplistic algorithm of this stream cipher [7–9,11,12,16] as well as there has been several proposals of new stream ciphers by applying several tweaks and modifications on the algorithm of RC4 [6,8,10,15–17,22] to remove the known weaknesses of the original stream cipher. Needless to say, all the proposed modifications were also the subject of some distinguishing attacks [2–5,13,18,19]. In 2013, Vanhoef et al. also showed some practically verifiable vulnerabilities present in WPA-TKIP [21]. Practical plaintext recovery attacks on RC4 in SSL/TLS were proposed by AlFardan et al. [1] in 2013 and Vanhoef and Piessens [20] in 2015.

© Springer International Publishing AG 2016
O. Dunkelman and S.K. Sanadhya (Eds.): INDOCRYPT 2016, LNCS 10095, pp. 305–321, 2016.
DOI: 10.1007/978-3-319-49890-4_17

In the response to these results, usage of RC4 has drastically decreased, especially in TLS, and major companies such as Google and Mozilla have officially removed the RC4 from web browsers in early 2016. Furthermore, RC4 use in TLS has been depreciated by RFC 7465 [14].

1.1 Description of the RC4 Stream Cipher

The RC4 stream cipher runs in two phases. The first phase is known as Key Scheduling Algorithm or KSA and the second phase is known as Pseudo-Random keystream Generation Algorithm or the PRGA. In the first phase, an array S is initialized with the elements $0, 1, \ldots, N$ where $N = 256$ and following a certain number of steps, permutation of $\{1, 2, \ldots, N\}$ is derived using a secret key K of size l bytes (typically $l = 16$). In the second phase known as the PRGA, the cipher runs for as many iterations as are needed to generate output bytes of the keystream. The input of the PRGA is the permutation derived after the KSA phase. The keystreams hence produced are XOR-ed with the plaintext to produce the corresponding ciphertext.

Algorithm 1. KSA

Input: S, K
Output: Permutation of S

---

for $i = 0$ to $N - 1$ **do**
  $S[i] = i$;
  $j = 0$;
end

for $i = 0$ to $N - 1$ **do**
  $j = (j + S[i]) + K[i \mod l])$
    $\mod N$;
  Swap($S[i]$,$S[j]$);
end

Algorithm 2. PRGA

Input: Permutation of S
Output: 1-byte output

$i = 0; j = 0$;
while Keystream is required **do**
  $i = (i + 1) \mod N$;
  $j = (j + S[i]) \mod N$;
  Swap($S[i]$,$S[j]$);
  Output$= S[(S[i] + S[j])$
    $\mod N]$;
end

1.2 Our Contribution and Organization of the Paper

In [20], authors present a list of biases present in the output bytes of the RC4 stream cipher. Most of these biases were based on the joint distribution of successive and non-successive keystream bytes and were reported following thorough experimental evaluations, however there were no theoretical proofs and explanations provided for them. The authors also categorized the biases as key-length dependent biases and the biases independent of key-length. In this paper, we prove most of the key-length independent output byte biases giving detailed theoretical explanations. Furthermore, we prove a number of significant biases based on the joint distribution of non-consecutive output byte pairs and three consecutive output bytes. In Table 1, we provide the list of biases presented in [20] and the biases unearthed by us during the analysis.

Table 1. Proved biases

#	Event	Observed probability	Source	Section
1	$\Pr[Z_3 = 4, Z_4 = -1, Z_5 = 4]$	$2/N^3$	New	2.1
2	$\Pr[Z_4 = 5, Z_5 = -1, Z_6 = -1]$	$2/N^3$	New	2.2
3	$\Pr[Z_1 = Z_4]$	$\frac{1}{N}(1 + \frac{0.7}{N})$	[20]	3.1
4	$\Pr[Z_1 = 0, Z_2 = 1]$	$\frac{1}{N^2}(1 - \frac{4}{N})$	[20]	4.1
5	$\Pr[Z_1 = 0, Z_2 = 2]$	$\frac{1}{N^2}(1 - \frac{5}{N})$	[20]	4.1
6	$\Pr[Z_1 = 0, Z_2 = 3, 4, \ldots]$	$\frac{1}{N^2}(1 - \frac{3}{N})$	[20]	4.1
7	$\Pr[Z_1 = x, Z_2 = 1], \forall x > 0$	$\frac{1}{N^2}(1 - \frac{2}{N})$	[20]	4.2
8	$\Pr[Z_1 = x, Z_2 = 258 - x], \forall x > 0$	$\frac{1}{N^2}(1 + \frac{1}{N})$	[20]	4.3
9	$\Pr[Z_1 = 257 - X, Z_X = 0], X = 2, 3, \ldots, 256$	$\frac{1}{N^2}(1 + \frac{\beta}{N})$	[20]	3.2
10	$\Pr[Z_1 = 257 - X, Z_X = X], X = 2, 3, \ldots, 256$	$\frac{1}{N^2}(1 + \frac{\beta}{N})$	[20]	3.3
11	$\Pr[Z_1 = 257 - X, Z_X = 257 - X], X = 2, 3, \ldots, 256$	$\frac{1}{N^2}(1 - \frac{\beta}{N})$	[20]	3.4
12	$\Pr[Z_1 = X - 1, Z_X = 1], X = 4, 5, \ldots, 256$	$\frac{1}{N^2}(1 + \frac{\beta}{N})$	[20]	3.5
13	$\Pr[Z_3 = 131, Z_{131} = 3]$	$\frac{1}{N^2}(1 + \frac{0.6}{N})$	[20]	2.3
14	$\Pr[(Z_{w256}, Z_{w256+2}) = (128, 0)], w \geq 1$	$\frac{1}{N^2}(1 + \frac{1}{N})$	[20]	4.4

2 Proofs of Biases Present in Non-consecutive Bytes and Consecutive Triple Bytes

In this section we provide proofs of biases present within joint distribution of non-consecutive bytes. The experimental values of these biases were listed in Vanhoef's paper [20] without any theoretical proofs. In the upcoming subsections, we will analyze the events leading to the biases present within these non-consecutive bytes. We will also provide proofs of the extended results found by us on the biases within the joint distribution of three consecutive bytes, during the course of theoretically proving the Vanhoef's results.

Some Notations: Let S denote the RC4 state consisting of random permutation of the elements $\{0, 1, 2, \ldots, 255\}$. Let S_r denote the RC4 state at the r-th round of the PRGA. Let Z_r denote the output byte generated after the PRGA round r.

2.1 Biased Probability of the Triplet $Z_3 = 4$, $Z_4 = 255$ and $Z_5 = 4$

In [20] it is mentioned that the probability of non-consecutive bytes (Z_3 and Z_5) being equal to 4 is biased and that the probability is given by the value $\Pr[Z_3 = 4, Z_5 = 4] \approx \frac{1}{N^2}(1 + \frac{1}{N})$ where $N = 256$. We present the following Lemma 1 to describe the events leading to the given bias. Interestingly, we found out that the same event leads to the biased probability of the given triple bytes. We also prove that the probability of $Z_3 = 4$, $Z_4 = 255$ and $Z_5 = 4$ is $\frac{2}{N^3}$ which is double compared to the probability in ideal case. Hence it represents a very huge bias in the joint distribution of the given triple bytes.

Lemma 1. *Let $S_0[1] = 4$, $S_0[2] = 1$ and $S_0[3] = 255$, then the output bytes Z_3 and Z_5 always ends up having the value 4. Moreover, the value of Z_4 is always equal to 255.*

Proof. We refer to the first round of the PRGA. At the first round, the public index $i = 0 + 1 = 1$ and the secret index $j = 0 + S_0[1] = 4$. After the swap operation $S_1[1] = X$ and $S_1[4] = 4$ where $X \notin \{1, 4, 255\}$. In the second round, $i = 2$ and $j = 5$. After the swap operation, $S_2[2] = Y$ and $S_2[5] = 1$ where again $Y \notin \{1, 4, 255\}$. In the third round, $i = 3$ and $j = 255 + 5 = 4$. Following the swap operation, $S_3[3] = 4$ and $S_3[4] = 255$. The value of third output byte, or Z_3 after the third PRGA round is given as

$$Z_3 = S_3[S_3[3] + S_3[4]] = S_3[4 + 255] = S_3[3] = 4$$

In the fourth round of the PRGA, $i = 4$ and $j = 4 + S_3[4] = 3$. After the swap operation, $S_4[4] = 4$ and $S_4[3] = 255$. The value of fourth output byte, or Z_4 after the fourth PRGA round is given as

$$Z_4 = S_4[S_4[4] + S_4[3]] = S_4[4 + 255] = S_4[3] = 255$$

In the fifth round, $i = 5$ and since the 5th location is not involved in swaps after round 2, $j = 3 + S_4[5] = 3 + S_2[5] = 4$. Following the swap operation, $S_5[5] = 4$ and $S_5[4] = 1$. The value of fifth output byte, or Z_5 after the fifth PRGA round is given as

$$Z_5 = S_5[S_5[5] + S_5[4]] = S_5[4 + 1] = 4$$

The following Theorem provides the proof of our extended result related to the given triple byte bias.

Theorem 1. *The probability of Z_3, Z_4 and Z_5 being equal to 4, 255 and 4 is given by the equation $\Pr[Z_3 = 4, Z_4 = 255, Z_5 = 4] \approx \frac{2}{N^3}$.*

Proof. Let E denote the event "$S_0[1] = 4$, $S_0[2] = 1$ and $S_0[3] = 255$". The probability of the event E can be given as $\frac{(N-3)!}{N!} \approx \frac{1}{N^3}$. According to Lemma 1, probability of Z_3, Z_4 and Z_5 being 4, 255 and 4 under the occurrence of event E is 1. By standard randomness assumptions supported by computer experiments, $\Pr[Z_3 = 4, Z_4 = 255, Z_5 = 4|E^c] = \frac{1}{N^3}$ where E^c denotes the compliment of the event E. Therefore the final probability can be given as

$$\begin{aligned}
\Pr[Z_3 = 4, Z_4 = 255, Z_5 = 4] &= \Pr[Z_3 = 4, Z_4 = 255, Z_5 = 4|E] \cdot \Pr[E] + \\
&\quad \Pr[Z_3 = 4, Z_4 = 255, Z_5 = 4|E^c] \cdot \Pr[E^c] \\
&= 1 \cdot \frac{1}{N^3} + \frac{1}{N^3} \cdot (1 - \frac{1}{N^3}) \\
&\approx \frac{2}{N^3}.
\end{aligned}$$

For an ideal cipher, the probability $\Pr[Z_3 = 4, Z_4 = 255, Z_5 = 4]$ should be only $\frac{1}{N^3}$, so we can see that in RC4, this probability is twice that of an ideal cipher. We now state the following theorem from [11], which outlines the number of output samples required to distinguish two distributions X and Y.

Theorem 2 *(Mantin-Shamir [11]). Let X, Y be distributions, and suppose that the event e happens in X with probability p and in Y with probability $p(1 + q)$. Then for small p and q, $O\left(\frac{1}{pq^2}\right)$ samples suffice to distinguish X from Y with a constant probability of success.*

Distinguishing from Random Sources: Let X be the probability distribution of Z_3, Z_4, Z_5 in an ideal random stream, and let Y be the probability distribution of Z_3, Z_4, Z_5 in streams produced by RC4 for randomly chosen keys. Let the event e denote $Z_3 = 4, Z_4 = 255, Z_5 = 4$, which occurs with probability of $\frac{1}{N^3}$ in X and $\frac{2}{N^3}$ in Y. By using the Theorem 2 with $p = \frac{1}{N^3}$ and $q = 1$, we can conclude that we need about 2^{24} output samples to reliably distinguish the two distributions.

2.2 Biased Probability of the Triplet $Z_4 = 5$, $Z_5 = 255$ and $Z_6 = 255$

Following similar techniques discussed in the previous Subsection, we will show the events which leads to this biased probability. We will also show that the probability $\Pr[Z_4 = 5, Z_5 = 255, Z_6 = 255]$ is again $\frac{2}{N^3}$. Please note that in [20], the experimental value of $\Pr[Z_4 = 5, Z_6 = 255]$ was listed and it was again around $\frac{1}{N^2}(1 + \frac{1}{N})$. However, we will be presenting proofs of the extended result found by us on the given triplet.

Lemma 2. *Let $S_0[1] = 5$, $S_0[2] = 255$ and $S_0[3] = 2$, then the output bytes Z_4, Z_5 and Z_6 always ends up having the values 5, 255 and 255.*

Proof. We refer to the first round of the PRGA. At the first round, the public index $i = 0 + 1 = 1$ and the secret index $j = 0 + S_0[1] = 5$. After the swap operation $S_1[1] = X$ and $S_1[5] = 5$ where $X \notin \{2, 5, 255\}$. In the second round, $i = 2$ and $j = 5 + 255 = 4$. After the swap operation, $S_2[2] = Y$ and $S_2[4] = 255$ where $Y \notin \{2, 5, 255\}$. At the third round, $i = 3$ and $j = 6$. Following the swap operation, $S_3[3] = W$ and $S_3[6] = 2$ where $W \notin \{2, 5, 255\}$. At the fourth round, $i = 4$ and $j = 6 + 255 = 5$. After the swap operation, $S_4[4] = 5$ and $S_4[5] = 255$. The value of fourth output byte, or Z_4 after the fourth PRGA round is given as

$$Z_4 = S_4[S_4[4] + S_4[5]] = S_4[5 + 255] = S_4[4] = 5$$

In the fifth round of the PRGA, $i = 5$ and $j = 5 + 255 = 4$. After the swap operation, $S_5[5] = 5$ and $S_5[4] = 255$. The value of fifth output byte, or Z_5 after the fifth PRGA round is given as

$$Z_5 = S_5[S_5[4] + S_5[5]] = S_5[5 + 255] = S_5[4] = 255$$

In the sixth round, $i = 6$ and $j = 6$. Since both the indices are same, there will be no swap in this round and $S_6[6] = 2$. The value of sixth output byte, or Z_6 after the sixth PRGA round is given as

$$Z_6 = S_6[S_6[6] + S_6[6]] = S_6[4] = 255$$

Theorem 3. *The probability of Z_4, Z_5 and Z_6 being equal to 5, 255 and 255 is given by the equation* $\Pr[Z_4 = 5, Z_5 = 255, Z_6 = 255] \approx \frac{2}{N^3}$.

Proof. Let E denote the event "$S_0[1] = 5$, $S_0[2] = 255$ and $S_0[3] = 2$". The probability of the event E can be given as $\frac{(N-3)!}{N!} \approx \frac{1}{N^3}$. According to Lemma 2, probability of Z_4, Z_5 and Z_6 being 5, 255 and 255 under the occurrence of event E is 1. By standard randomness assumptions supported by computer experiments, $\Pr[Z_4 = 5, Z_5 = 255, Z_6 = 255|E^c] = \frac{1}{N^3}$ where E^c denotes the compliment of the event E. Therefore the final probability can be given as

$$
\begin{aligned}
\Pr[Z_4 = 5, Z_5 = 255, Z_6 = 255] &= \Pr[Z_4 = 5, Z_5 = 255, Z_6 = 255|E] \cdot \Pr[E] + \\
&\quad \Pr[Z_4 = 5, Z_5 = 255, Z_6 = 255|E^c] \cdot \Pr[E^c] \\
&= 1 \cdot \frac{1}{N^3} + \frac{1}{N^3} \cdot (1 - \frac{1}{N^3}) \\
&\approx \frac{2}{N^3}.
\end{aligned}
$$

The probability of this triplet again is $\frac{2}{N^3}$ which is twice as the probability in case of the ideal cipher. This brings a scope of broadcast attack on RC4 based on these triple byte biases. We follow the similar lines as given in previous Subsection to reliably distinguish the probability distribution of Z_4, Z_5, Z_6 in an ideal random stream from the distribution of Z_4, Z_5, Z_6 in streams produced by RC4 for randomly chosen keys. It is easy to deduce that about 2^{24} samples are required in order to distinguish the two distributions.

We attempted to find more consecutive triple bytes leading to similar highly biased probabilities by analyzing the keystream bytes further using similar approach, but couldn't find biases strong enough as in the previously described couple of cases. Rather the biases in further triple consecutive bytes were very weak.

2.3   Bias of $Z_3 = 131$ and $Z_{131} = 3$

In [20], the biased probability of $Z_3 = 131$ and $Z_{131} = 3$ is given as $\Pr[Z_3 = 131, Z_{131} = 3] = \frac{1}{N^2}(1 + \frac{0.6}{N})$. With the help of following lemma and theorems, we provide the theoretical proof of this bias.

Theorem 4. *The probability of $Z_3 = 131$ and $Z_{131} = 3$ is given by the equation* $\Pr[Z_3 = 131, Z_{131} = 3] \approx \frac{1}{N^2}(1 + \frac{0.6}{N})$.

Proof. Let the event "$S_0[1] = 131$, $S_0[2] = 128$ and $S_{130}[j] = 3$ and $j \neq 131$ for rounds $r = 4, 5, \ldots, 130$" be denoted by E. The state transitions can then be described as follows.

Looking at the first round of the PRGA, the public index $i = 1$ and the secret index $j = 0 + S_0[1] = 131$. Following the swap operation, $S_1[1] = X$ where $X \notin \{131, 128\}$ and $S_1[131] = 131$. In the second round, $i = 2$ and $j = 131 + S_1[2] = 3$. After the swap operation, $S_2[2] = Y$ where $Y \notin \{131, 128\}$ and $S_2[3] = 128$. In the third round, $i = 3$ and $j = 3 + S_2[3] = 131$. After the swap operation, $S_3[3] = 131$ and $S_3[131] = 128$. The value of the third output byte can then be calculated as,

$$Z_3 = S_3[S_3[3] + S_3[131]] = S_3[131 + 128] = S_3[3] = 131$$

Now, if the value of the secret index j is never 131 for the next 126 rounds and considering that the value of $S_{130}[j] = 3$, then in the 131st round we have $S_{131}[j] = 128$ and $S_{131}[131] = 3$. Now, the value of 131st output byte can be calculated as,

$$Z_{131} = S_{131}[S_{131}[131] + S_{131}[j]] = S_{131}[3 + 128] = S_{131}[131] = 3$$

The probability of $S_0[1] = 131$, $S_0[2] = 128$ is $\frac{1}{N^2}$. Probability of $S_{130}[j] = 3$ is $\frac{1}{N}$. Finally the probability of $j \neq 131$ for rounds $r = 4, 5, \ldots, 130$ is $(1 - \frac{1}{N})^{126}$, which is close to 0.6. Considering these sub-events independent, we have $\Pr[E] \approx \frac{0.6}{N^3}$. The probability of $Z_3 = 131$ and $Z_{131} = 3$ under the occurrence of E is 1. By standard randomness assumptions supported by computer experiments, $\Pr[Z_3 = 131, Z_{131} = 3|E^c] = \frac{1}{N^2}$ where E^c denotes the compliment of the event E. Therefore the final probability can be given as

$$\begin{aligned} \Pr[Z_3 = 131, Z_{131} = 3] &= \Pr[Z_3 = 131, Z_{131} = 3|E] \cdot \Pr[E] + \\ &\quad \Pr[Z_3 = 131, Z_{131} = 3|E^c] \cdot \Pr[E^c] \\ &= 1 \cdot \frac{0.6}{N^3} + \frac{1}{N^2} \cdot (1 - \frac{0.6}{N^3}) \\ &\approx \frac{1}{N^2} + \frac{0.6}{N^3}. \end{aligned}$$

3   Proofs of Biases Influenced by Z_1

In this section, we provide the proofs of biases where the output byte Z_1 influence all initial 256 keystream bytes along with the biased equality $Z_1 = Z_4$.

3.1   Bias in the Equality $Z_1 = Z_4$

In [20], it is mentioned that the probability of $Z_1 = Z_4$ is positively biased and has the value around $\frac{1}{N}(1 + \frac{0.7}{N})$. In the light of the following lemma and theorem, we provide detailed proof of this probability being biased.

Theorem 5. *The probability of $Z_1 = Z_4$ is given by the equation $\Pr[Z_1 = Z_4] \approx \frac{1}{N}(1 + \frac{0.7}{N})$.*

Proof. Let the event "$S_0[1] = 2$, $S_0[2] \notin \{-1, 0, 1\}$ and $S_0[3]$ equals $N - 3$ or $N - 5$" be denoted by E. Consider initially the case when $S_0[3] = N - 3$ and then the following state transitions.

Referring to the first round of the PRGA, i is incremented and takes the value 1, and $j = 2$. After the swap operation, $S_1[1] = X$ (X is the initial value in index location 2), here we need to impose the added condition that $X \notin \{N - 1, 0, 1\}$

the reason for which will become apparent shortly. We also have $S_1[2] = 2$. The value of output byte Z_1 is then calculated as

$$Z_1 = S_1[S_1[1] + S_1[2]] = S_1[X + 2]$$

In the second and third round, the values of i, j, $S[i]$ and $S[j]$ change as following,

1. $i = 2$, $j = 2 + S_1[2] = 4$, $S_2[2] = Y$ where $Y \notin \{2, N - 3\}$, $S_2[4] = 2$.
2. $i = 3$, $j = 4 + S_2[3] = 1$, $S_3[3] = X$, $S_3[1] = N - 1$.

In the fourth round, $i = 4$ and $j = 1 + S_3[4] = 3$. After the swap operation, $S_4[4] = 2$ and $S_4[3] = X$. The output byte Z_4 is now calculated as,

$$Z_4 = S_4[S_4[4] + S_4[3]] = S_4[X + 2]$$

If $X \notin \{N - 1, 0, 1\}$, none of the indices i, j in the first 4 rounds would have touched the value $X + 2$, and so the value at index location $X + 2$ never gets swapped out. Hence we can conclude that the output bytes Z_1 and Z_4 are always same under the given conditions.

The case $S_0[3] = N - 5$ can be dealt with similarly. Following is the list of state transitions in the first 4 rounds.

1. $i = 1$, $j = 0 + S_0[1] = 2$, $S_1[1] = X$, $S_1[2] = 2$ and $Z_1 = S_1[X + 2]$
2. $i = 2$, $j = 2 + S_1[2] = 4$, $S_2[2] = Y$ where $Y \notin \{2, N - 3\}$, $S_2[4] = 2$.
3. $i = 3$, $j = 4 + S_2[3] = N - 1$, $S_3[3] = W$, $S_3[N - 1] = N - 5$.
4. $i = 4$, $j = N - 1 + S_3[4] = 1$, $S_4[1] = 2$, $S_4[4] = X$ and $Z_4 = S_4[X + 2]$

Again we have $Z_1 = Z_4$ if the value in index $X + 2$ does not get swapped out at any time. It is easy to see that $X \notin \{N - 3, N - 1, 0, 1\}$ ensures that. The probability of $\Pr[E] \approx \frac{2}{N^2}$. The probability of $Z_1 = Z_4$ under the occurrence of E is approximately $1 - \frac{3}{N}$ if $S_0[3] = N - 3$, and around $1 - \frac{4}{N}$ if $S_0[3] = N - 5$. Thus, we have

$$\Pr[Z_1 = Z_4 | E] = \frac{1}{2}(1 - \frac{3}{N}) + \frac{1}{2}(1 - \frac{4}{N}) = (1 - \frac{7}{2N})$$

However we experimentally observed that $\Pr[Z_1 = Z_4 | E^c]$ is non-uniform and has the value close to $\frac{1}{N} - \frac{1.3}{N^2}$ where E^c denotes the compliment of the event E. We don't have an exact analytical reasoning of why this non-uniformity occurs, and in that respect the proof is incomplete. But we do identify the event largely responsible for the positive bias of this event. Therefore, based on the experimental evaluation, we give the final probability as

$$\begin{aligned}\Pr[Z_1 = Z_4] &= \Pr[Z_1 = Z_4 | E] \cdot \Pr[E] + \\ &\quad \Pr[Z_1 = Z_4 | E^c] \cdot \Pr[E^c] \\ &= (1 - \frac{7}{2N}) \cdot \frac{2}{N^2} + (\frac{1}{N} - \frac{1.3}{N^2}) \cdot (1 - \frac{2}{N^2}) \\ &\approx \frac{1}{N} + \frac{0.7}{N^2}.\end{aligned}$$

3.2   Bias in $Z_1 = 257 - X$ and $Z_X = 0$

The joint distribution of the output bytes $Z_1 = 257 - X$ and $Z_X = 0$ where $X = \{2, 3, \ldots, 256\}$ is positively biased. In the following theorem, we provide the proof of this biased probability.

Theorem 6. *The probability of $Z_1 = 257 - X$ and $Z_X = 0$ is given by the equation* $\Pr[Z_1 = 257 - X, Z_X = 0] \approx \frac{1}{N^2}(1 + \frac{\beta}{N})$ *where* $\beta = (1 - \frac{1}{N})^{X-2}$.

Proof. Let the event E denote "$S_0[1] = X$, $S_0[X] = N + 1 - X$, $j_X = Y$, $S_0[Y] = 0$ and $S_1[X]$ not being swapped out from round 2 to round $i = X - 1$", then we have the following transitions.

In round one, $i = 1$ and $j = X$. After the swap operation, $S_1[1] = N + 1 - X$ and $S_1[X] = X$. The output byte Z_1 can now be calculated as

$$Z_1 = S_1[S_1[1] + S_1[X]] = S_1[1] = N + 1 - X$$

In order to get $Z_X = 0$, we don't need the value of $S_1[X]$ being swapped out from round 2 to round $i = X - 1$. The probability of the value of $S_1[X]$ not being swapped out from round 2 to round $i = X - 1$, is given as $(1 - \frac{1}{N})^{X-2} = \beta$. In round X, $i = X$ and $j_X = Y$. After the swap operation, $S_X[X] = 0$ and $S_X[Y] = X$. The output byte Z_X is then calculated as

$$Z_X = S_X[S_X[X] + S_X[Y]] = S_X[X] = 0$$

The probability of the event E is around $\frac{\beta}{N^3}$. We have $\Pr[Z_1 = 257 - X, Z_X = 0|\mathsf{E}] = 1$. Due to standard randomness assumptions, we also know $\Pr[Z_1 = 257 - X, Z_X = 0|\mathsf{E}^c] = \frac{1}{N^2}$. Therefore using the Bayes' Theorem, the total probability can be given as

$$
\begin{aligned}
\Pr[Z_1 = 257 - X, Z_X = 0] &= \Pr[Z_1 = 257 - X, Z_X = 0|\mathsf{E}] \cdot \Pr[\mathsf{E}] + \\
&\quad \Pr[Z_1 = 257 - X, Z_X = 0|\mathsf{E}^c] \cdot \Pr[\mathsf{E}^c] \\
&= 1 \cdot \frac{\beta}{N^3} + \frac{1}{N^2} \cdot (1 - \frac{\beta}{N^3}) \\
&\approx \frac{1}{N^2} + \frac{\beta}{N^3}.
\end{aligned}
$$

3.3   Bias in $Z_1 = 257 - X$ and $Z_X = X$

The joint distribution of the output bytes $Z_1 = 257 - X$ and $Z_X = X$ where $X = \{2, 3, \ldots, 256\}$ is also positively biased and in the following theorem, we provide the proof of this biased probability.

Theorem 7. *The probability of $Z_1 = 257 - X$ and $Z_X = X$ is given by the equation* $\Pr[Z_1 = 257 - X, Z_X = X] \approx \frac{1}{N^2}(1 + \frac{\beta}{N})$ *where* $\beta = (1 - \frac{1}{N})^{X-2}$.

Proof. Let the event E denote "$S_0[1] = X$, $S_0[X] = N + 1 - X$, $j_X = 1$ and $S_1[X]$ not being swapped out from round 2 to round $i = X - 1$", then we have the following transitions.

In round one, $i = 1$ and $j = X$. After the swap operation, $S_1[1] = N + 1 - X$ and $S_1[X] = X$. The output byte Z_1 can now be calculated as

$$Z_1 = S_1[S_1[1] + S_1[X]] = S_1[1] = N + 1 - X$$

The probability of the value of $S_1[X]$ not being swapped out from round 2 to round $i = X - 1$, is given as $(1 - \frac{1}{N})^{X-2} = \beta$. In round X, $i = X$ and $j_X = 1$. After the swap operation, $S_X[X] = N + 1 - X$ and $S_X[1] = X$. The output byte Z_X is then calculated as

$$Z_X = S_X[S_X[X] + S_X[1]] = S_X[1] = X$$

The probability of the event E is also around $\frac{\beta}{N^3}$. We have $\Pr[Z_1 = 257 - X, Z_X = X | E] = 1$. Due to standard randomness assumptions, we also know $\Pr[Z_1 = 257 - X, Z_X = X | E^c] = \frac{1}{N^2}$. Therefore using the Bayes' Theorem, the total probability can be given as

$$
\begin{aligned}
\Pr[Z_1 = 257 - X, Z_X = X] &= \Pr[Z_1 = 257 - X, Z_X = X | E] \cdot \Pr[E] + \\
&\quad \Pr[Z_1 = 257 - X, Z_X = X | E^c] \cdot \Pr[E^c] \\
&= 1 \cdot \frac{\beta}{N^3} + \frac{1}{N^2} \cdot (1 - \frac{\beta}{N^3}) \\
&\approx \frac{1}{N^2} + \frac{\beta}{N^3}.
\end{aligned}
$$

3.4   Bias in $Z_1 = 257 - X$ and $Z_X = 257 - X$

The joint distribution of the output bytes $Z_1 = 257 - X$ and $Z_X = 257 - X$ where $X = \{2, 3, \ldots, 256\}$ is negatively biased and the probability of this event is around $\frac{1}{N^2}(1 - \frac{\beta}{N})$ where $\beta = (1 - \frac{1}{N})^{X-2}$.

Theorem 8. *The probability of $Z_1 = 257 - X$ and $Z_X = 257 - X$ is given by the equation $\Pr[Z_1 = 257 - X, Z_X = 257 - X] \approx \frac{1}{N^2}(1 - \frac{\beta}{N})$.*

Proof. Since the bias is negative, we will consider the event in which it is impossible to have Z_1 and Z_X both equals to $257 - X$. Let the event E denote "$S_0[1] = X$, $S_0[X] = N + 1 - X$, $j_X \neq 1$ and $S_1[X]$ not being swapped out from round 2 to round $i = X - 1$", then we have the following transitions.

In round one, $i = 1$ and $j = X$. After the swap operation, $S_1[1] = N + 1 - X$ and $S_1[X] = X$. The output byte Z_1 can now be calculated as

$$Z_1 = S_1[S_1[1] + S_1[X]] = S_1[1] = N + 1 - X$$

The probability of the value of $S_1[X]$ not being swapped out from round 2 to round $i = X - 1$, is given as $(1 - \frac{1}{N})^{X-2} = \beta$. In round X, $i = X$ and $j_X = Y$

where $Y \neq 1$. After the swap operation, $S_X[X] = Z$ (say) and $S_X[Y] = X$. The output byte Z_X is then calculated as

$$Z_X = S_X[S_X[X] + S_X[Y]] = S_X[X + Z]$$

Since $Y \neq 1$, this ensures that $Z_X \neq N + 1 - X$

The probability of the event E is around $\frac{\beta}{N}(1 - \frac{1}{N}) = m$ (say). We have $\Pr[Z_1 = 257 - X, Z_X = 257 - X|E] = 0$. Due to standard randomness assumptions, we also know $\Pr[Z_1 = 257 - X, Z_X = 257 - X|E^c] = \frac{1}{N^2}$. Therefore using the Bayes' Theorem, the total probability can be given as

$$\begin{aligned}
\Pr[Z_1 = 257 - X, Z_X = 257 - X] &= \Pr[Z_1 = 257 - X, Z_X = 257 - X|E] \cdot \Pr[E] + \\
&\quad \Pr[Z_1 = 257 - X, Z_X = 257 - X|E^c] \cdot \Pr[E^c] \\
&= 0 \cdot m + \frac{1}{N^2} \cdot (1 - m) \\
&\approx \frac{1}{N^2} - \frac{\beta}{N^3}.
\end{aligned}$$

3.5   Bias in $Z_1 = X - 1$ and $Z_X = 1$

This condition doesn't hold when $X = 1$. When X is 2, it falls in the one of the categories of biases listed separately in [20]. When $X = 3$, the bias becomes negligible as j_3 cannot be 1. From $X = 4$ onwards, the given joint distribution is positively biased and the probability of this event is around $\frac{1}{N^2}(1 + \frac{\beta}{N})$.

Theorem 9. *The probability of $Z_1 = X - 1$ and $Z_X = 1$ is given by the equation $\Pr[Z_1 = X - 1, Z_X = 1] \approx \frac{1}{N^2}(1 + \frac{\beta}{N})$ where $(1 - \frac{1}{N})^{X-3} = \beta$.*

Proof. Let the event E denote "$S_0[1] = 1$, $S_0[2] = X - 1$, $j_X = 1$ and $S_2[X]$ not being swapped out from round 3 to round $i = X - 1$", then we have the following transitions.
In round one, $i = 1$ and $j = 1$. Since both the indices are same, no swap happens in the first round. The output byte Z_1 can now be calculated as

$$Z_1 = S_1[S_1[1] + S_1[1]] = S_1[2] = X - 1$$

Note that in round two, $i = 2$ and $j = 1 + X - 1 = X$. After swap, $S_2[2] = Y$ (say) and $S_2[X] = X - 1$. We need the value of $S_2[X]$ not being swapped out from round 3 to round $i = X - 1$. The probability of this event can be given as $(1 - \frac{1}{N})^{X-3} = \beta$.
In round X, $i = X$ and $j_X = 1$. After the swap operation, $S_X[X] = 1$ and $S_X[1] = X - 1$. The output byte Z_X is then calculated as

$$Z_X = S_X[S_X[X] + S_X[1]] = S_X[X] = 1$$

The probability of the event E is around $\frac{\beta}{N^3}$. We have $\Pr[Z_1 = X - 1, Z_X = 1|E] = 1$. Due to standard randomness assumptions, we also know $\Pr[Z_1 = X - 1, Z_X = 1|E^c] = \frac{1}{N^2}$. Therefore using the Bayes' Theorem, the total probability can be given as

$$
\begin{aligned}
\Pr[Z_1 = X - 1, Z_X = 1] &= \Pr[Z_1 = X - 1, Z_X = 1|E] \cdot \Pr[E] + \\
&\quad \Pr[Z_1 = X - 1, Z_X = 1|E^c] \cdot \Pr[E^c] \\
&= 1 \cdot \frac{\beta}{N^3} + \frac{1}{N^2} \cdot \left(1 - \frac{\beta}{N^3}\right) \\
&\approx \frac{1}{N^2} + \frac{\beta}{N^3}.
\end{aligned}
$$

4 Proofs of Consecutive Bytes Biases and Long-Term Biases

In this section, we will prove that the joint distribution of consecutive output bytes Z_1 and Z_2 is biased positively and negatively for certain values of Z_1 and Z_2. Furthermore, we will prove a long-term bias in the output bytes Z_{w256} and Z_{w256+2} where $w \geq 1$. We will show that the probability of this event is given by the equation $\Pr[(Z_{w256}, Z_{w256+2}) = (128, 0)] \approx \frac{1}{N^2} + \frac{1}{N^3}$.

4.1 The Biased Consecutive Output Bytes $Z_1 = 0$ and $Z_2 = x$

The joint distribution of $Z_1 = 0$ and $Z_2 = x$ for $x \neq 0$ is negatively biased. The bias varies for different values of x which will be explained in the following theorem.

Theorem 10. *The probabilities of $Z_1 = 0$ and $Z_2 = x$ for $x \neq 0$ are given by*

$x = 1$: $\Pr[Z_1 = 0, Z_2 = 1] \approx \frac{1}{N^2}(1 - \frac{4}{N})$
$x = 2$: $\Pr[Z_1 = 0, Z_2 = 2] \approx \frac{1}{N^2}(1 - \frac{5}{N})$
$x = U$: $\Pr[Z_1 = 0, Z_2 = U] \approx \frac{1}{N^2}(1 - \frac{3}{N})$ *where $U = 3, 4, \ldots$.*

Proof. Let the cases $Z_1 = 0$ and $Z_2 = x$ be denoted by $\mathcal{A}$ and $\mathcal{B}$. Since the biases are negative, we will be looking into the events which would make either $\mathcal{A}$ or $\mathcal{B}$ impossible to happen.

When $x = 1$. Let the event E_1 denote the case when "$S_0[1] = 0$". In the first round of the PRGA, $i = 1$ and $j = S_0[1] = 0$. Then we have $S_1[0] = 0$ and $S_1[1] = Y$ (say). The first output byte Z_1 is given as $S_1[Y] \neq 0$.

Let the event E_2 denote the case when $S_0[1] = 1$ and $S_0[2] \neq 0$. In the first PRGA round we have, $i = 1$ and $j = 1$. Since no swaps happen in this round we have $Z_1 = S_1[2] \neq 0$.

Let E_A denotes the both E_1 and E_2 combined. Therefore we have probability $\Pr[E_A] \approx \frac{2}{N} - \frac{1}{N^2}$. We know that $\Pr[\mathcal{A}|E_A] = 0$ and $\Pr[\mathcal{A}|E_A^c] = \frac{1}{N}$. Therefore the total probability of $\mathcal{A}$ is $\Pr[\mathcal{A}] \approx \frac{1}{N} - \frac{2}{N^2}$.

Let the event E_3 denote the case when "$S_0[1] \neq 1$ and $S_0[2] = 0$". In the first round of the PRGA, $i = 1$ and $j = S_0[1] = Q$. After the swap, we have $S_1[1] = R$ (say) and $S_1[Q] = Q$. In the second round we have $i = 2$ and $j = Q$. After the swap, we get $S_2[2] = Q$ and $S_2[Q] = 0$. The output byte Z_2 is given as $S_2[Q] = 0$.

Let the event E_4 denote the case when "$S_0[2] = 1$". In the first round of the PRGA, $i = 1$ and $j = S_0[1] = Z$. After the swap, we have $S_1[1] = W$ (say) and $S_1[Z] = Z$ (say). In the second round we have $i = 2$ and $j = Z + 1$. After the swap, we get $S_2[2] = T$ (say) and $S_2[Z + 1] = 1$. The output byte Z_2 is given as $S_2[T + 1]$. Since $T \neq Z$, we have $Z_2 \neq 1$.

Let $E_\mathcal{B}$ denotes the both E_3 and E_4 combined then the probability of $E_\mathcal{B}$ is again $\frac{2}{N} - \frac{1}{N^2}$. Following similar approach as case $\mathcal{A}$, we have $\Pr[\mathcal{B}] \approx \frac{1}{N} - \frac{2}{N^2}$.

Considering the cases $\mathcal{A}$ and $\mathcal{B}$ independent of one another, we have $\Pr[\mathcal{A} \cdot \mathcal{B}] = \Pr[\mathcal{A}] \cdot \Pr[\mathcal{B}] \approx \frac{1}{N^2} - \frac{4}{N^3}$.

When $x = 2$. For $x = 2$, we will follow the similar approach to find events which will make the cases $\mathcal{A}$ or $\mathcal{B}$ impossible to happen. Notably, there are three same events for $x = 2$ which lead to the biases when $x = 1$. The events E_1 and E_2 used previously plays the same role in the biased probability of the case $\mathcal{A}$. The event E_3 directly makes $Z_2 = 0$ and hence can be considered one of the events when $Z_2 \neq 2$. There are two more events which make $Z_2 \neq 2$.

Let the event E' be "$S_0[2] = 2$". In the first round of the PRGA, we have $i = 1$ and $j = S_0[1] = X$ (say). After the swap operation we get $S_1[1] = Y$ (say) and $S_1[X] = X$. In the second round, we have $i = 2$ and $j = X + 2$. After the swap operation, we have $S_2[2] = W$ (say) and $S_2[X + 2] = 2$. Then the second output byte Z_2 is $S_2[W + 2]$. Since $X \neq W$, this ensures $Z_2 \neq 2$.

Let the event E'' be "$S_0[1] = 2$". In the first round of the PRGA, we have $i = 1$ and $j = S_0[1] = 2$. After the swap operation we get $S_1[1] = P$ (say) and $S_1[2] = 2$. In the second round, we have $i = 2$ and $j = 4$. After the swap operation, we have $S_2[2] = Q$ (say) and $S_2[4] = 2$. Then the second output byte Z_2 is $S_2[Q + 2]$. Since $Q \neq 2$, this ensures $Z_2 \neq 2$.

Hence, when $x = 2$, we have the events E_3, E' and E'' which makes the case $\mathcal{B}$ impossible to happen. Let $E'_\mathcal{B}$ denotes combination of these three events. Therefore we have $\Pr[E'_\mathcal{B}] \approx \frac{3}{N} - \frac{1}{N^2}$. Therefore probability of case $\mathcal{B}$ can now be given as $\Pr[\mathcal{B}] \approx \frac{1}{N} - \frac{3}{N^2}$.

Again considering the cases $\mathcal{A}$ and $\mathcal{B}$ independent of one another, we have $\Pr[\mathcal{A} \cdot \mathcal{B}] = \Pr[\mathcal{A}] \cdot \Pr[\mathcal{B}] \approx \frac{1}{N^2} - \frac{5}{N^3}$.

When $x = U$. For $x = U$, the biased probability is caused by the events E_1, E_2 and E_3 which were explained earlier when $x = 1$. Following the similar approaches, the probability of the joint distribution can be given as $\Pr[\mathcal{A} \cdot \mathcal{B}] = \Pr[\mathcal{A}] \cdot \Pr[\mathcal{B}] \approx \frac{1}{N^2} - \frac{3}{N^3}$.

4.2   The Biased Consecutive Output Bytes $Z_1 = x$ and $Z_2 = 1$

The joint distribution of $Z_1 = x$ and $Z_2 = 1$ (for $x > 0$) is biased negatively with the probability $\frac{1}{N^2} - \frac{2}{N^3}$. In [20], it was mentioned that the probability of this event is biased positively, but during the course of our analysis, we found the event to be biased negatively. We think there may have been some possible typing error in [20]. The following theorem describes the events resulting in the biased probability of this distribution.

Theorem 11. *The probability of $Z_1 = x$ and $Z_2 = 1$ is given by the equation* $\Pr[Z_1 = x, Z_2 = 1] \approx \frac{1}{N^2}(1 - \frac{2}{N})$.

Proof. Let us again denote the cases $Z_1 = x$ and $Z_2 = 1$ by $\mathcal{A}$ and $\mathcal{B}$. In the previous subsection, we discussed several events which made the previous cases impossible to happen. The event E_3 described in the previous subsection makes the output byte Z_2 directly equal to 0. Therefore the event E_3 makes the case $\mathcal{B}$ impossible.

Let us denote the event E_x given by "$S_0[1] = x$ and $S_0[x] \neq 0$". Under this event, in PRGA round one, we have $i = 1$ and $j = S_0[1] = x$. After the swap operation we have $S_1[1] = y$ (say) and $S_1[x] = x$. Therefore the first output byte Z_1 can be given by $S_1[S_1[1] + S_1[x]] = S_1[x + y]$. We know that y cannot be equal to 0 and it ensures that under this condition, Z_1 cannot be x.

We have $\Pr[E_3] \approx \frac{1}{N} - \frac{1}{N^2}$ and $\Pr[E_x] \approx \frac{1}{N} - \frac{1}{N^2}$ and under these events, we accordingly have $\Pr[\mathcal{B}|E_3]$ and $\Pr[\mathcal{A}|E_x]$ equal to 0. Therefore the final probabilities of each $\Pr[\mathcal{A}]$ and $\Pr[\mathcal{B}]$ is around $\frac{1}{N} - \frac{1}{N^2}$.

Considering the cases $\mathcal{A}$ and $\mathcal{B}$ independent of one another, we have $\Pr[\mathcal{A} \cdot \mathcal{B}] = \Pr[\mathcal{A}] \cdot \Pr[\mathcal{B}] \approx \frac{1}{N^2} - \frac{2}{N^3}$.

4.3 The Biased Consecutive Output Bytes $Z_1 = x$ and $Z_2 = 258 - x$

In this subsection, we prove that the joint distribution of the output bytes $Z_1 = x$ and $Z_2 = 258 - x$ (for $x > 0$) is positively biased. The following theorem describes the biased probability of the given output bytes.

Theorem 12. *The probability of $Z_1 = x$ and $Z_2 = 258 - x$ is given by the equation* $\Pr[Z_1 = x, Z_2 = 258 - x] \approx \frac{1}{N^2}(1 + \frac{1}{N})$.

Proof. Let the event E denote "$S_0[1] = 1$, $S_0[2] = x$, $S_0[x+1] = 258 - x$", then we have the following transitions.

In the first round, $i = 1$ and $j = S_0[1] = 1$. Since both the indices are same, it results in no swaps. The value of Z_1 is then calculated as

$$Z_1 = S_1[S_1[1] + S_1[1]] = S_1[2] = x$$

In the next round, $i = 2$ and $j = 1 + x$. After the swap operation, $S_2[2] = 258 - x$ and $S_2[1 + x] = x$. The output byte Z_2 is calculated as

$$Z_2 = S_2[S_2[2] + S_2[1 + x]] = S_2[258] = S_2[2] = 258 - x$$

The probability of the event E is around $\frac{1}{N^3}$. We have $\Pr[Z_1 = x, Z_2 = 258 - x|E] = 1$. Due to standard randomness assumptions, we also know

$\Pr[Z_1 = x, Z_2 = 258 - x | E^c] = \frac{1}{N^2}$. Therefore using the Bayes' Theorem, the total probability can be given as

$$\begin{aligned}
\Pr[Z_1 = x, Z_2 = 258 - x] &= \Pr[Z_1 = x, Z_2 = 258 - x | E] \cdot \Pr[E] + \\
&\quad \Pr[Z_1 = x, Z_2 = 258 - x | E^c] \cdot \Pr[E^c] \\
&= 1 \cdot \frac{1}{N^3} + \frac{1}{N^2} \cdot (1 - \frac{1}{N^3}) \\
&\approx \frac{1}{N^2} + \frac{1}{N^3}.
\end{aligned}$$

4.4 Long-Term Bias in Output Bytes Z_{w256} and Z_{w256+2}

The joint distribution of the output bytes Z_{w256} and Z_{w256+2} where $w \geq 1$ are biased positively and persists in long-term. The theoretical analysis of the biased probability is given below.

Theorem 13. *The probability of Z_{w256} and Z_{w256+2} being equal to the values 128 and 0 is given by the equation* $\Pr[(Z_{w256}, Z_{w256+2}) = (128, 0)] \approx \frac{1}{N^2} + \frac{1}{N^3}$.

Proof. Let that after the completion of round $w256 - 1$, we have $S_{w256-1}[0] = \frac{N}{2}$, $S_{w256-1}[2] = 0$ and $j_{w256-1} = \frac{N}{2}$. Let us denote this event as E. Then we have the following transitions in next 3 rounds.

1. $i_{w256} = w256 - 1 + 1 = 0$, $j_{w256} = \frac{N}{2} + S_{w256-1}[0] = 0$, $Z_{w256} = S_{w256}[\frac{N}{2} + \frac{N}{2}] = S_{w256}[0] = \frac{N}{2}$.
2. $i_{w256+1} = 0 + 1 = 1$, $j_{w256+1} = 0 + S_{w256}[1] = X$ (say). After the swap, $S_{w256+1}[1] = Y$ (say) and $S_{w256+1}[X] = X$.
3. $i_{w256+2} = 2$, $j_{w256+2} = X + S_{w256+1}[2] = X$. After the swap, $S_{w256+2}[2] = X$ and $S_{w256+2}[X] = 0$, $Z_{w256+2} = S_{w256+2}[X + 0] = S_{w256+2}[X] = 0$.

The probability of the event E is around $\frac{1}{N^3}$. The probability $\Pr[(Z_{w256}, Z_{w256+2}) = (128, 0) | E] = 1$. Also $\Pr[(Z_{w256}, Z_{w256+2}) = (128, 0) | E^c] = \frac{1}{N^2}$. Therefore the total probability comes to

$$\begin{aligned}
\Pr[(Z_{w256}, Z_{w256+2}) = (128, 0)] &= \Pr[(Z_{w256}, Z_{w256+2}) = (128, 0) | E] \cdot \Pr[E] + \\
&\quad \Pr[(Z_{w256}, Z_{w256+2}) = (128, 0) | E^c] \cdot \Pr[E^c] \\
&= 1 \cdot \frac{1}{N^3} + \frac{1}{N^2} \cdot (1 - \frac{1}{N^3}) \\
&\approx \frac{1}{N^2} + \frac{1}{N^3}.
\end{aligned}$$

5 Conclusion

In this paper, we have tried to theoretically explain and prove numerous biases present in the keystream of RC4 that were experimentally evaluated by the authors of [20] without any theoretical explanations. Furthermore, we have also unearthed couple of strong significant biases present in the joint distribution of 3 consecutive output bytes. These biases are huge and have twice the probability compared to the probability in the ideal cases. There are still a number of biases mentioned in [20] including biases in consecutive output bytes and single byte biases, both key length dependent, which are yet to be proved and seems as an interesting area of research.

References

1. AlFardan, N.J., Bernstein, D.J., Paterson, K.G., Poettering, B., Schuldt, J.C.N.: On the security of RC4 in TLS. In: USENIX Security Symposium 2013, pp. 305–320 (2013)
2. Banik, S., Sarkar, S., Kacker, R.: Security analysis of the RC4+ stream cipher. In: Paul, G., Vaudenay, S. (eds.) INDOCRYPT 2013. LNCS, vol. 8250, pp. 297–307. Springer, Heidelberg (2013). doi:10.1007/978-3-319-03515-4_20
3. Banik, S., Jha, S.: Some security results of the RC4+ stream cipher. Secur. Commun. Netw. 8(18), 4061–4072 (2015)
4. Banik, S., Jha, S.: How not to combine RC4 states. In: Chakraborty, R.S., Schwabe, P., Solworth, J. (eds.) SPACE 2015. LNCS, vol. 9354, pp. 95–112. Springer, Heidelberg (2015). doi:10.1007/978-3-319-24126-5_6
5. Banik, S., Isobe, T.: Cryptanalysis of the full Spritz stream cipher. In: Peyrin, T. (ed.) FSE 2016. LNCS, vol. 9783, pp. 63–77. Springer, Heidelberg (2016). doi:10.1007/978-3-662-52993-5_4
6. Gong, G., Gupta, K.C., Hell, M., Nawaz, Y.: Towards a general RC4-like keystream generator. In: Feng, D., Lin, D., Yung, M. (eds.) CISC 2005. LNCS, vol. 3822, pp. 162–174. Springer, Heidelberg (2005). doi:10.1007/11599548_14
7. Isobe, T., Ohigashi, T., Watanabe, Y., Morii, M.: Full plaintext recovery attack on broadcast RC4. In: Moriai, S. (ed.) FSE 2013. LNCS, vol. 8424, pp. 179–202. Springer, Heidelberg (2014). doi:10.1007/978-3-662-43933-3_10
8. Lv, J., Zhang, B., Lin, D.: Distinguishing attacks on RC4 and a new improvement of the cipher. Cryptology ePrint Archive: Report 2013/176
9. Maitra, S.: Four Lines of Design to Forty Papers of Analysis: The RC4 Stream Cipher. http://www.isical.ac.in/~indocrypt/indo12.pdf
10. Maitra, S., Paul, G.: Analysis of RC4 and proposal of additional layers for better security margin. In: Chowdhury, D.R., Rijmen, V., Das, A. (eds.) INDOCRYPT 2008. LNCS, vol. 5365, pp. 27–39. Springer, Heidelberg (2008). doi:10.1007/978-3-540-89754-5_3
11. Mantin, I., Shamir, A.: A practical attack on broadcast RC4. In: Matsui, M. (ed.) FSE 2001. LNCS, vol. 2355, pp. 152–164. Springer, Heidelberg (2002). doi:10.1007/3-540-45473-X_13
12. Maximov, A., Khovratovich, D.: New state recovery attack on RC4. In: Wagner, D. (ed.) CRYPTO 2008. LNCS, vol. 5157, pp. 297–316. Springer, Heidelberg (2008). doi:10.1007/978-3-540-85174-5_17
13. Maximov, A.: Two linear distinguishing attacks on VMPC and RC4A and weakness of RC4 family of stream ciphers. In: Gilbert, H., Handschuh, H. (eds.) FSE 2005. LNCS, vol. 3557, pp. 342–358. Springer, Heidelberg (2005). doi:10.1007/11502760_23
14. Papov, A.: Prohibiting RC4 cipher suites. In: Internet Engineering Task Force (IETF). https://tools.ietf.org/html/rfc7465

15. Paul, G., Maitra, S., Chattopadhyay, A.: Quad-RC4: merging four RC4 states towards a 32-bit stream cipher. IACR Cryptology eprint Archive 2013:572 (2013)
16. Paul, S., Preneel, B.: A new weakness in the RC4 keystream generator and an approach to improve the security of the cipher. In: Roy, B., Meier, W. (eds.) FSE 2004. LNCS, vol. 3017, pp. 245–259. Springer, Heidelberg (2004). doi:10.1007/978-3-540-25937-4_16
17. Rivest, R.L., Schuldt, J.C.N.: Spritz—a spongy RC4-like stream cipher and hash function. https://people.csail.mit.edu/rivest/pubs/RS14.pdf
18. Sarkar, S.: Further non-randomness in RC4, RC4A and VMPC. Crypt. Commun. **7**(3), 317–330 (2015)
19. Tsunoo, Y., Saito, T., Kubo, H., Shigeri, M., Suzaki, T., Kawabata, T.: The most efficient distinguishing attack on VMPC and RC4A. In: SKEW 2005. http://www.ecrypt.eu.org/stream/papers.html
20. Vanhoef, M., Piessens, F.: All your biases belong to us: breaking RC4 in WPA-TKIP and TLS. In: 24th USENIX Security Symposium 2015, pp. 97–112 (2015)
21. Vanhoef, M., Piessens, F.: Practical verification of WPA-TKIP vulnerabilities. In: ASIACCS 2013, Proceedings of the 8th ACM SIGSAC Symposium on Information, Computer and Communications Security, pp. 427–436 (2013)
22. Zoltak, B.: VMPC one-way function and stream cipher. In: Roy, B., Meier, W. (eds.) FSE 2004. LNCS, vol. 3017, pp. 210–225. Springer, Heidelberg (2004). doi:10.1007/978-3-540-25937-4_14

Practical Low Data-Complexity Subspace-Trail Cryptanalysis of Round-Reduced PRINCE

Lorenzo Grassi[1]([✉]) and Christian Rechberger[1,2]

[1] IAIK, Graz University of Technology, Graz, Austria
{lorenzo.grassi,christian.rechberger}@iaik.tugraz.at
[2] DTU Compute, Technical University of Denmark, Lyngby, Denmark

Abstract. Subspace trail cryptanalysis is a very recent new cryptanalysis technique, and includes differential, truncated differential, impossible differential, and integral attacks as special cases.

In this paper, we consider PRINCE, a widely analyzed block cipher proposed in 2012. After the identification of a 2.5 rounds subspace trail of PRINCE, we present several (truncated differential) attacks up to 6 rounds of PRINCE. This includes a very practical attack with the lowest data complexity of only 8 plaintexts for 4 rounds, which co-won the final round of the PRINCE challenge in the 4-round chosen-plaintext category. The attacks have been verified using a C implementation.

Of independent interest, we consider a variant of PRINCE in which ShiftRows and MixLayer operations are exchanged in position. In particular, our result shows that the position of ShiftRows and MixLayer operations influences the security of PRINCE. The same analysis applies to follow-up designs inspired by PRINCE.

Keywords: PRINCE · Subspace trails cryptanalysis · Invariant subspace attack · Truncated differential attack · Practical attack · MANTIS

1 Introduction

The area of lightweight cryptography involves ciphers with low implementation costs, adequate for use in smart devices that have very limited resources (regarding memory, computing power, battery supply). Lightweight ciphers are designed in order to ensure a high level of security, even in the presence of tight constraints, that is they should be designed as a trade-off between security, cost of implementation and performance.

One of the most analyzed recent lightweight block ciphers is PRINCE [6]. The structure was designed in order to have efficient instantaneously encryption of a given plaintext, i.e. the entire encryption and decryption process should take place within the shortest possible delay, using little chip area. Follow-up designs (e.g. [2,3,5]) were inspired by PRINCE.

PRINCE has already gained a lot of attention from the academic community, and some interesting cryptanalysis have been published. Most of the earlier

© Springer International Publishing AG 2016
O. Dunkelman and S.K. Sanadhya (Eds.): INDOCRYPT 2016, LNCS 10095, pp. 322–342, 2016.
DOI: 10.1007/978-3-319-49890-4_18

attacks came with very high time and data complexity. In order to encourage more practically relevant cryptanalysis, "The PRINCE Challenge"[1] was organized, which started in (middle) 2014 and recently concluded with its third round. The challenge involved two settings: a chosen-plaintext scenario and the known-plaintext one. Since the competition aims at finding practical attacks, submissions must respect some initial restrictions regarding data, time and memory complexity: a particular emphasis is on *restricting the amount of data* (plaintext) that is available to the attacker.

Studying practical attacks on round-reduced versions of ciphers is motivated in many ways, see e.g. [7] for a survey of such reasons. A recent example is a creative attack on a full version of a 2nd-round CAESAR candidate ELmD [4] which in turn relies on an attack on AES reduced to 6 rounds. As use-cases of PRINCE are particularly sensitive to the choice of the number of rounds (due to latency constraints) it is very interesting to understand how much security can be at most hoped for when rounds are reduced.

In this paper, we present truncated differential attacks on reduced PRINCE, derived in a natural way exploiting so-called "subspace trails" of reduced-versions of PRINCE. The subspace trail framework was recently introduced in [12] as generalization of the invariant-subspace attack [15,16], and found already application in the cryptanalysis of AES [12]. While we describe applications in various settings, we focus on 4-round reduced PRINCE. The result improves upon all earlier results and has the lowest data complexity while still being entirely practical and practically verified. *This attack also co-won the final round of the PRINCE challenge in the 4-round chosen-plaintext category.*

As a second important aspect, we study the security of PRINCE when the rounds are slightly modified. Without going into the details here already, a round of PRINCE is very similar to an AES one, with the main difference that the ShiftRows operation is computed after the MixLayer one (instead of before). We show that the order of these two operations influences the security of PRINCE, and we show a possible way to overcome this problem. The same analysis applies to other encryption schemes that follow the same design of PRINCE, as the low-latency tweakable block cipher MANTIS [5] presented at CRYPTO 2016.

Review of Attacks on PRINCE. Known cryptanalysis of PRINCE includes theoretical attacks and observations on round-reduced and full PRINCE, and also a number of practical attacks on round-reduced versions. Here we review those most relevant to our work.

Derbez and Perrin described in [8] attacks based on a Meet-in-the-Middle approach, applicable (theoretically) up to 10 rounds of the algorithm. In [17], Morawiecki introduced attack relying on Integral and Higher-Order-Differential Cryptanalysis, up to 7 rounds[2]. This attack is based on a 3.5 round distinguisher

---

[1] https://www.emsec.rub.de/research/research_startseite/prince-challenge/.

[2] Table 1 of [17] contains an error about the data complexity and the time complexity for the Integral Attack on 4 rounds. The correct values for this attack (as also confirmed by the author of [17]) are those reported in Table 1 of this paper.

on PRINCE with one active nibble. Starting from this work, Posteuca and Negara [18] found a 4.5 round integral distinguisher for PRINCE which needs three (not arbitrary) active nibbles instead of one.

Due to the involution structure of PRINCE, a modified version of a differential attack was presented in [1]. Instead of choosing pairs of plaintexts with a known difference and studying its propagation through the encryption process (as in a classical differential attack), authors are able to recover the key using the difference among the nibbles of the plaintexts and of the respective ciphertexts (the nibbles are in the same positions). A related work about truncated differentials [14] has been presented in [21], which showed the existence of 5- and 6-round truncated differential distinguishers.

Our Contribution. We describe practical key-recovery attacks based on subspace trails of PRINCE which resemble truncated differentials, and we analyze in details the security of PRINCE-like ciphers focusing on the order of ShiftRows and MixLayer operations.

We base our work on the *Subspace Trail Cryptanalysis*, a technique that was recently introduced in [12]. Starting from [12], in Sect. 3 we investigate the behavior of subspaces in PRINCE. At a high level, we fix a subspace of plaintexts that maintain predictable properties after repeated applications of a key-variant round function. In other words, we identify (constant dimensional) subspace trails, that is a coset of a plaintext subspace that encrypts to proper subspaces of the state space over several rounds.

In Sect. 4 we present an *"equivalent"* version of PRINCE (with respect to the attacks we consider), which allows a better understanding of the design of this encryption scheme. As we have already mentioned, a round of PRINCE is very similar to an AES one, with the main difference that the ShiftRows operation is computed after the MixLayer one. Our analysis shows that if these two operations are exchanged of position (to have something similar to AES), the attacks present in literature can usually cover more rounds with the same (or even less) complexity. As example, for this modified version it is possible to set up a subspace trail that covers one more round. Thus, we present how to modify the middle-rounds of PRINCE in order to obtain a version equivalent to the original one (also from the security point of view) and where the ShiftRows operation is computed before the MixLayer one. Similar analysis applies also to other encryption schemes that follows the (same) design of PRINCE. In particular, a detailed analysis for the MANTIS encryption scheme is presented in Appendix A. Finally, we highlight that this problem arises only for PRINCE-like ciphers, i.e. the security of AES-like cipher is not influenced by the position of the ShiftRows operation with respect to the MixLayer, while PRINCE-like ciphers are.

In the following sections, we use the found subspace trails as starting points to set up competitive key recovery attacks to round-reduced PRINCE. In particular, we present two different truncated differential key-recovery attacks on 3 rounds of PRINCE. The idea - described at the beginning of Sect. 5 and in details in App. E of [11] - is simple. Assume to fix a coset of a particular subspace $\mathcal{C}$ of the

Table 1. Comparison table of attacks on 4-round PRINCE. These are the four central rounds, that is the middle rounds, one round before and one round after. Data complexity is measured in number of required chosen plaintexts (CP). Time complexity is measured in round-reduced PRINCE encryption equivalents (E). Memory complexity is measured in plaintexts (64 bits).

Technique	Data (CP)	Computation (E)	Memory	Reference
Trunc. Diff. (EE)	$8 = 2^3$	$2^{18.25}$	**Small**	Sect. 5
Bit-pattern integral	$48 = 2^{5.6}$	2^{22}	Small	[17]
(Pre-computed) integral	$64 = 2^6$	$2^{7.4}$	Small	[19]
Integral	$160 = 2^{7.32}$	$2^{9.32}$	Small	[17]
Trunc. Diff. (EB)	$430 = 2^{8.75}$	$2^{8.15}$	**Small**	[11] **(App. G)**
Diff./Logic	2^{10}	$5\,\mathrm{sec}$	$\ll 2^{27}$	[8]
Differential	2^{32}	$2^{56.26}$	2^{48}	[1]

(EE: Extension at End - EB: Extension at Beginning)

plaintexts space. After 2.5 rounds, each element of a (fixed) coset of C belongs to a coset of another particular subspace M, i.e. a coset of C is mapped into a coset of M after 2.5 rounds. Equivalently, if two elements belong to the same coset of C, after 2.5 rounds they belong to the same coset of M independently of the secret key. Thus, the key of the final round must satisfy the condition that, given two ciphertexts (whose plaintexts belong to the same coset of C), they belong to the same coset of M half round before. As main result, we show that *a truncated differential attack that exploits relationships among the nibbles is (much) more powerful than one that works independently on each nibble.*

In Sect. 5, we show how to extend this attack to 4 rounds by adding one round at the end. This attack needs only 8 chosen plaintexts and it is the best one from the point of view of the data complexity (the computational cost is also very competitive), improving previous results of a factor 6 for the data complexity and of a factor 2^4 for the computational cost. All these attacks have been verified using a C/C++ implementation. A comparison of all known state of art of attacks on PRINCE and our attacks is given in Table 1.

It is also possible to extend the attack on 3 rounds at the beginning (see App. G of [11]) which leads to higher data but lower time complexity. Using both the extension at the end and at the beginning, it is possible to attack 5- and 6- rounds of PRINCE (see App. H of [11]).

Practical Verification of 3- and 4-Rounds Attacks. We practically verified all the 3 rounds attacks described in this paper and the 4 rounds attack described in Sect. 5, using a C/C++ implementation[3]. For all the attacks, the full key recovery takes a fraction of a second on a desktop PC. We also practical verified

---

[3] The source code is available at https://github.com/Krypto-iaik/PRINCE_Attacks.

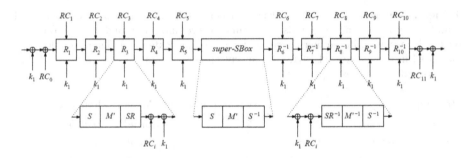

Fig. 1. A scheme of the PRINCEcore cipher.

some of the attacks (e.g. the square one) present in literature against the modified versions of PRINCE presented in Sect. 4.

2   Description of PRINCE

PRINCE [6] is a lightweight cipher with a state size of 64 bits - the 64-bits state of PRINCE can be visualized as a 4×4-matrix, where every cell represents a nibble - and a key length of 128 bits. It is based on the so-called FX construction [13], where one part of the key is used for a core cipher F, which contains the major encryption process, and the remaining parts are used for whitenings before and after the core: $FX_{k,k_1,k_2} = k_2 \oplus F_k(x \oplus k_1)$. First, the 128-bit key k is split into two 64-bit words (i.e. $k = (k_0||k_1)$), and then it is expanded into 192 with a simple linear transformation: $(k_0||k_0'||k_1) := (k_0||(k_0 \ggg 1) \oplus (k_0 \lll 63)||k_1)$. The 64-bit subkeys k_0 and k_0' are used as whitening keys to the underlying block cipher called PRINCEcore, while the 64-bit key k_1 is used for the core.

The core cipher "PRINCEcore" is a substitution-permutation network composed of 12 rounds (see Fig. 1). Every round in PRINCE consists of an S-Box layer, a Linear layer, a ShiftRows operation, a key addition and the addition of a round constant:

- **S-Box Layer**: Every nibble in the internal state is replaced by using a 4×4-bit S-Box, which has algebraic degree 3 and which is differential 4-uniform.
- **Linear Layer M′**: In the linear layer, the state is multiplied by an involutive 64×64-matrix, a kind of equivalent of MixColumns in AES. More precisely, two 16×16 submatrices $\hat{M}^{(0)}$ and $\hat{M}^{(1)}$ are arranged on the diagonal of a bigger matrix, where every submatrix affects a 16-bit chunk x_i of the 64-bit state $x = (x_1||x_2||x_3||x_4)$:

$$M' \cdot x = (\hat{M}^{(0)} \cdot x_1||\hat{M}^{(1)} \cdot x_2||\hat{M}^{(1)} \cdot x_3||\hat{M}^{(0)} \cdot x_4).$$

- **ShiftRows Operation SR**: Equal to the one in the AES cipher.
- **A bit-wise XOR with a round constant** RC_i, for $i = 0, \ldots, 11$.
- **A bit-wise XOR with the secret key** k_1.

In the last 5 rounds (the backward rounds), the order of operations is inverse with respect to the first 5 rounds (the forward rounds), where only the round constants differ. The middle rounds consist of three key-less operations: an S-Box layer, a matrix multiplication with M' and an inverse S-Box layer. Since the matrix M' is self-inverting (i.e. $M' = M'^{-1}$), the same linear layer M' operation is used in forward and backward rounds. Like AES, the combination of matrix multiplication and shifting provides full diffusion after only two rounds. Moreover, the varying round constants RC_i supplement the round transformation in order to prevent slide attacks. The difference between $RC_i \oplus RC_{11-i}$ is always equal to a constant value α. Since the round constants satisfy $RC_i \oplus RC_{11-i} = \alpha$ and since M' is an involution, the core cipher has the so called α-reflection property, i.e. the core cipher is such that the inverse of PRINCEcore parametrized with k is equal to PRINCEcore parametrized with $k \oplus \alpha$: $D_{(k_0||k_0'||k_1)}(\cdot) = E_{(k_0'||k_0||k_1 \oplus \alpha)}(\cdot)$.

For the following, we use the term "PRINCE-like cipher" to denote a cipher with middle rounds, r forward rounds and r backwards rounds, and which has the α-reflection property - examples are MANTIS [5] or QARMA [2].

Notation. In our attack, we suppose that the 64-bit state is organized as a 16×1 array of nibbles, and we use the notation $[z]$ to denote the nibble in position z (the z-th nibble is in row $r = z \bmod 4$ and in column $c = (z-r)/4$). We denote by R one round of PRINCE, while we denote i rounds by $R^{(i)}$ (without distinction between the forward and the backward direction). To simplify the notation we denote by *super-SBox* the middle rounds, by $\hat{k}$ the key of the final round and by $\tilde{k}$ the key of the first round, that is:

$$super\text{-}SBox(\cdot) = \text{S-Box}^{-1} \circ M' \circ \text{S-Box}(\cdot), \quad \hat{k} := k_1 \oplus k_0' \oplus \alpha, \quad \tilde{k} := k_1 \oplus k_0.$$

We attack round-reduced variants of PRINCE. In case of an even number of rounds, we keep the symmetry of the cipher.

3   Subspace Trails

Let F denote a round function in a iterative block cipher and let $V \oplus a$ denote a coset of a vector space V. Then if $F(V \oplus a) = V \oplus a$ we say that $V \oplus a$ is an *invariant coset* of the subspace V for the function F. This concept can be generalized to *trails of subspace*.

Definition 1. *Let* $(V_1, V_2, \ldots, V_{r+1})$ *a set of* $r + 1$ *subspaces with* $\dim(V_i) \leq \dim(V_{i+1})$. *If for each* $i = 1, \ldots, r+1$ *and for each* $a_i \in V_i^{\perp}$, *there exists (unique)* $a_{i+1} \in V_{i+1}^{\perp}$ *such that* $F(V_i \oplus a_i) \subseteq V_{i+1} \oplus a_{i+1}$, *then* $(V_1, V_2, \ldots, V_{r+1})$ *is a subspace trail of length* r *for the function* F. *If the previous relation holds with equality, then the trail is called a constant-dimensional subspace trail.*

We refer to [12] for more details about the concept of subspace trails. Our treatment here is however meant to be self-contained.

In the following, we present two subspace trails for 2.5 rounds of PRINCE. The first one is composed of the middle rounds (without the final S-Box^{-1}) and 1 round before it, while the second one is composed of the middle rounds and 0.5 round after it. In particular, this second one is composed of an *invariant subspace* of the middle rounds. All the proofs of the theorems and of the propositions of this section can be found in App. A of [11].

3.1 Subspaces of PRINCE

In this section, we define the subspaces of PRINCE, analogous to those of AES presented in [12]. For the following, let $E = \{e[0], \ldots, e[15]\}$ denote the unit vectors of $\mathbb{F}_{2^4}^{16}$ (e.g. e_i has a single 1 in position i). Moreover, we recall that given a generic subspace X, two different cosets $X \oplus a$ and $X \oplus b$ (i.e. $a \neq b$) are *equivalent* if and only if $a \oplus b \in X$.

For each $i = 0, \ldots, 3$, let $\mathcal{C}_i$ the *column subspace* of dimension 16 defined as:

$$\mathcal{C}_i = \langle e[4 \cdot i], e[4 \cdot i + 1], e[4 \cdot i + 2], e[4 \cdot i + 3] \rangle. \tag{1}$$

For instance, $\mathcal{C}_0$ correspond to matrix representation:

$$\mathcal{C}_0 = \left\{ \begin{bmatrix} x & 0 & 0 & 0 \\ z & 0 & 0 & 0 \\ w & 0 & 0 & 0 \\ y & 0 & 0 & 0 \end{bmatrix} \middle| \forall x, y, z, w \in \mathbb{F}_{2^4} \right\} \equiv \begin{bmatrix} x & 0 & 0 & 0 \\ z & 0 & 0 & 0 \\ w & 0 & 0 & 0 \\ y & 0 & 0 & 0 \end{bmatrix}.$$

For each $i = 0, \ldots, 3$, let $\mathcal{D}_i$ the *diagonal subspace* and $\mathcal{ID}_i$ the *inverse-diagonal subspace* - both of dimension 16 - defined as:

$$\mathcal{D}_i = SR(\mathcal{C}_i), \qquad \mathcal{ID}_i = SR^{-1}(\mathcal{C}_i) \tag{2}$$

For instance, $\mathcal{D}_0$ and $\mathcal{ID}_0$ correspond to matrix representations:

$$\mathcal{D}_0 \equiv \begin{bmatrix} x & 0 & 0 & 0 \\ 0 & 0 & 0 & y \\ 0 & 0 & w & 0 \\ 0 & z & 0 & 0 \end{bmatrix}, \qquad \mathcal{ID}_0 \equiv \begin{bmatrix} x & 0 & 0 & 0 \\ 0 & z & 0 & 0 \\ 0 & 0 & w & 0 \\ 0 & 0 & 0 & y \end{bmatrix}.$$

Finally, let $\mathcal{M}_i$ the *mixed subspace* and $\mathcal{IM}_i$ the *inverse-mixed subspace* - both of dimension 16 - defined as

$$\mathcal{M}_i := M'(\mathcal{D}_i), \qquad \mathcal{IM}_i := M'(\mathcal{ID}_i) \tag{3}$$

For instance, $\mathcal{M}_0$ and $\mathcal{IM}_0$ correspond to matrix representations:

$$\mathcal{M}_0 \equiv \begin{bmatrix} \alpha_3(x) & \alpha_3(z) & \alpha_0(w) & \alpha_2(y) \\ \alpha_2(x) & \alpha_2(z) & \alpha_3(w) & \alpha_1(y) \\ \alpha_1(x) & \alpha_1(z) & \alpha_2(w) & \alpha_0(y) \\ \alpha_0(x) & \alpha_0(z) & \alpha_1(w) & \alpha_3(y) \end{bmatrix}, \qquad \mathcal{IM}_0 \equiv \begin{bmatrix} \alpha_3(x) & \alpha_1(z) & \alpha_0(w) & \alpha_0(y) \\ \alpha_2(x) & \alpha_0(z) & \alpha_3(w) & \alpha_3(y) \\ \alpha_1(x) & \alpha_3(z) & \alpha_2(w) & \alpha_2(y) \\ \alpha_0(x) & \alpha_2(z) & \alpha_1(w) & \alpha_1(y) \end{bmatrix}$$

where $\alpha_i(\cdot)$ are defined as

$$\alpha_i(x) = x \wedge (0x2^i \oplus 0xf), \tag{4}$$

and where $\wedge$ is the *and (logic) operator*.

Let $I \subseteq \{0, 1, 2, 3\}$. Subspaces $\mathcal{C}_I, \mathcal{D}_I, \mathcal{ID}_I, \mathcal{M}_I$ and $\mathcal{IM}_I$ are defined as:

$$\mathcal{C}_I = \bigoplus_{i \in I} \mathcal{C}_i, \quad \mathcal{D}_I = \bigoplus_{i \in I} \mathcal{D}_i, \quad \mathcal{ID}_I = \bigoplus_{i \in I} \mathcal{ID}_i, \quad \mathcal{M}_I = \bigoplus_{i \in I} \mathcal{M}_i, \quad \mathcal{IM}_I = \bigoplus_{i \in I} \mathcal{IM}_i.$$

Note that $\mathcal{C}_I$ is an *invariant subspace* for the middle-rounds of PRINCE, that is for each $a \in \mathcal{C}_I^\perp$, there exists unique $b \in \mathcal{C}_I^\perp$ such that

$$\text{S-Box}^{-1} \circ M' \circ \text{S-Box}(\mathcal{C}_I \oplus a) = \mathcal{C}_I \oplus b. \tag{5}$$

As noticed in [20] and in [8], the middle rounds of PRINCE have 2^{32} *fixed points* (x is a fixed point of a function f iff $f(x) = x$). In particular, in [8] authors showed that if a plaintext of PRINCE "corresponds" to a fixed point of the middle rounds, then the encryption scheme is much simplified, since the 4 center rounds - minus the first and the last key addition - become a simple S-Box layer. However, no key-recovery attack has been showed: due to the presence of the secret key, it is not possible to choose a priori the plaintexts in order to satisfy the previous requirement. In our case, we show how to exploit the invariant subspace (that is, set of points instead of a single point) in order to mount powerful attacks on reduced-round PRINCE.

3.2 Subspace Trails of PRINCE

In the following, we present several subspace trails for round-reduced PRINCE.

Subspace Trail for 1+1.5 rounds of PRINCE. Let $R^{(1+1.5)}(\cdot)$ defined as:

$$R^{(1+1.5)}(\cdot) := M' \circ \text{S-Box} \circ R \circ ARK(\cdot), \tag{6}$$

i.e. the middle rounds without the final S-Box (denoted by "1.5") and the previous round (denoted by "1").

Theorem 1. *Let $I \subseteq \{0, 1, 2, 3\}$. For each $a \in \mathcal{C}_I^\perp$, there exists unique $b \in \mathcal{M}_I^\perp$ such that $R^{(1+1.5)}(\mathcal{C}_I \oplus a) = \mathcal{M}_I \oplus b$, where b depends on a and on the secret key. Equivalently:*

$$\text{Prob}(R^{(1+1.5)}(x) \oplus R^{(1+1.5)}(y) \in \mathcal{M}_I \mid x \oplus y \in \mathcal{C}_I) = 1. \tag{7}$$

This means that a coset of $\mathcal{C}_I$ is mapped into a coset of $\mathcal{M}_I$ after 2.5 rounds:

$$\mathcal{C}_I \oplus a \xrightarrow{R \circ ARK(\cdot)} \mathcal{D}_I \oplus b \xrightarrow{M' \circ \text{S-Box}(\cdot)} \mathcal{M}_I \oplus c.$$

Thus, a subspace trail for $1+1.5$ rounds of PRINCE is composed by the subspaces $\{\mathcal{C}_I, \mathcal{D}_I, \mathcal{M}_I\}$. Since S-Box($\mathcal{M}_I$) is mapped into a subspace of dimension 64 (that is all the space), it is not possible to extend the found subspace trail anymore. Moreover, observe that if X is a generic subspace, $X \oplus a$ is a coset of X and if x and y are two elements of the (same) coset $X \oplus a$, then $x \oplus y \in X$. This justifies the probability (7).

Subspace Trail for $2+0.5$ rounds of PRINCE. Let $R^{(2+0.5)}(\cdot)$ defined as:

$$R^{(2+0.5)}(\cdot) := M' \circ SR^{-1} \circ ARK \circ \text{super-SBox} \circ ARK(\cdot), \qquad (8)$$

i.e. the middle rounds ("2") and the linear part of the next round ("0.5").

Theorem 2. *Let $I \subseteq \{0, 1, 2, 3\}$. For each $a \in \mathcal{C}_I^\perp$, there exists unique $b \in \mathcal{IM}_I^\perp$ such that $R^{(2+0.5)}(\mathcal{C}_I \oplus a) = \mathcal{IM}_I \oplus b$, where b depends on a and on the secret key. Equivalently:*

$$\text{Prob}(R^{(2+0.5)}(x) \oplus R^{(2+0.5)}(y) \in \mathcal{IM}_I \,|\, x \oplus y \in \mathcal{C}_I) = 1. \qquad (9)$$

This means that a coset of $\mathcal{C}_I$ is mapped into a coset of $\mathcal{IM}_I$ after 2.5 rounds:

$$\mathcal{C}_I \oplus a \xrightarrow{\text{super-SBox} \circ ARK(\cdot)} \mathcal{C}_I \oplus b \xrightarrow{M' \circ SR^{-1} \circ ARK(\cdot)} \mathcal{IM}_I \oplus c.$$

Thus, a subspace trail for $2+0.5$ rounds of PRINCE is composed by the subspaces $\{\mathcal{C}_I, \mathcal{IM}_I\}$.

4   An "Equivalent" Representation of PRINCE

In this section, we present an "*equivalent*" representation of PRINCE from the point of view of the security. The PRINCEcore round is very similar to the AES round. The major difference between them is that in a PRINCE round the MixLayer operation is performed before the ShiftRows operation, while in an AES round is the opposite.

In order to better understand the PRINCE algorithm, we evaluate the security of a version of PRINCE - called in the following PRINCE' - where these two linear operations are exchanged in position, both in the forward and in the backward rounds. First of all, in this case it is possible to set up a subspace trail for 3.5 rounds of PRINCE' (i.e. one more round than original PRINCE):

$$\mathcal{ID}_I \oplus a \xrightarrow{R \circ ARK(\cdot)} \mathcal{C}_I \oplus b \xrightarrow{\text{super-SBox}(\cdot)} \mathcal{C}_I \oplus c \xrightarrow{M' \circ SR^{-1} \circ ARK(\cdot)} \mathcal{IM}_I \oplus d, \quad (10)$$

where $I \subseteq \{0, 1, 2, 3\}$, using the property that $\mathcal{C}_I$ is an invariant subspace for the middle rounds. The proof follows immediately by the definition of the subspaces and by the order of the ShiftRows and of the MixLayer operations.

Also due to the following cryptanalysis of PRINCE' against the most popular attacks present in literature, we can conclude that this version of PRINCE is weaker than the original one (as the designers, we don't consider the related key attacks for this security analysis):

- *Differential/Linear Cryptanalysis:* For the original PRINCE, *"any differential characteristic and any linear-trail over 4 consecutive rounds of PRINCE has at least 16 active S-Boxes"* (see App. C of [6] for more details). For the modified version PRINCE' and using the same argumentation given in [6], the number of active S-Boxes over 4 consecutive rounds is at least 12 instead of 16.
- *Square Attack:* For the original PRINCE, the balanced property holds for 4.5 rounds (2 forward rounds + middle rounds + 1 backward rounds) starting with three input active nibbles which lie on the same column (see [18] for more details). For PRINCE', the balanced property holds for 5.5 rounds (2 forward rounds + middle rounds + 2 backward rounds) starting with a single input active nibble (result practical verified).
- *Meet-in-the-Middle Attack:* The Meet-in-the-Middle Attacks presented in [8] are not influenced by the positions of the MixLayer and of the ShiftRows operations, that is there are analogous meet-in-the-middle attacks for this modified version similar to the ones presented for the original PRINCE.

As a consequence, the position of the ShiftRows and of the MixLayer operations influences the security of this encryption scheme.

A version of PRINCE - called in the following PRINCE$^\sharp$ - with the same security of the original one and where ShiftRows and MixLayer operations are ordered as in AES can be obtained by changing the original *middle-rounds* of PRINCE with the following one:

$$middle\text{-}rounds(x) = \text{S-Box}^{-1} \circ SR^{-1} \circ M' \circ SR \circ \text{S-Box}(x), \tag{11}$$

and with some more slight modifications, as we show in details in the following. By definition of PRINCE:

$$p \xrightarrow{ARK(\cdot)} \xrightarrow{ARK \circ M' \circ SR \circ \text{S-Box}(\cdot)} \cdots \xrightarrow{ARK \circ M' \circ SR \circ \text{S-Box}(\cdot)} \underbrace{\xrightarrow{\text{S-Box}^{-1} \circ M' \circ \text{S-Box}(\cdot)}}_{middle\text{-}rounds}$$

$$\xrightarrow{\text{S-Box}^{-1} \circ SR^{-1} \circ M' \circ ARK(\cdot)} \cdots \xrightarrow{\text{S-Box}^{-1} \circ SR^{-1} \circ M' \circ ARK(\cdot)} \xrightarrow{ARK(\cdot)} c.$$

Exchanging S-Box, ARK and SR operations, and applying a SR^{-1} operation on the plaintext and a SR on the ciphertext, one obtains:

$$p' \xrightarrow{ARK'(\cdot)} \xrightarrow{ARK' \circ SR \circ M' \circ \text{S-Box}(\cdot)} \cdots \xrightarrow{ARK' \circ SR \circ M' \circ \text{S-Box}(\cdot)} \underbrace{\xrightarrow{\text{S-Box}^{-1} \circ SR \circ M' \circ SR^{-1} \circ \text{S-Box}(\cdot)}}_{middle\text{-}rounds}$$

$$\xrightarrow{\text{S-Box}^{-1} \circ M' \circ SR^{-1} \circ ARK''(\cdot)} \cdots \xrightarrow{\text{S-Box}^{-1} \circ M' \circ SR^{-1} \circ ARK''(\cdot)} \xrightarrow{ARK''(\cdot)} c',$$

where $ARK'(\cdot) := \cdot \oplus SR^{-1}(k)$, $ARK''(\cdot) := \cdot \oplus SR(k)$, $p' = SR^{-1}(p)$ and $c' = SR(c)$. PRINCE$^\sharp$ is an equivalent representation of PRINCE, where ShiftRows operation is performed before the MixColumns one (as in AES) and where the *super-SBox* operation is a little modified, for the cost of 2 additional ShiftRows operations. With respect to the original PRINCE, a (slight) different key schedule is used and SR^{-1} (respectively SR) is applied on p (respectively on c).

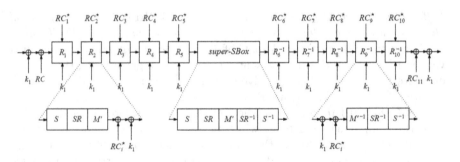

Fig. 2. A scheme of the PRINCEcore of the cipher PRINCE*, analogous (but not completely equivalent - see the text for details) to the original PRINCE.

In this equivalent representation, a ShiftRows operation (respectively Inverse-SR) is applied to the key in the forward rounds (respectively backwards). Thus, we finally define another version of PRINCE - called in the following PRINCE* - depicted in Fig. 2, which is analogous to the original PRINCE but not completely equivalent, since *no* ShiftRows operation (respectively Inverse-SR) is applied to the key in the forward rounds (respectively backward). Thus, we claim that PRINCE* has the same security of the original PRINCE. Note that the order of ShiftRows and MixLayer operations of PRINCE* is the same of AES. In App. C of [11], we present a related key attack that exploits this equivalent version of PRINCE.

Conclusion and AES-Like Ciphers. Our analysis can be applied in a natural way to other PRINCE-like ciphers, as for example the MANTIS encryption scheme [5] - see Appendix A for details and [9], where a similar but independent analysis with analogous results and conclusion has been proposed - or the QARMA block cipher family [2].

By our analysis, *the order of the MixLayer and the ShiftRows operations influences the security of PRINCE-like ciphers.* Thus, it seems advisable to use only one of the two following options for future designs of PRINCE-like schemes:

- the middle-rounds as in the original PRINCE cipher (that is, S-Box$^{-1} \circ M' \circ$ S-Box$(\cdot)$) and the MixLayer computed before (respectively after) the ShiftRows in the forward (respectively backward) rounds;
- the middle rounds as in the PRINCE* cipher (that is, S-Box$^{-1} \circ SR^{-1} \circ M' \circ SR \circ$ S-Box$(\cdot)$) and the ShiftRows computed before (respectively after) the MixLayer in the forward (respectively backward) rounds.

Finally, we emphasize that this analysis holds due to the particular structure of PRINCE-like cipher. In particular, consider key-recovery attacks that are independent of the key-schedule[4], excluding related-key attacks. For an AES-like cipher (with r identical rounds and without middle rounds), the position of the ShiftRows operation with respect to the MixLayer one does not influence the security. Indeed, consider the AES encryption scheme (where the final MixColumns operation can also be omitted):

$$p \xrightarrow{ARK} \xrightarrow{ARK \circ MC \circ SR \circ \text{S-Box}(\cdot)} \ldots \xrightarrow{ARK \circ MC \circ SR \circ \text{S-Box}(\cdot)} c.$$

Changing the position of SR, ARK and S-Box, and applying a ShiftRows operation to the ciphertexts (note that this is a linear operation, so it doesn't influence the security of the encryption scheme), one obtains:

$$SR(p) \xrightarrow{ARK''} \xrightarrow{ARK'' \circ SR \circ MC \circ \text{S-Box}(\cdot)} \ldots \xrightarrow{ARK'' \circ SR \circ MC \circ \text{S-Box}(\cdot)} c,$$

where $ARK''(\cdot) = \cdot \oplus SR(k)$ and k is the secret key. It follows that an equivalent version of AES - called for consistency AES* - defined as

$$p \xrightarrow{ARK} \xrightarrow{ARK \circ SR \circ MC \circ \text{S-Box}(\cdot)} \ldots \xrightarrow{ARK \circ SR \circ MC \circ \text{S-Box}(\cdot)} c,$$

where ShiftRows operation is computed after the MixColumns one, has the same security of the original version.

5 Truncated Differential Attack on 4 Rounds of PRINCE

Using the first 2.5 rounds subspace trail presented in previous section, it is possible to set up an attack on 3 rounds of PRINCE:

$$p \xrightarrow{R(\cdot)} q \xrightarrow{M' \circ \text{S-Box}(\cdot)} s \xrightarrow{\text{S-Box}^{-1}(\cdot)} c,$$

where the plaintexts are chosen in the same coset of $\mathcal{C}_I$, and the states s belong in the same coset of $\mathcal{M}_I$. Briefly, given a pair of ciphertexts (c^1, c^2), the idea is to find the final key using the condition S-Box$(c^1 \oplus \hat{k}) \oplus$ S-Box$(c^2 \oplus \hat{k}) \in \mathcal{M}_I$. When the full key $\hat{k}$ has been found, to find k_1 the idea is to use plaintexts in the same coset of $\mathcal{D}_I$, to decrypt the corresponding ciphertexts and to find the key k_1 using the condition that S-Box$(q^1 \oplus k_1) \oplus$ S-Box$(q^2 \oplus k_1) \in SR(\mathcal{M}_I)$, where $q := super\text{-}SBox^{-1}(c) \oplus R_1$. The attack is presented in details in App. E of [11], and here we focus on the attack on 4 rounds of PRINCE (giving all the details).

---

[4] We observe that attacks that exploit the key-schedule can be affected by the order of linear operations. To better highlight this fact, we refer to the analysis done in [10] about the effect of the omission of the final MixColumns operation. While in general key-recovery attacks are not influenced by the presence of the last MixColumns operation, some of the attacks that exploit it (e.g. Meet-in-the-Middle attacks) are affected, since a different key schedule can affect the amount of key material that has to be guessed in key-recovery attacks (also in the standard single-key model). In a similar way, the same analysis holds also when the positions of the MixColumns and ShiftRows operations are exchanged.

To attack 4 rounds, a possibility is to extend the attack on 3 rounds (the middle rounds and one round before) at the end. Consider four rounds of PRINCE:

$$p \xrightarrow{R(\cdot)} \hat{p} \xrightarrow{M' \circ \text{S-Box}(\cdot)} q \xrightarrow{\text{S-Box}^{-1}(\cdot)} s \xrightarrow{R^{-1}(\cdot)} c,$$

where $p \in \mathcal{C}_I \oplus a$ (for $a \in \mathcal{C}_I^\perp$ fixed). Given a pair of plaintexts/ciphertexts (where the plaintexts belong to the same coset of $\mathcal{C}_I$), the idea is simply to guess the key of the final round, to decrypt partially one round, and to find the key of the third round such that the two corresponding texts belong to the same coset of $\mathcal{M}_I$. That is, if $\hat{k}$ is a candidate of the final key (as we show in the following, the attacker must test all the possibilities) and if $R_{\hat{k}}(\cdot)$ denotes the final round with key $\hat{k}$, given a pair (c^1, c^2) the right key k_0 must satisfy the condition:

$$\text{S-Box}(R_{\hat{k}}(c^1) \oplus k_1 \oplus RC_1 \oplus a) \oplus \text{S-Box}(R_{\hat{k}}(c^2) \oplus k_1 \oplus RC_1 \oplus a) \in \mathcal{M}_I.$$

Candidates $\hat{k}$ and k_1 of the key must be tested checking if this condition is satisfied for other pairs of ciphertexts. To find them, the idea is to work independently on each column of $\mathcal{M}_I$. For the following, we limit to the case $I = \{0\}$.

First Step of the Attack. Let c^1 and c^2 two ciphertexts such that the corresponding plaintexts belong to the same coset of $\mathcal{C}_0$, that is $p^1 \oplus p^2 \in \mathcal{C}_0$. As for the attack on 3 rounds, the idea is to work independently on each column of the key, due to the fact that the columns of $\mathcal{M}_0$ depend on different and independent variables. Initially the attacker guesses 1 column (that is 4 nibbles) of the final key, as for example $\hat{k}[0], \hat{k}[1], \hat{k}[2]$ and $\hat{k}[3]$, and she uses them to partially decrypt c^1 and c^2, that is she computes 4 nibbles of $s^1 := R_{\hat{k}}(c^1)$ and of $s^2 := R_{\hat{k}}(c^2)$. Note that the attacker cannot guess 4 arbitrary nibbles of the final key but an entire column, since she has to compute the Linear Layer M'. Moreover, since the attacker can not impose any restriction/condition on the final key, she has to repeat the next steps for each values of these four nibbles of the final key, which are $(2^4)^4 = 2^{16}$ possible values in total.

Due to the ShiftRows operation, after one-round of decryption these four nibble belong to different column. To find four nibbles of k_0, the attacker must work independently on each nibble (note that since they lie on different columns, no relationship holds among them, due to the definition of $\mathcal{M}_I$). For example, using the definition of $\mathcal{M}_0$, the nibble $k_1[0]$ has to satisfy the condition:

$$(\text{S-Box}(s^1[0] \oplus k_1[0] \oplus RC_2[0]) \oplus \text{S-Box}(s^2[0] \oplus k_1[0] \oplus RC_2[0])) \wedge \text{0x8} = 0, \quad (12)$$

and similar conditions hold for the nibbles $k_1[7], k_1[10]$ and $k_1[13]$. To find these 4 nibbles, the attacker needs at least four different pairs of chosen ciphertexts (each of these conditions involves only one bit - it is satisfied with prob. 2^{-1}).

Since these found 4 nibbles of k_1 (which are $k_1[0], k_1[7], k_1[10], k_1[13]$) depend on the 4 guessed nibbles of $\hat{k}$ (which are $\hat{k}[0], \ldots, \hat{k}[3]$), for each combination of the first column of $\hat{k}$, the attacker finds on average one combination of the 4 nibbles of k_1, that is 2^{16} in total. To discover the right combination, the attacker has to test these values using other pairs of ciphertexts, that is given other pairs of ciphertexts $(c^\star, c')$ she has to check if the corresponding texts $(q^\star, q')$ - where $q := \text{S-Box}(s \oplus k_1 \oplus RC_2)$ - belong to the same coset of $\mathcal{M}_0$. Since each condition involves one bit and since there are 2^{16} combinations for the 4 nibbles of $\hat{k}$ and k_1, she needs at least other four different pairs to check the found values (since $2^{16} \times (2^{-4})^4 = 1$). Thus the attacker needs at least eight different pairs for this first step. To save memory, a good idea is to check immediately the found values of $\hat{k}$ and k_1 with other pairs of ciphertexts: in this way, the attacker doesn't need to store anything.

Second Step of the Attack. When the attacker has found 1 nibble for each column of k_1, the idea is to use the relationships that hold among the nibbles of the same column to discover the other nibbles of the key much faster than working on each nibble independently by the others.

As before, the attacker guesses one column (e.g. the second one) of $\hat{k}$, and decrypts partially the pair of ciphertexts. In order to find other 4 nibbles of k_1 (one per column), the idea is to use the relationships that hold among the nibbles of the same column given in Theorem 7 - App. E of [11], and not to work independently on each nibble. For example, the attacker can find the nibble of the first column $k_1[1]$ using the relationship:

$$(\text{S-Box}(s^1[0] \oplus k_1[0] \oplus RC_2[0]) \oplus \text{S-Box}(s^2[0] \oplus k_1[0] \oplus RC_2[0])) \wedge \text{0xb}$$
$$= (\text{S-Box}(s^1[1] \oplus k_1[1] \oplus RC_2[1]) \oplus \text{S-Box}(s^2[1] \oplus k_1[1] \oplus RC_2[1])) \wedge \text{0x7},$$

where the left part of the equation is known ($k_1[0]$ is already known). Analogous relationships hold for the other nibbles and for the other columns. Observe that since these relationships involve more than one bit, in this second step the attacker needs a lower number of pairs of ciphertexts to discover the right key.

As before, the attacker has to repeat the step for each possible values of the second column of $\hat{k}$, and to test the found values against other plaintexts/ciphertexts pairs in order to eliminate wrong candidates (remember that she finds on average one candidate of four nibbles of k_1 for each of the 2^{16} guess values of $\hat{k}$). The same procedure is used for the third and for the fourth columns of $\hat{k}$. For example, for each guess value of the third column of $\hat{k}$, the relationships that the nibble $k_1[2]$ have to satisfy are $\tilde{s}[2] \wedge \text{0xb} = \tilde{s}[1] \wedge \text{0xd}$ and $\tilde{s}[2] \wedge \text{0x4} = \tilde{s}[0] \wedge \text{0x4}$, where $\tilde{s}[i] := \text{S-Box}(s^1[i] \oplus k_1[i] \oplus RC_2[i]) \oplus \text{S-Box}(s^2[i] \oplus k_1[i] \oplus RC_2[i])$.

For completeness, note that also in this second step the attacker can work independently on each nibble, but the total computational cost would be higher.

Estimation of the Data Complexity. Our implementation shows that if only eight pairs of ciphertexts are used for the first step, then some false positive candidates of the key pass the test. To avoid this problem, one possibility is to use more pairs of ciphertexts, making the filter stronger[5].

---

[5] We emphasize that the right key is always found. We use more plaintexts only to discard false positives that pass the test.

In the following, we first try to give a theoretical estimation of the number of pairs necessary to eliminate almost all the false positive candidates of the key. In the first step of the attack, the problem of the false positives arises when $R_{\hat{k}}(c^1)[i] = R_{\hat{k}}(c^2)[i]$ for a certain $i = 0, 7, 10, 13$. For example, note that if $R_{\hat{k}}(c^1)[0] = R_{\hat{k}}(c^2)[0]$ in Eq. (12), then each possible value of $k_1[0]$ passes the test. Thus, a first estimation can be done by calculating the minimum number of pairs such that there exist at least eight pairs with 4 different nibbles.

Before to continue, an important observation has to be done. Given n texts, it is possible to construct $n \cdot (n-1)/2$ different pairs, but actually only $n-1$ pairs are useful for the attack. Consider for example three texts t^1, t^2, t^3 and the corresponding pairs (t^1, t^2), (t^1, t^3), (t^2, t^3). If $k_1[0]$ satisfies the condition (12) for the pairs $(t^1[0], t^2[0])$ and $(t^1[0], t^3[0])$, then it automatically satisfies this condition also for the pair $(t^2[0], t^3[0])$[6]. Thus, only two pairs are really useful for the attack. More generally, given n texts, only $n-1$ pairs are useful for the attack - for the following we suppose that one text is in common for all the pairs. As shown in details in App. F of [11], the probability that only the right key is found using 9 chosen plaintexts is about $(0.0604)^4$. By calculation (we refer to App. F of [11] for more details), the attacker needs at least 16 plaintexts in order to have a good probability of success, which is approximately 73%.

Actually, this is only a rough approximation, since an important aspect is not taken into account. For each guess of the first column of $\hat{k}$, the attacker is able to find 4 nibbles of k_1. Then, she checks these candidates of the keys using other texts. Note that it is sufficient that one nibble of k_1 doesn't pass this test to conclude that the guess value of $\hat{k}$ is wrong, independently by the other three nibbles of k_1. Moreover, it is also possible that wrong key candidates found at the first step are detected and eliminated in the second step of the attack. Thus, as our implementation shows, a lower number of texts (with respect to that predicted by our theoretical model) turns out to be sufficient for the attack. In particular, we found that 12 chosen plaintexts (instead of 16) - i.e. 11 pairs - are sufficient with (very) high probability. Working in a similar way, it turns out that only 6 chosen plaintexts - i.e. 5 pairs - are sufficient for the second step.

The number of plaintexts chosen for the first step (i.e. 12) allows to eliminate almost all the false candidates of the key. For the second step of the attack, a lower number of plaitexts (i.e. 6) is sufficient to recover the entire key. As a consequence, in the second step the attacker doesn't use and wastes a lot of information about the (chosen) plaintexts/ciphertexts pairs.

To improve the attack, the idea is to reduce the total number of chosen plaintexts used for the attack, and to use all the available plaintexts/ciphertexts pairs also in the second step. That is, the idea is to reduce the number of chosen plaintexts used for the first step and to increase this number for the second

---

[6] Note that: [S-Box($t^2[0] \oplus k_1[0]$) $\oplus$ S-Box($t^3[0] \oplus k_1[0]$)] $\wedge$ 0x8 = [S-Box($t^2[0] \oplus k_1[0]$) $\oplus \oplus$ S-Box($t^1[0] \oplus k_1[0]$) $\oplus$ S-Box($t^1[0] \oplus k_1[0]$) $\oplus$ S-Box($t^3[0] \oplus k_1[0]$)] $\wedge$ 0x8 = 0.

step - we assume that these two numbers are equal. As a consequence, in the first step of the attack more false positive candidates pass the test, but they are soon detected in the second step, thanks to the higher number of pairs used. Moreover, note that in the second step the false candidates are detected much faster than in the first one, since the probability that the relationships are satisfied is much lower (remember that they involve an higher number of bits).

Our implementation shows that 8 chosen plaintexts (i.e. 7 pairs) are sufficient for this mode of the attack (note that we use 8 chosen plaintexts both in the first step and in the second one), and that the total computational cost is approximately unchanged (with respect to the previous mode).

Estimation of the Computational Cost. The computational cost of the first step can be estimated as follows. Given 5 chosen ciphertexts, the computational cost to calculate 4 nibbles of $R^{-1}(c)$ is $8 \times 4 = 2^5$ S-Box look-ups. As shown in detail in App. E.1 of [11], the cost to find one nibble of k_1 working independently on each nibble is $2^{5.46}$ S-Box look-ups (thus for 4 nibbles, the cost is $4 \times 2^{5.46} = 2^{7.46}$). The cost to check candidates of $\hat{k}$ and of k_1 against other pairs of ciphertexts can be estimated by $4 \times 2 \times 8 = 2^6$ S-Box look-ups. Thus, to find the right combination and to do the requested check, the total computational cost for the first step is well approximated by 2^{16} (possible values) $\times (4 \cdot 2^{5.46} + 8 \times 4 + 2^6) \simeq 2^{24.1}$ S-Box looks ups.

Some optimizations allow to improve the computational cost of the attack. For example, for each guessed value of $\hat{k}$, the attacker should focus on a single nibble of k_1, e.g. $k_1[0]$. In this way, it is possible to eliminate wrong candidates simply checking the found candidates of $k_1[0]$ and of the column of $\hat{k}$ against all the available pairs of texts before to work on the other three nibbles of k_1. It follows that it is sufficient to consider only these survived combinations of the first column of $\hat{k}$ (instead of all the 2^{16} possible values) in order to find the other 3 nibbles of k_1. The computational cost to find the remaining 3 nibbles of k_1 becomes negligible compared to the cost to find $k_1[0]$, and the total computational cost can be approximated by $2^{16} \times (2^{5.46} + 2^5 + 2^6) \simeq 2^{23.1}$ S-Box look ups.

The computational cost for the second step can be computed in a similar way. As shown in an analogous case in App. E.2 of [11], the cost to find one nibble of k_1 is of $2^{4.6}$ S-Box look-ups (given another nibble of the same column and exploiting the relationship among the nibbles). Thus, the total cost for this step can be approximated by 3 (columns) $\times 2^{16} \times (2^{4.6}$ (cost of the subspace attack - single equivalence) $+ 4 \times 8$ (check) $+ 2^5$ (partial decryption)) $\simeq 2^{23.9}$ S-Box look-ups, using the same optimizations as before.

The total computational cost can be approximated by $2^{23.9} + 2^{23.1} \simeq 2^{24.25}$ S-Box look-ups, that is $2^{18.25}$ four-rounds encryption, and the attacker needs only 8 different chosen plaintexts (that belong to the same coset of $\mathcal{C}_0$).

Data: 7 ciphertexts pairs (c^1, c^i) where $i = 2, \ldots, 8$, whose corresponding plaintexts belong in the same coset of $\mathcal{C}_0$

Result: Secret Key $\hat{k}$ and k_1.

for *all* 2^{16} *possible combinations of* $(\hat{k}[0], \hat{k}[1], \hat{k}[2], \hat{k}[3])$ **do**

    decrypt one round: $s^i[j] = $ S-Box$(c^i \oplus \hat{k}[j])$ $\forall i = 1, \ldots, 8$ and $\forall j = 0, \ldots, 3$;

    for $k_1[0]$ *from* 0 *to* $2^4 - 1$ **do**

        check if for each possible pairs (s^1, s^i) where $i = 2, \ldots, 8$:

        [S-Box$(s^1[0] \oplus k_1[0] \oplus RC_2[0]) \oplus$ S-Box$(s^i[0] \oplus k_1[0] \oplus RC_2[0])] \wedge$ 0x8 $= 0$;

        If not satisfied, **then** next value (i.e. next $k_1[0]$ or/and $(\hat{k}[0], \ldots, \hat{k}[3])$);

        else

            identify candidates for $k_1[0]$ and $(\hat{k}[0], \ldots, \hat{k}[3])$;

            use these candidates of the first column of $\hat{k}$, to find candidates of $k_1[7], k_1[10], k_1[13]$ - use the same algorithm described for $k_1[0]$ and work independently on each nibble;

        end

    end

end

for *all candidates of* $k_1[0], k_1[7], k_1[10], k_1[13]$ *and* $(\hat{k}[0], \ldots, \hat{k}[3])$ **do**

    for *all* 2^{16} *possible combinations of* $(\hat{k}[4], \hat{k}[5], \hat{k}[6], \hat{k}[7])$ **do**

        decrypt one round: $s^i[j] = $ S-Box$(c^i \oplus \hat{k}[j])$ $\forall i = 1, \ldots, 8$ and $\forall j = 4, \ldots, 7$;

        for $k_1[1]$ *from* 0 *to* $2^4 - 1$ **do**

            check if for each possible pairs (s^1, s^i) where $i = 2, \ldots, 8$:

            [S-Box$(s^1[0] \oplus k_1[0] \oplus RC_2[0]) \oplus$ S-Box$(s^i[0] \oplus k_1[0] \oplus RC_2[0]))] \wedge$ 0xb $=$ [S-Box$(s^1[1] \oplus k_1[1] \oplus RC_2[1]) \oplus$ S-Box$(s^i[1] \oplus k_1[1] \oplus RC_2[1])] \wedge$ 0x7;

            If not satisfied, **then** next value (i.e. next $k_1[1]$ or/and $(\hat{k}[4], \ldots, \hat{k}[7])$ or/and $k_1[0]$ or/and $(\hat{k}[0], \ldots, \hat{k}[3])$);

            else

                identify candidates for $k_1[1]$ and $(\hat{k}[4], \ldots, \hat{k}[7])$;

                use these candidates of the second column of $\hat{k}$, to find candidates of $k_1[4], k_1[11], k_1[14]$ - use the same algorithm described for $k_1[1]$ and exploit the relationships among the nibbles;

            end

        end

    end

end

Repeat this second step for the third and for the fourth column of $\hat{k}$;

return *Secret Key* $\hat{k}$ *and* k_1.

Algorithm 1: *Truncated differential attack on 4 rounds of PRINCE - extension at the end.* For simplicity, this pseudo-code is not completely optimized as described in the text.

Acknowledgements. The work in this paper has been partially supported by the Austrian Science Fund (project P26494-N15).

A   MANTIS Encryption Scheme: Subspace Trail Cryptanalysis

MANTIS encryption scheme [5] is a low-latency tweakable block cipher proposed at CRYPTO 2016. The starting point used by the designer for this encryption scheme is a PRINCE-like encryption scheme, keeping the entire design symmetric around the middle (to have the α-reflection property). In order to improve the security, the PRINCE-round has been replaced by the MIDORI-round function. This simple change results in a cipher with improved latency and improved security compared to PRINCE. Note that in contrast to PRINCE, the PermuteCells operation is performed before the MixLayer one.

MANTIS$_r$ has a 64-bit block length and works with a 128-bit key ($k = k_0 || k_1$ with 64-bit subkeys k_0, k_1) and 64-bit tweak T. The parameter r specifies the number of rounds of one half of the cipher. As PRINCE, MANTIS is based on the FX-construction and thus applies whitening keys before and after applying its core components (the whitening keys are generated in the same way as for PRINCE). Every round $R^i(\cdot)$ in MANTIS is defined as

$$R^i(\cdot) = M \circ P(h^i(T) \oplus k_1 \oplus RC_i \oplus \text{S-Box}(\cdot)),$$

for $i = 0, \ldots, r$, where[7]:

- **S-Box layer**: Every byte in the internal state is replaced by using the involutory 4×4-bit MIDORI S-Box;
- **A bit-wise XOR with the (full) round tweakey state** $h^i(T) \oplus k_1$, for $i = 0, \ldots, r$, where T is the tweak and h^i is the tweak permutation;
- **PermuteCells Operation P**: The cells of the internal state are permuted according to the MIDORI permutation;
- **MixColumns M**: Each column of the cipher internal state array is multiplied by the MixColumns binary matrix of MIDORI M (we recall that $M = M^{-1}$):
- **A bit-wise XOR with the key** k_1 **and a round constant** RC_i.

As for PRINCE, in the last r rounds the order of operations is inverse with respect to the first r rounds, where only the round constants differ. Moreover, the middle rounds consist of three key-less operations: an S-Box layer, a matrix multiplication with M and an inverse S-Box layer. Finally, as PRINCE, MANTIS has the α-reflection property, that is $D_{(k_0||k_0'||k_1)}(\cdot, T) = E_{(k_0'||k_0||k_1 \oplus \alpha)}(\cdot, T)$. Thus, our results presented in Sect. 4 can be applied on MANTIS.

Subspace Trail of MANTIS. Proceeding as for PRINCE, we first identify analogous subspace trails for MANTIS. The column, diagonal and mixed subspaces are defined exactly as the ones defined for PRINCE in Sect. 3.1, but their representations are a little different (expect for the column space).

---

[7] We refer to [5] and [3] for a complete description of the S-Box, the PermuteCells and the MixColumns operations.

For instance, $\mathcal{D}_0 = P(\mathcal{C}_0)$, $\mathcal{ID}_0 = P^{-1}(\mathcal{C}_0)$, $\mathcal{M}_0 = M(\mathcal{D}_0)$ and $\mathcal{IM}_0 = M(\mathcal{ID}_0)$ correspond to matrix representations:

$$
\mathcal{D}_0 \equiv \begin{bmatrix} x & 0 & 0 & 0 \\ 0 & 0 & y & 0 \\ 0 & 0 & 0 & z \\ 0 & w & 0 & 0 \end{bmatrix} \quad
\mathcal{ID}_0 \equiv \begin{bmatrix} x & 0 & 0 & 0 \\ 0 & y & 0 & 0 \\ 0 & 0 & z & 0 \\ 0 & 0 & 0 & w \end{bmatrix} \quad
\mathcal{M}_0 \equiv \begin{bmatrix} 0 & w & y & z \\ x & w & 0 & z \\ x & w & y & 0 \\ x & 0 & y & z \end{bmatrix} \quad
\mathcal{IM}_0 \equiv \begin{bmatrix} 0 & w & y & z \\ x & 0 & y & z \\ x & w & 0 & z \\ x & w & y & 0 \end{bmatrix}.
$$

Let $I \subseteq \{0,1,2,3\}$. Since $\mathcal{C}_I$ is an invariant subspace for the middle rounds, note that it is possible to set up a subspace trail for 3.5 rounds of MANTIS:

$$
\mathcal{ID}_I \oplus a \xrightarrow{R \circ ARK(\cdot)} \mathcal{C}_I \oplus b \xrightarrow{super\text{-}SBox(\cdot)} \mathcal{C}_I \oplus c \xrightarrow{M' \circ SR^{-1}(\cdot)} \mathcal{IM}_I \oplus d.
$$

A More Secure Version of MANTIS. As for PRINCE, we consider a version of MANTIS where the MixColumns and the PermuteCells operations are exchanged in positions - called for the following MANTIS*. In this version, the rounds of MANTIS* are defined similar of the PRINCE ones, where the Mix-Columns operation is performed before (resp. after) the PermuteCells one in the forward (resp. backwards) rounds.

As first consequence, in this case it is only possible to set up a subspace trail for 2.5 rounds (similar to PRINCE), that is $\mathcal{C}_I \oplus a \xrightarrow{R(\cdot)} \mathcal{D}_I \oplus b \xrightarrow{M \circ S\text{-}Box(\cdot)} \mathcal{M}_I \oplus c$ or $\mathcal{C}_I \oplus a \xrightarrow{super\text{-}SBox(\cdot)} \mathcal{C}_I \oplus b \xrightarrow{M \circ SR^{-1}(\cdot)} \mathcal{IM}_I \oplus c$.

Moreover, *"as one round of MANTIS is almost identical to one round in MIDORI, most of the security analysis can simply be copied from the latter"* (see Sect. 6.3 of [5]). By our analysis of Sect. 4 and since MIDORI [3] is an AES-like cipher, its security is not influenced by the positions of the MixColumns and of the PermuteCells operations. Thus, the version of MIDORI - called for consistency MIDORI* - in which the MixColumns operation is performed before the PermuteCells operation has the same security of the original one.

Due to previous considerations and since the analysis done for PRINCE in Sect. 4 also applies on MANTIS as well, we can claim that *MANTIS** (i.e. the version of MANTIS in which MixColumns and PermuteCells are exchanged in positions) *is more secure than the original version proposed by [5] with respect to the attack vectors considered in this paper*. Note that this claim is also justified by the fact that authors didn't consider related-key attacks in order to evaluate the security of MANTIS, and that its key schedule is linear (in particular, there is no key-schedule since all the subkeys are equal to the whitening key).

For completeness and following our analysis of Sect. 4, we defined another version of MANTIS - called in the following MANTIS$'$, such that MANTIS$'$ is identical to the original MANTIS excepted for the middle rounds, defined as

$$
middle\text{-}rounds(\cdot) = \text{S-Box}^{-1} \circ P^{-1} \circ M \circ P \circ \text{S-Box}(\cdot).
$$

As for MANTIS*, we can claim that MANTIS$'$ is more secure than the original version proposed by [5], and that it has the same security of MANTIS*. For completeness, a similar but independent analysis is proposed in [9], which leads to analogous results and conclusions.

References

1. Abed, F., List, E., Lucks, S.: On the Security of the Core of PRINCE Against Biclique and Differential Cryptanalysis. Cryptology ePrint Archive, Report 2016/712 (2016)
2. Avanzi, R.: The QARMA Block Cipher Family - Almost MDS Matrices Over Rings With Zero Divisors, Nearly Symmetric Even-Mansour Constructions With Non-Involutory Central Rounds, and Search Heuristics for Low-Latency S-Boxes. Cryptology ePrint Archive, Report 2016/444 (2016)
3. Banik, S., Bogdanov, A., Isobe, T., Shibutani, K., Hiwatari, H., Akishita, T., Regazzoni, F.: Midori: a block cipher for low energy. In: Iwata, T., Cheon, J.H. (eds.) ASIACRYPT 2015. LNCS, vol. 9453, pp. 411–436. Springer, Heidelberg (2015). doi:10.1007/978-3-662-48800-3_17
4. Bay, A., Ersoy, O., Karakoç, F.: Universal Forgery and Key Recovery Attacks on ELmD Authenticated Encryption Algorithm. Cryptology ePrint Archive, Report 2016/640 (2016). To appear at Asiacrypt 2016
5. Beierle, C., Jean, J., Kölbl, S., Leander, G., Moradi, A., Peyrin, T., Sasaki, Y., Sasdrich, P., Sim, S.M.: The SKINNY family of block ciphers and its low-latency variant MANTIS. In: Robshaw, M., Katz, J. (eds.) CRYPTO 2016. LNCS, vol. 9815, pp. 123–153. Springer, Heidelberg (2016). doi:10.1007/978-3-662-53008-5_5
6. Borghoff, J., Canteaut, A., Güneysu, T., Kavun, E.B., Knezevic, M., Knudsen, L.R., Leander, G., Nikov, V., Paar, C., Rechberger, C., Rombouts, P., Thomsen, S.S., Yalçın, T.: PRINCE – a low-latency block cipher for pervasive computing applications. In: Wang, X., Sako, K. (eds.) ASIACRYPT 2012. LNCS, vol. 7658, pp. 208–225. Springer, Heidelberg (2012). doi:10.1007/978-3-642-34961-4_14
7. Bouillaguet, C., Derbez, P., Dunkelman, O., Fouque, P., Keller, N., Rijmen, V.: Low-data complexity attacks on AES. IEEE Trans. Inf. Theory **58**(11), 7002–7017 (2012)
8. Derbez, P., Perrin, L.: Meet-in-the-middle attacks and structural analysis of round-reduced PRINCE. In: Leander, G. (ed.) FSE 2015. LNCS, vol. 9054, pp. 190–216. Springer, Heidelberg (2015). doi:10.1007/978-3-662-48116-5_10
9. Dobraunig, C., Eichlseder, M., Mendel, F.: Key recovery for MANTIS-5. Cryptology ePrint Archive, Report 2016/754 (2016)
10. Dunkelman, O., Keller, N.: The effects of the omission of last round's MixColumns on AES. Inf. Process. Lett. **110**(8–9), 304–308 (2010)
11. Grassi, L., Rechberger, C.: Practical low data-complexity subspace-trail cryptanalysis of round-reduced PRINCE. IACR Cryptology ePrint Archive (2016)
12. Grassi, L., Rechberger, C., Rønjom, S.: Subspace trail cryptanalysis and its applications to AES. Cryptology ePrint Archive, Report 2016/592 (2016)
13. Kilian, J., Rogaway, P.: How to protect DES against exhaustive key search. In: Koblitz, N. (ed.) CRYPTO 1996. LNCS, vol. 1109, pp. 252–267. Springer, Heidelberg (1996). doi:10.1007/3-540-68697-5_20
14. Knudsen, L.R.: Truncated and higher order differentials. In: Preneel, B. (ed.) FSE 1994. LNCS, vol. 1008, pp. 196–211. Springer, Heidelberg (1995). doi:10.1007/3-540-60590-8_16
15. Leander, G., Abdelraheem, M.A., AlKhzaimi, H., Zenner, E.: A cryptanalysis of PRINTCIPHER: the invariant subspace attack. In: Rogaway, P. (ed.) CRYPTO 2011. LNCS, vol. 6841, pp. 206–221. Springer, Heidelberg (2011). doi:10.1007/978-3-642-22792-9_12

16. Leander, G., Minaud, B., Rønjom, S.: A generic approach to invariant subspace attacks: cryptanalysis of Robin, iSCREAM and Zorro. In: Oswald, E., Fischlin, M. (eds.) EUROCRYPT 2015. LNCS, vol. 9056, pp. 254–283. Springer, Heidelberg (2015). doi:10.1007/978-3-662-46800-5_11

17. Morawiecki, P.: Practical Attacks on the Round-reduced PRINCE. Cryptology ePrint Archive, Report 2016/245 (2016)

18. Posteuca, R., Negara, G.: Integral Cryptanalysis of Round-Reduced PRINCE Cipher. Proceedings of the Romanian Academy, Series A **16**, 265–270 (2015)

19. Raddum, H., Rasoolzadeh, S.: Faster Key Recovery Attack on Round-Reduced PRINCE. Cryptology ePrint Archive, Report 2016/828 (2016). To appear at Light-Sec 2016

20. Soleimany, H., Blondeau, C., Yu, X., Wu, W., Nyberg, K., Zhang, H., Zhang, L., Wang, Y.: Reflection cryptanalysis of PRINCE-like ciphers. J. Crypt. **28**(3), 718–744 (2013)

21. Zhao, G., Sun, B., Li, C., Su, J.: Truncated differential cryptanalysis of PRINCE. Secur. Commun. Netw. **8**(16), 2875–2887 (2015)

Foundations

On Negation Complexity of Injections, Surjections and Collision-Resistance in Cryptography

Douglas Miller, Adam Scrivener, Jesse Stern,
and Muthuramakrishnan Venkitasubramaniam[✉]

University of Rochester, Rochester, NY, USA
muthuv@cs.rochester.edu

Abstract. Goldreich and Izsak (Theory of Computing, 2012) initiated the research on understanding the role of negations in circuits implementing cryptographic primitives, notably, considering one-way functions and pseudo-random generators. More recently, Guo, Malkin, Oliveira and Rosen (TCC, 2015) determined tight bounds on the minimum number of negations gates (i.e., negation complexity) of a wide variety of cryptographic primitives including pseudo-random functions, error-correcting codes, hardcore-predicates and randomness extractors.

We continue this line of work to establish the following results:

1. First, we determine tight lower bounds on the negation complexity of collision-resistant and target collision-resistant hash-function families.
2. Next, we examine the role of injectivity and surjectivity on the negation complexity of one-way functions. Here we show that,
   (a) Assuming the existence of one-way injections, there exists a monotone one-way injection. Furthermore, we complement our result by showing that, even in the worst-case, there cannot exist a monotone one-way injection with constant stretch.
   (b) Assuming the existence of one-way permutations, there exists a monotone one-way surjection.
3. Finally, we show that there exists list-decodable codes with monotone decoders.

In addition, we observe some interesting corollaries to our results.

Keywords: Monotone boolean circuits · One-way functions · Collision-resistant hash-functions · Injections · Surjections

1 Introduction

A boolean circuit C is *monotone* if it comprises only of fanin-2 AND and OR gates. Monotone circuits have been extensively studied in complexity theory, where one of the fundamental goals is to establish lower bounds on circuit size, and learning theory [1,5,6,10,17]. In the context of cryptography, Goldreich and Izsak began exploring whether cryptographic primitives can be implemented

© Springer International Publishing AG 2016
O. Dunkelman and S.K. Sanadhya (Eds.): INDOCRYPT 2016, LNCS 10095, pp. 345–363, 2016.
DOI: 10.1007/978-3-319-49890-4_19

via monotone circuits [8]. They showed that, assuming the existence of one-way functions, there exists a one-way function implementable by monotone circuits. On the negative side, they proved that pseudo-random generators cannot be implemented using monotone circuits. More recently, Guo, Malkin, Oliveira and Rosen, inspired by a long series of works [2–4,7,12,13,15,16] in complexity theory, initiated the study of the *negation complexity* of realizing cryptographic primitives. Loosely speaking, the *negation complexity* of a primitive is the minimum number of negation gates required in any circuit implementing that primitive.

Markov [12] showed that any boolean function on n-bit inputs can be realized using a circuit with at most $\lceil \log(n+1) \rceil$ negation gates. A more efficient version was proved by Fischer [7], which, in particular, implies that any polynomial-time computable boolean function can be realized using a polynomial-size monotone circuit with $\lceil \log(n+1) \rceil$ negation gates. In essence, the negation complexity of any arbitrary polynomial-time computable function is $O(\log n)$. In [9], quite surprisingly, they show that this is *tight* for various primitives such as pseudorandom functions, error-correcting codes, randomness-extractors and generic hard-core predicates. Interestingly, a new technique was required to establish each lower-bound.

Our first motivating question concerns the negation complexity of collision-resistant hash-functions. Various notions of collision-resistance have been used in cryptographic constructions. Standard hash-functions, such as MD5 and SHA, are referred to as collision-resistant hash-functions (CRH), where the security game requires an adversary to find a pair of colliding inputs given a function picked uniformly from the family. A weaker form of collision-resistance that can be realized from one-way functions, referred to as universal one-way hash-functions (or target collision-resistant functions (TCR)) [14] require the adversary to produce a target for which it needs to find a collision before seeing the description of the hash-function. Yet another related primitive is second-preimage resistant hash-functions (SPR) where it is infeasible for any adversary to find a collision for a randomly chosen hash-function on a uniformly chosen input. Our first motivating question is:

What is the negation complexity of collision-resistance?

Another interesting result presented in the work of Guo et al. [9] proves the impossibility of monotone one-way permutations.[1] The main result shows that any monotone circuit that implements a permutation must be of the form where the output bits are a permutation of the input bits, in essence, making them easily invertible with probability 1. The result of Guo et al. and Goldreich and Izsak show a gap in what kind of one-way functions are achievable using monotone circuits. Our second motivating question is:

Do there exist polynomial-sized one-way surjections or one-way injections with monotone circuit implementations?

---

[1] A permutation is a length-preserving function that is both *injective* (i.e., one-to-one) and *surjective* (i.e., onto).

1.1 Our Results

Our first result establishes optimal bounds on the negation complexity of CRHs and TCRs.

Theorem 1 (Informal). *Collision-resistant hash-functions and Target Collision-resistant hash-functions require* $\theta(\log n)$ *negation gates.*

While we resolve the question for CRHs and TCRs, the problem remains open for Second Pre-Image Resistant functions. Interestingly, since TCRs and SPRs are equivalent,[2] any result on the negation complexity of SPR could reveal something about the negation complexity of universal hash-functions and, consequently, the XOR function.

Our second result explores whether injectivity or surjectivity influences the negation complexity of constructing one-way functions. We answer in the affirmative that, if we relax one of these conditions from a one-way permutation, it is indeed possible to construct monotone one-way functions. More precisely, we prove the following theorems.

Theorem 2 (Informal). *Assume the existence of one-way surjections. Then there exists a (poly-sized) one-way surjection that is computable by a monotone circuit.*

Theorem 3 (Informal). *Assume the existence of one-way injections. Then there exists a (poly-sized) one-way injection that is computable by a monotone circuit.*

We remark that if we start with a one-way permutation $f : \{0,1\}^n \rightarrow \{0,1\}^n$, then we can construct a one-way injection $g : \{0,1\}^n \rightarrow \{0,1\}^{3n}$ and a one-way surjection $h : \{0,1\}^n \rightarrow \{0,1\}^{n-2\log n}$ that are both monotone.

We also complement our injectivity result by showing that there do not exist one-way injections with constant stretch, where stretch refers to the difference in lengths of the output and input. More formally, we prove the following theorem:

Theorem 4 (Informal). *There exists an algorithm that, given oracle access to an injective monotone function* $f : \{0,1\}^n \rightarrow \{0,1\}^m$, *where* $m - n = O(1)$, *can invert* f *in polynomial time.*

This can be viewed as a generalization of the result of [9] where they give an algorithm only when $m = n$. We remark that not only can the algorithm invert an arbitrary element in the range of the function, it can also determine whether an element is in the range of the function.

In the work of Guo et al. [9], they show that error-correcting codes are highly non-monotone. Here we extend their result to list-decodable codes. This result follows techniques from [9] and is presented in the full version.

---

[2] A TCR can be constructed from an SPR by computing a universal hash-function (1-wise independent) on the input before feeding it to the SPR function, namely, masking the inputs with a random key.

1.2 Our Techniques

We remark that for most of our results we rely on different techniques. We believe that understanding the negation complexity of cryptography is a fundamental problem, and that our work sheds light on how different properties of functions influence the negation complexity of cryptographic primitives. We briefly mention some of our techniques below.

Collision-Resistant Hash-Functions: Establishing that collision-resistant hash-function requires negation gates follows using the pigeon-hole principle. Consider any monotone function with m-bit outputs: In any $m + 1$ chain of inputs $x^1, \ldots, x^{m+1}$ (i.e. $x^i \preceq x^{i+1}$)[3] which has strictly increasing Hamming weights there must exist a pair of consecutive inputs that collide. This is because each bit of the output can change at most once in the sequence and there are only m output bits. We can then conclude by using the fact that hash-functions are compressing, i.e. $m < n$ and a chain can be easily constructed by simply flipping input bits one at a time from 0 to 1 starting from 0^n. Following ideas of [9], using a theorem of Markov [12], we also extend this to show that if the collision-resistant function is highly-compressing (namely by polynomially many bits) then it must require $\log n$ negations.

One-Way Monotone Surjections: Our technique for constructing one-way monotone surjections follows the idea of Goldreich and Izsak [8] for constructing one-way monotone functions. Let $\mathsf{ham}(x)$ denote the Hamming weight of the string x. Recall that in their construction, starting from any one-way function $f : \{0,1\}^n \to \{0,1\}^m$, they consider a function f' that behaves as follows:

- On inputs x such that $\mathsf{ham}(x) < \frac{n}{2}$, f' assumes the value 0^m.
- On inputs x such that $\mathsf{ham}(x) = \frac{n}{2}$, f' assumes the value $f(x)$.
- On inputs x such that $\mathsf{ham}(x) > \frac{n}{2}$, f' assumes the value 1^m.

They then show that this function can be implemented using a monotone circuit. Since the middle slice occupies at least $\frac{1}{\sqrt{n}}$ fraction of the total inputs, f' is a weak one-way function. Then a strong monotone one-way function can be obtained via standard parallel repetition.

Suppose we start with a function f that is surjective and apply the construction specified above, the resulting function will no longer be surjective. Instead, we rely on a an error correcting code G with a certain property. More precisely, G maps 'balanced' (i.e., contains the same number of 0s as 1s) strings of length $n + O(\log n)$ to strings of length n surjectively. An example of one such code is the Knuth code [11]. Given such codes, we can augment Goldreich's construction to get a monotone surjective one-way function as follows: We consider a function f' that takes as input $n + O(\log n)$ inputs and first applies G to its input, followed by f on balanced inputs while the rest of the inputs are defined just as before. It follows from the definition that this function is surjective and weak one-way.

---

[3] We write $a \preceq b$, if for any i, i^{th} bit of a is 1 implies that the ith bit of b is 1.

One-Way Monotone Injections: The technique for our one-way monotone injection construction can be explained using the same construction as above. Note that, in this construction, inputs of Hamming weight smaller than $n/2$ all map to the same output. Similarly, inputs with Hamming greater than $n/2$ also map to the same value. The main idea here to achieve injection is to get rid of these collisions.

Towards this, we add two blocks of n-bits to the output where the purpose of the first block is to handle inputs of Hamming weight smaller than $n/2$ and the other for those with Hamming weight greater than $n/2$. The result then follows by showing that the following functions f_{lower} and f_{upper} can be built using monotone circuits:

- $f_{\text{lower}}(x) = x$ when $\text{ham}(x) < |x|/2$ and $f_{\text{lower}}(x) = 1^n$ otherwise.
- $f_{\text{upper}}(x) = x$ when $\text{ham}(x) > |x|/2$ and $f_{\text{upper}}(x) = 0^n$ otherwise.

On a high-level this idea follows the approach of [8] and we present the formal statement and proof in Sect. 5.

A Deterministic Algorithm to Invert Monotone Injection: One of our main technical contributions and novel approach is in proving our lower bound on injections. Here we give an explicit deterministic algorithm that allows inversion of any injections with constant stretch. In fact, we provide a general analysis where the run-time of the algorithm depends on the input length and stretch.

At a high-level, the idea is that given a target $b = f(s)$ of a function $f : \{0,1\}^n \rightarrow \{0,1\}^m$, we systematically break f down into a number of different cases, each of which is permutation-like, in the sense that most of the output bits are permutations of the input bits. More specifically, we reduce f to a piecewise function of restrictions of f (over domains where certain input bits are held constant) where fixing one additional bit of the input fixes exactly one bit of the output. At that point we use the correspondence between input and output bits to find pre-images for all but the stretch bits of y, which are brute-forced. We iterate through the cases until we find one that produces a preimage for y, or we exhaust all cases, in which case no preimage can exist.

The main trick of the proof is the reduction of f into a polynomial (assuming $m - n = O(1)$) number of cases. Essentially, this is done by searching for an input bit that causes more than one output bit to be fixed when set to 1 (or 0), and recursing into both the case where we set it to one and set it to zero. Then, by a combinatorial argument, we show that this will only recurse polynomially many times. In particular, this helps us characterize an important property of functions of super-contstant stretch (that is, $f : \{0,1\}^n \mapsto \{0,1\}^{n+\omega(1)}$) that are potentially hard to solve: they will very often exhibit the worst-case behavior where fixing an input bit to 0 fixes only one output bit to 0, but fixing it to 1 fixes multiple output bits to 1, or vice versa. Functions like this are thus ideal candidates for one-way monotone injections of small stretch.

2   Preliminaries

For some $x, y \in \{0,1\}^n$, we write $x \preceq y$ if $x_i \leq y_i$ for all $i \in [n]$. By definition, a Boolean function $f : \{0,1\}^n \to \{0,1\}$ is monotone iff $f(x) \leq f(y)$ whenever $x \preceq y$ and a function $g : \{0,1\}^n \to \{0,1\}^m$ is monotone iff every output bit of g is a monotone Boolean function. If x is a binary string, let $\mathsf{ham}(x)$ be the *Hamming weight* of x, i.e. the amount of 1s that appear in x. A chain $X = (x^1, \ldots, x^t)$ is a monotone sequence of strings over $\{0,1\}^n$, i.e., $x^i \leq x^{i+1}$ for every $i \in [1, t-1]$. Define an increasing chain as a chain for which all $x^j \neq x^k$ when $j \neq k$.

Let $f : \{0,1\}^n \to \{0,1\}$ be a Boolean function, and let $X = (x^1, x^2, \ldots, x^n)$ be a chain. Then, define $a(f, X)$ to be the largest set of indexes $\{0 \leq i_0, \ldots, i_m \leq n-1\}$ such that $f(x^{i_j}) \neq f(x^{i_j+1})$ for every $j \in [0, m-1]$. Furthermore, define $a(f) = max_X(a(f, X))$ where X is a chain to be the *alternating complexity* of f.

The following result was shown by Markov [12]:

Theorem 5. *Let $f : \{0,1\}^n \to \{0,1\}^m$ be a Boolean function computed by a circuit with at most t negations. Then $a(f) \in O(2^t)$.*

2.1   One-Way Functions

Definition 1. *Let $f : \{0,1\}^* \to \{0,1\}^*$ be a polynomial-time computable function. f is (strong) one-way if for every PPT machine A, there exists a negligible function $\nu(\cdot)$ such that*

$$\Pr[x \leftarrow \{0,1\}^n; y = f(x) : A(1^n, y) \in f^{-1}(f(x))] \leq \nu(n)$$

A function $f \{0,1\}^n \to \{0,1\}^m$ is said to be *injective* if for any $x, y \in \{0,1\}^n$, $x \neq y \implies f(x) \neq f(y)$. f is said to be *surjective* if for any $z \in \{0,1\}^m$, there is an $x \in \{0,1\}^n$ such that $f(x) = z$. f is said to be a weak one-way function if there exists a polynomial p such that for all adversaries A the probability with which it can invert the function is at most $\frac{1}{p(n)}$ for sufficiently large n where the probability is over a randomly chosen input x and the random coins of A.

Definition 2 (Exponentially-hard one-way functions). *Let $f : \{0,1\}^* \to \{0,1\}^*$ be a polynomial-time computable function. f is exponentially-hard one-way if there exists some c such that for every probabilistic adversary A that runs in $O(2^{cn})$ time, there exists a negligible function $\nu(\cdot)$ such that*

$$\Pr[x \leftarrow \{0,1\}^n; y = f(x) : A(1^n, y) \in f^{-1}(f(x))] \leq \nu(n)$$

Definition 3 (Regular one-way functions). *Let $f : \{0,1\}^* \to \{0,1\}^*$ be a one-way function. f is regular if there exists a function $\alpha : N \to N$ such that for every $n \in N$ and every $x \in \{0,1\}^n$ we have: $|f^{-1}(f(x))| = \alpha(n)$.*

We assume that the regularity $\alpha(\cdot)$ of a function f is not known (i.e. not polynomial-time computable). Without loss of generality, we assume the one-way function is length preserving i.e. $f(\{0,1\}^n) \subseteq \{0,1\}^n$.

2.2   Target Collision-Resistant Hash-Function Families

Definition 4. *Let $\mathcal{G} = \{g_k\}_{k \in \mathcal{K}}$ be a family of functions where each function g_k goes from $\{0,1\}^{n+\ell}$ to $\{0,1\}^n$. We say that $\mathcal{G}$ is a* Collision-Resistant Hash-Function Family *if (i) the functions g_k are efficiently computable and (ii) for every efficient adversary A, the probability that A succeeds in the following game is negligible in n:*

- *Choose $k \leftarrow \mathcal{K}$*
- *Let $x, x' \leftarrow A(1^n, k)$*
- *A succeeds if $x \neq x'$ and $g_k(x) = g_k(x')$*

Definition 5. *Let $\mathcal{G} = \{g_k\}_{k \in \mathcal{K}}$ be a family of functions where each function g_k goes from $\{0,1\}^{n+\ell}$ to $\{0,1\}^n$. We say that $\mathcal{G}$ is a* Universal One-Way Hash-Function Family *if (i) the functions g_k are efficiently computable and (ii) for every efficient adversary A, the probability that A succeeds in the following game is negligible in n:*

- *Let $(x, \sigma) \leftarrow A(1^n)$ where σ is some state information output by A*
- *Choose $k \leftarrow \mathcal{K}$*
- *Let $x' \leftarrow A(\sigma, k)$*
- *A succeeds if $x \neq x'$ and $g_k(x) = g_k(x')$*

Universal One-Way Hash-Function Families [14] as defined above enjoy the property of target collision-resistance. The related notion of *Second Preimage Resistance* follows the same security game with the exception that x and k are uniformly chosen and handed to the adversary instead of allowing the adversary to first choose x before seeing the key. It is well-known how to construct UOWHFs from second preimage resistant families.

2.3   Error-Correcting and Balanced Codes

Let $E : \{0,1\}^n \to \{0,1\}^m$ be a polynomial-time computable function. Given strings $y, y' \in \{0,1\}^m$, define $\Delta(y, y') = \frac{\sum_{i=1}^m y_i \neq y_i'}{m}$. We say E is γ-*error-correcting* if for any pair of distinct inputs $x, x' \in \{0,1\}^n$, $\Delta(E(x), E(x')) \geq \gamma$.

Definition 6. *Let $f : \{0,1\}^n \to \{0,1\}^m$ be a polynomial-time computable function. f is a* balanced code *if every element in the image of f is balanced, i.e. $\lceil \frac{m}{2} \rceil$ of its bits are 1 and $\lfloor \frac{m}{2} \rfloor$ of its bits are 0.*

In addition, we define $\mathsf{BAL}_n = \{x \in \{0,1\}^n | x \text{ is balanced}\}$, and O_n to be the strict order on BAL_n that is inherited from the normal lexicographic order on $\{0,1\}^n$. For example, if $n = 4$, $O_n = 0011, 0101, 0110, 1001, 1010, 1100$. Finally, if w is a binary string, let w^k represent w whose first k bits are flipped and the rest of the bits remain the same.

We are interested in constructing a surjective, poly-time computable function $G : \mathsf{BAL}_n \to \{0,1\}^{n-2\log n}$ (assuming n is a power of 2 for simplicity). This is known as a decoder for a balanced Knuth Code [11].

3   Negation Complexity of Collision-Resistance

In this section, we prove our results regarding the negation complexity of collision-resitant hash-functions. We start with a warm-up Lemma that shows that for any compressing (unkeyed) function that is monotone a collision can be found easily by uniform adversaries.

Lemma 1. *Let $g : \{0,1\}^n \to \{0,1\}^m$ be any monotone function such that $m < n$. Then there exists a uniform adversary A that can output $x, x' \in \{0,1\}^n$ such that $x \neq x'$ and $g(x) = g(x')$.*

Proof. On a high-level the proof will demonstrate that every increasing chain has a collision. Consider an algorithm A that does the following:

- Choose an arbitrary increasing chain $X = (x^1, \ldots, x^{n+1})$ over $\{0,1\}^n$ and compute $g(x^i)$ for all i. Suppose there exists an i such that $g(x^i) = g(x^{i+1})$. If so, output (x^i, x^{i+1}). Otherwise output fail.

For all pairs $g(x^i)$ and $g(x^{i+1})$, the montonicity of g and the definition of an increasing chain implies that, either there exists at least one position in the bits of $g(x^i)$ that was a 0 turns to a 1 in $g(x^{i+1})$, or $g(x^i) = g(x^{i+1})$. There are n possible values for i and a maximum of m possible positions that can be flipped from 0 to 1. Furthermore, monotonicity implies that each position can flip at most once. Since $n > m$, by the pigeonhole principle, there must exist at least one such pair that has no bit that flips and thus, for this pair, $g(x^i) = g(x^{i+1})$ and $x^i \neq x^{i+1}$.

Corollary 1. *There does not exist a family of collision-resistant hash-functions that is computable by monotone circuits.*

Next, we extend the proof to rule out monotone implementations of target collision-resistant functions. Recall that in the security game for target collision-resistance, the adversary needs to pick a target x for which it is required to find a collision before it receives the description of the hash-function.

Theorem 6. *There does not exist a family of target collision-resistant hash-functions that is computable by monotone circuits.*

Assume, for contradiction, that there exists a family of target collision-resistant hash-functions $\{g_k\}_{k \in \mathcal{K}}$ where g_k is a function from $\{0,1\}^n$ to $\{0,1\}^m$ with $m < n$ that can be computed using a monotone circuit of polynomial size. From Lemma 1, we know that for any compressing function, and therefore for any g_k, every increasing chain of inputs has at least one collision.

Pick an arbitrary increasing chain $X = (x^1, \ldots, x^{n+1})$. Since every function g_k has a collision in this chain, using a standard averaging argument we can conclude that there exists at least one index in $[n]$, say i^*, such that with probability at least $\frac{1}{n}$ over the functions $k \in \mathcal{K}$, $g_k(x^{i^*}) = g_k(x^{i^*+1})$. Consider an adversary A that picks an input uniformly from the chain X and submits it as a target. Upon receiving the function g_k, it outputs (x^i, x^{i+1}) if $g_k(x^i) = g_k(x^{i+1})$. Otherwise, it outputs fail.

It follows that A picks x^{i^*} with probability at least $\frac{1}{n}$ and when this happens, it succeeds in outputting a collision with probability at least $\frac{1}{n}$. Therefore, A succeeds in outputting a collision with non-negligible probability, which is a contradiction.

Next, we strengthen our bound by proving that both CRHs and TCRs are highly non-monotone.

Lemma 2. *Let* $g : \{0,1\}^n \to \{0,1\}^m$ *be an arbitrary function that can be implemented by a circuit with* $t = (1 - \epsilon)\log n$ *gates. Then there exists a constant* $c > 0$ *such that, if* $m < cn^\epsilon$, *then there exists an adversary* A *that the can output* $x, x' \in \{0,1\}^n$ *such that* $x \neq x'$ *and* $g(x) = g(x')$.

Proof. We will rely on Markov's theorem (cf. Theorem 5) that shows that for any boolean function f on n-bit inputs that can be computed by a circuit with at most t gates, $a(f) = O(2^t)$. Recall that $a(f) = max_\mathcal{X} a(f, \mathcal{X})$ where $\mathcal{X}$ is any non-decreasing chain over $\{0,1\}^n$ and $a(f, \mathcal{X})$ denotes the number of times the output of f changes when iterating through the chain.

We prove that any increasing chain will have consecutive elements whose outputs are identical under g. Then we can rely precisely on the same adversary as in Lemma 1 that examines the elements in an arbitrary increasing chain and outputs if it finds a collision on any consecutive elements in the chain.

Let $X = (x^1, \ldots, x^{n+1})$ be an arbitrary increasing chain over $\{0,1\}^n$. As before, we have that every pair of consecutive elements either collide under g or differ by at least one position in the output. By Markov's theorem we know that each position in the output can change at most $c'2^t = c'n^{1-\epsilon}$ times. Hence if $n > mc'n^{1-\epsilon}$ then there must be a collision by the pigeonhole principle. By fixing $c = \frac{1}{c'}$ and observing that $m < \frac{1}{c'}n^\epsilon$, the lemma follows.

Using the same techniques as before we obtain the following corollary.

Corollary 2. *Let* $\{g_k\}_{k \in \mathcal{K}}$ *be a family of collision-resistant hash-functions or target collision-resistant functions where* g_k *is a function from* n *bits to* m *bits such that* $m < cn^\epsilon$ *and computable by a circuit with* t *gates. Then* $t \geq (1-\epsilon)\log n$.

4   One-Way Monotone Surjections

In this section, we provide our construction of one-way monotone surjections. First we require the following definitions:

Definition 7. *If* $f : \{0,1\}^n \mapsto \{0,1\}^m$ *is a function, let the* $k - cut$ *of* f *be the function* $f^{(k)} : \{0,1\}^n \mapsto \{0,1\}^m$ *such that*

$$f^{(k)}(x) = \begin{cases} 0^m & : \text{ham}(x) < k \\ f(x) & : \text{ham}(x) = k \\ 1^m & : \text{ham}(x) > k \end{cases}$$

We remark that in literature this sometimes referred to as k-slice.

Definition 8. *If x is a binary string and k is an integer, the* threshold function *of threshold* k *is the function T_k such that*

$$T_k(x) = \begin{cases} 0 : ham(x) < k \\ 1 : ham(x) \geq k \end{cases}$$

Definition 9. *If x is a binary string, let $x^{i \to 0}$ stand for the binary string produced by changing the ith bit in x to 0.*

Proposition 1. *If x is a binary string and $ham(x) = k$, then $T_k(x^{i \to 0}) = \neg x_i$.*

Proof. If $x_i = 1$, then $x^{i \to 0}$ has Hamming weight $k - 1$. Thus, $T_k(x^{i \to 0}) = 0$. If $x_i = 0$, changing the ith bit to 0 does not change x, and so it does not change the Hamming weight. Thus, $T_k(x^{i \to 0}) = 1$.

We will rely on the following proposition (from [8]) that states that $f^{(k)}$ is computable by a monotone circuits.

Proposition 2. *If $f : \{0,1\}^n \mapsto \{0,1\}^m$ is a function computable in polynomial time, then for any k, the $k - cut$ of f, $f^{(k)}$, is computable by a polynomial-sized monotone circuit.*

Theorem 7. *If one-way permutations exist, then one-way surjections exist with $O(\log n)$ compression which can be computed by a polynomial-sized monotone circuit.*

Proof. In this proof, we follow the convention that, if w is a binary string, then w^k will represent the string obtained by flipping the first k bits of w and keeping the rest of the bits remain the same. Under this notation, we have that $w^{(k)(k)} = w$.

The following function $G : \mathsf{BAL}_n \to \{0,1\}^{n-2\log n}$ will be useful in proving this theorem:

```
1: function G(s):
2:     u ← first 2 log n bits of s
3:     if u has Hamming weight of log n then
4:         k ← position of u in O_{2 log n} for BAL_{2 log n}
5:         w ← last n − 2 log n bits of s
6:         return w^k
7:     else
8:         return 0^{n−2 log n}
```

Claim. $G : \mathsf{BAL}_n \to \{0,1\}^{n-2\log n}$ is surjective and computable in polynomial time.

Proof. First, we prove that G is surjective. Towards proving this, we prove the following Claim.

Claim. Suppose $w \in \{0,1\}^{n-2\log n}$. Then, there exists a $k \in \{0,\ldots,n-2\log n\}$ such that w^k has Hamming weight $\frac{|w|}{2}$.

Proof. To see this, let $\mathrm{ham}(w) = l$. Observe that w^0 has Hamming weight l, and $w^{|w|}$ has Hamming weight $|w| - l$. Furthermore, $w^0, w^1, w^2, \ldots$ is a sequence of bit strings such that $\mathrm{ham}(w^i) = \mathrm{ham}(w^{i+1}) \pm 1$. Thus, the Hamming weights of $w^0, w^1, w^2, \ldots, w^{|w|}$ hit every natural number (inclusively) in between $\mathrm{ham}(w^0) = l$ and $\mathrm{ham}(w^{|w|}) = |w| - l$. In this set of numbers lies $\frac{|w|}{2}$. This is because, if $l > \frac{|w|}{2}$, $|w| - l < |w| - \frac{|w|}{2} = \frac{|w|}{2}$, and a similar argument goes for when $l < \frac{|w|}{2}$, and is trivial when $l = \frac{|w|}{2}$. Thus, there is a k such that $\mathrm{ham}(w^k) = \frac{|w|}{2}$.

Now, to prove that G is surjective, it suffices to show the following:

Claim. If u is the kth string in $\mathsf{BAL}_{2\log n}$ according to $O_{2\log n}$, $G(uw^k) = w$.

Proof. First, we show that the kth string in $\mathsf{BAL}_{2\log n}$ indeed exists. It suffices to show that $|\mathsf{BAL}_{2\log n}| \geq |w|$, as we are never required to use $k = |w|$, as if $w^{|w|}$ has even Hamming weight, so does w^0. Thus we only consider $k \in [0, \ldots, |w|-1]$. First,

$$|\mathsf{BAL}_{2\log n}| = \binom{2\log n}{\log n} \geq \frac{2^{2\log n - 1}}{\sqrt{\log n}}$$

by Stirling's approximation. Then, we have

$$\frac{2^{2\log n - 1}}{\sqrt{\log n}} = \frac{n^2}{2\sqrt{\log n}} \geq \frac{n^2}{2\log n}$$

Finally,

$$\frac{n^2}{2\log n} \geq n - 2\log n \iff \frac{\frac{n^2}{2} - n\log n + 2\log^2 n}{\log n} \geq 0$$

And for sufficiently large n, the numerator on the left hand side is greater than 1, and $\frac{1}{\log n} \geq 0$ for $n > 1$.

And so, for sufficiently large n,

$$|\mathsf{BAL}_{2\log n}| \geq |w|$$

Claim. $uw^k \in \mathsf{BAL}_n$.

Proof. $\mathrm{ham}(u) = \log n$ since $2\log n$ is even, and thus

$$\mathrm{ham}(uw^k) = \mathrm{ham}(u) + \mathrm{ham}(w^k) = \frac{2\log n}{2} + \frac{|w^k|}{2} = \frac{2\log n + |w^k|}{2} = \frac{n}{2}$$

Then, upon inspection of G, we see that since u has even Hamming weight, G will compute k, the position of u in $O_{2\log n}$. This is the same k as in w^k, by definition of u. Then, G will return $w^{(k)(k)} = w$.

Let G be the function based on Knuth's balanced code [11] described in Sect. 2.3. Recall that G is computable in polynomial time and surjective. Using this function G, we will now construct the following function, assuming one-way permutations exist. Let $f : \{0,1\}^{n-2\log n} \mapsto \{0,1\}^{n-2\log n}$ be a one-way permutation. Let $f' : \{0,1\}^n \mapsto \{0,1\}^{n-2\log n}$ be the function $(f \circ G)^{(\frac{n}{2})}$.

Claim. f' is monotone and surjective.

Proof. Given any $x \in \{0,1\}^{n-2\log n}$, there is a $y \in \{0,1\}^{n-2\log n}$ such that $f(y) = x$ (since f is surjective). Then, since G is surjective, there is a $z \in \mathsf{BAL}_n$ such that $G(z) = y$. Thus, $f(G(z)) = x$. Then, since $z \in \mathsf{BAL}_n$, $\mathrm{ham}(z) = \frac{n}{2}$. And so $f'(z) = f(G(z)) = x$. Furthermore, f' is known as the $\frac{n}{2}$-cut of $f \circ G$, and we have already proven that any k-cut of a polynomial-time algorithm has a polynomial-size monotone circuit that computes it.

It suffices to show that f' is weak one-way, as by a simple extension we can construct a strong one-way function from f'. Again, following [8], we can show that f' is weak.

Lemma 3. *f' is a weak one-way function.*

Assume that adversary D' inverts $f'(U_n)$ with probability at least $1 - \frac{1}{n^2} + \frac{1}{n^3}$. Now, construct adversary D that, on input $y \in f'(U_n)$, returns $G(D'(y))$.

It follows from Claims 4 and 4the for every string $w \in 0, 1^{n-2\log n}$ there exists at least one value for $k <= n - 2\log n$ such that $G(uw^k) = w$ where u is the kth string in some ordering of strings in $\mathsf{BAL}_{2\log n}$. However, there could be more than one input for which G outputs w. We will pick exactly one preimage w.r.t. G for every string $w \in 0, 1^{n-2\log n}$ and form a set S. The weak one-wayness of f can be concluded as follows:

1. G is injective and surjective from S to $0, 1^{n-2\log n}$.
2. For any string y that is not the all 0s or all 1s string, if D' inverts y successfully under f' as x', then $G(x')$ gives a preimage of y under f. That is, D inverts y wrt f.
3. Let $A(n)$ be the probability that x' sampled from $0, 1^n$ is in S. Since S occupies $1/n^2$ fraction of $0, 1^n$, it holds that $A(n) = 1/n^2$.

$$\Pr[x \leftarrow \{0,1\}^{n-2\log n} : D(f(x)) \in f^{-1}(f(x))]$$
$$= \Pr[x \leftarrow \{0,1\}^{n-2\log n} : D'(f(x)) \in f'^{(-1)} \text{ (Using 2)}$$
$$= \Pr[x' \leftarrow \{0,1\}^n : D'(f'(x')) \in f'^{-1}(f'(x'))|x \in S] \text{ (Using 1)}$$
$$= \frac{1}{\Pr[x' \in S]} \Pr[x' \leftarrow \{0,1\}^n : D'(f'(x')) \in f'^{-1}(f'(x')) \wedge x \in S]$$
$$= \frac{1}{\Pr[x' \in S]} (\Pr[x' \leftarrow \{0,1\}^n : D'(f'(x')) \in f'^{-1}(f'(x'))]$$
$$\qquad - \Pr[x' \leftarrow \{0,1\}^n : D'(f'(x')) \in f'^{-1}(f'(x')) \wedge x \notin S])$$
$$>= n^2(1 - A(n) + \frac{1}{n^3} - (1 - A(n)) \text{ (Using 3)}$$
$$= \frac{1}{n}$$

which is non-negligible and a contradiction to the fact that f is (strong) one-way.

Therefore, f' is weak. We conclude the proof of Theorem 7 by observing that the standard amplification of weak one-way functions to strong one-way functions via repetition preserves both the monotonicity and surjectivity.

Corollary 3. *If a one-way surjection* $f_S : \{0,1\}^{n-2\log n} \to \{0,1\}^m$ *exists, then a one-way monotone surjection* $g_S : \{0,1\}^n \to \{0,1\}^m$ *exists.*

Proof. By replacing the one-way permutation used in the preceding proof with f_S, and using the fact that the injectivity of the permutation is not used in the proof, this corollary follows.

Corollary 4. *If strong exponentially hard one-way permutations exist, weak one-way monotone surjections exist from n bits to $n - O(\log\log(n))$ bits.*

Proof. Suppose some f_n is a sequence of strong exponentially hard one-way permutations. Let $d \in \omega(\log n)$. Now, via a compexity leveraging argument, we see that there exists a strong one-way permutation g_n on d-bits that is strong one-way with respect to polynomial-time adversaries. By the previous theorem, this means that there exists a monotone weak one-way surjection from d bits to $d - \log d$ bits. Append $n - d$ input and output bits (each input mapping directly to the corresponding output bit, so as to maintain surjectivity) to get a monotone weak one-way surjection from n bits to $n - \log d$ which will be $n - O(\log\log(n))$ as desired by setting d appropriately.

5   One-Way Monotone Injections

Theorem 8. *If strong one-way permutations exist, then one-way monotone injections exist. In particular, if strong one-way permutations exist, then strong one-way monotone injections exist from n bits to $3n$ bits.*

Proof. Suppose there exists some one-way family of permutations $f_n : \{0,1\}^n \mapsto \{0,1\}^n$ that have a polynomial bound $Q(n)$ on their circuit size. Then, let $T_{i,n} : \{0,1\}^n \mapsto \{0,1\}$ be defined as the function such that $T_i(x)$ is true if and only if x has at least i of its n bits set to 1. First, we show that T_i can be calculated via a polynomial-sized circuit through the following recursive construction: let $T_0(x) = 1$ for all x, and for $i > 0$, let $T_i = \bigvee_{j=1}^{n}(x_j \wedge T_{i-1}(x \text{ with bit } j \text{ set to zero}))$. Recall that if f_n has circuit size bounded by $Q(n)$, then by De Morgan's laws, we can still construct a monotone circuit $f'_n : \{0,1\}^n \times \{0,1\}^n \mapsto \{0,1\}^n$ with circuit size $\leq Q(n)$ such that $f'_n(x, \neg x) = f_n(x)$ for all x. If we set $H_{i,n}(x) = T_{\lfloor \frac{n}{2} \rfloor, n}(x \text{ with bit } i \text{ set to } 0)$, then note that for x with Hamming weight $\lfloor \frac{n}{2} \rfloor$, $f'_n(x, H_{1,n}(x) \| H_{2,n}(x) \| \cdots \| H_{n,n}(x)) = f_n(x)$. Then, defining $F_n(x) = f'_n(x, H_{1,n}(x) \| H_{2,n}(x) \| \cdots \| H_{n,n}(x))$, consider the following constructed family:

$$g_n(x) = F_n(x) \| \left(x \wedge [T_{\lfloor \frac{n}{2} \rfloor + 1, n}(x)]^n \right) \| \left(x \vee [T_{\lfloor \frac{n}{2} \rfloor, n}(x)]^n \right)$$

Note that the concatenation of the H functions has a polynomial bound in n. Thus, we can construct g_n in a polynomial-size monotone circuit.

Claim. g_n is injective.

Proof. For inputs x with Hamming weight less than $\lfloor \frac{n}{2} \rfloor$, the last n bits of g_n equals x, and so within this range, there cannot be any collisions. A similar proof shows that there are no collisions among inputs with Hamming weight greater than $\lfloor \frac{n}{2} \rfloor$.

In addition, there are no collisions between inputs of Hamming weight $\overline{\lfloor \frac{n}{2} \rfloor}$, as F_n is injective.

There are also no collisions between inputs of Hamming weight less than $\lfloor \frac{n}{2} \rfloor$ and those with Hamming weight greater than $\lfloor \frac{n}{2} \rfloor$, because the third section of the latter consists of all 1s, and the third section of the former cannot have Hamming weight greater than or equal to $\lfloor \frac{n}{2} \rfloor$.

Finally, there are no collisions between inputs of Hamming weight less than $\lfloor \frac{n}{2} \rfloor$ and Hamming weight equal to $\lfloor \frac{n}{2} \rfloor$, as the third section of the former cannot have Hamming weight greater than or equal to $\lfloor \frac{n}{2} \rfloor$, and the third section of the latter consists of all 1s. A similar proof shows that there are no collisions between inputs with Hamming weight equal to $\lfloor \frac{n}{2} \rfloor$ and inputs with Hamming weight greater than $\lfloor \frac{n}{2} \rfloor$.

Now, recall that there are $\Omega(1/\sqrt{n})$ fraction of inputs with Hamming weight $\lfloor \frac{n}{2} \rfloor$. Then, since the first n bits of $g_n = f_n$ for inputs with Hamming weight $\lfloor \frac{n}{2} \rfloor$, the first n bits of $g_n = f_n$ for $\Omega(1/\sqrt{n})$ fraction of inputs. Therefore, g_n is $\Omega(1/\sqrt{n})$ hard. g_n can then be amplified by a simple extension to obtain strong one-way injections.

Corollary 5. *If a one-way injection $f_I : \{0,1\}^n \to \{0,1\}^m$ exists, then a one-way monotone injection $g_I : \{0,1\}^n \to \{0,1\}^{m+2n}$ exists.*

Proof. Since the preceding proof does not require that the one-way function be surjective, the proof is equivalent. Furthermore, the output of the constructed function is m bits concatenated by two sequences of n bits each.

As a simple corollary, we can construct regular length-preserving functions:

Corollary 6. *If one-way injections exist, regular length-preserving one-way monotone functions exist.*

Proof. By the preceding theorem, we can construct a family of monotone weak one-way functions. $g_n : \{0,1\}^n \mapsto \{0,1\}^{3n}$. Then, construct a family of functions $h_n : \{0,1\}^{3n} \mapsto \{0,1\}^{3n}$ such that

$$h_n(x) = g_n(\text{last n bits of x})$$

Then, h_n is 2^{2n}-regular, as for any element y in the image of h_n, there is exactly one element $x' \in \{0,1\}^n$ such that $g(x') = y$. Then, there are exactly 2^{2n} elements in the domain of h_n whose last n bits are equal to x', thus each of these maps to y. Now, if there was an adversary D and a polynomial q such that h_n can be inverted with probability $1 - A(n) + \frac{1}{q(n)}$, an adversary

$$D'(y) = \text{last } n \text{ bits of } D(y)$$

would invert g_n with probability $1 - A(n) + \frac{1}{q(n)}$ as well, thus contradicting the fact that g_n is weak one-way. Thus, h_n is weak one-way, and can be extended to a strong one-way function.

Using a complexity leveraging argument, we obtain the following corollary assuming the existence of exponentially hard one-way functions.

Corollary 7. *If strong exponentially hard one-way permutations exist, strong one-way monotone injections exist from n bits to $n + \omega(\log(n))$ bits.*

We need the following lemma to prove our corollary. We will prove the corollary using the following lemma and then prove the lemma.

Lemma 4. *If $f_n : \{0,1\}^n \mapsto \{0,1\}^{m(n)}$ is sequence of strong exponentially hard one-way functions, then for any function $d \in \omega(\log(n))$, the sequence of functions $g_n : \{0,1\}^{d(n)} \mapsto \{0,1\}^{m(d(n))}$ defined as $g_n(x) = f_{d(n)}(x)$ is strong one-way with respect to poly(n).*

Proof. Suppose some f_n is a sequence of strong exponentially hard one-way permutations. Let $d(n) \in \omega(\log(n))$. Now, by the lemma, there exists a strong one-way sequence of permutations g_n on $d(n)/2$ bits that is strong one-way with respect to poly(n). By the previous theorem, this means there exists a strong one-way monotone injection from $d(n)/2$ bits to $1.5*d(n)$ bits. Append $n-d(n)/2$ input and output bits (each input mapping directly to the corresponding output bit, so as to maintain injectivity) to get a strong one-way monotone injection from n bits to $n + d(n)$ bits, as desired.

Proof of Lemma 4. Suppose that given some f_n exponentially hard (specifically, $O(2^{cn})$ hard) and some $d \in \omega(\log(n))$, there exists some probabilistic adversary A that can invert g_n with non-negligible probability in (assuming n is sufficiently large) poly(n) time. However, this means that A can invert f_n in $Bn^p = B2^{\log(n)p}$ time. Since f_n is exponentially hard, this must be greater than $k2^{cd(n)}$. Since $d \in \omega(\log(n))$, this will be false for large enough n, a contradiction. By contradiction, g_n is a strong one-way function with respect to poly(n).

6 Negation Complexity of Some One-Way Injections

One of our main technical contributions is presented in this section. We show that an arbitrary injective one-way function f from n bits to m bits that is computable by monotone circuits can be inverted in deterministic time $O(poly(n,m)\binom{m}{m-n}2^{m-n})$.

We say a function f is *perfectly invertible* in P if there is a deterministic polynomial-time adversary that, with only oracle access to any f and given any $x \in$ codomain(f), outputs the preimage $f^{-1}(x)$ or indicates that x is not in the range of f.

Theorem 9. *There exists an inverter I that, for any $k > 0$, perfectly inverts in P the monotone injections $\{0,1\}^n \mapsto \{0,1\}^m$ in deterministic time $O(poly(n,m)\binom{m}{m-n}2^{m-n})$, where $m = n + k$.*

In order to prove the theorem, we will require the following proposition that follows immediately from an injectivity argument:

Proposition 3. *Given any monotone injection, and any input bit of that function, restricting an input bit to zero (or one) will necessarily restrict at least one output bit to zero (or one). In other words, for any monotone injection $f : \{0,1\}^n \mapsto \{0,1\}^m$, where $f(b_1 b_2 \cdots b_n) = b'_1 b'_2 \cdots b'_m$, then for all $1 \leq t \leq n$, there exists a t_1 such that $b'_{t_1} = 1$ whenever b_t is 1, and a t_0 such that $b'_{t_0} = 0$ whenever b_t is 0.*

Proof. The inverter I consists of two main parts, the reduce method, and the solve method. The adversary begins at the reduce method, and calls the reduce method recursively, until it reaches the base case, at which point the solve method is run.

The reduce method operates on strings in $\{0,1,\star\}^*$, which should be interpreted as follows: the $\star$ bits are considered "free" and may vary throughout the scope of the variables, whereas the 0 and 1 bits are fixed, and will not change. For convenience, we define a couple functions on $\{0,1,\star\}^*$: let $z(y)$ denote the string obtained from y where all positions with a $\star$ in y are replaced with zero, and similarly let $w(y)$ denote the string where all $\star$ bits in y are replaced with ones.

The inverter I operates as follows:

Let x be the input, which has m bits. I calls reduce$(x, \star^n, \star^m)$, beginning a recursive procedure. The procedure reduce$(x, y, \star^m)$ takes three inputs: the target $x \in \{0,1\}^m$ (which is unused, except insofar as it is passed onto the final solve method), the input bit configuration $y \in \{0,1,\star\}^n$ and the output bit configuration, $p \in \{0,1,\star\}^m$. The fixed bits in y should be interpreted as bits which are assumed to be consistent with the preimage of x, and the fixed bits in p should be interpreted as bits that are constant over all possible settings of the free input bits. We will maintain as a recursive invariant in calls to reduce that the difference between the number of free bits in p and the free bits in y is at most $m - n$.

The reduce method is as follows:

- Define f' as the restriction of f such that the domain is the subset of $\{0,1\}^n$ that only disagree with y on its free bits, and the range is the subset of $\{0,1\}^m$ that only disagree with p on its free bits. Note that f' is a monotone injection. For each free bit in y, indexed by t, let $a_t \in \{0,1\}^n$ be the string that has only position t set to one, $0^{t-1}10^{n-t}$, and let b_t be the complement of a_t. Let $r_t = (a_t \vee z(y))$ and $s_t = (b_t \wedge w(y))$ where $\vee$ and $\wedge$ are computed bitwise. Now, consider ham$(f'(r_t))$ and $l - ham(f'(s_t))$. By Proposition 3, both quantities are ≥ 1. If both are $= 1$ for all t, then as a base case, return the result of solve(x, y, p). Otherwise, in the case that one of the quantities is ≥ 2 for some t, branch into two recursive calls: reduce$(x, a_t \vee y, r_t \vee p)$ and reduce$(x, b_t \wedge y, s_t \wedge p)$, where the bitwise $\vee$ and $\wedge$ operations are extended in the natural way, so that $0 \vee \star = \star$, $1 \vee \star = 1$, $\star \vee \star = \star$, $0 \wedge \star = 0$, $1 \wedge \star = \star$, and $\star \wedge \star = \star$. If either call returns a preimage for x, return that preimage; otherwise, if both branches indicate no solution, return that there is no preimage.

Observe that when the solve method is called from the reduce method, it holds that all of the free bits in the input y, when restricted to 0, set exactly one free bit in the output p to 0, and when restricted to 1, set exactly one free bit in the output p to 1 (since, otherwise we recurse further). Let π, τ be partial functions from free bits in y to the free bits in p such that on input a position in the input indicate which output bit is set to zero and one, respectively. Now, by injectivity, we know that π and τ must themselves be injective, because otherwise, setting two different free input bits would result in the same output.

The solve method is as follows:

- Suppose there are k free input bits. First, discover the π and τ permutations by simply testing each free input bit, and seeing which free output bit corresponds to that input. Consider the images of π and τ, P and T. We know the number of free output bits is limited, so $|P \cup T| \le k + (m - n)$, and $k = |P| = |T|$. So, $m + k - n \ge |P \cup T| = |P| + |T| - |P \cap T| = 2k - |P \cap T|$, meaning that $k - |P \cap T| \le (m - n)$. For each bit $b \in P \cap T$, examine the corresponding bit in x; if x has a zero at b, it must be the case that a preimage of x has $T^{-1}(b)$ set to zero, and if x has a one at b, a preimage of x has $P^{-1}(b)$ set to one. Thus, $|P \cap T|$ of the free bits can be deduced, leaving at most $m - n$ remaining free bits. Perform a brute-force search over all possible configurations of these bits, returning a preimage of x if one is found, or indicating no such preimage exists.

It is clear that the solve method runs in $O(m2^{m-n})$ steps. Next, we count the maximum number of times reduce is called. Let $T(i, j)$ be the number of times reduce is called, where i is the number of free input bits and j is the number of free output bits minus the number of free input bits. Note that in the worst case, $T(i, j) = 1 + T(i-1, j) + T(i-1, j-1)$, since in at least one of the paths, more than one outbit bit is set by restricting an input bit. The recursion will necessarily stop as soon as either $i = 0$ or $j = 0$. Equivalently this quantity is bounded by the number of non-backtracking paths between corners on a $n \times (m - n)$ rectangular grid, which is $\binom{m}{m-n}$. Each call takes only $O(nm)$ time, and so the total running time for the whole method is $O\left(nm^2 \binom{m}{m-n} 2^{m-n}\right)$, which for constant $m - n$ is polynomial in n.

7 Some Corollaries to Our Injection Lower Bound

We present some interesting (and immediate) corollaries to this theorem in this section and provide the proofs in the full version.

Corollary 8. *For any constant c, for any OWI with constant stretch, for all inputs of Hamming weight less than c [assuming n is sufficiently large], at least one negation gate must not trigger, and for all inputs of Hamming weight more than $m - c$, at least one negation gate must trigger.*

Corollary 9. *Any monotone function f that has a constant number of collisions (in other words, $|\{(x,y)|x \neq y, f(x) = f(y)\}|$ is constant) is perfectly invertible in P.*

Corollary 10. *Any monotone injection from $\{0,1\}^n \mapsto \{0,1\}^{n+O(\log n)}$ can be inverted in time $n^{O(\log n)}$*

Recall that Corollary 7 shows that that assuming existence exponential-time hard one-way permutations we can get a monotone injection with $\omega(\log n)$ stretch secure against polynomial-time adversaries. While still a small gap remains, this gives an indication that we cannot significantly improve our results.

Corollary 11. *There is an inverter that, for any constants k, g, perfectly inverts in P any monotone injection $f : \{0,1\}^n \mapsto \{0,1\}^{n+k}$ with g negation gates at the bottom (meaning the negation gates are over only input bits).*

Corollary 12. *OWIs (one-way injections) with constant stretch and negation gates only at the bottom must have a super-constant number of negation gates.*

Proof. This is the contrapositive of the previous corollary.

Corollary 13. *OWIs with constant stretch must either have a super-constant number of negation gates, or a negation gate above a subcircuit of superconstant size.*

Corollary 14. *OWIs in NC_0 with constant width require a super-constant number of negation gates.*

References

1. Amano, K., Maruoka, A.: A superpolynomial lower bound for a circuit computing the clique function with at most (1/6) log log n negation gates. SIAM J. Comput. **35**(1), 201–216 (2005)
2. Beals, R., Nishino, T., Tanaka, K.: More on the complexity of negation-limited circuits. In: Proceedings of the Twenty-Seventh Annual ACM Symposium on Theory of Computing, Las Vegas, Nevada, USA, 29 May–1 June 1995, pp. 585–595 (1995)
3. Beals, R., Nishino, T., Tanaka, K.: On the complexity of negation-limited Boolean networks. SIAM J. Comput. **27**(5), 1334–1347 (1998)
4. Blais, E., Canonne, C.L., Oliveira, I.C., Servedio, R.A., Tan, L.: Learning circuits with few negations. CoRR abs/1410.8420 (2014)
5. Blum, A., Burch, C., Langford, J.: On learning monotone Boolean functions. In: 39th Annual Symposium on Foundations of Computer Science, FOCS 1998, Palo Alto, California, USA, 8–11 November 1998, pp. 408–415 (1998)
6. Buresh-Oppenheim, J., Kabanets, V., Santhanam, R.: Uniform hardness amplification in NP via monotone codes. Electron. Colloquium Comput. Complex. (ECCC) **13**(154) (2006)
7. Fischer, M.J.: The complexity of negation-limited networks - a brief survey. In: Brakhage, H. (ed.) GI-Fachtagung 1975. LNCS, vol. 33, pp. 71–82. Springer, Heidelberg (1975). doi:10.1007/3-540-07407-4_9

8. Goldreich, O., Izsak, R.: Monotone circuits: one-way functions versus pseudorandom generators. Theory Comput. **8**(1), 231–238 (2012)
9. Guo, S., Malkin, T., Oliveira, I.C., Rosen, A.: The power of negations in cryptography. In: Dodis, Y., Nielsen, J.B. (eds.) TCC 2015. LNCS, vol. 9014, pp. 36–65. Springer, Heidelberg (2015). doi:10.1007/978-3-662-46494-6_3
10. Karchmer, M., Wigderson, A.: Monotone circuits for connectivity require superlogarithmic depth. In: Proceedings of the 20th Annual ACM Symposium on Theory of Computing, Chicago, Illinois, USA, 2–4 May 1988, pp. 539–550 (1988)
11. Knuth, D.E.: Efficient balanced codes. IEEE Trans. Inf. Theory **32**(1), 51–53 (1986)
12. Markov, A.A.: On the inversion complexity of a system of functions. J. ACM **5**(4), 331–334 (1958)
13. Morizumi, H.: Limiting negations in non-deterministic circuits. Theor. Comput. Sci. **410**(38–40), 3988–3994 (2009)
14. Naor, M., Yung, M.: Universal one-way hash functions and their cryptographic applications. In: Proceedings of the 21st Annual ACM Symposium on Theory of Computing, Seattle, Washigton, USA, 14–17 May 1989, pp. 33–43 (1989)
15. Santha, M., Wilson, C.B.: Limiting negations in constant depth circuits. SIAM J. Comput. **22**(2), 294–302 (1993)
16. Sung, S.C., Tanaka, K.: Limiting negations in bounded-depth circuits: an extension of Markov's theorem. Inf. Process. Lett. **90**(1), 15–20 (2004)
17. Tardos, É.: The gap between monotone and non-monotone circuit complexity is exponential. Combinatorica **8**(1), 141–142 (1988)

Implicit Quadratic Property of Differentially 4-Uniform Permutations

Theo Fanuela Prabowo[✉] and Chik How Tan

Temasek Laboratories, National University of Singapore,
5A Engineering Drive 1, #09-02, Singapore 117411, Singapore
{tsltfp,tsltch}@nus.edu.sg

Abstract. Substitution box (S-box) is an important component of block ciphers for providing nonlinearity. It is often constructed from differentially 4-uniform permutation. In this paper, we examine all (to the best of our knowledge) the differentially 4-uniform permutations that are known in the literature and determine whether they are implicitly quadratic. We found that all of them are implicitly quadratic, making them vulnerable to algebraic attack [10,12–14]. This leads to an open question of whether there exists a differentially 4-uniform permutation over $\mathbb{F}_{2^n}$ that is not implicitly quadratic. We provide a partial answer to this question by solving it for the special cases of $n = 11$ and $n = 13$.

Keywords: Cryptography · S-box · Implicitly quadratic functions · Implicit quadratic property · Differentially 4-uniform permutations · Algebraic attack

1 Introduction

Substitution box (S-box) is an important component of block ciphers as it is often the only nonlinear component of a block cipher. Thus, choosing a good S-box is very important to ensure the security of a block cipher. The S-box of AES, which uses the affine transform of the Inverse function on $\mathbb{F}_{2^8}$, is chosen as to maximize resistance against various major attacks such as linear, differential, and higher order differential attack.

Unfortunately, due to its simple algebraic description in $\mathbb{F}_{2^8}$, the S-box of AES is vulnerable to algebraic attack. It is well known that 39 linearly independent multivariate quadratic equations over $\mathbb{F}_2$ can be derived from the S-box of AES (see [8,9,11]). One can then perform algebraic attack on AES by solving a system of quadratic equations, e.g. using the XL algorithm [12], XSL algorithm [13], Gröbner basis algorithm [14], etc.

It is noted that there is still no known practical attack on full-round AES. However, the property that there exist some quadratic equations satisfied by the input and output bits of the S-box opens up some vulnerabilities. An $n \times n$ S-box (i.e. (n, n)-function) having such property is called implicitly quadratic. The implicit quadratic property of (n, n)-function has been studied in several other

© Springer International Publishing AG 2016
O. Dunkelman and S.K. Sanadhya (Eds.): INDOCRYPT 2016, LNCS 10095, pp. 364–379, 2016.
DOI: 10.1007/978-3-319-49890-4_20

papers. [9] (and later extended in [8]) studied the number of linearly independent multivariate quadratic equations that can be derived from some power functions (i.e. functions of the form x^e for some integer e) of certain types, including the Inverse, Gold, and Kasami functions. However, there are some inaccuracies in their results as pointed out by [11]. Apart from that, [11] also determined the exact number of multivariate quadratic equations from the Inverse function on $\mathbb{F}_{2^n}$. [19] then expanded and generalized the technique used in [11] to study the implicit quadratic property for power functions more generally.

Most of the previous studies focused on the power functions. However, there are numerous other classes of functions that are suitable to be used as an S-box. Differentially 4-uniform permutations are candidates for a good S-box as they possess good resistance against differential attack. In this paper, we examine all the differentially 4-uniform permutations that are known in the literature (not limited to power functions) and determine whether they are implicitly quadratic.

The rest of the paper is organized as follows. In Sect. 2, a precise definition of implicitly quadratic functions is given. In Sect. 3, we introduce some useful tools to determine whether a given function is implicitly quadratic. We then examine all the known differentially 4-uniform permutations and determine whether they are implicitly quadratic in Sect. 4. In Sect. 5, we pose an open question of whether there exist differentially 4-uniform permutations over $\mathbb{F}_{2^n}$ that are not implicitly quadratic. We also present a solution to this question for the special cases when $n = 11$ and $n = 13$. Finally, we conclude the paper in Sect. 6.

2   Preliminaries

Let $n \geq 1$ be integer. We denote the finite field of 2 elements and of 2^n elements by $\mathbb{F}_2$ and $\mathbb{F}_{2^n}$ respectively. Let $\mathbb{F}_2[t]$ be the univariate polynomial ring over $\mathbb{F}_2$. Suppose $p(t) \in \mathbb{F}_2[t]$ is an irreducible polynomial of degree n. We identify the finite field $\mathbb{F}_{2^n}$ with $\mathbb{F}_2[t]/(p(t))$. Note that $\mathbb{F}_{2^n}$ and $\mathbb{F}_2^n$ are isomorphic as vector spaces over $\mathbb{F}_2$. We define a linear isomorphism $\varphi : \mathbb{F}_{2^n} \to \mathbb{F}_2^n$ by $\varphi(t^{n-1}) := (1, 0, 0, \cdots, 0)$, $\varphi(t^{n-2}) := (0, 1, 0, 0, \cdots, 0), \cdots, \varphi(1) := (0, 0, \cdots, 0, 1)$.

Definition 1. A function $\widetilde{F} : \mathbb{F}_2^n \to \mathbb{F}_2^n$ is implicitly quadratic if there exists $q \in \mathbb{F}_2[t_1, t_2, \cdots, t_{2n}]/(t_1^2 + t_1, t_2^2 + t_2, \cdots, t_{2n}^2 + t_{2n})$ with $\deg(q) = 2$ such that for all $(x_1, x_2, \cdots, x_n) \in \mathbb{F}_2^n$, if $(y_1, y_2, \cdots, y_n) := \widetilde{F}(x_1, x_2, \cdots, x_n)$, then $q(x_1, \cdots, x_n, y_1, \cdots, y_n) = 0$.

Any such quadratic polynomial q is called an implicit quadratic equation (IQE) satisfied by $\widetilde{F}$.

Let V be a vector space over $\mathbb{F}_2$ of dimension n. Then V is isomorphic to $\mathbb{F}_2^n$. Let $\phi : V \to \mathbb{F}_2^n$ be a linear isomorphism. Given a function $F : V \to V$, we define $\psi_\phi(F) : \mathbb{F}_2^n \to \mathbb{F}_2^n$ by $\psi_\phi(F) := \phi \circ F \circ \phi^{-1}$.

Definition 2. Let V be an n-dimensional $\mathbb{F}_2$-vector space and $\phi : V \to \mathbb{F}_2^n$ be a linear isomorphism. A function $F : V \to V$ is implicitly quadratic if the corresponding function $\psi_\phi(F) : \mathbb{F}_2^n \to \mathbb{F}_2^n$ is implicitly quadratic.

Lemma 1. *The definition above is independent of ϕ, that is, if $\phi_1, \phi_2 : V \to \mathbb{F}_2^n$ are linear isomorphisms, then $\psi_{\phi_1}(F)$ is implicitly quadratic if and only if $\psi_{\phi_2}(F)$ is implicitly quadratic. Furthermore, the number of IQE satisfied by $\psi_{\phi_1}(F)$ is the same as the number of IQE satisfied by $\psi_{\phi_2}(F)$.*

Proof. Let $\widetilde{F_1} := \phi_1 \circ F \circ \phi_1^{-1}$ be implicitly quadratic. We show that $\widetilde{F_2} := \phi_2 \circ F \circ \phi_2^{-1}$ is implicitly quadratic as well.

Since $\widetilde{F_1}$ is implicitly quadratic, there exists a quadratic polynomial q such that for any $x' \in \mathbb{F}_2^n$, if $y' := \widetilde{F_1}(x') = \phi_1 \circ F \circ \phi_1^{-1}(x')$, then $q(x', y') = 0$. Let $x \in \mathbb{F}_2^n$ and $y := \widetilde{F_2}(x) = \phi_2 \circ F \circ \phi_2^{-1}(x)$. Then $\phi_1 \circ \phi_2^{-1}(y) = \phi_1 \circ F \circ \phi_1^{-1} \circ \phi_1 \circ \phi_2^{-1}(x)$. In other words, we have $\phi_1 \circ \phi_2^{-1}(y) = \widetilde{F_1}(\phi_1 \circ \phi_2^{-1}(x))$. Thus, we have $q(\phi_1 \circ \phi_2^{-1}(x), \phi_1 \circ \phi_2^{-1}(y)) = 0$.

Since $\phi_1 \circ \phi_2^{-1} : \mathbb{F}_2^n \to \mathbb{F}_2^n$ is a linear isomorphism, then each component of $\phi_1 \circ \phi_2^{-1}(x)$ (resp. $\phi_1 \circ \phi_2^{-1}(y)$) is a linear combination of the components of x (resp. y). So, expanding $q(\phi_1 \circ \phi_2^{-1}(x), \phi_1 \circ \phi_2^{-1}(y)) = 0$ gives a quadratic polynomial q' in terms of the components of x and y. Hence, $\widetilde{F_2}$ is implicitly quadratic.

We now show that the number of IQE satisfied by $\widetilde{F_1}$ is the same as the number of IQE satisfied by $\widetilde{F_2}$. Let Q (resp. Q') be the set of all IQE satisfied by $\widetilde{F_1}$ (resp. $\widetilde{F_2}$). Define $\tau : Q \to Q'$ by $\tau(q) := q'$, where $q'(t_1, \cdots, t_{2n}) := q(\phi_1 \circ \phi_2^{-1}(t_1, \cdots, t_n), \phi_1 \circ \phi_2^{-1}(t_{n+1}, \cdots, t_{2n}))$. For any $q \in Q$, we have $\tau(q) \in Q'$. So $\tau : Q \to Q'$ is well-defined. Define another function $\tau' : Q' \to Q$ by $\tau'(q') := q$, where $q(t_1, \cdots, t_{2n}) := q'(\phi_2 \circ \phi_1^{-1}(t_1, \cdots, t_n), \phi_2 \circ \phi_1^{-1}(t_{n+1}, \cdots, t_{2n}))$. Note that τ and τ' are inverses of each other. So τ is bijective, and thus $|Q| = |Q'|$. $\square$

3   Main Tools

In this section, we present some tools to determine whether a given function $\widetilde{F} : \mathbb{F}_2^n \to \mathbb{F}_2^n$ is implicitly quadratic.

Using the same setting as in Definition 1, let $(x_1, \cdots, x_n) \in \mathbb{F}_2^n$ and $(y_1, \cdots, y_n) := \widetilde{F}(x_1, \cdots, x_n)$. Define

$$\mathbf{u}_{(x_1, \cdots, x_n)} := (1, x_1, \cdots, x_n, x_1 x_2, \cdots, x_{n-1} x_n, y_1, \cdots, y_n,$$
$$y_1 y_2, \cdots, y_{n-1} y_n, x_1 y_1, x_1 y_2, \cdots, x_n y_n).$$

Note that $\mathbf{u}_{(x_1, \cdots, x_n)} \in \mathbb{F}_2^{L_n}$, where $L_n := 1 + 2\left[n + \binom{n}{2}\right] + n^2 = 2n^2 + n + 1$. Suppose

$$M_{\widetilde{F}} := \begin{pmatrix} \mathbf{u}_{(0,0,\cdots,0)} \\ \mathbf{u}_{(0,0,\cdots,1)} \\ \vdots \\ \mathbf{u}_{(1,1,\cdots,1)} \end{pmatrix}.$$

For any matrix M, we recall that its nullspace is defined to be the vector space $\{\mathbf{v} | M\mathbf{v} = 0\}$. We denote the nullspace of M and the nullity of M (that is, the dimension of its nullspace) by Nullspace(M) and Nullity(M) respectively.

We define Θ : {IQE satisfied by $\widetilde{F}$} $\rightarrow$ Nullspace($M_{\widetilde{F}}$) as follows. Any IQE satisfied by $\widetilde{F}$ is of the form

$$a+\sum_{i=1}^{n}b_i x_i+\sum_{i=1}^{n-1}\sum_{j=i+1}^{n}c_{i,j}x_i x_j+\sum_{i=1}^{n}d_i y_i+\sum_{i=1}^{n-1}\sum_{j=i+1}^{n}e_{i,j}y_i y_j+\sum_{i=1}^{n}\sum_{j=1}^{n}f_{i,j}x_i y_j = 0,$$

(1)

where a, b_i, $c_{i,j}$, d_i, $e_{i,j}$, and $f_{i,j}$ are coefficients belonging to $\mathbb{F}_2$. We define the function Θ by sending IQE of the form (1) to the vector $(a, b_1, \cdots, b_n, c_{1,2}, \cdots, c_{n-1,n}, d_1, \cdots, d_n, e_{1,2}, \cdots, e_{n-1,n}, f_{1,1}, \cdots, f_{n,n})^T$.

Theorem 1. *The function Θ: {IQE satisfied by $\widetilde{F}$} $\rightarrow$ Nullspace($M_{\widetilde{F}}$) is a well-defined linear isomorphism.*

Proof. Let q be an IQE satisfied by $\widetilde{F}$ of the form (1), and suppose $\mathbf{v} := \Theta(q)$. Note that (1) is equivalent to $\mathbf{u}_{(x_1,\cdots,x_n)}\mathbf{v} = 0$ for all $(x_1, \cdots, x_n) \in \mathbb{F}_2^n$. This can be combined into one matrix equation $M_{\widetilde{F}}\,\mathbf{v} = 0$, which holds if and only if $\mathbf{v} \in$ Nullspace($M_{\widetilde{F}}$). Thus, q is an IQE of the form (1) satisfied by $\widetilde{F}$ if and only if $\mathbf{v} \in$ Nullspace($M_{\widetilde{F}}$). This shows that Θ is well-defined and surjective. As it is clear that Θ is linear and injective, we conclude that Θ is a well-defined linear isomorphism. $\qquad\square$

As a corollary of Theorem 1, we have the following:

Corollary 1. *The number of linearly independent IQE satisfied by $\widetilde{F}$ is equal to Nullity($M_{\widetilde{F}}$). In particular, the function $\widetilde{F}$ is implicitly quadratic if and only if Nullity($M_{\widetilde{F}}$) $\neq 0$.*

Remark 1. Given any function $\widetilde{F} : \mathbb{F}_2^n \rightarrow \mathbb{F}_2^n$, Corollary 1 gives a method to determine whether $\widetilde{F}$ is implicitly quadratic. The quantity Nullity($M_{\widetilde{F}}$) can be computed easily by Gaussian elimination. Moreover, Gaussian elimination also gives a basis for the nullspace of $M_{\widetilde{F}}$, which can then be converted to IQE satisfied by $\widetilde{F}$ via the one-to-one correspondence Θ. This method of obtaining IQE satisfied by $\widetilde{F}$ is similar to that of [1].

Lemma 2. *Let $1 \leq n \leq 6$ be integer. Then any function $\widetilde{F} : \mathbb{F}_2^n \rightarrow \mathbb{F}_2^n$ is implicitly quadratic.*

Proof. Note that the matrix $M_{\widetilde{F}}$ is a $2^n \times L_n$ matrix, where $L_n = 2n^2 + n + 1$. Since $2^n < L_n$ for $1 \leq n \leq 6$, we see that Rank($M_{\widetilde{F}}$) $\leq 2^n < L_n$. So, by Rank-Nullity Theorem [18], we have Nullity($M_{\widetilde{F}}$) $= L_n -$ Rank($M_{\widetilde{F}}$) > 0. Hence, by Corollary 1, $\widetilde{F}$ is implicitly quadratic. $\qquad\square$

Remark 2. In view of the above lemma, we shall henceforth only consider the case $n \geq 7$.

Lemma 3 [5]. *Suppose $\widetilde{F_1}, \widetilde{F_2}: \mathbb{F}_2^n \rightarrow \mathbb{F}_2^n$ are CCZ-equivalent. Then $\widetilde{F_1}$ is implicitly quadratic if and only if $\widetilde{F_2}$ is implicitly quadratic. Moreover, the number of IQE satisfied by $\widetilde{F_1}$ is the same as the number of IQE satisfied by $\widetilde{F_2}$.*

4   Examining the Known Differentially 4-Uniform Permutations

In this section, we are only interested to examine the known differentially 4-uniform permutations over $\mathbb{F}_{2^n}$ for n even as $n \times n$ S-box with n even is commonly used in block ciphers.

4.1   Functions Constructed by Primary Construction

There are 5 classes of primarily-constructed differentially 4-uniform permutations (bijective (n, n)-functions). These are listed in Table 1. The Inverse, Gold and Kasami functions are already known to be implicitly quadratic [8,9,11]. In this subsection, we show that the other functions are also implicitly quadratic.

Table 1. Primarily-constructed differentially 4-uniform permutations over $\mathbb{F}_{2^n}$

	Functions	Conditions	Ref
Gold	x^{2^i+1}	$n = 2k$, k is odd and $\gcd(n, i) = 2$	[15]
Kasami	$x^{2^{2i}-2^i+1}$	$n = 2k$, k is odd and $\gcd(n, i) = 2$	[16]
Inverse	x^{-1} $(0^{-1} := 0)$	n is even	[20]
Bracken-Leander	$x^{2^{2m}+2^m+1}$	$n = 4m$ and m is odd	[3]
Binomial	$\alpha x^{2^s+1} + \alpha^{2^m} x^{2^{-m}+2^{m+s}}$	$n = 3m$, m even, $m/2$ odd, $\gcd(n, s) = 2$, $3\mid m+s$ and α is a primitive element of $\mathbb{F}_{2^n}$	[4]

In the last column of Table 2, we list some equations derived from $y = F(x)$, where F is a primarily-constructed differentially 4-uniform permutation. Note that the IQE satisfied by $\psi_\varphi(F)$ can be obtained by applying φ to any of the derived equations listed on the table and looking at the components. Thus, using Table 2, we have just shown that all of the primarily-constructed differentially 4-uniform permutations are implicitly quadratic.

As the precise number of IQE satisfied by the Inverse function is known and will be useful to us, we recall the following well-known result.

Lemma 4 [11]. *Let* $I : \mathbb{F}_{2^n} \rightarrow \mathbb{F}_{2^n}$ *be the Inverse function. Then,* $Nullity(M_{\psi_\varphi(I)}) = 5n - 1$.

4.2   Functions Constructed by Switching Method

A number of differentially 4-uniform permutations have been constructed via the switching method applied to the Inverse function. In the following, we list the known differentially 4-uniform permutations of $\mathbb{F}_{2^n}$ constructed by switching method for n even.

Table 2. Derived equations from primarily-constructed permutations

	Functions	Derived equations
Gold	x^{2^i+1}	$y = x^{2^i+1}$ $xy^{2^i} = x^{2^{2i}}y$ $xy^{2^i} = x^{2^{2i}}y$ $xy^{2^i} = x^{2^{2i}}y$
Kasami	$x^{2^{2i}-2^i+1}$	$x^{2^i}y = x^{2^{2i}+1}$ $y^{2^i+1} = x^{2^{3i}+1}$
Inverse	x^{-1} $(0^{-1} := 0)$	$xy = 1$ (if $x \neq 0$), $xy = 0$ (if $x = 0$) $x^2y = x$ $x^4y = x^3$ $xy^2 = y$ $xy^4 = y^3$
Bracken-Leander	$x^{2^{2m}+2^m+1}$	$xy^{2^m} = x^{2^{3m}}y$
Binomial	$\alpha x^{2^s+1} + \alpha^{2^m}x^{2^{-m}+2^{m+s}}$	$y^{2^m} = \alpha^{2^m}x^{2^{m+s}+2^m} + \alpha^{2^{2m}}x^{1+2^{2m+s}}$

- Qu-Tan-Tan-Li [25]: $x^{-1}+Tr(x^{-d}+(x^{-1}+1)^d)$, where $d = 2^n - 2$, or $3(2^t+1)$ for $2 \leq t \leq n/2 - 1$.

- Qu-Tan-Li-Gong [26]: $x^{-1} + g(x)$, where g is some Boolean function.

- Peng-Tan [21]:

$$F(x) = \begin{cases} x^{-1} + 1 & \text{if } x \in T, \\ x^{-1} & \text{if } x \in \mathbb{F}_{2^n} \setminus T, \end{cases}$$

where $T \subset \mathbb{F}_{2^n}$. As there are at least $2^{2^{n-2}-1}$ such T, it is not possible to list them here. For more details on T, please refer to [21].

- Peng-Tan (II) [22]:

$$F(x) = \begin{cases} \beta(x+1)^{-1} + \alpha & \text{if } x \in \mathbb{F}_{2^d}, \\ x^{-1} & \text{if } x \in \mathbb{F}_{2^n} \setminus \mathbb{F}_{2^d}, \end{cases}$$

where α, β, d, n satisfy any of the following conditions:
(1) $\alpha \in \mathbb{F}_{2^d}$, $\beta = 1$, d is even, or
(2) $\alpha = \beta = 1$, d is odd, or
(3) $\alpha = 0$, $\beta = 1$, $d = 1, 3$, $n/2$ is odd, or
(4) $\alpha, \beta \in \mathbb{F}_{2^d}$, $Tr(\beta^{-1}) = 1$, n/d is odd.

- Peng-Tan-Wang [23]:

$$F(x) = \begin{cases} (\gamma x)^{-1} & \text{if } x \in U, \\ x^{-1} & \text{if } x \in \mathbb{F}_{2^n} \setminus U, \end{cases}$$

where $\gamma \in \mathbb{F}_{2^n}$ and U is a union of some cosets of the cyclic group $\langle \gamma \rangle$. Please refer to [23] for more details on the set U.

– Zha-Hu-Sun (I) [29]:

$$F(x) = \begin{cases} x^{-1} + 1 & \text{if } x \in S, \\ x^{-1} & \text{if } x \in \mathbb{F}_{2^n} \setminus S, \end{cases}$$

where S satisfies any of the following conditions:
(1) $S = \mathbb{F}_{2^{k_1}} \cup \mathbb{F}_{2^{k_2}}$; k_1, k_2 are even; $k_1 | n$; $k_2 | n$, or
(2) $S = \mathbb{F}_{2^3} \cup \mathbb{F}_{2^{k_1}}$; k_1 is even; $k_1 | n$; $\gcd(3, k_1) = 1$; $6 | n$; $\frac{n}{6}$ is odd.

– Zha-Hu-Sun (II) [28]:

$$F(x) = \begin{cases} x^{-1} + \alpha & \text{if } x \in \mathbb{F}_{2^d}, \\ x^{-1} & \text{if } x \in \mathbb{F}_{2^n} \setminus \mathbb{F}_{2^d}, \end{cases}$$

where $\alpha \in \mathbb{F}_{2^d}$; $d | n$; d is even, or $d = 1, 3$; $n/2$ is odd.

– Zha-Hu-Sun (III) [28]:

$$F(x) = \begin{cases} \beta x^{-1} + \alpha & \text{if } x \in \mathbb{F}_{2^d}, \\ x^{-1} & \text{if } x \in \mathbb{F}_{2^n} \setminus \mathbb{F}_{2^d}, \end{cases}$$

where $\alpha, \beta \in \mathbb{F}_{2^d}$; $Tr(\frac{1}{\alpha}) = 1$; $d | n$; d is even; $\frac{n}{d}$ is odd.

– Tang-Carlet-Tang [27]:

$$F(x) = \begin{cases} (x+1)^{-1} & \text{if } x \in T, \\ x^{-1} & \text{if } x \in \mathbb{F}_{2^n} \setminus T, \end{cases}$$

where T satisfies:
(1) if $x \in T$, then $x + 1 \in T$, and
(2) if $x \in T$, then $Tr(\frac{1}{x}) = Tr(\frac{1}{x+1}) = 1$.

Remark 3. The Qu-Tan-Tan-Li function [25] and Qu-Tan-Li-Gong function [26] can be expressed in the following form

$$F(x) = \begin{cases} x^{-1} + 1 & \text{if } x \in S, \\ x^{-1} & \text{if } x \in \mathbb{F}_{2^n} \setminus S, \end{cases}$$

where $S = \{x \in \mathbb{F}_{2^n} \mid Tr(x^{-d} + (x^{-1} + 1)^d) = 1\}$ with $d = 2^n - 2$ or $d = 3(2^t + 1)$ for $2 \le t \le n/2 - 1$ for the Qu-Tan-Tan-Li function [25]; and $S = \{x \in \mathbb{F}_{2^n} \mid g(x) = 1\}$ for the Qu-Tan-Li-Gong function [26].

In this subsection, we will show that all the functions listed above are implicitly quadratic.

Theorem 2. *Let $F : \mathbb{F}_{2^n} \to \mathbb{F}_{2^n}$ be any implicitly quadratic function with $N :=$ Nullity$(M_{\psi_\varphi(F)}) > 2n$. Suppose U is a subset of $\mathbb{F}_{2^n}$ and $\alpha \in \mathbb{F}_{2^n} \setminus \{0\}$. Define $G : \mathbb{F}_{2^n} \to \mathbb{F}_{2^n}$ by*

$$G(x) := \begin{cases} F(x) + \alpha & \text{if } x \in U, \\ F(x) & \text{if } x \in \mathbb{F}_{2^n} \setminus U. \end{cases}$$

Then G is implicitly quadratic and Nullity$(M_{\psi_\varphi(G)}) \geq N - 2n$.

Proof. Suppose $\{u_1, \cdots, u_{n-1}, \alpha\}$ is a basis for $\mathbb{F}_{2^n}$. Define a linear isomorphism $\phi : \mathbb{F}_{2^n} \to \mathbb{F}_2^n$ by $\phi(u_1) := (1, 0, \cdots, 0)$, $\phi(u_2) := (0, 1, 0, \cdots, 0)$, $\cdots$, $\phi(u_{n-1}) := (0, \cdots, 0, 1, 0)$, $\phi(\alpha) := (0, \cdots, 0, 1)$. Let $\widetilde{F} := \psi_\phi(F)$ and $\widetilde{G} := \psi_\phi(G)$.

As $\phi(\alpha) = (0, 0, \cdots, 0, 1)$, we see that $M_{\widetilde{G}}$ is obtained from $M_{\widetilde{F}}$ by modifying only the columns corresponding to the terms involving y_n. There are $2n$ such terms: $y_n, y_1 y_n, \cdots, y_{n-1} y_n, x_1 y_n, \cdots, x_n y_n$. Thus, $M_{\widetilde{G}}$ is obtained from $M_{\widetilde{F}}$ by modifying $2n$ columns. Therefore the number of linearly independent columns in $M_{\widetilde{G}}$ is at most $2n$ more than the number of linearly independent columns in $M_{\widetilde{F}}$. In other words, Rank$(M_{\widetilde{G}}) \leq$ Rank$(M_{\widetilde{F}}) + 2n$. Applying Rank-Nullity Theorem and simplifying, we have Nullity$(M_{\widetilde{G}}) \geq$ Nullity$(M_{\widetilde{F}}) - 2n$. By Lemma 1, we have $N =$ Nullity$(M_{\widetilde{F}})$ and Nullity$(M_{\psi_\varphi(G)}) =$ Nullity$(M_{\widetilde{G}})$. So, Nullity$(M_{\psi_\varphi(G)}) =$ Nullity$(M_{\widetilde{G}}) \geq$ Nullity$(M_{\widetilde{F}}) - 2n = N - 2n > 0$. Therefore, by Corollary 1, $\widetilde{G}$ (and hence G) is implicitly quadratic. $\qquad\square$

By Lemma 4 and Theorem 2, we deduce the following corollary:

Corollary 2. *Let $n \geq 7$ be an integer and U be any subset of $\mathbb{F}_{2^n}$. Then the function $F : \mathbb{F}_{2^n} \to \mathbb{F}_{2^n}$ given by*

$$F(x) := \begin{cases} x^{-1} + 1 & \text{if } x \in U, \\ x^{-1} & \text{if } x \in \mathbb{F}_{2^n} \setminus U, \end{cases}$$

is implicitly quadratic. Moreover, Nullity$(M_{\psi_\varphi(F)}) \geq 3n - 1$.

Corollary 3. *The Zha-Hu-Sun (I) function [29], the Qu-Tan-Tan-Li function [25], the Qu-Tan-Li-Gong function [26] and the Peng-Tan function [21] are implicitly quadratic.*

Proof. This is true by Corollary 2 and Remark 3. $\qquad\square$

Corollary 4. *The Zha-Hu-Sun (II) function [28] is implicitly quadratic.*

Proof. This is a consequence of Lemma 4 and Theorem 2. $\qquad\square$

Corollary 5. *Let F be the Tang-Carlet-Tang function [27]. Then F is implicitly quadratic.*

Proof. Let F^{-1} be the composite inverse of F. By Corollary 2, F^{-1} is implicitly quadratic. Moreover, as F and F^{-1} are CCZ-equivalent, we see by Lemma 3 that F is implicitly quadratic. $\qquad\square$

Now it remains to prove that the Zha-Hu-Sun (III) function [28], Peng-Tan (II) function [22], and Peng-Tan-Wang function [23] are implicitly quadratic. In order to do that, we need the following theorem.

Theorem 3. *Let S be a proper subfield of $\mathbb{F}_{2^n}$. Suppose $F : \mathbb{F}_{2^n} \to \mathbb{F}_{2^n}$ is a function such that $x \cdot F(x) \in S$ for all $x \in \mathbb{F}_{2^n}$. Then F is implicitly quadratic.*

Proof. Let $\{v_1, v_2, \cdots, v_d\}$ be a basis for S over $\mathbb{F}_2$. We can extend it to a basis $\{v_1, \cdots, v_d, v_{d+1}, \cdots, v_n\}$ for $\mathbb{F}_{2^n}$. We define a linear isomorphism $\phi : \mathbb{F}_{2^n} \to \mathbb{F}_2^n$ by $\phi(v_1) := (1, 0, \cdots, 0)$, $\phi(v_2) := (0, 1, 0, \cdots, 0)$, $\cdots$, $\phi(v_n) := (0, \cdots, 0, 1)$. We shall show that $\widetilde{F} := \psi_\phi(F)$ is implicitly quadratic. Let $x_1, \cdots, x_n, y_1, \cdots, y_n$ be variables. Suppose $x := \phi^{-1}(x_1, \cdots, x_n)$ and $y := \phi^{-1}(y_1, \cdots, y_n)$. Let $(s_1, s_2, \cdots, s_n) := \phi(xy)$.

We claim that each component of $\phi(xy)$ is quadratic in terms of $x_1, \cdots, x_n$, $y_1, \cdots, y_n$. Suppose $(x_1', x_2', \cdots, x_n') := \varphi(x) = \varphi \circ \phi^{-1}(x_1, x_2, \cdots, x_n)$ and $(y_1', y_2', \cdots, y_n') := \varphi(y) = \varphi \circ \phi^{-1}(x_1, x_2, \cdots, x_n)$. Thus, each of the x_i' (resp. y_i') is a linear combination of $x_1, \cdots, x_n$ (resp. $y_1, \cdots, y_n$). Note that each component of $\varphi(xy)$ is quadratic in terms of $x_1', \cdots, x_n', y_1', \cdots, y_n'$. Since $\varphi \circ \phi^{-1}$ is linear, we see that any component of $\varphi(xy)$ is also quadratic in terms of $x_1, \cdots, x_n, y_1, \cdots, y_n$. Moreover, as $\phi \circ \varphi^{-1}$ is linear, we conclude that any component of $\phi(xy) = \phi \circ \varphi^{-1}(\varphi(xy))$ is quadratic in terms of $x_1, \cdots, x_n, y_1, \cdots, y_n$.

Now, if we let $y = F(x)$, then we have $xy \in S$. As $xy \in S$ and $\{v_1, v_2, \cdots, v_d\}$ is a basis for S, the last $(n - d)$ components of $\phi(xy)$ must all be zero. So, $\widetilde{F}$ satisfies the following $(n - d)$ IQE: $s_{d+1} = 0$, $s_{d+2} = 0, \cdots, s_n = 0$. Thus, $\widetilde{F}$ is implicitly quadratic. Hence, F is implicitly quadratic. $\square$

Corollary 6. *The Zha-Hu-Sun (III) function [28] and the Peng-Tan (II) function [22] are implicitly quadratic.*

Proof. Let F_1 and F_2 be the Zha-Hu-Sun (III) function and the Peng-Tan (II) function respectively. Note that

$$x \cdot F_1(x) = \begin{cases} \beta + \alpha x & \text{if } x \in \mathbb{F}_{2^d}, \\ 1 & \text{if } x \in \mathbb{F}_{2^n} \setminus \mathbb{F}_{2^d}, \end{cases}$$

and

$$x \cdot F_2(x) = \begin{cases} \beta x(x+1)^{-1} + \alpha x & \text{if } x \in \mathbb{F}_{2^d}, \\ 1 & \text{if } x \in \mathbb{F}_{2^n} \setminus \mathbb{F}_{2^d}. \end{cases}$$

For any $x \in \mathbb{F}_{2^d}$, we have $\beta + \alpha x, \beta x(x+1)^{-1} + \alpha x \in \mathbb{F}_{2^d}$ as $\alpha, \beta \in \mathbb{F}_{2^d}$ and $\mathbb{F}_{2^d}$ is closed under addition, multiplication, and taking inverse. For any $x \in \mathbb{F}_{2^n} \setminus \mathbb{F}_{2^d}$, we have $x \cdot F_1(x) = x \cdot F_2(x) = 1$. Thus, $x \cdot F_1(x), x \cdot F_2(x) \in \mathbb{F}_{2^d}$ for all $x \in \mathbb{F}_{2^n}$. Hence, by Theorem 3, F_1 and F_2 are implicitly quadratic. $\square$

Corollary 7. *Let F be the Peng-Tan-Wang function [23]. Then F is implicitly quadratic.*

Proof. First note that if $\langle \gamma \rangle = \mathbb{F}_{2^n}$, then $F(x) = \gamma^{-1}x^{-1}$ for all $x \in \mathbb{F}_{2^n}$. In this case, F is affine equivalent to the Inverse function, and so is implicitly quadratic by Lemma 3. Now suppose that $\langle \gamma \rangle \neq \mathbb{F}_{2^n}$. Then $\mathbb{F}_2(\gamma)$ is a proper subfield of $\mathbb{F}_{2^n}$. Note that $x \cdot F(x) = \begin{cases} \gamma^{-1} & \text{if } x \in U, \\ 1 & \text{if } x \in \mathbb{F}_{2^n} \setminus U. \end{cases}$

Thus, $x \cdot F(x) \in \mathbb{F}_2(\gamma)$ for any $x \in \mathbb{F}_{2^n}$. Hence, by Theorem 3, F is implicitly quadratic. □

4.3   Functions Constructed by Expansion

In 2013, Carlet et al. [7] introduced a method to construct differentially 4-uniform permutations over $\mathbb{F}_{2^{n+1}}$ from known permutations over $\mathbb{F}_{2^n}$. The construction is as follows.

- Carlet et al. [7]: Let $n \geq 5$ be an odd integer. For any element $\alpha \in \mathbb{F}_{2^n} \setminus \{0,1\}$ such that $Tr(\alpha) = Tr(1/\alpha) = 1$, define an $(n+1, n+1)$-function F as follows:

$$F(x_1, \cdots, x_n, x_{n+1}) := \begin{cases} \left(\frac{1}{x'}, g(x')\right) & \text{if } x_{n+1} = 0, \\ \left(\frac{\alpha}{x'}, g(\frac{x'}{\alpha}) + 1\right) & \text{if } x_{n+1} = 1, \end{cases}$$

where $x' \in \mathbb{F}_{2^n}$ is identified with $(x_1, \cdots, x_n) \in \mathbb{F}_2^n$ and g is an arbitrary Boolean function defined on $\mathbb{F}_{2^n}$. Note that $\frac{1}{0}$ is defined as 0.

In the following, we show that the above function is implicitly quadratic.

Theorem 4. *Suppose $\alpha \in \mathbb{F}_{2^n} \setminus \mathbb{F}_2$ and $g_1, g_2 : \mathbb{F}_{2^n} \to \mathbb{F}_2$ are Boolean functions. Let $I : \mathbb{F}_{2^n} \to \mathbb{F}_{2^n}$ be the Inverse function. Define the function $F : \mathbb{F}_{2^n} \times \mathbb{F}_2 \to \mathbb{F}_{2^n} \times \mathbb{F}_2$ by*

$$F(x_0, x) := \begin{cases} (g_1(x), I(x)) & \text{if } x_0 = 0, \\ (g_2(x), \alpha I(x)) & \text{if } x_0 = 1. \end{cases}$$

Then F is implicitly quadratic.

Proof. Let $F_2 : \mathbb{F}_{2^n} \to \mathbb{F}_{2^n}$ be defined by $F_2(x) := \alpha I(x)$. Suppose $\widetilde{F_1} := \psi_\varphi(I)$, $\widetilde{F_2} := \psi_\varphi(F_2)$, $g_1' := g_1 \circ \varphi^{-1}$, and $g_2' := g_2 \circ \varphi^{-1}$. We define $\widetilde{F} : \mathbb{F}_2^{n+1} \to \mathbb{F}_2^{n+1}$ by

$$\widetilde{F}(x_0, x_1, \cdots, x_n) := \begin{cases} (g_1'(x_1, \cdots, x_n), \widetilde{F_1}(x_1, \cdots, x_n)) & \text{if } x_0 = 0, \\ (g_2'(x_1, \cdots, x_n), \widetilde{F_2}(x_1, \cdots, x_n)) & \text{if } x_0 = 1. \end{cases}$$

Let M be the matrix obtained from $M_{\widetilde{F}}$ by removing all $(4n+3)$ columns corresponding to the terms involving x_0 or y_0. Note that $\text{Nullity}(M) \leq \text{Nullity}(M_{\widetilde{F}})$. By Corollary 1 and the previous inequality, in order to show that F is implicitly quadratic, it suffices to show that $\text{Nullity}(M) > 0$.

Observe that $M = \begin{pmatrix} M_{\widetilde{F_1}} \\ M_{\widetilde{F_2}} \end{pmatrix}$. Thus, $v \in \text{Nullspace}(M)$ if and only if $v \in \text{Nullspace}(M_{\widetilde{F_1}}) \cap \text{Nullspace}(M_{\widetilde{F_2}})$. In the following, we are going to show that $\text{Nullspace}(M_{\widetilde{F_1}}) \cap \text{Nullspace}(M_{\widetilde{F_2}})$ is nonzero by finding some IQE that are satisfied by both $\widetilde{F_1}$ and $\widetilde{F_2}$.

We first find some IQE satisfied by $\widetilde{F_1}$. Let $x_1, \cdots, x_n, y_1, \cdots, y_n$ be variables and $x := \varphi^{-1}(x_1, \cdots, x_n)$ and $y := \varphi^{-1}(y_1, \cdots, y_n)$. Suppose $(\alpha_1, \cdots, \alpha_n) := \varphi(\alpha)$ and $(s_1, \cdots, s_n) := \varphi(xy)$. Note that each s_i is quadratic in terms of $x_1, \cdots, x_n, y_1, \cdots, y_n$. From the equation $y = I(x)$, we see that

$$xy = \begin{cases} 1 & \text{if } x \neq 0, \\ 0 & \text{if } x = 0. \end{cases}$$

Applying φ to the equation above gives

$$(s_1, s_2, \cdots, s_n) = \begin{cases} (0, 0, \cdots, 0, 1) & \text{if } x \neq 0, \\ (0, 0, \cdots, 0, 0) & \text{if } x = 0. \end{cases}$$

Thus, $\widetilde{F_1}$ satisfies the $(n-1)$ IQE: $s_1 = 0, s_2 = 0, \cdots, s_{n-1} = 0$.

Now we find some IQE satisfied by $\widetilde{F_2}$. Consider the equation $y = \alpha x^{-1}$. Then

$$xy = \begin{cases} \alpha & \text{if } x \neq 0, \\ 0 & \text{if } x = 0. \end{cases}$$

Applying φ to the equation above, we have

$$(s_1, s_2, \cdots, s_n) = \begin{cases} (\alpha_1, \alpha_2, \cdots, \alpha_{n-1}, \alpha_n) & \text{if } x \neq 0, \\ (0, 0, \cdots, 0, 0) & \text{if } x = 0. \end{cases}$$

Suppose $\{i_1, i_2, \cdots, i_k\} := \{i \in \{1, 2, \cdots, n-1\} | \alpha_i = 0\}$ and $\{j_1, j_2, \cdots, j_l\} := \{j \in \{1, 2, \cdots, n-1\} | \alpha_j = 1\}$. Then $s_{i_1} = 0, s_{i_2} = 0, \cdots, s_{i_k} = 0$ are k IQE satisfied by $\widetilde{F_2}$ (and so these k IQE are satisfied by both $\widetilde{F_1}$ and $\widetilde{F_2}$). Moreover, the following $(l-1)$ IQE are also satisfied by both $\widetilde{F_1}$ and $\widetilde{F_2}$: $s_{j_1} + s_{j_2} = 0, s_{j_1} + s_{j_3} = 0, \cdots, s_{j_1} + s_{j_l} = 0$. Thus, there are at least $k + (l-1) = (n-2)$ IQE satisfied by both $\widetilde{F_1}$ and $\widetilde{F_2}$. Hence, $\text{Nullity}(M) = \dim(\text{Nullspace}(M)) = \dim(\text{Nullspace}(M_{\widetilde{F_1}}) \cap \text{Nullspace}(M_{\widetilde{F_2}})) \geq n - 2 > 0$. $\square$

Recently, Perrin et al. [24] introduced the so-called butterfly structure, which is a $2n$-bit mapping obtained by concatenating two bivariate functions over $\mathbb{F}_{2^n}$. Such butterflies have two CCZ-equivalent representations: one is a quadratic non-bijective function (denoted V_e^α) and one is a degree $n+1$ permutation (denoted H_e^α) as described in the following.

- Perrin et al. [24]: Let $\alpha \in \mathbb{F}_{2^n}$ and e be an integer such that the mapping $x \mapsto x^e$ is a permutation on $\mathbb{F}_{2^n}$. The Butterfly Structures are the functions on $(\mathbb{F}_{2^n})^2$ defined as follows:
  • the open butterfly H_e^α is defined by

$$\mathsf{H}_e^\alpha(x, y) := \left((y + \alpha((x + y^e)^{1/e} + \alpha y))^e + ((x + y^e)^{1/e} + \alpha y)^e, (x + y^e)^{1/e} + \alpha y \right),$$

- the closed butterfly V_e^α is defined by

$$V_e^\alpha(x, y) := ((x + \alpha y)^e + y^e, (y + \alpha x)^e + x^e).$$

It was shown in [24] that the differential uniformity of these functions is at most 4 when n, e, and α satisfy any of the following conditions:

(1) n is odd, $e = 3 \cdot 2^t$ for some integer t, and $\alpha \in \mathbb{F}_{2^n} \setminus \{0, 1\}$; or
(2) $n = 6$, $e = 5$, α is a certain element of $\mathbb{F}_{2^n}$; or
(3) n is odd, $e = 2^{2k} + 1$ for some integer k, $\alpha = 1$.

However [24] also stated that in any of these cases, the function V_e^α is quadratic (and hence is implicitly quadratic as well). This is because the left and right side of $V_e^\alpha(x, y)$ are equal to $(x + \alpha y)^e + y^e$ and $(y + \alpha x)^e + x^e$ respectively, both of which are quadratic in any of the three cases (note that the mapping $x \mapsto x^e$ is quadratic in any of these cases). As V_e^α and H_e^α are CCZ-equivalent, by Lemma 3, we see that the differentially 4-uniform permutation H_e^α is also implicitly quadratic.

4.4   Functions Constructed by Contraction

In [6], Carlet presented a method to construct differentially 4-uniform permutations over $\mathbb{F}_{2^n}$ by using APN permutations over $\mathbb{F}_{2^{n+1}}$. The construction is as follows.

- Carlet [6]: Let $c \equiv n \bmod 2$, $S = \{x \in \mathbb{F}_{2^{n+1}} \mid Tr(x) = 0\}$, $\alpha \in \mathbb{F}_{2^{n+1}}$ such that $Tr(\alpha) = 1$. Define $F(x) = x + \frac{1}{x + \alpha + c} + (\frac{1}{x + \alpha + c})^2$. Identify S with $\mathbb{F}_{2^n}$, then $F(x)|_S$ is a permutation over $\mathbb{F}_{2^n}$.

In this subsection, we show that the above function is implicitly quadratic.

Theorem 5. *Let V be an $\mathbb{F}_2$-vector space of dimension n and U be a vector subspace of V of dimension $d < n$. Suppose $F : V \to V$ is implicitly quadratic and $F(U) \subseteq U$. Then $F|_U : U \to U$ is implicitly quadratic.*

Proof. Let $\{v_1, \cdots, v_d\}$ be a basis for U. We can extend it to a basis $\{v_1, \cdots, v_d, v_{d+1}, \cdots, v_n\}$ for V. Define a linear isomorphism $\phi : U \to \mathbb{F}_2^d$ by $\phi(v_1) := (1, 0, \cdots, 0)$, $\phi(v_2) := (0, 1, 0, \cdots, 0), \cdots, \phi(v_d) := (0, \cdots, 0, 1)$. Similarly, define another linear isomorphism $\overline{\phi} : V \to \mathbb{F}_2^n$ by $\overline{\phi}(v_1) := (1, 0, \cdots, 0)$, $\overline{\phi}(v_2) := (0, 1, 0, \cdots, 0), \cdots, \overline{\phi}(v_n) := (0, \cdots, 0, 1)$.

As F is implicitly quadratic, then $\psi_{\overline{\phi}}(F)$ is also implicitly quadratic. So, there exists a quadratic polynomial q such that $q(x_1, \cdots, x_n, y_1, \cdots, y_n) = 0$. Define a quadratic polynomial q' in $2d$ variables by $q'(x_1, \cdots, x_d, y_1, \cdots, y_d) := q(x_1, \cdots, x_d, \underbrace{0, \cdots, 0}_{n-d \text{ times}}, y_1, \cdots, y_d, \underbrace{0, \cdots, 0}_{n-d \text{ times}})$. Then q' is an IQE satisfied by $\psi_\phi(F|_U)$. Thus, $F|_U : U \to U$ is also implicitly quadratic. $\qquad\square$

Lemma 5. *Suppose $c \in \{0,1\}$ and $\alpha \in \mathbb{F}_{2^n}$. Define $F : \mathbb{F}_{2^n} \to \mathbb{F}_{2^n}$ by $F(x) := x + \frac{1}{x+\alpha+c} + \left(\frac{1}{x+\alpha+c}\right)^2$. Then F is implicitly quadratic.*

Proof. Note that from $y = F(x)$, we can derive

$$(x + \alpha + c)^2(y + \alpha + c) + (x + \alpha + c)^3 + (x + \alpha + c) = \begin{cases} 1 & \text{if } x \neq \alpha + c, \\ 0 & \text{if } x = \alpha + c. \end{cases}$$

We can obtain $(n - 1)$ IQE satisfied by $\psi_\varphi(F)$ by applying φ to the above equation and looking at the components. Thus, F is implicitly quadratic. $\square$

Combining Theorem 5 and Lemma 5, we see that the function constructed by Carlet in [6] is implicitly quadratic.

5   An Open Question and Its Partial Answer

All of the differentially 4-uniform permutations we have examined so far are implicitly quadratic. This naturally leads us to pose the following open question.

Open Question. Let $n \geq 7$ be integer. Do there exist differentially 4-uniform permutations over $\mathbb{F}_{2^n}$ which are not implicitly quadratic? If so, then how to construct them?

In attempt to solve the open question, we performed an exhaustive search on all power functions (functions of the form x^e for some integer e) on $\mathbb{F}_{2^n}$ for $7 \leq n \leq 13$. We found that there is no differentially 4-uniform power function on $\mathbb{F}_{2^7}, \mathbb{F}_{2^8}, \mathbb{F}_{2^9}$, and $\mathbb{F}_{2^{12}}$ that is not implicitly quadratic. We managed to find some non-implicitly quadratic differentially 4-uniform power functions on $\mathbb{F}_{2^{10}}$. For $e \in \{87, 174, 237, 315, 348, 369, 423, 453, 474, 555, 630, 669, 696, 723, 738, 789, 846, 873, 906, 948\}$, the functions x^e on $\mathbb{F}_{2^{10}}$ are non-implicitly quadratic differentially 4-uniform functions. However, none of them is a permutation.

Fortunately, we successfully found some non-implicitly quadratic differentially 4-uniform permutations on $\mathbb{F}_{2^{11}}$ and $\mathbb{F}_{2^{13}}$, solving the open question for the special cases of $n = 11$ and $n = 13$. Let $E_1 := \bigcup_{i=0}^{10}\{109 \cdot 2^i \pmod{2^{11} - 1}, 695 \cdot 2^i \pmod{2^{11} - 1}\}$ and $E_2 := \bigcup_{i=0}^{10}\{251 \cdot 2^i \pmod{2^{11} - 1}, 367 \cdot 2^i \pmod{2^{11} - 1}\}$. For any $e \in E_1 \cup E_2$, the function x^e on $\mathbb{F}_{2^{11}}$ is a differentially 4-uniform permutation which is not implicitly quadratic. We also computed the differential spectrum and nonlinearity of these power functions. For $e \in E_1$, the differential spectrum of x^e is [2433883, 1420618, 337755], while for $e \in E_2$, the differential spectrum of x^e is [2568985, 1150414, 472857].[1] The nonlinearity of x^e for any $e \in E_1 \cup E_2$ is 960. This is slightly less than the nonlinearity of the Inverse function on $\mathbb{F}_{2^{11}}$, which equals to 980.

For the case $n = 13$, we found that for any $e \in \bigcup_{i=0}^{12}\{303 \cdot 2^i \pmod{2^{13} - 1}, 947 \cdot 2^i \pmod{2^{13} - 1}\}$, the function x^e on $\mathbb{F}_{2^{13}}$ is a differentially 4-uniform

---

[1] The differential spectrum $[a, b, c]$ represents the multi-set in which 0 appears a times, 2 appears b times and 4 appears c times.

permutation that is not implicitly quadratic. The differential spectrum of these power functions is $[37703173, 25244662, 4152837]$, while the nonlinearity is 3968, slightly less than the nonliearity of the Inverse function on $\mathbb{F}_{2^{13}}$ which equals to 4006.

Extending our exhaustive search on all power functions on $\mathbb{F}_{2^n}$ for $7 \leq n \leq 13$, we remark that it was shown in Table 1 of [2], that for $14 \leq n \leq 24$, any power function on $\mathbb{F}_{2^n}$ which is a differentially 4-uniform permutation is CCZ-equivalent to either the Inverse function, the Kasami function, the Gold function, or the Bracken-Leander function. We have seen in Subsect. 4.1 that all of these functions are implicitly quadratic. Thus, by Lemma 3, we conclude that for any $14 \leq n \leq 24$, there is no power function on $\mathbb{F}_{2^n}$ that is both differentially 4-uniform permutation and not implicitly quadratic.

6 Conclusion

In this paper, we studied the implicit quadratic property for S-boxes constructed from differentially 4-uniform permutations (bijective (n,n)-functions). If an (n,n)-function used as an S-box in some ciphers is implicitly quadratic, then one can derive some quadratic equations to express the S-box. As a result, the cipher is vulnerable to algebraic attack [10,12–14]. Thus, the property of being not implicitly quadratic is a desired property for a good S-box.

It is desirable to find a differentially 4-uniform permutation that is not implicitly quadratic as such function has a good resistance against both algebraic and differential attack. We have examined all (to the best of our knowledge) the known differentially 4-uniform permutations over $\mathbb{F}_{2^n}$. We proved that all of them are implicitly quadratic, except for the Li-Wang's function [17].[2] This leads to an open question of whether there exist non-implicitly quadratic differentially 4-uniform permutations over $\mathbb{F}_{2^n}$ and how to construct them if they exist. We found some power functions on $\mathbb{F}_{2^{11}}$ and $\mathbb{F}_{2^{13}}$ which are non-implicitly quadratic differentially 4-uniform permutations, solving the question for the special cases of $n = 11$ and $n = 13$. However, solving the open question for the general case remains an interesting problem to be explored for future research.

References

1. Biryukov, A., De Cannière, C.: Block ciphers and systems of quadratic equations. In: Johansson, T. (ed.) FSE 2003. LNCS, vol. 2887, pp. 274–289. Springer, Heidelberg (2003). doi:10.1007/978-3-540-39887-5_21
2. Blondeau, C., Canteaut, A., Charpin, P.: Differential properties of power functions. Int. J. Inf. Coding Theory **1**(2), 149–170 (2010)
3. Bracken, C., Leander, G.: A highly nonlinearity differentially 4-uniform power mapping that permutes fields of even degree. Finite Fields Appl. **16**(4), 231–242 (2010)

---

[2] We have tested the Li-Wang functions [17] in Magma for $n \leq 12$. The computation shows that all of them are implicitly quadratic. This suggests that it is likely that all the Li-Wang functions are implicitly quadratic.

4. Bracken, C., Tan, C.H., Tan, Y.: Binomial differentially 4-uniform permutations with high nonlinearity. Finite Fields Appl. **18**(3), 537–546 (2012)

5. Budaghyan, L., Carlet, C., Pott, A.: New class of almost bent and almost perfect nonlinear polynomials. IEEE Trans. Inf. Theory **52**(3), 1141–1152 (2006)

6. Carlet, C.: On known and new differentially uniform functions. In: Parampalli, U., Hawkes, P. (eds.) ACISP 2011. LNCS, vol. 6812, pp. 1–15. Springer, Heidelberg (2011). doi:10.1007/978-3-642-22497-3_1

7. Carlet, C., Tang, D., Tang, X., Liao, Q.: New construction of differentially 4-uniform bijections. In: Lin, D., Xu, S., Yung, M. (eds.) Inscrypt 2013. LNCS, vol. 8567, pp. 22–38. Springer, Heidelberg (2014). doi:10.1007/978-3-319-12087-4_2

8. Cheon, J.H., Lee, D.H.: Quadratic equations from APN power functions. IEICE Trans. Fundam. **E89-A**(1), 1–9 (2006)

9. Cheon, J.H., Lee, D.H.: Resistance of S-boxes against algebraic attacks. In: Roy, B., Meier, W. (eds.) FSE 2004. LNCS, vol. 3017, pp. 83–93. Springer, Heidelberg (2004). doi:10.1007/978-3-540-25937-4_6

10. Cid, C., Murphy, S., Robshaw, M.: Algebraic Aspects of the Advanced Encryption Standard. Springer, Heidelberg (2006)

11. Courtois, N.T., Debraize, B., Garrido, E.: On exact algebraic [non-]immunity of S-boxes based on power functions. In: Batten, L.M., Safavi-Naini, R. (eds.) ACISP 2006. LNCS, vol. 4058, pp. 76–86. Springer, Heidelberg (2006). doi:10.1007/11780656_7

12. Courtois, N., Klimov, A., Patarin, J., Shamir, A.: Efficient algorithms for solving overdefined systems of multivariate polynomial equations. In: Preneel, B. (ed.) EUROCRYPT 2000. LNCS, vol. 1807, pp. 392–407. Springer, Heidelberg (2000). doi:10.1007/3-540-45539-6_27

13. Courtois, N.T., Pieprzyk, J.: Cryptanalysis of block ciphers with overdefined systems of equations. In: Zheng, Y. (ed.) ASIACRYPT 2002. LNCS, vol. 2501, pp. 267–287. Springer, Heidelberg (2002). doi:10.1007/3-540-36178-2_17

14. Faugère, J.C.: A new efficient algorithm for computing Grobner bases without reduction to zero (F5). In: ISSAC 2002, pp. 75–83. ACM, New York (2002)

15. Gold, R.: Maximal recursive sequences with 3-valued recursive cross-correlation functions (corresp.). IEEE Trans. Inf. Theory **14**(1), 154–156 (1968)

16. Kasami, T.: The weight enumerators for several classes of subcodes of the 2nd order binary reed-muller codes. Inf. Control **18**(4), 369–394 (1971)

17. Li, Y.Q., Wang, M.S.: Constructing differentially 4-uniform permutations over $\mathbb{F}_{2^{2m}}$ from quadratic APN permutations over $\mathbb{F}_{2^{2m+1}}$. Des. Codes Cryptogr. **72**, 249–264 (2014)

18. De Meyer, C.: Matrix Analysis and Applied Linear Algebra. SIAM, Philadelphia (2000)

19. Nawaz, Y., Gupta, K.C., Gong, G.: Algebraic immunity of S-boxes based on power mappings, analysis and construction. IEEE Trans. Inf. Theory **55**(9), 4263–4273 (2009)

20. Nyberg, K.: Differentially uniform mappings for cryptography. In: Helleseth, T. (ed.) EUROCRYPT 1993. LNCS, vol. 765, pp. 55–64. Springer, Heidelberg (1994). doi:10.1007/3-540-48285-7_6

21. Peng, J., Tan, C.H.: New explicit constructions of differentially 4-uniform permutations via special partitions of $\mathbb{F}_{2^{2k}}$. Finite Fields Appl. **40**, 73–89 (2016)

22. Peng, J., Tan, C.H.: New differentially 4-uniform permutations by modifying the inverse function on subfields. Cryptogr. Commun. doi:10.1007/s12095-016-0181-x

23. Peng, J., Tan, C.H., Wang, Q.: A new family of differentially 4-uniform permutations over $\mathbb{F}_{2^{2k}}$ for odd k. Sci. China Math. **59**(6), 1221–1234 (2016)

24. Perrin, L., Udovenko, A., Biryukov, A.: Cryptanalysis of a theorem: decomposing the only known solution to the big APN problem. In: Robshaw, M., Katz, J. (eds.) CRYPTO 2016. LNCS, vol. 9815, pp. 93–122. Springer, Heidelberg (2016). doi:10.1007/978-3-662-53008-5_4

25. Qu, L.J., Tan, Y., Tan, C.H., Li, C.: Constructing differentially 4-uniform permutations over $\mathbb{F}_{2^{2k}}$ via the switching method. IEEE Trans. Inf. Theory **59**(7), 4675–4686 (2013)

26. Qu, L.J., Tan, Y., Li, C., Gong, G.: More constructions of differentially 4-uniform permutations on $\mathbb{F}_{2^{2k}}$. Des. Codes Cryptogr. **78**(2), 391–408 (2016)

27. Tang, D., Carlet, C., Tang, X.: Differentially 4-uniform bijections by permuting the inverse function. Des. Codes Cryptogr. **77**(1), 117–141 (2015)

28. Zha, Z.B., Hu, L., Sun, S.W.: Constructing new differentially 4-uniform permutations from the inverse function. Finite Fields Appl. **25**, 64–78 (2014)

29. Zha, Z.B., Hu, L., Sun, S.W., Shan, J.Y.: Further results on differentially 4-uniform permutations over $\mathbb{F}_{2^{2m}}$. Sci. China Math. **58**(7), 1577–1588 (2015)

Secret Sharing for mNP: Completeness Results

Mahabir Prasad Jhanwar[1]([⊠]) and Kannan Srinathan[2]

[1] Ashoka University, Sonepat, India
mahavir.jhawar@gmail.com
[2] IIIT Hyderabad, Hyderabad, India

Abstract. We show completeness results for secret sharing schemes realizing mNP access structures. We begin by proposing a new, Euclidean-type, division technique for access structures. Using this new technique we obtain several results in characterizing access structures for efficient (unconditionally secure) secret sharing schemes:

- We show a useful transformation that achieves efficient schemes for complex access structures using schemes realizing simple access structures.
- We show that, assuming every access structure in P ∩ mono admits efficient secret sharing, the existence of an efficient secret sharing for an access structure in mNP that is also complete for mNP under Karp/Levin *monotone-reductions* implies secret sharing schemes for all of mNP.
- We finally improve upon the above completeness result by obtaining the same under *ordinary* Karp/Levin reductions.

1 Introduction

Secret sharing schemes enable a dealer, holding a secret piece of information, to distribute this secret among a set $\mathcal{P}_n = \{P_1, \ldots, P_n\}$ of n players such that only some predefined authorized subsets of players can reconstruct the secret from their shares. The (monotone) collection $\Gamma_n \subseteq 2^{\mathcal{P}_n}$ of authorized sets that can reconstruct the secret is called an access structure. The security of a secret sharing scheme requires that any unauthorized set B of players, i.e., $B \notin \Gamma_n$, pulling its shares together and attempt to reconstruct the secret should fail with high probability. Consequently, the security is termed unconditional (computational) if the players are computationally unbounded (computationally bounded).

A secret sharing scheme realizing an access structure Γ_n over n players is termed size-efficient, if the total length of the n shares is polynomial in n; semi-efficient, if the share distribution is computable in $\mathsf{poly}(n)$ time; and efficient, if both share distribution and reconstruction are computable in $\mathsf{poly}(n)$ time. The notions of semi-efficiency and efficiency are stronger than size-efficiency.

A major problem in this field is the characterization of access structures in terms of secret sharing schemes that they admit, where the security and efficiency of the later is measured as a combination of the following:

- Unconditional/computational security, and
- size-efficiency/semi-efficiency/efficiency.

© Springer International Publishing AG 2016
O. Dunkelman and S.K. Sanadhya (Eds.): INDOCRYPT 2016, LNCS 10095, pp. 380–390, 2016.
DOI: 10.1007/978-3-319-49890-4_21

For concrete characterization, now onwards, we use the term *access structure* for referring to an infinite family of access structures $\Gamma = \{\Gamma_n\}_{n\in\mathbb{N}}$ (for every n, Γ_n is an access structure over $\mathcal{P}_n$) and the term "scheme realizing Γ" for referring to an infinite family of secret sharing schemes $\{\Pi_n\}_{n\in\mathbb{N}}$ such that for every n, Π_n realizes Γ_n.

Associating sets $A \subseteq \mathcal{P}_n$ with there characteristic vectors $x_A \in \{0,1\}^n$, we can define a language $L_\Gamma \subseteq \{0,1\}^*$ associated with an access structure $\Gamma = \{\Gamma_n\}_{n\in\mathbb{N}}$. Namely, $L_\Gamma = \cup_{n=1}^{\infty}\{x_A \in \{0,1\}^n \mid A \in \Gamma_n\}$. An access structure $\Gamma = \{\Gamma_n\}_{n\in\mathbb{N}}$ is said to be in the complexity class P ∩ mono if the associated language $L_\Gamma \in$ P ∩ mono. The Γ is said to be in mNP if $L_\Gamma \in$ mNP.

The question of access structures characterization has been widely studied. The extensive work in this area can be divided under the following two category of security: unconditional and computational. The most general class of access structures with known characterization results under them are given below.

– **Unconditional Security**
  • **P ∩ mono:** It has been extensively studied whether there exists efficient secret sharing schemes for every access structures in P ∩ mono? In fact, it is wide open if the same is true for all of mP - the class of access structure strictly contained in P ∩ mono. With several schemes realizing different classes of access structures [6–8,11,12,16], the most general class of access structures in P∩mono that admit efficient perfect secret sharing are those that can be described by a polynomial-size monotone span program [13].
  • **mNP:** The question of obtaining unconditionally secure efficient schemes for access structures in mNP was met with an impossibility result. Steven Rudich observed that if NP ≠ coNP, then for Hamiltonian access structure in NP there exists no semi-efficient secret-sharing scheme (specifically, schemes with perfect privacy) [4].
– **Computational Security**
  • **P ∩ mono:** It is known that the whole of mP admit efficient secret sharing schemes that are computationally secure - assuming that one-way functions exists [4,17].
  • **mNP:** Komargodski, Naor and Yogev [14] showed semi-secret sharing schemes for all of mNP (and therefore cover all of P ∩ mono), where the reconstruction algorithm is polynomial-time if the NP-witnesses for the authorized sets are given. Their scheme assumes existence of witness encryption [9] for whole of NP and one-way functions.

1.1 Our Results

An important corollary of the main result of Komargodski, Naor and Yogev [14] is the following completeness theorem for secret sharing schemes realizing mNP access structures:

Theorem 1 [14]. *Assume that one-way functions exists. Then existence of an efficient computational secret sharing for an access structure in* mNP *that is also complete for* mNP *under Karp/Levin reductions implies efficient computational secret sharing scheme for every access structure in* mNP.

The above theorem was established using the following two results:

- A secret sharing scheme for an access structure $\Gamma = \{\Gamma_n\}_{n \in \mathbb{N}}$ implies witness encryption for the associated language L_Γ.
- *Completeness theorem of witness encryption*: Using standard Karp/Levin reductions between NP-complete languages, one can transform a witness encryption for a single NP-complete language to a witness encryption scheme for any other language in NP.

Beside one-way functions, the completeness result in Theorem 1, therefore, is obtained based on the existence of witness encryption which in turn relies on strong computational assumptions related to indistinguishability obfuscation [2,3].

In this paper we obtain such completeness results for mNP access structures assuming that efficient secret sharing schemes exists for access structures in P∩mono. More importantly, *our completeness results hold under reductions with unconditional security*. As a corollary, our completeness results also partially resolve the following problem that was left open in [14]: Is there a way that can use secret sharing scheme for access structures in P ∩ mono to achieve secret sharing scheme for access structures in mNP?

In particular, this paper makes the following important contributions:

- Our foremost contribution lies in defining a new Euclidean-type division technique for access structures. Namely, for a given pair of access structures (more like a pair of *dividend* and *divisor*), this new technique distill a list of access structures, possibly simpler then dividend and divisor (more like a *remainder*). Unlike the ordinary Euclidean division for numbers, the remainder access structures are not fixed and choosing them carefully is of great importance as it allows for simplified reductions among schemes realizing these access structures.
- We next illustrate the usefulness of our proposed division property by describing a transformation that achieves efficient secret sharing scheme for a given access structure using secret sharing schemes for appropriately defined divisor and remainder access structures.
- The above transformation helps us to achieve our first completeness theorem: Namely we show that, assuming access structures in P ∩ mono admit efficient secret sharing, the existence of an efficient secret sharing for an access structure in mNP that is also complete for mNP under Karp/Levin *monotone-reductions* implies secret sharing schemes for all of mNP.
- The above completeness theorem is obtained for NP-completeness under monotone-reductions. Removing the later restriction proved to be an important achievement of our work. A clever construction of remainder access structures helped us to obtain our second completeness theorem: Namely we show, assuming access structures in P ∩ mono admit efficient secret sharing, the existence of an efficient secret sharing for an access structure in mNP implies efficient secret sharing for all of mNP.

2   Preliminaries

2.1   Access Structure and Its Complexity

Let $\mathcal{P}_n \overset{\text{def}}{=} \{P_1, \dots, P_n\}$ be a set of n players. A collection $\Gamma \subseteq 2^{\mathcal{P}_n}$ of subsets of $\mathcal{P}_n$ is called *monotone increasing* if, $A \in \Gamma$ and $A \subseteq B \subseteq \mathcal{P}_n$ implies $B \in \Gamma$. A collection $\Gamma' \subseteq 2^{\mathcal{P}_n}$ is called *monotone decreasing* if, $A \in \Gamma'$ and $B \subseteq A$ implies $B \in \Gamma'$.

Definition 1 (Access Structure). *An access structure on $\mathcal{P}_n$ is a tuple (Γ_n, Γ'_n), where $\Gamma_n, \Gamma'_n \subseteq 2^{\mathcal{P}_n}$, such that*

- *Γ_n is monotone increasing; Γ'_n is monotone decreasing, and*
- *$\Gamma_n \cap \Gamma'_n = \emptyset$.*

For an access structure (Γ_n, Γ'_n), the collection Γ'_n is often called an *adversary access structure*. We call an access structure complete if, the adversary access structure Γ'_n complements Γ_n in full. We consider only complete access structures in this paper and they are simply denoted by Γ_n.

Definition 2 (Complete Access Structure). *An access structure (Γ_n, Γ'_n) is called complete if, $\Gamma'_n = 2^{\mathcal{P}_n} \backslash \Gamma_n$, i.e., $\Gamma_n \cup \Gamma'_n = 2^{\mathcal{P}_n}$.*

An access structure Γ_n can be freely identified with its characteristic Boolean function $f_{\Gamma_n} : \{0,1\}^n \to \{0,1\}$. To each set $A \subseteq \mathcal{P}_n$ associate a unique (characteristic vector) $v^A = (v_1^A, \dots, v_n^A) \in \{0,1\}^n$ as follows: for every j in $1 \le j \le n$, $v_j^A = 1$ iff $P_j \in A$. Define, $D_{\Gamma_n} = \{v^A \mid A \in \Gamma_n\} \subseteq \{0,1\}^n$.

Definition 3 (Associated Boolean function). *For access structure Γ_n, the corresponding boolean function $f_{\Gamma_n} : \{0,1\}^n \to \{0,1\}$ is defined as follows: for $x \in \{0,1\}^n$, $f_{\Gamma_n}(x) = 1$ iff $x \in D_{\Gamma_n}$.*

Clearly, the boolean function f_{Γ_n} is monotone. Associating access structures Γ_n with their boolean functions f_{Γ_n}, we can associate a language $L_\Gamma \subseteq \{0,1\}^*$ to a family of access structures $\Gamma = \{\Gamma_n\}_{n \in \mathbb{N}}$.

Definition 4 (Associated Language). *For an access structure $\Gamma = \{\Gamma_n\}_{n \in \mathbb{N}}$, the corresponding language $L_\Gamma \subseteq \{0,1\}^*$ is defined as follows: $L_\Gamma = \{x \in \{0,1\}^* \mid f_{\Gamma_{|x|}}(x) = 1\}$, where $|x|$ denotes the length of the binary string x.*

For any access structure $\Gamma = \{\Gamma_n\}_{n \in \mathbb{N}}$, the corresponding language L_Γ is clearly in the complexity class mono - the class of monotone languages.

Definition 5 (Access Structure Complexity). *An access structure $\Gamma = \{\Gamma_n\}_{n \in \mathbb{N}}$ is said to be*

1. *in $\mathsf{P} \cap$ mono if $L_\Gamma \in \mathsf{P} \cap$ mono,*
2. *in $\mathsf{NP} \cap$ mono if $L_\Gamma \in \mathsf{NP} \cap$ mono.*

It is a well known fact that, $\mathsf{P} \cap$ mono $\neq$ mP [1,15], where the complexity class mP denotes languages that admit monotone circuits of polynomial-size; but $\mathsf{NP} \cap$ mono $=$ mNP [10], where mNP denotes the class of languages accepted by polynomial-size monotone non-deterministic circuits. We will refer to access strutures in $\mathsf{NP} \cap$ mono by mNP access structures.

2.2   Secret Sharing

An n-party secret sharing scheme involves $n + 1$ players: A dealer $\mathcal{D}$, a set $\mathcal{P}_n = \{P_1, \ldots, P_n\}$ of n participants, and an access structure Γ_n over $\mathcal{P}$. A secret sharing scheme for an arbitrary Γ_n allows the dealer to *distribute shares* of a *secret value* such that

- **Privacy**: any unauthorized set $B \subseteq \mathcal{P}$ of participants, i.e., $B \notin \Gamma_n$, must not obtain any information on the secret from their collective shares.
- **Reconstructability**: any authorized coalitions $A \subseteq \mathcal{P}$ of participants, i.e., $A \in \Gamma_n$, must always reconstruct the secret from their collective shares.

Definition 6 (Secret Sharing). *An n-party secret sharing for an access structure Γ_n over $\mathcal{P}_n = \{P_1, \ldots, P_n\}$ is a tuple $\Pi = (\mathsf{Share}, \mathsf{Rec}, \Sigma, \Sigma_1, \ldots, \Sigma_n)$ such that the following holds:*

- *Algorithms*
  - $\mathsf{Share}.\Pi$: *The share distribution algorithm $\mathsf{Share}.\Pi$ is a probabilistic algorithm that, on input $s \in \Sigma$ returns $(\mathsf{Sh}_1, \ldots, \mathsf{Sh}_n) \xleftarrow{\$} \mathsf{Share}.\Pi(s)$, where $\mathsf{Sh}_i \in \Sigma_i$, $1 \le i \le n$.*
  - $\mathsf{Rec}.\Pi$: *The secret reconstruction algorithm $\mathsf{Rec}.\Pi$ is a deterministic algorithm that on input $(\sigma_1, \ldots, \sigma_n) \in \prod_{i=1}^{n}(\Sigma_i \cup \{*\})$ returns a value $\sigma \leftarrow \mathsf{Rec}.\Pi(\sigma_1, \ldots, \sigma_n)$ where $\sigma \in \Sigma \cup \{\perp\}$. The distinguished symbols $*$ and $\perp$ have the following meanings: $\sigma_i = *$ means the ith share is missing, and $\perp \leftarrow \mathsf{Rec}.\Pi(\sigma_1, \ldots, \sigma_n)$ indicates that the algorithm is unable to recover the underlying secret.*
- *Property*
  - **Correctness:** *For every authorized set of players $A \subseteq \mathcal{P}_n$, i.e., $A \in \Gamma_n$, and for every $s \in \Sigma$, we have*

$$\mathsf{Rec}\big(\mathsf{Share}.\Pi(s)_A\big) = s \tag{1}$$

  where $\mathsf{Share}.\Pi(s)_A$ restricts the n length vector $(\mathsf{Sh}_1, \ldots, \mathsf{Sh}_n) \xleftarrow{\$} \mathsf{Share}.\Pi(s)$ to its A-entries, i.e., $\mathsf{Share}.\Pi(s)_A = \{\mathsf{Sh}_i\}_{P_i \in A}$.
  - **Security:** *The security of a secret sharing scheme is measured by the maximum probability with which a adversary $\mathcal{A}$ can win the following privacy game - PrivacySS.*

The game is played between the dealer $\mathcal{D}$ and an adversary $\mathcal{A}$ as follows:

1. *$\mathcal{A}$ first picks a pair of secrets $s_0, s_1 \in S$, and gives them to $\mathcal{D}$.*
2. *$\mathcal{D}$ chooses a random bit $b \in \{0, 1\}$ and executes $\mathsf{Share}.\Pi(s_b)$.*
3. *$\mathcal{A}$ queries shares of a set of participants $B \subseteq \mathcal{P}$ such that $B \notin \Gamma_n$.*
4. *$\mathcal{A}$ outputs a guess b' for b using the shares $\mathsf{Share}.\Pi(s_b)_B$.*

The adversary is said to win the game if $b' = b$. We measure its success as

$$Adv^{\mathsf{PrivacySS}}(\mathcal{A}) = 2 \cdot \Pr[b' = b] - 1.$$

$$
\boxed{
\begin{array}{l}
\Sigma \ni s_0, s_1 \leftarrow \mathcal{A}; \\
b \xleftarrow{\$} \{0,1\}; \\
(\mathsf{Sh}_1, \ldots, \mathsf{Sh}_n) \xleftarrow{\$} \mathsf{Share}.\Pi(s_b); \\
\Gamma_n \not\ni B \leftarrow \mathcal{A}; \\
\{0,1\} \ni b' \leftarrow \mathcal{A}\big(\mathsf{Share}.\Pi(s_b)_B\big)
\end{array}
}
$$

Fig. 1. PrivacySS: The Privacy Game

Definition 7 (Privacy). *A secret sharing scheme is said to have:*

* Perfect-Privacy, *when $\mathcal{A}$ is unbounded and $\mathsf{Adv}^{\mathsf{PrivacySS}}(\mathcal{A}) = 0$*
* ϵ-Statistical Privacy, *when $\mathcal{A}$ is unbounded and $\mathsf{Adv}^{\mathsf{PrivacySS}}(\mathcal{A}) < \epsilon$, where $\epsilon > 0$.*
* Computational-Privacy, *when $\mathcal{A}$ is a probabilistic polynomial time (PPT) algorithm and $\mathsf{Adv}^{\mathsf{PrivacySS}}(\mathcal{A}) < \eta(k)$, where $\eta(\cdot)$ is a negligible function, and k denotes the underlying security parameter of the scheme[1].*

- **Efficiency:** *Different measure of efficiency is used in the secret sharing literature. A secret sharing scheme Π is termed*

* *Size Efficient, if the total length of the n shares is polynomial in n.*
* *Semi Efficient, if the share distribution algorithm $\mathsf{Share}.\Pi$ is computable in $\mathsf{poly}(n)$ time.*
* *Efficient, if both $\mathsf{Share}.\Pi$ and $\mathsf{Rec}.\Pi$ are computable in $\mathsf{poly}(n)$ time.*

Definition 8 (Secret Sharing for Languages). *A family of secret sharing schemes $\Pi = \{\Pi_n\}_{n \in \mathbb{N}}$ is said to realize $\Gamma = \{\Gamma_n\}_{n \in \mathbb{N}}$ if for every $n \in \mathbb{N}$, Π_n realizes Γ_n. Then Π is also called a secret sharing scheme for the corresponding language L_Γ (see Definition 4).*

Consequently, $\Pi = \{\Pi_n\}_{n \in \mathbb{N}}$ realizing $\Gamma = \{\Gamma_n\}_{n \in \mathbb{N}}$ is said to be (size/semi) efficient if for every $n \in \mathbb{N}$, Π_n realizing Γ_n is (size/semi) efficient.

In the following, all the secret sharing schemes that we will present are both efficient and have perfect privacy.

3   A Division Property for Access Structures

For $n, m \in \mathbb{N}$, consider the following access structures:

- Γ_n - an access structure over $\mathcal{P}_n = \{P_1, \ldots, P_n\}$
- Δ_m - an access structure over $\mathcal{Q}_m = \{Q_1, \ldots, Q_m\}$, and
- for every i in $1 \le i \le m$, $\Gamma_n^{(i)}$ - an access structure over $\mathcal{P}_n$.

---

[1] In this setting, the instantiations of n, $|\Sigma|$, $\mathsf{Share}.\Pi$, $\mathsf{Rec}.\Pi$ and so on, admits an additional parameter k.

Definition 9. *We say* $\Gamma_n \bmod \Delta_m \stackrel{\text{def}}{=} \{\Gamma_n^{(1)}, \ldots, \Gamma_n^{(m)}\}$ *if, for every* $A \subseteq \mathcal{P}_n$ *the set* $A \bmod \Delta_m \stackrel{\text{def}}{=} \{Q_i \in \mathcal{Q}_m \mid A \in \Gamma_n^{(i)}\} \subseteq \mathcal{Q}_m$ *satisfies the following property:*

$$A \in \Gamma_n \iff A \bmod \Delta_m \in \Delta_m \tag{2}$$

The division property in Definition 9 closely resembles the ordinary Euclidean division for integers, where Γ_n is dividend, Δ_m is divisor, and remainder is formed by the list of access structures $\{\Gamma_n^{(1)}, \ldots, \Gamma_n^{(m)}\}$. Clearly, the size (the number of authorized sets) of each $\Gamma_n^{(i)}$ is at most that of Γ_n. We will later see the importance of obtaining smaller size (and therefore simpler) $\Gamma_n^{(i)}$'s.

4   A Transformation

Theorem 2. *Let* $\Gamma_n, \Gamma_n^{(1)}, \ldots, \Gamma_n^{(m)}$ *be access structures on* $\mathcal{P}_n$, *and* Δ_m *be an access structure on* $\mathcal{Q}_m$ *such that* $\Gamma_n \bmod \Delta_m = \{\Gamma_n^{(1)}, \ldots, \Gamma_n^{(m)}\}$. *Assume*

1. $\Pi_{\Delta_m} = (\text{Share.}\Pi_{\Delta_m}, \text{Rec.}\Pi_{\Delta_m})$ *is a perfect secret sharing scheme realizing* Δ_m, *and*
2. *for every* i *in* $1 \leq i \leq m$, $\Pi_{\Gamma_n^{(i)}} = (\text{Share.}\Pi_{\Gamma_n^{(i)}}, \text{Rec.}\Pi_{\Gamma_n^{(i)}})$ *is a perfect secret sharing realizing* $\Gamma_n^{(i)}$

then there exists Π_{Γ_n} - *a perfect secret sharing scheme realizing* Γ_n.

Proof: The secret sharing scheme Π_{Γ_n} can be described as follows:

– Share.Π_{Γ_n}: The share distribution algorithm distributes a secret s among players in $\mathcal{P}_n = \{P_1, \ldots, P_n\}$ as follows:
  - Compute $(s_1, \ldots, s_m) \xleftarrow{\$} \text{Share.}\Pi_{\Delta_m}(s)$
  - For every i in $1 \leq i \leq m$, compute $(s_{i1}, \ldots, s_{in}) \xleftarrow{\$} \text{Share.}\Pi_{\Gamma_n^{(i)}}(s_i)$

  The player P_j, for every j in $1 \leq j \leq n$, gets the following share:

$$P_j \leftarrow (s_{1j}, s_{2j}, \ldots, s_{mj})$$

– Rec.Π_{Γ_n}: For every authorized set $A \in \Gamma_n$, the players in A pull together their respective shares and reconstruct the secret as follows. Let $A \bmod \Delta_m = \{Q_{i_1}, \ldots, Q_{i_r}\} \subseteq \mathcal{Q}_m$, for some r in $1 \leq r \leq m$. By the definition of $A \bmod \Delta_m$, $A \in \Gamma_n^{(i_j)}$, j in $1 \leq j \leq r$, and therefore players in A reconstruct intermediate shares s_{i_j}'s using reconstruction algorithm Rec.$\Pi_{\Gamma_n^{(i_j)}}$'s respectively. As $A \bmod \Delta_m$ is in Δ_m, the secret is finally reconstructed by computing $s \leftarrow \text{Rec.}\Pi_{\Delta_m}(s_{i_1}, \ldots, s_{i_r})$.

– Privacy: Secret is perfectly hidden from the combined shares of any unauthorized set $A' \notin \Gamma_n$. Let $A' \bmod \Delta_m = \{Q_{i_1}, \ldots, Q_{i_u}\}$ and it does not belongs to Δ_m. The players in A' can compute intermediate shares s_{i_j}'s, $1 \leq j \leq u$, of the secret s. But these shares $\{s_{i_1}, \ldots, s_{i_u}\}$ will not reveal any information (perfectly hidden) about s as $\{Q_{i_1}, \ldots, Q_{i_u}\} \notin \Delta_m$.

5   Completeness Under Monotone-Reductions

Theorem 3. *Assume access structures in* $\mathsf{P} \cap \mathsf{mono}$ *admit efficient secret sharing. Then existence of an efficient secret sharing for an access structure in* mNP *that is also complete for* mNP *under Karp/Levin monotone-reductions implies secret sharing schemes for all of* mNP.

Proof: Let $\Delta = \{\Delta_m\}_{m \in \mathbb{N}}$ be an access structure in mNP that is also complete for mNP under monotone-reductions and suppose it admits an efficient secret sharing scheme. Consider an arbitrary access structure $\Gamma = \{\Gamma_n\}_{n \in \mathbb{N}}$ from mNP. We now show, for every $n \in \mathbb{N}$, Γ_n admits an efficient secret sharing scheme. For any fix n, there exists (completeness of Δ) an $m \in \mathbb{N}$ such that Γ_n is monotone-reducible to Δ_m, i.e., there exists a polynomial time computable *monotone* function $K_R : 2^{\mathcal{P}_n} \to 2^{\mathcal{Q}_m}$ such that the following holds:

$$\forall A \subseteq \mathcal{P}_n, A \in \Gamma_n \iff K_R(A) \in \Delta_m. \tag{3}$$

Define, for every i in $1 \le i \le m$, an access structure $\Gamma_n^{(i)}$ over $\mathcal{P}_n$ as follows:

$$\text{For } i \in [m], \Gamma_n^{(i)} = \{A \subseteq \mathcal{P}_n \mid Q_i \in K_R(A)\}. \tag{4}$$

The theorem follows by proving the following claims (see Theorem 2):

Claim 1: Each $\Gamma_n^{(i)}$ is in $\mathsf{P} \cap \mathsf{mono}$, $1 \le i \le m$
Claim 2: $\Gamma_n \bmod \Delta_m = \{\Gamma_n^{(1)}, \ldots, \Gamma_n^{(m)}\}$.

Proof of Claim 1: We first show $\Gamma_n^{(i)}$ is monotone, i.e., for every $A, B \subseteq \mathcal{P}_n$ with $\Gamma_n^{(i)} \ni A \subseteq B$, we show $B \in \Gamma_n^{(i)}$. Firstly, $Q_i \in K_R(A)$ as $A \in \Gamma_n^{(i)}$. Secondly, the monotone property of K_R map implies $K_R(A) \subseteq K_R(B)$. These two mean that $Q_i \in K_R(B)$, implying B belongs to $\Gamma_n^{(i)}$.

We now show $\Gamma_n^{(i)}$ is in P. For any set $A \subseteq \mathcal{P}_n$, $A \in \Gamma_n^{(i)}$ iff $Q_i \in K_R(A)$. But, K_R is a polynomial time computable function and therefore computing $K_R(A)$ is efficient, implying $\Gamma_n^{(i)}$ is in P.

Proof of Claim 2: We now prove $\Gamma_n \bmod \Delta_m = \{\Gamma_n^{(1)}, \ldots, \Gamma_n^{(m)}\}$, i.e., for every $A \subseteq \mathcal{P}_n$, $A \in \Gamma_n$ iff $A \bmod \Delta_m \in \Delta_m$. But

$$\begin{aligned}
A \bmod \Delta_m &= \{Q_i \in \mathcal{Q}_m \mid A \in \Gamma_n^{(i)}\} \\
&= \{Q_i \in \mathcal{Q}_m \mid Q_i \in K_R(A)\} \\
&= K_R(A)
\end{aligned}$$

Therefore, for every set $A \subseteq \mathcal{P}_n$

$$A \in \Gamma_n \overset{eqn-3}{\iff} K_R(A) \in \Delta_m$$
$$\iff A \bmod \Delta_m \in \Delta_m$$

This completes the proof.

6   Completeness Without Monotone-Reductions

Theorem 4. *Assume access structures in* P ∩ mono *admit efficient secret sharing. Then existence of an efficient secret sharing for an access structure in* mNP *that is also complete for* mNP *under ordinary (not necessarily monotone) Karp/Levin reductions implies efficient secret sharing for all those* $\Gamma = \{\Gamma_n\}_{n \in \mathbb{N}} \in$ mNP *that satisfy the following: for every n there exists a* $k_n \in \mathbb{N}$ *such that* $\Gamma_n = B_{k_n} \cup \{A \subseteq \mathcal{P}_n \mid |A| \geq k_n + 1\}$, *where* B_{k_n} *is a subset of* $A_{k_n} \stackrel{\text{def}}{=} \{A \subseteq \mathcal{P}_n \mid |A| = k_n\}$.

Proof: Let $\Delta = \{\Delta_m\}_{m \in \mathbb{N}}$ be an access structure in mNP that is also complete and it admits an efficient secret sharing scheme. Consider an arbitrary access structure $\Gamma = \{\Gamma_n\}_{n \in \mathbb{N}}$ from mNP satisfying the following: for every n there exists a $k_n \in \mathbb{N}$ such that $\Gamma_n = B_{k_n} \cup \{A \subseteq \mathcal{P}_n \mid |A| \geq k_n + 1\}$, where B_{k_n} is a subset of A_{k_n}, the set of all k_n-size subsets of $\mathcal{P}_n$. We now show that Γ_n admits efficient secret sharing scheme for every $n \in \mathbb{N}$. For any fix n, there exists (completeness of Δ) $m \in \mathbb{N}$ such that Γ_n is Karp/Levin reducible to Δ_m, i.e., there exists a polynomial time computable function $K_R : 2^{\mathcal{P}_n} \to 2^{\mathcal{Q}_m}$ with the following property:

$$\forall A \subseteq \mathcal{P}_n, A \in \Gamma_n \iff K_R(A) \in \Delta_m. \tag{5}$$

We now define, for every i in $1 \leq i \leq m$, an access structure $\Gamma_n^{(i)}$ on $\mathcal{P}_n$ as follows:

$$\Gamma_n^{(i)} = \{A \subseteq \mathcal{P}_n \mid Q_i \in K_R(A) \wedge |A| = k_n\} \cup \{A \subseteq \mathcal{P}_n \mid |A| \geq k_n + 1\} \tag{6}$$

It is easy to see that, for every i in $1 \leq i \leq m$, $\Gamma_n^{(i)}$ is in P∩mono. To prove the theorem, it suffices to show (by Theorem 2) that $\Gamma_n \bmod \Delta_m = \{\Gamma_n^{(1)}, \ldots, \Gamma_n^{(m)}\}$, i.e., for every $A \subseteq \mathcal{P}_n$, $A \in \Gamma_n$ iff $A \bmod \Delta_m \in \Delta_m$. We consider the following exhaustive cases.

- $|A| < k_n$: Clearly, $A \notin \Gamma_n$ and $A \bmod \Delta_m = \emptyset \notin \Delta_m$, and therefore $A \in \Gamma_n$ iff $A \bmod \Delta_m \in \Delta_m$ holds true.
- $|A| \geq k_n + 1$: In this case, $A \in \Gamma_n$ and $A \bmod \Delta_m = \mathcal{Q}_m \in \Delta_m$, and therefore $A \in \Gamma_n$ iff $A \bmod \Delta_m \in \Delta_m$ holds true.
- $|A| = k$: Finally, in this case

$$
\begin{aligned}
A \bmod \Delta_m &= \{Q_i \in \mathcal{Q}_m \mid A \in \Gamma_n^{(i)}\} \\
&= \{Q_i \in \mathcal{Q}_m \mid (Q_i \in K_R(A) \wedge |A| = k_n) \vee (|A| \geq k_n + 1)\} \\
&= \{Q_i \in \mathcal{Q}_m \mid Q_i \in K_R(A)\} \\
&= K_R(A)
\end{aligned}
$$

Hence, $A \in \Gamma_n \overset{eqn-5}{\iff} K_R(A) \in \Delta_m \iff A \bmod \Delta_m \in \Delta_m$.

Corollary 1. *Assume access structures in* P∩mono *admit efficient secret sharing. Then existence of an efficient secret sharing for an access structure in* mNP *implies efficient secret sharing for all of* mNP.

Proof: It suffices (by Theorem 4) to prove the following: the class of access structures $\Gamma = \{\Gamma_n\}_{n\in\mathbb{N}} \in \text{mNP}$ as described in Theorem 4 cover whole of mNP. This follows by a technique developed in [5]. We now show access structures in mNP are in one-one correspondence with access structures of the type described in Theorem 4.

Let $\hat{\Gamma} = \{\hat{\Gamma}_n\}_{n\in\mathbb{N}}$ be an arbitrary access structure in mNP. For every $n \in \mathbb{N}$, we now define, based on $\hat{\Gamma}_n$, an access structure $\tilde{\Gamma}_{2n}$. First identify $\hat{\Gamma}_n$ with the set $L_{\hat{\Gamma}_n} \subseteq \{0,1\}^n$. Now define $\tilde{\Gamma}_{2n}$ over a set of $2n$ players $\tilde{\mathcal{P}}_{2n} = \{P_{i,b}\}_{1\le i\le n; b\in\{0,1\}}$:

$$\tilde{\Gamma}_{2n} = B_n \cup \{A \subseteq \tilde{\mathcal{P}}_{2n} \mid |A| \ge n+1\}$$

where the collection B_n consists of precisely the following n-size subsets of $\tilde{\mathcal{P}}_{2n}$: for every $x = (x_1, \ldots, x_n) \in L_{\hat{\Gamma}_n}$, the set $\{P_{1,x_1}, \ldots, P_{n,x_n}\}$ is in B_n. Clearly, the complexity of checking whether a set $A \subseteq \tilde{\mathcal{P}}_{2n}$ is in $\tilde{\Gamma}_{2n}$ is exactly the complexity of deciding the membership in $L_{\hat{\Gamma}_n}$. However $L_{\hat{\Gamma}} = \{L_{\hat{\Gamma}_n}\}_{n\in\mathbb{N}}$ is in mNP (as $\hat{\Gamma} \in \text{mNP}$) and so $\tilde{\Gamma} = \{\tilde{\Gamma}_{2n}\}_{n\in\mathbb{N}}$ is in mNP. Finally, $\tilde{\Gamma} = \{\tilde{\Gamma}_{2n}\}_{n\in\mathbb{N}}$ is clearly of the type described in Theorem 4. This proves the corollary.

References

1. Alon, N., Boppana, R.B.: The monotone circuit complexity of boolean functions. Combinatorica **7**(1), 1–22 (1987)
2. Barak, B., Goldreich, O., Impagliazzo, R., Rudich, S., Sahai, A., Vadhan, S., Yang, K.: On the (im)possibility of obfuscating programs. In: Kilian, J. (ed.) CRYPTO 2001. LNCS, vol. 2139, pp. 1–18. Springer, Heidelberg (2001). doi:10.1007/3-540-44647-8_1
3. Barak, B., Goldreich, O., Impagliazzo, R., Rudich, S., Sahai, A., Vadhan, S.P., Yang, K.: On the (im)possibility of obfuscating programs. J. ACM **59**(2), 6 (2012)
4. Beimel, A.: Secret-sharing schemes: a survey. In: Chee, Y.M., Guo, Z., Ling, S., Shao, F., Tang, Y., Wang, H., Xing, C. (eds.) IWCC 2011. LNCS, vol. 6639, pp. 11–46. Springer, Heidelberg (2011). doi:10.1007/978-3-642-20901-7_2
5. Beimel, A., Ishai, Y.: On the power of nonlinear secrect-sharing. In: IEEE Conference on Computational Complexity, pp. 188–202 (2001)
6. Benaloh, J., Leichter, J.: Generalized secret sharing and monotone functions. In: Goldwasser, S. (ed.) CRYPTO 1988. LNCS, vol. 403, pp. 27–35. Springer, Heidelberg (1990). doi:10.1007/0-387-34799-2_3
7. Blakley, G.: Safeguarding cryptographic keys. In: AFIPS National Computer Conference, vol. 48, pp. 313–317 (1979)
8. Brickell, E.F.: Some ideal secret sharing schemes. In: Quisquater, J.-J., Vandewalle, J. (eds.) EUROCRYPT 1989. LNCS, vol. 434, pp. 468–475. Springer, Heidelberg (1990). doi:10.1007/3-540-46885-4_45
9. Garg, S., Gentry, C., Sahai, A., Waters, B.: Witness encryption and its applications. In: Boneh, D., Roughgarden, T., Feigenbaum, J. (eds.) STOC, pp. 467–476. ACM (2013)
10. Grigni, M., Sipser, M.: Monotone complexity. In: LMS Workshop on Boolean Function Complexity, vol. 169, pp. 57–75. Cambridge University Press (1992)

11. Ito, M., Saito, A., Nishizeki, T.: Secret sharing scheme realizing general access structure. Electron. Commun. Jpn. (Part III: Fundam. Electron. Sci.) **72**(9), 56–64 (1989)
12. Jackson, W.-A., Martin, K.M.: Cumulative arrays and geometric secret sharing schemes. In: Seberry, J., Zheng, Y. (eds.) AUSCRYPT 1992. LNCS, vol. 718, pp. 48–55. Springer, Heidelberg (1993). doi:10.1007/3-540-57220-1_51
13. Karchmer, M., Wigderson, A.: On span programs. In: Structure in Complexity Theory Conference, pp. 102–111 (1993)
14. Komargodski, I., Naor, M., Yogev, E.: Secret-sharing for NP. In: Sarkar, P., Iwata, T. (eds.) ASIACRYPT 2014. LNCS, vol. 8874, pp. 254–273. Springer, Heidelberg (2014). doi:10.1007/978-3-662-45608-8_14
15. Razborov, A.A.: Lower bounds on the monotone complexity of some Boolean functions. Dokl. Akad. Nauk SSSR **281**, 798–801 (1985). English translation in Sov. Math. Doklady **31**, 354–357 (1985)
16. Shamir, A.: How to share a secret. Commun. ACM **22**(11), 612–613 (1979)
17. Vinod, V., Narayanan, A., Srinathan, K., Rangan, C.P., Kim, K.: On the power of computational secret sharing. In: Johansson, T., Maitra, S. (eds.) INDOCRYPT 2003. LNCS, vol. 2904, pp. 162–176. Springer, Heidelberg (2003). doi:10.1007/978-3-540-24582-7_12

New Cryptographic Constructions

Receiver Selective Opening Security from Indistinguishability Obfuscation

Dingding Jia[1,2,3(✉)], Xianhui Lu[1,2,3], and Bao Li[1,2,3]

[1] State Key Laboratory of Information Security,
Institute of Information Engineering, CAS, Beijing, China
{ddjia,xhlu,lb}@is.ac.cn
[2] Data Assurance and Communication Security Research Center,
CAS, Beijing, China
[3] University of Chinese Academy of Sciences, Beijing, China

Abstract. In this paper we study public key encryptions secure against RSO (receiver selective opening) attacks. To do so, we exploit the puncturable property of several existing CCA secure schemes that employs the "all-but-one" technique, use an indistinguishability obfuscator to wrap up the decryption circuit and set the obfuscated circuit as the secret key. Concretely, our first construction is from lossy trapdoor functions; our second construction is a bit encryption from puncturable pseudo-random functions and is secure against chosen ciphertext attacks simultaneously.

Keywords: Indistinguishability obfuscation · Receiver selective opening · Public key encryption

1 Introduction

The notion of selective opening attacks is firstly considered in the multi-party computation scenario [6]. It studies security for the uncorrupted parties, when messages of different parties are correlated and some parties are corrupted with internal randomness revealed. In 2009, Bellare, Hofheinz and Yilek introduced the formal definition of selective opening in the PKE (public key encryption) scenario [3], and numerous works on this topic appeared ever since then [1,2,13, 14,17–19,21–26,30].

Compared with the ordinary IND-CPA/CCA (indistinguishability against chosen plaintext/ciphertext attacks) security, security in the selective opening case is more complicated, for the reason that the opening of the randomness allows the adversary to verify the correspondence between the ciphertext and the message. In [3] they formulated the definition of selective opening in two styles: one is the indistinguishability-based style, which we will call IND-SO; the other is the simulation-based style, which we will call SIM-SO. Relations among

This work is Supported by the National Basic Research Program of China (973 project) (No. 2013CB338002), the National Nature Science Foundation of China (No. 61502484, No. 61379137, No. 61572495).

O. Dunkelman and S.K. Sanadhya (Eds.): INDOCRYPT 2016, LNCS 10095, pp. 393–410, 2016.
DOI: 10.1007/978-3-319-49890-4_22

IND-SO, SIM-SO and the ordinary IND security have been well studied recently [1, 2, 4, 14, 21–23, 30].

Depending on the attack scenario, selective opening has been considered in two flavors. In the SSO (sender selective opening) setting, there are many senders, the attacker can corrupt some of them and get the messages together with the encryption randomness. In the RSO (receiver selective opening) setting, there are many receivers, the attacker can corrupt some of them and get the messages together with the corresponding secret keys for decryption (here we assume that the randomness for the key generation algorithm is erased once key pairs are generated). One may notice that the encryption randomness in the SSO setting and the secret key in the RSO setting play similar roles in some sense.

Constructions in the SSO Setting. The first feasibility results for selective opening security are in the sender setting and leverage an interesting relation with LE (lossy encryption) [3]. They proved that LE implies IND-SSO-CPA security, and LE that supports efficient opening implies SIM-SSO-CPA security. Several constructions of LE were promoted in [3, 19]. Approaches to achieving SSO and CCA security simultaneously are various and more primitives are involved, such as ABM-TDF (all-but-many lossy trapdoor functions), XAC (cross authenticated code) [13, 18, 19, 25, 26].

Constructions in the RSO Setting. Despite there are many constructions in the SSO setting, constructions in the RSO setting are relatively less. Though existence for RSO security schemes is implied by the existence of non-committing encryption [5, 6, 8, 11, 20], Nielsen [29] showed that multi-message SIM-RSO security in the standard model without erasures is impossible for key pairs of constant length[1]. To circumvent this, Canetti *et al.* gave a framework to transform a single-message SIM-RSO-CPA secure construction to multi-message secure construction via a key-evolving system [7]. They also gave concrete SIM-RSO-CPA constructions for single-message case [7]. In 2015, Hazay *et al.* [21] introduced a primitive called tNCER (tweaked non-committing encryptions for receivers) which implies IND-RSO-CPA security and proposed several instantiations.

Motivation. In this work, we focus on constructions of SIM-RSO security. While in the IND-RSO security notion, the message distributions are restricted to be efficiently conditional resamplable, security notions for SIM-RSO does not suffer from such restrictions. And the SIM-RSO security implies IND-RSO security. In a nutshell, a scheme is SIM-RSO secure if for any adversary who can see a sequence of public keys and ciphertexts, part of which will be opened with secret keys later, there exists a simulator that can compute the same output without seeing the ciphertexts and secret keys. While SIM-RSO security requires the opening of secret keys, one may notice that the IND-CCA security requires answering decryption queries, which needs information about the secret keys too.

---

[1] Note that Nielsen's bound is only effective for SIM-RSO security; for IND-RSO setting, security for single-message case and multi-message case is equivalent, which can be easily proved via a hybrid argument as that for ordinary IND-CPA security.

A natural question is whether there exist approaches to transforming IND-CCA security to SIM-RSO security.

There are mainly two approaches to constructing practical IND-CCA secure PKE in the standard model: one employs the universal HPS [9], which we call the Cramer-Shoup type; the other employs an ABO (all-but-one) technique [31–34], which we call the ABO type. In the security proof, to answer decryption queries correctly, the simulator should hold some private informtion about the secret key. For the Cramer-Shoup type this private information is the original secret key, so it is natural to open the secret key together with the predetermined message, schemes of this kind can fit into the tNCER paradigm[2]. For the ABO type the simulator just holds an ABO key, with which the simulator can answer all decryption queries except for those in some relation with the challenge ciphertext, which makes it not able to open secret keys. Since schemes of the ABO type admit instantiations from search problems, such as factoring, and search problems encompass a larger class of intractable problems than decisional assumptions, it is worthwhile to study how to obtain the SIM-RSO security from the ABO type schemes. In this paper we try to convert IND-CCA secure schemes of the ABO type to achieve SIM-RSO security.

As Matsuda and Hanaoka [28] pointed out, schemes of the ABO type satisfy a decryption puncturable property. Sahai and Waters [33] in 2014 and Garg et al. [15] in 2013 proposed methods to apply $i\mathcal{O}$ (indistinguishability obfuscation) to puncturable programs and obtained deniable encryptions and several other primitives, such as PKE, signatures, injective hash functions, etc.. Generally speaking, $i\mathcal{O}$ assures that the obfuscation of any two distinct (equal-size) circuits that implement identical functionality be computationally indistinguishable. In this paper we will use $i\mathcal{O}$ to handle the decryption puncturable property and achieve SIM-RSO security.

Contributions. In this paper we use $i\mathcal{O}$ to convert two IND-CCA secure schemes to be SIM-RSO secure. That is, we set the secret key to be the obfuscated decryption circuit. Firstly, we transform the construction from LTDF and ABO-TDF in [31] to achieve SIM-RSO-CPA security. Secondly, we transform the bit encryption from puncturable PRF (pseudo-random function) in [33] to achieve SIM-RSO-CCA security. The main observation is to handle the decryption puncturable property with an $i\mathcal{O}$ and set the obfuscated circuit as the secret key[3]. In the following are some technique overviews.

Since $i\mathcal{O}$ only assures indistinguishability of circuits that of equal functionality, but for encryption schemes of the ABO type, the ABO key cannot decrypt the challenge ciphertext, one problem for using $i\mathcal{O}$ is to assure that the simulated decryption algorithm outputs the same result for the "one challenge ciphertext. Our resolution is to hardwire the challenge ciphertext and the matching answer as the constant information into the $i\mathcal{O}$. For ciphertexts different from

---

[2] Note that tNCER is IND-RSO secure, and can achieve SIM-RSO security only if a fake ciphertext can be opened to any message with a secret key efficiently.

[3] Similar technique has been used recently in other works [8,10]. Our work is independent to that.

the challenge ciphertext, it could be decrypted by the simulated key without the embedded constant information.

For the construction based on LTDF [31], we use the function value of LTDF as the branch tag of the ABO-TDF, i.e. the ciphertext is of form $(c_1 = f(x), c_2 = g(c_1, x), c_3 = h(x) \oplus m)$, where f is an LTDF, g is an ABO-TDF, h is a pairwise independent hash function, sk is set to be the obfuscation of the circuit that takes (c_1, c_2, c_3) as input and outputs $m = c_3 \oplus h(x)$ where $x = f^{-1}(c_1)$ if equations $c_1 = f(x), c_2 = g(c_1, x)$ hold. In the security proof, we puncture the key at challenge ciphertext c^*, and hardwire the mapping $c^* \rightarrow m^*$ in order to preserve the input-output behavior, then we switch f and g to lossy according to the indistinguishability of LTDF and ABO-TDF, thus $h(x)$ and c_3 are randomly distributed, which is irrelevant to m and can be simulated with a dummy message. With similar technique, we can modify schemes from ABO-XHPS (all-but-one extractable hash proof system) [34] and one-way functions for correlated product input [32] to be SIM-RSO-CPA secure respectively.

For the concrete CCA secure bit encryption from puncturable PRF in [33], the ciphertext is of form $(c_1 = PRG(r), c_2 = PRF_1(K_1, c_1) \oplus m, c_3 = PRF_2(K_2, c_1 \| c_2))$. Since c_3 verifies c_1 and c_2 simultaneously, the transformed construction, in which we set sk to be the obfuscation of the circuit that hardwired (K_1, K_2), takes (c_1, c_2, c_3) as input and outputs $m = c_2 \oplus PRF_1(K_1, c_1)$ if $c_3 = PRF_2(K_2, c_1 \| c_2)$, can achieve SIM-RSO-CCA security. In the security proof, we puncture the key at challenge ciphertext $c_1^*, (c_1^* \| 0, c_1^* \| 1)$, and hardwire the mapping $c^* \rightarrow m^*$ in order to preserve the input-output behavior, then we switch c_2^*, c_3^* to random according to the pseudorandom property of puncturable PRF, thus c_2^* is irrelevant to m and can be simulated with a dummy message. It is easy to see that correctness of decryption is preserved. We state that the construction is meaningful for two reasons: firstly, it is better to take minimalist approach outside of the strong tool "obfuscation", such as one-way functions; secondly, the construction achieves a stronger security against chosen ciphertext attacks.

As far as we know, except for those from non-committing encryption, current RSO secure constructions only consider cases in which key generation randomness is erased after key pairs are generated. We leave it as an open problem to build RSO secure schemes that withstand learning the whole generation randomness and which are not non-committing encryptions.

Organization. The rest of the paper is organized as follows: in Sect. 2 we give some preliminaries and definitions, in Sect. 3 we give two constructions and prove their SIM-RSO security and Sect. 4 is the conclusion of the whole paper.

2   Preliminaries and Definitions

2.1   Preliminaries

Notations. In this paper we use PPT to represent probabilistic polynomial time for short. Let $[n]$ be the set of $\{1, 2, ..., n\}$. $a \leftarrow A$ is used to denote choosing a random element from A when A is a set, and to denote picking an element

Experiment. $Exp_{\text{real}}^{\text{sim-rso-cca}}(\mathcal{A})$:	Experiment. $Exp_{\text{ideal}}^{\text{sim-rso-cca}}(\mathcal{S})$:
$state = \epsilon$ $(\boldsymbol{pk}, \boldsymbol{sk}) := (pk_i, sk_i)_{i\in[n]} \leftarrow_R Setup(1^\lambda)$ $(state, dist) \leftarrow \mathcal{A}^{Dec(\cdot,\cdot)}(\boldsymbol{pk}, state)$ $\boldsymbol{m} \leftarrow dist$ $\boldsymbol{c}^* \leftarrow Enc(\boldsymbol{pk}, \boldsymbol{m})$ $(state, I) \leftarrow \mathcal{A}^{Dec(\cdot,\cdot)}(\boldsymbol{c}, state)$ $output_A \leftarrow \mathcal{A}^{Dec(\cdot,\cdot)}(\boldsymbol{sk}_I, \boldsymbol{m}_I, state)$ return $(\boldsymbol{m}, output_A)$	$state = \epsilon$ $(state, dist) \leftarrow \mathcal{S}(1^\lambda, state)$ $\boldsymbol{m} \leftarrow dist$ $(state, I) \leftarrow \mathcal{S}(state)$ $output_S \leftarrow \mathcal{S}(\boldsymbol{m}_I, state)$ return $(\boldsymbol{m}, output_S)$

Fig. 1. SIM-RSO-CCA security

according to A when A is a distribution. We use the lower case boldface to denote vectors. $Enc(\boldsymbol{pk}, \boldsymbol{m}) := (Enc(pk_1, m_1), ..., Enc(pk_n, m_n))$ when $\boldsymbol{pk}, \boldsymbol{m}$ are vectors of dimension n. The min-entropy of a distribution $\mathcal{X}$ over domain $\mathcal{D}$ is defined as $\text{H}_\infty(\mathcal{X}) = max_{a\in\mathcal{D}}(-log \Pr[\mathcal{X} = a])$. The statistical distance of two distributions $\mathcal{X}$ and $\mathcal{Y}$ over a common domain $\mathcal{D}$ is defined as $SD(\mathcal{X}, \mathcal{Y}) = \frac{1}{2}\sum_{a\in\mathcal{D}} |\Pr[\mathcal{X} = a] - \Pr[\mathcal{Y} = a]|$.

2.2 Security Definitions

A PKE scheme consists of the following algorithms:

Keygen: the key generation algorithm takes as input a security parameter λ and outputs a public key pk and a secret key sk. $Keygen(1^\lambda) \rightarrow (pk, sk)$.

Enc: the encryption algorithm takes as input the public key pk, a message m in the message space $\mathbb{M}$, and outputs a ciphertext c. $Enc(pk, m) \rightarrow c$.

Dec: the decryption algorithm takes the secret key sk and a ciphertext c as input and outputs a message m or $\perp$. $Dec(sk, c) \rightarrow m$ or $\perp$.

Correctness. A PKE scheme satisfies correctness, if $Dec(sk, Enc(pk, m)) = m$ for all $(pk, sk) \leftarrow Keygen(1^\lambda), m \in \mathbb{M}$.

Security. Here we give the simulation based security against receiver selective opening chosen ciphertext attacks (SIM-RSO-CCA) of a PKE scheme as in [21]. Here we modify the output of the Experiment from $(\boldsymbol{m}, output_A, I, dist)$ $/(\boldsymbol{m}, output_S, I, dist)$ to $(\boldsymbol{m}, output_A)$ $/(\boldsymbol{m}, output_S)$, since $(I, dist)$ is chosen by the adversary/simulator, hence can be included in $output_A/output_S$.

Note that in $Exp_{\text{real}}^{\text{sim-rso-cca}}(\mathcal{A})$, the decryption query is of the form (c, j), and is answered by $Dec(sk_j, c)$, and it is required that $c \neq c_j^*$. And after receiving $\boldsymbol{sk}_I$, it is required that the decryption query (c, j) satisfies $j \notin I$. The advantage of a distinguisher $\mathcal{D}$ with binary output is defined as $Adv_{\mathcal{D}}^{\text{sim-RSO-CCA}} = \left| \Pr[D(Exp_{\text{real}}^{\text{sim-rso-cca}}(\mathcal{A})) = 1] - \Pr[D(Exp_{\text{ideal}}^{\text{sim-rso-cca}}(\mathcal{S})) = 1] \right|$. When omitting the decryption oracle, the above experiment gives the definition of IND-RSO-CPA security.

Definition 1 (IND-RSO-CCA/CPA Security). *A PKE scheme is SIM-RSO-CCA secure if for any PPT adversary $\mathcal{A}$, there exists a PPT simulator $\mathcal{S}$, such that for any PPT distinguisher D, $Adv_D^{SIM\text{-}RSO\text{-}CCA}$ is negligible in λ. And it is SIM-RSO-CPA secure if for any PPT adversary $\mathcal{A}$, there exists a PPT simulator $\mathcal{S}$, such that for any PPT distinguisher D, $Adv_D^{SIM\text{-}RSO\text{-}CPA}$ is negligible in λ.*

For simplicity here we use the security definition that the corruption is one-shot, there is also a slightly stronger definition which allows for promoting corruption queries $i \in I$ adaptively [1]. In fact our constructions satisfy the stronger definition.

2.3 Indistinguishability Obfuscation

Intuitively, an indistinguishability obfuscator keeps the functionality unchanged for a circuit, and other information about the circuit is computationally protected. In the following we show the formal definition as in [15,33].

Definition 2 (Indistinguishability Obfuscator $(i\mathcal{O})$). *A uniform PPT machine $i\mathcal{O}$ is called an indistinguishability obfuscator for a circuit class $\{\mathcal{C}_\lambda\}$ if the following conditions hold:*

- *for all security parameter λ, for all $C \in \mathcal{C}_\lambda$, for all inputs x, it satisfies that*

$$\Pr[C'(x) = C(x) : C' \leftarrow i\mathcal{O}(1^\lambda, C)] = 1.$$

- *For any PPT algorithms $Samp, \mathcal{D}$, if there exists a negligible function $\alpha(\cdot)$ such that: if $\Pr[\forall x, C_0(x) = C_1(x) : (C_0, C_1, \sigma) \leftarrow Samp] > 1 - \alpha(1^\lambda)$, then*

$$\left| \Pr[\mathcal{D}(\sigma, i\mathcal{O}(1^\lambda, C_0)) = 1] - \Pr[\mathcal{D}(\sigma, i\mathcal{O}(1^\lambda, C_1)) = 1] \right| \leq \alpha(1^\lambda),$$

where $(C_0, C_1, \sigma) \leftarrow Samp$.

If the circuit class $\{\mathcal{C}_\lambda\}$ is of size at most $poly(1^\lambda)$, then the above obfuscator is called an indistinguishability obfuscator for polynomial-size circuits.

2.4 Puncturable Pseudo-random Functions

The notion of puncturable PRFs was introduced by Sahai and Waters [33]. In a puncturable PRF, a punctured key is computed from a normal key and some punctured points. When a punctured key is given out, function can be correctly evaluated on inputs that are not punctured; however, on inputs that are punctured, the outputs are pseudo-random.

Definition 3. *A puncturable family of PRFs mapping from $n(1^\lambda)$ to $m(1^\lambda)$ consists of three algorithms $(Key_F, Puncture_F, Eval_F)$, where $(Key_F, Eval_F)$ are as normal function description and $Puncture_F$ computes a punctured key K_S from K and a set S. It satisfies the following properties:*

Functionality preserved under puncturing. *For every PPT adversary $\mathcal{A}$ that outputs a set $S \subset \{0,1\}^{n(1^\lambda)}$ and every $x \in \{0,1\}^{n(1^\lambda)} \setminus S$,*

$$\Pr[F(K,x) = F(K(S),x) : K \leftarrow Key_F, K(S) = Puncture_F(K,S)] = 1.$$

Pseudo-random on punctured points. *For every PPT adversary $\mathcal{A}$, $Adv_{\mathcal{A}}^{pr} = |2\Pr[Exp^{pr}(\mathcal{A}) = 1] - 1|$ is negligible, where $Exp^{pr}(\mathcal{A})$ is defined as follows (Fig. 2.):*

2.5 LTDFs and ABO-TDFs

A collection of LTDFs consists of two indistinguishable families of functions. Functions in one family are injective, while functions in the other family are lossy. Concretely,

Definition 4 (LTDFs). *A collection of (n,l)-LTDFs is a 4-tuple PPT algorithms (S_0, S_1, F, F^{-1}) such that:*

Sampling a lossy function: *$S_0(1^\lambda)$ outputs a function index $s \in \{0,1\}^*$. The algorithm $F(s,\cdot)$ computes a function $f_s : \{0,1\}^n \mapsto \{0,1\}^*$, whose image size is at most 2^{n-l}.*

Sampling an injective function with its trapdoor: *$S_1(1^\lambda)$ outputs a pair $(s,t) \in \{0,1\}^* \times \{0,1\}^*$. The algorithm $F(s,\cdot)$ computes a function $f_s : \{0,1\}^n \mapsto \{0,1\}^*$, and $F^{-1}(t,\cdot)$ computes its inverse, which satisfies that for every $x \in \{0,1\}^n$, $F^{-1}(t, F(s,x)) = x$.*

Hard to distinguish injective from lossy: *For any PPT adversary $\mathcal{A}$, $Adv_{\mathcal{A}}^{LTDF}$ is negligible, where*

$$Adv_{\mathcal{A}}^{LTDF} = \Pr[\mathcal{A}(s : s \leftarrow S_0(1^\lambda)) = 1] - \Pr[\mathcal{A}(s : (s,t) \leftarrow S_1(1^\lambda)) = 1].$$

An ABO-TDF (all-but-one lossy trapdoor function) is associated with a branch set B. The sampling algorithm takes $b^* \in B$ as input, and outputs a

---

Experiment. $Exp^{pr}(\mathcal{A})$:

$b \leftarrow_R \{0,1\}$
$(S \subset \{0,1\}^{n(1^\lambda)}, st) \leftarrow \mathcal{A}$
$K \leftarrow Key_F$
$K(S) = Puncture_F(K,S)$
$Z_0 \leftarrow \{0,1\}^{m(1^\lambda)|S|}$
$Z_1 \leftarrow F(K,S) (= \text{concatenation of } F(K,x) \text{ for } x \in S)$
$b' \leftarrow \mathcal{A}(K(S), Z_b, st)$
if $b = b'$, outputs 1, else outputs 0

Fig. 2. puncturable PRF

description s_0 of the function $G(\cdot,\cdot,\cdot)$ together with a trapdoor t_0. The function has the property that for any branch $b \neq b^*$ the function $G(s_0,b,\cdot)$ is injective, and the function $G(s_0,b^*,\cdot)$ is lossy. For security, it requires that it is hard to distinguish a function description generated by different branches.

Definition 5 (ABO-TDF [12,31]). *A collection of (n,l)-ABO-TDFs consists of 4-tuple PPT algorithms (B,S,G,G^{-1}) such that:*

Sampling a branch: $B(1^\lambda)$ *outputs a branch $b \in \{0,1\}^v$.*

Sampling a function: *For every b output by $B(1^\lambda)$, the algorithm $S(1^\lambda,b)$ outputs a pair $(s_0,t_0) \in \{0,1\}^* \times \{0,1\}^*$ consists of a function index s_0 and a trapdoor t_0.*

Evaluation of lossy and injective branches: *For every b^* produced by $B(1^\lambda)$ and for every (s_0,t_0) produced by $S(1^\lambda,b^*)$, the algorithm $G(s_0,b^*,\cdot)$ computes a function $g_{s_0,b^*} : \{0,1\}^n \mapsto \{0,1\}^*$, whose image size is at most 2^{n-l}; for $b \neq b^*$, the algorithm $G(s_0,b,\cdot)$ computes an injective function $g_{s_0,b} : \{0,1\}^n \mapsto \{0,1\}^*$ and G^{-1} computes its inverse, which satisfies that for every $x \in \{0,1\}^n$, $G^{-1}(t_0,b,G(s_0,b,x)) = x$.*

Security: *For any PPT adversary $\mathcal{A}$ that outputs (b_0,b_1), $Adv_{\mathcal{A}}^{ABO-TDF}$ is negligible, where*

$$Adv_{\mathcal{A}}^{ABO-TDF} = \Pr[\mathcal{A}((b_0,b_1,s_0) : (s_0,t_0) \leftarrow S(1^\lambda,b_0)) = 1] - \Pr[\mathcal{A}((b_0,b_1,s_1) :$$

$$(s_1,t_1) \leftarrow S(1^\lambda,b_1)) = 1].$$

LTDFs and ABO-TDFs are equivalent for appropriate choice of parameters [31], and they can be constructed from decisional assumptions, like DDH assumption [31], quadratic residuosity assumption, composite residuosity assumption, d-linear assumption, LWE assumption, etc. [12]. Some of these constructions are slightly lossy, but using the method in [27], one can amplify the lossiness.

2.6 Randomness Extractor

In this work, we use pairwise independent hash function as a randomness extractor as in [31], the leftover hash lemma states that for a random variable X, as long as the difference between the min-entropy of X and the output length is large enough, the output $h(X)$ is statistically close to uniform.

Lemma 1 (Leftover Hash Lemma [16]). *Let $X \in \mathcal{X}$ be a random variable where $H_\infty(X) \geq \kappa$. Let H be a family of pairwise independent hash functions with domain $\mathcal{X}$ and range $\{0,1\}^l$. Then for $h \leftarrow H$ and $U_l \leftarrow \{0,1\}^l$, it satisfies that*

$$SD((h,h(X)),(h,U_l)) \leq 2^{(l-\kappa)/2}.$$

3   Constructions

3.1   CPA Secure Construction from LTDF

Here we describe a SIM-RSO-CPA secure construction converted from the PKE construction in [31]. Let (S_0, S_1, F, F^{-1}) be a collection of (v, l_1)-LTDFs, without loss of generality, we assume that the image of F is $\{0,1\}^v$. Let H be a family of pairwise independent hash functions from $\{0,1\}^v$ to $\{0,1\}^l$. Let (B, S, G, G^{-1}) give a collection of (v, l_0)-ABO-TDFs having branches $B = \{0,1\}^v$ (which contains the image of F). The message space is $\{0,1\}^l$.

As in [31], it requires that the parameters satisfy that $(v-l_1)+(v-l_0) \leq v-\kappa$ for some $\kappa = w(\log v)$, and $l \leq \kappa - 2\lg(1/\epsilon)$ for a negligible ϵ.

Keygen: It first generates an injective trapdoor function: $(s, t) \leftarrow S_1(1^\lambda)$ and an ABO-TDF having lossy branch $0^v : (s_0, t_0) \leftarrow S(1^\lambda, 0^v)$. Then it chooses a pairwise independent hash function $h \leftarrow H$. After that, it creates an obfuscation of the program Decrypt of Fig. 3, and pad the size to be the maximum of itself and program Decrypt* of Fig. 4. Finally erase the randomness that is used for key generation. The public key $pk = (s, s_0, h)$ and the secret key sk is the obfuscated program Decrypt.

Enc: It takes as input pk and chooses $x \in \{0,1\}^v$ uniformly at random, computes $c_1 = F(s, x)$, then it computes $c_2 = G(s_0, c_1, x)$ and $c_3 = h(x) \oplus m$. The ciphertext $c = (c_1, c_2, c_3)$.

Dec: It runs the obfuscated program of Decrypt on input c and takes the output as the decryption.

Constants: $(pk = (s, s_0, h), t)$
Input: $(c_1, c_2, c_3) \in \{0,1\}^v \times \{0,1\}^* \times \{0,1\}^l$.

1. compute $x = F^{-1}(t, c_1)$;
2. verify if $c_1 = F(s, x)$ and $c_2 = G(s_0, c_1, x)$;
3. If the equations hold, output $m = c_3 \oplus h(x)$; else reject.

Fig. 3. Program Decrypt

Constants: $(pk = (s, s_0, h), t_0, c_1^*, c_2^*, c_3^*, m^*)$.
Input: $(c_1, c_2, c_3) \in \{0,1\}^v \times \{0,1\}^* \times \{0,1\}^l$.

1. if $(c_1, c_2) = (c_1^*, c_2^*)$, output $m = c_3 \oplus c_3^* \oplus m^*$;
2. if $c_1 = c_1^*, c_2 \neq c_2^*$, output $\perp$;
2. if $c_1 \neq c_1^*$, compute $x = G^{-1}(t_0, c_1, c_2)$;
3. verify if $c_1 = F(s, x)$ and $c_2 = G(s_0, c_1, x)$;
4. if the equations hold, output $m = c_3 \oplus h(x)$, otherwise output $\perp$.

Fig. 4. Program Decrypt*

Theorem 1. *If the $i\mathcal{O}$ is a secure indistinguishability obfuscation, (S_0, S_1, F, F^{-1}) is a collection of (v, l_1)-LTDFs, (B, S, G, G^{-1}) is a collection of (v, l_0)-ABO-TDFs, H is a family of pairwise independent hash functions from $\{0, 1\}^v$ to $\{0, 1\}^l$, then the above scheme is SIM-RSO-CPA secure.*

Proof. To prove the security of the above scheme, we define a sequence of games whereby no PPT adversary can tell the difference between two consecutive games. Let n be the total number of receivers.

$Game_0$: the real security game. The challenger $\mathcal{C}$ generates n injective keys $(s_i, t_i) \leftarrow S_1(1^\lambda)$ for $i = 1, ..., n$ and n ABO-TDFs $(s_{0i}, t_{0i}) \leftarrow S(1^\lambda, 0^v)$, n pairwise independent hash functions $h_i \leftarrow H$. The public keys are set as $\{pk_i = (s_i, s_{0i}, h_i)\}_{i \in [n]}$, the secret keys are set as the obfuscation of the program Decrypt.
  - When $\mathcal{A}$ promotes a query $dist$, the challenger picks $m^* \leftarrow dist$, chooses random $x_i^* \in \{0, 1\}^v$ for $i \in [n]$, computes $(c_{1i}^* = F(s_i, x_i^*), c_{2i}^* = G(s_{0i}, c_{1i}^*, x_i^*), c_{3i}^* = h(x_i^*) \oplus m_i^*)$ and sends $c = \{(c_{1i}^*, c_{2i}^*, c_{3i}^*)\}_{i \in [n]}$ to $\mathcal{A}$.
  - When $\mathcal{A}$ promotes a set $I \subset [n]$, the challenger responds with (m_I^*, sk_I).
$Game_1$: The same as $Game_0$ except that secret keys are changed to be the obfuscation of the program Decrypt* with constants $(pk, t_0, c_1^*, c_2^*, c_3^*, m^*)$. Later we will show that $Game_1$ and $Game_0$ are indistinguishable according to the security of $i\mathcal{O}$.
$Game_2$: The same as $Game_1$ except that s_i's are generated via $s_i \leftarrow S_0(1^\lambda)$. It is easy to see that the probability in distinguishing $Game_2$ and $Game_1$ is bounded by $nAdv^{LTDF}$.
$Game_3$: The same as $Game_2$ except that s_{0i}'s are generated via $s_{0i} \leftarrow S(1^\lambda, c_{1i}^*)$. It is easy to see that the probability in distinguishing $Game_3$ and $Game_2$ is bounded by $nAdv^{ABO-TDF}$.
$Game_4$: The same as $Game_3$ except that c_{3i}^*s are changed to be randomly chosen. We will prove that $Game_3$ and $Game_4$ are statistically indistinguishable and it is easy to show that for all PPT adversary $\mathcal{A}$, there exists a simulator $\mathcal{S}$ that proceeds the same as the challenger in $Game_4$ and hence can output what $\mathcal{A}$ outputs, so $Adv_4 = 0$.

Let $W_i = \Pr[D(Exp_{real}^{sim\text{-}rso\text{-}cpa}(\mathcal{A})) = 1]$ in $Game_i$. Next we prove that the probability between consecutive games is negligibly close.

Lemma 2. *For any adversary $\mathcal{A}$, $W_1 - W_0$ is negligible, assuming $i\mathcal{O}$ is an indistinguishability obfuscator.*

Proof. To prove the lemma, we define intermediate games $Game_{0,i}$ for $i = 1, ..., n$.

$Game_{0,i}$: The same as $Game_{0,i-1}$ except that sk_i is the obfuscation of program Decrypt* (Fig. 4) instead of the program Decrypt. $Game_{0,0}$ is $Game_0$.

Then we prove that $Game_{0,i}$ and $Game_{0,i-1}$ are indistinguishable.

$\mathcal{B}$ generates n injective keys $(s_i, t_i) \leftarrow S_1(1^\lambda)$ for $i = 1, ..., n$ and n ABO-TDFs $(s_{0i}, t_{0i}) \leftarrow S(1^\lambda, 0^v)$, n pairwise independent hash functions $h_i \leftarrow H$. The public keys are set as $\{pk_i = (s_i, s_{0i}, h_i)\}_{i \in [n]}$. On receiving a query $dist$, $\mathcal{B}$ picks $\boldsymbol{m}^* \leftarrow dist$ and randomly $\boldsymbol{x}^*$, computes $(c_{1i}^* = F(s_i, x_i^*), c_{2i}^* = G(s_{0i}, c_{1i}^*, x_i^*)$, $c_{3i}^* = h(x_i^*) \oplus m_i^*)$ and creates two circuits $\mathcal{C}_1$ of Fig. 3 with constants (pk_i, t_i) and $\mathcal{C}_2$ of Fig. 4 with constants $(pk_i, t_{0i}, x_i^*, c_{1i}^*, c_{2i}^*, c_{3i}^*, m_i^*)$, sends both circuits to the $i\mathcal{O}$ and sets the output as sk_i. For $l < i$, sk_l is set as the obfuscation of program Decrypt* (Fig. 4); for $l > i$, sk_l is set as the obfuscation of program Decrypt (Fig. 3).

When $\mathcal{A}$ requires openness, $\mathcal{B}$ responds with $(\boldsymbol{m}_I^*, \boldsymbol{sk}_I)$.

Finally $\mathcal{B}$ outputs what $\mathcal{A}$ outputs.

Note that when $\mathcal{B}$ receives an obfuscation of program Decrypt, the above game perfectly simulates $Game_{0,i-1}$; when $\mathcal{B}$ receives an obfuscation of program Decrypt*, the above game perfectly simulates $Game_{0,i}$.

Next we analysis that for every input $(c = (c_1, c_2, c_3), i)$, the output of program Decrypt and Decrypt* are identical. clearly, when $c_1 \neq c_{1i}^*$, both circuits compute the same x since $F(s_i, \cdot)$ and $G(s_{0i}, c_{1i}, \cdot)$ are both injective; when $c_1 = c_{1i}^*$, verification holds iff $c_2 = G(s_{0i}, c_1, x_i^*) = c_2^*$, and the decryption $m = c_3 \oplus c_{3i}^* \oplus m_i^*$. □

Lemma 3. *For any PPT adversary $\mathcal{A}$, $W_4 - W_3$ is negligible.*

Proof. In $Game_3$, $F(s_i, \cdot)$ and $G(s_{0i}, c_{1i}^*, \cdot)$ are both lossy, and the output value of (c_{1i}^*, c_{2i}^*) can take at most $2^{v-l_1+v-l_0}$ values, so for the unopened ciphertexts, the conditional min-entropy of x_i^* is bounded by

$$H_\infty(x_i^* | c_{1i}^*, c_{2i}^*) \geq v - (v - l_1 + v - l_0) \geq \kappa.$$

Then by leftover hash lemma, $(pk_i, c_{1i}^*, c_{2i}^*, h(x_i^*))$ is statistical close to $(pk_i, c_{1i}^*, c_{2i}^*, u_i)$, where u_i is randomly distributed over $\{0,1\}^l$, hence $(pk_i, c_{1i}^*, c_{2i}^*, h(x_i^*) \oplus m_i^*)$ is irrelevant to m_i^* information theoretically. □

3.2 CCA Secure Construction from Puncturable PRFs (for Bit Encryption)

Let PRG be a pseudo-random generator mapping $\{0,1\}^\lambda$ to $\{0,1\}^{2\lambda}$, F_1 be a puncturable PRF mapping $\{0,1\}^{2\lambda}$ to $\{0,1\}$, F_2 be a puncturable PRF mapping $\{0,1\}^{2\lambda+1}$ to $\{0,1\}^\lambda$. The scheme is described as follows:

Keygen: It first picks random K_1 and K_2 for F_1 and F_2 separately, then it creates an obfuscation of the program Encrypt of Fig. 5. The size of the program is padded to be the maximum of itself and Encrypt* of Fig. 6. It also creates an obfuscation of the program Decrypt of Fig. 7. And pad the size to be the maximum of itself and program Decrypt* of Fig. 8. The public key pk is the obfuscated program Encrypt and the secret key sk is the obfuscated program Decrypt. Finally erase the randomness used in the key generation algorithm.

Constants: PRF keys K_1, K_2.
Input: message $m \in \{0,1\}$, randomness $r \in \{0,1\}^{\lambda}$.

1. compute $t = PRG(r)$.
2. output $c = (c_1 = t, c_2 = F_1(K_1, t) \oplus m, c_3 = F_2(K_2, c_1 \| c_2))$.

Fig. 5. Program Encrypt

Constants: Punctured PRF keys $K_1(\{t^*\}), K_2(\{t^*\|0, t^*\|1\})$.
Input: message $m \in \{0,1\}$, randomness $r \in \{0,1\}^{\lambda}$.

1. compute $t = PRG(r)$.
2. output $c = (c_1 = t, c_2 = F_1(K_1, t) \oplus m, c_3 = F_2(K_2, c_1 \| c_2))$.

Fig. 6. Program Encrypt*

Constants: PRF keys K_1, K_2.
Input: $(c_1, c_2, c_3) \in \{0,1\}^{2\lambda} \times \{0,1\} \times \{0,1\}^{\lambda}$.

1. verify if $c_3 = F_2(K_2, c_1 \| c_2)$.
2. if the equation holds, output $m = F_1(K_1, c_1) \oplus c_2$, otherwise output $\perp$.

Fig. 7. Program Decrypt

Enc: It takes $m \in \{0,1\}$ as input and picks a random value $r \in \{0,1\}^{\lambda}$, then it runs the obfuscated program of pk on inputs (m, r) and takes the output as the ciphertext.

Dec: It runs the obfuscated program of sk on input c and takes the output as the decryption.

Theorem 2. *If the $i\mathcal{O}$ is an indistinguishability obfuscation, PRG is a secure pseudo-random generator, F_1, F_2 are secure punctured PRFs, then the above scheme is SIM-RSO-CCA secure.*

Proof. To prove the security of the above scheme, we define a sequence of games whereby no PPT adversary can tell the difference between two consecutive games. Let q denote the number of decryption queries that the adversary makes during the whole game, n be the total number of receivers.

Game$_0$: the real security game. The challenger $\mathcal{C}$ picks n key pairs (K_{1i}, K_{2i}) for $i = 1, ..., n$ for the puncturable PRFs F_1, F_2. The public keys are set as the obfuscation of the program Encrypt and later sent to the adversary $\mathcal{A}$, the secret keys are set as the obfuscation of the program Decrypt.

– When $\mathcal{A}$ promotes a query $dist$, the challenger picks $\boldsymbol{m}^* \leftarrow dist$, chooses random $r_i^* \in \{0,1\}^{\lambda}$ for $i \in [n]$, computes $c_{1i}^* = t_i^* = PRG(r_i^*)$, $c_{2i}^* =$

Constants: Punctured PRF keys $K_1(\{t^*\}), K_2(\{t^*\|0, t^*\|1\})$ and $t^*, c_2^*, \beta_0, \beta_1, m^*$.
Input: $(c_1, c_2, c_3) \in \{0,1\}^{2\lambda} \times \{0,1\} \times \{0,1\}^{\lambda}$.

1. if $c_1 = t^*$ and $c_3 \neq \beta_{c_2}$, output $\perp$.
2. if $c_1 = t^*$ and $c_3 = \beta_{c_2}$, output $c_2 \oplus c_2^* \oplus m$.
3. otherwise, verify if $c_3 = F_2(K_2, c_1\|c_2)$.
4. if the equation holds, output $m = F_1(K_1, c_1) \oplus c_2$, otherwise output $\perp$.

Fig. 8. Program Decrypt*

$F_1(K_{1i}, t_i^*) \oplus m_i$, $c_{3i}^* = F_2(K_{2i}, c_{1i}^*\|c_{2i}^*)$ and sends $\boldsymbol{c}^* = \{(c_{1i}^*, c_{2i}^*, c_{3i}^*)\}_{i \in [n]}$
to $\mathcal{A}$.

- When $\mathcal{A}$ promotes a decryption query (c, j), the challenger first checks that $c \neq c_j^*$ and rejects otherwise, then it verifies whether the equation $c_3 = F_2(K_{2j}, c_1\|c_2)$ holds, if it holds, it responds with $m = F_1(K_{1j}, c_1) \oplus c_2$, otherwise it just rejects.
- When $\mathcal{A}$ promotes a set $I \subset [n]$, the challenger responds with $(\boldsymbol{m}_I^*, \boldsymbol{sk}_I)$.
- Finally, $\mathcal{A}$ outputs $output_A$.

$Game_1$: The same as $Game_0$ except that t_i^*'s are chosen randomly from $\{0,1\}^{2\lambda}$. It is easy to see that the probability in distinguishing $Game_1$ and $Game_0$ is bounded by $nAdv^{PRG}$.

$Game_2$: The same as $Game_1$ except that public keys are changed to be the obfuscation of the program Encrypt* with punctured keys $K_{1i}(t_i^*), K_{2i}(t_i^*\|0, t_i^*\|1)$. Later we will show that $Game_1$ and $Game_2$ are indistinguishable.

$Game_3$: The same as $Game_2$ except that secret keys are changed to be the obfuscation of the program Decrypt* with constants $K_{1i}(\{t_i^*\}), K_{2i}(\{t_i^*\|0, t_i^*\|1\})$ and $t_i^*, \beta_{0i}, \beta_{1i}, c_{2i}^*, m_i^*$, where $\alpha_i = F_1(K_{1i}, t_i^*)$, $\beta_{0i} = F_2(K_{2i}, t_i^*\|0)$, $\beta_{1i} = F_2(K_{2i}, t_i^*\|1)$. And $c_{2i}^* = \alpha_i \oplus m_i^*$, $c_{3i}^* = \beta_{c_{2i}^* i}$. Later we will show that $Game_3$ and $Game_2$ are indistinguishable according to the security of $i\mathcal{O}$.

$Game_4$: The same as $Game_3$ except that $\{\beta_{0i}, \beta_{1i}\}_{i \in [n]}$ are randomly chosen from $\{0,1\}^{\lambda}$. Later we will show that $Game_4$ and $Game_3$ are indistinguishable according to the pseudo-random property of $F_2(K_{2i}, \cdot)$.

$Game_5$: The same as $Game_4$ except that $\{c_{2i}^*\}_{i \in [n]}$ are randomly chosen from $\{0,1\}$. $Game_5$ and $Game_4$ are indistinguishable according to the pseudo-random property of $F_1(K_{1i}, \cdot)$. It is easy to show that for all PPT adversary $\mathcal{A}$, there exists a simulator $\mathcal{S}$ that proceeds the same as the challenger in $Game_5$ and hence can output what $\mathcal{A}$ outputs, so $Adv_5 = 0$.

Let $W_i = \Pr[D(Exp_{real}^{sim\text{-}rso\text{-}cca}(\mathcal{A})) = 1]$ in $Game_i$. Next we show that the probability between consecutive games is negligibly close.

Lemma 4. *For any PPT algorithm $\mathcal{A}$, $W_1 - W_0 \leq nAdv^{PRG}$.*

Proof. To prove the lemma, we define intermediate games $Game_{0,i}$ for $i = 1, ..., n$.

$Game_{0,i}$: The same as $Game_{0,i-1}$ except that t_i^* is chosen randomly from $\{0,1\}^{2\lambda}$. $Game_{0,0}$ is $Game_0$.

$\mathcal{B}$ receives $T \in \{0,1\}^{2\lambda}$ and its task is to decide whether there exists a $r \in \{0,1\}^{\lambda}$ s.t. $T = PRG(r)$ or not. $\mathcal{B}$ then proceeds as in $Game_{0,i-1}$ except that when creating the i-th challenge ciphertext, it sets $t_i^* = T$. Finally $\mathcal{B}$ outputs what $\mathcal{A}$ outputs.

Note that when T is in the image of PRG, the above game perfectly simulates $Game_{0,i-1}$; when T is randomly chosen from $\{0,1\}^{2\lambda}$, the above game perfectly simulates $Game_{0,i}$. $\square$

Lemma 5. *For any PPT adversary $\mathcal{A}$, $W_2 - W_1$ is negligible, assuming $i\mathcal{O}$ is an indistinguishability obfuscation.*

Proof. To prove the lemma, we define intermediate games $Game_{1,i}$ for $i = 1, ..., n$.

$Game_{1,i}$: The same as $Game_{1,i-1}$ except that pk_i is the obfuscation of program Encrypt* (Fig. 6) instead of the program Encrypt (Fig. 5). $Game_{1,0}$ is $Game_1$.

$\mathcal{B}$ chooses random key pairs (K_{1l}, K_{2l}) and $t_l^* \in \{0,1\}^{2\lambda}$ for $l = 1, ..., n$. Then it creates two circuits $\mathcal{C}_1$ of Fig. 5 with constants (K_{1i}, K_{2i}) and $\mathcal{C}_2$ of Fig. 6 with constants $(K_{1i}(t_i^*), K_{2i}(t_i^*\|0, t_i^*\|1))$, sends both circuits to the $i\mathcal{O}$ and sets the output as pk_i, other public keys are set as follows: for $l < i$, pk_l is the obfuscation of program Encrypt* with constants $K_{1l}(t_l^*), K_{2l}(t_l^*\|0, t_l^*\|1)$; for $l > i$, pk_l is the obfuscation of program Encrypt with constants (K_{1l}, K_{2l}). $\mathcal{B}$ then proceeds as in $Game_1$. Finally $\mathcal{B}$ outputs what $\mathcal{A}$ outputs.

Note that when $\mathcal{B}$ receives an obfuscation of program Encrypt, the above game perfectly simulates $Game_{1,i-1}$; when $\mathcal{B}$ receives an obfuscation of program Encrypt*, the above game perfectly simulates $Game_{1,i}$.

Next we analysis that for every input (m, r), the output of program Encrypt and Encrypt* are identical except with negligible probability. Since t_i^* is chosen randomly from $\{0,1\}^{2\lambda}$, with probability $1 - 1/2^{\lambda}$, $t_i^* \notin PRG(1^{\lambda})$, and when $t \neq t_i^*$, $F(K_1, t)$ and $F(K_2, t\|b)$ can be correctly computed. $\square$

Lemma 6. *For all PPT algorithm $\mathcal{A}$, $W_3 - W_2$ is negligible, assuming $i\mathcal{O}$ is an indistinguishability obfuscation.*

Proof. To prove the lemma, we define intermediate games $Game_{2,i}$ for $i = 1, ..., n$. $Game_{2,i}$ is the same as $Game_{2,i-1}$ except that sk_i is the obfuscation of program Decrypt* (Fig. 8) instead of the program Decrypt (Fig. 7). $Game_{2,0}$ is $Game_2$.

$\mathcal{B}$ chooses random key pairs (K_{1l}, K_{2l}) and $t_l^* \in \{0,1\}^{2\lambda}$ for $l = 1, ..., n$. Then it sets pk_l be the obfuscation of program Encrypt* with constants $(K_{1l}(t_l^*), K_{2l}(t_l^*\|0, t_l^*\|1))$. $\mathcal{B}$ then computes $c_{2i}^* = F_1(K_{1i}, t_i^*) \oplus m_i^*$, $\beta_{0i} = F_2(K_{2i}, t_i^*\|0)$, $\beta_{1i} = F_2(K_{2i}, t_i^*\|1)$ and creates two circuits $\mathcal{C}_1$ as in Fig. 7 with constants (K_{1i}, K_{2i}) and $\mathcal{C}_2$ as in Fig. 8 with constants $(K_{1i}(t_i^*), K_{2i}(t_i^*\|0, t_i^*\|1), t_i^*, \beta_{0i}, \beta_{1i}, c_{2i}^*, m_i^*)$, sends both circuits to the $i\mathcal{O}$ and sets the output as sk_i. For $l < i$, sk_l is set as the obfuscation of circuit Decrypt*; for $l > i$, sk_l is set as the obfuscation of circuit Decrypt.

When $\mathcal{A}$ promotes a decryption query, $\mathcal{B}$ responds with its secret key. When $\mathcal{A}$ promotes an encryption query, $\mathcal{B}$ sets $c_{1l}^* = t_l^*, c_{2l}^* = F_1(K_{1l}, t_l^*) \oplus m_l^*, c_{3l}^* = F_2(K_{2l}, c_{1l} \| c_{2l})$. When $\mathcal{A}$ requires openness, $\mathcal{B}$ responds with $(\boldsymbol{m_I^*}, \boldsymbol{sk_I})$.

Finally $\mathcal{B}$ outputs what $\mathcal{A}$ outputs.

Note that when $\mathcal{B}$ receives an obfuscation of program Decrypt, the above game perfectly simulates $Game_{2,i-1}$; when $\mathcal{B}$ receives an obfuscation of program Decrypt*, the above game perfectly simulates $Game_{2,i}$.

Next we analysis that for every input $(c = (c_1, c_2, c_3), j)$, the output of program Decrypt and Decrypt* are identical. clearly, when $c_1 \neq t_i^*$ or $j \neq i$, both circuits proceed identically; when $c_1 = t_i^*$ and $j = i$, verification holds iff $c_3 = F_2(K_2, c_1 \| c_2) = \beta_{c_2 i}$, and the decryption $m = c_2 \oplus F_1(K_1, c_1) = c_2 \oplus c_{2i}^* \oplus m_i^*$. $\square$

Lemma 7. *For all PPT algorithm $\mathcal{A}$, $W_4 - W_3 \leq nAdv_{F_2}^{PRF}$.*

Proof. To prove the lemma, we define intermediate games $Game_{3,i}$ for $i = 1, ..., n$. $Game_{3,i}$ is the same as $Game_{3,i-1}$ except that β_{0i}, β_{1i} are changed to be random. $Game_{3,0}$ is $Game_3$.

$\mathcal{B}$ chooses random key pairs K_{1l} for $l = 1, ..., n$ and K_{2l} for $l \neq i, l = 1, ..., n$ and $t_l^* \in \{0, 1\}^{2\lambda}$ for $l = 1, ..., n$. Then it sends $(t_i^* \| 0, t_i^* \| 1)$ to its challenge oracle and receives $(K_{2i}(t_i^* \| 0, t_i^* \| 1), \beta_{0i}, \beta_{1i})$ as response, it sets pk_l be the obfuscation of program Encrypt* with constants $(K_{1l}(t_l^*), K_{2l}(t_l^* \| 0, t_l^* \| 1))$ and sk_l be the obfuscation of program Decrypt* with constants $(K_{1l}(t_l^*), K_{2l}(t_l^* \| 0, t_l^* \| 1), t_l^*, \beta_{0l}, \beta_{1l}, c_{2l}^*, m_l^*)$, where $c_{2l}^* = F_1(K_{1l, t_l^*}) \oplus m_l^*$ and for $l < i$, (β_{0l}, β_{1l}) are randomly chosen from $\{0, 1\}^{2\lambda}$ and for $l > i$, $(\beta_{0l}, \beta_{1l}) = (F_2(K_{2l}, t_l^* \| 0), F_2(K_{2l}, t_l^* \| 1))$.

When $\mathcal{A}$ promotes a decryption query, $\mathcal{B}$ answers with secret keys. When $\mathcal{A}$ promotes an encryption query, $\mathcal{B}$ sets $c_{1l}^* = t_l^*, c_{2l}^*, c_{3l}^* = \beta_{c_{2l}^* l}$. When $\mathcal{A}$ requires openness, $\mathcal{B}$ responds with $(\boldsymbol{m_I^*}, \boldsymbol{sk_I})$.

Finally $\mathcal{B}$ outputs what $\mathcal{A}$ outputs.

Note that when $(\beta_{0i}, \beta_{1i}) = (F_2(K_{2i}, t_i^* \| 0), F_2(K_{2i}, t_i^* \| 1))$, the above game perfectly simulates $Game_{3,i-1}$; when (β_{0i}, β_{1i}) are randomly chosen, the above game perfectly simulates $Game_{3,i}$. $\square$

Lemma 8. *For all PPT algorithm $\mathcal{A}$, $W_5 - W_4 \leq nAdv_{F_1}^{PRF}$.*

Proof. To prove the lemma, we define intermediate games $Game_{4,i}$ for $i = 1, ..., n$. $Game_{4,i}$ is the same as $Game_{4,i-1}$ except that c_{2i}^* is changed to be random. $Game_{4,0}$ is $Game_4$.

$\mathcal{B}$ chooses random key pairs K_{1l} for $l \neq i, l = 1, ..., n$ and K_{2l}, $t_l^* \in \{0, 1\}^{2\lambda}$ for $l = 1, ..., n$. Then it sends t_i^* to its challenge oracle and receives $K_1(t_i^*), \alpha_i$ as response, it computes $c_{2i}^* \oplus m_i^*$ and chooses random c_{2l}^* for $l < i$ and computes $c_{2l}^* = F_1(K_{1l}, t_l^*) \oplus m_l^*$ for $l > i$, then it sets pk_l be the obfuscation of program Encrypt* with constants $(K_{1l}(t_l^*), K_{2l}(t_l^* \| 0, t_l^* \| 1))$ and sk_l be the obfuscation of program Decrypt* with constants $(K_{1l}(t_l^*), K_{2l}(t_l^* \| 0, t_l^* \| 1), t_l^*, \beta_{0l}, \beta_{1l}, c_{2l}^*, m_l^*)$, where β_{0l}, β_{1l} are randomly chosen from $\{0, 1\}^{\lambda}$.

When $\mathcal{A}$ promotes a decryption query, $\mathcal{B}$ answers with secret keys. When $\mathcal{A}$ promotes an encryption query, $\mathcal{B}$ sets $c_{1l}^* = t_l^*, c_{2l}^*, c_{3l}^* = \beta_{c_{2l}^* l}$. When $\mathcal{A}$ requires openness, $\mathcal{B}$ responds with $(\boldsymbol{m}_I^*, \boldsymbol{sk}_I)$.

Finally $\mathcal{B}$ outputs what $\mathcal{A}$ outputs.

Note that when $\alpha_i = F_1(K_{1i}, t_i^*)$, the above game perfectly simulates the real game of $Game_{4,i-1}$; when α_i is randomly chosen, the above game perfectly simulates the real game of $Game_{4,i}$.   □

4   Conclusion

In this paper we propose a method to convert IND-CCA secure schemes of the ABO type to achieve SIM-RSO security with the help of $i\mathcal{O}$. In concrete, we use indistinguishability obfuscator to wrap up the decryption circuit and set the obfuscated circuit as the secret key. As a result, we get an SIM-RSO-CPA secure construction from lossy trapdoor functions, and an SIM-RSO-CCA bit encryption from puncturable pseudo-random functions.

Acknowledgments. We are very grateful to anonymous reviewers for their helpful comments.

References

1. Bellare, M., Dowsley, R., Waters, B., Yilek, S.: Standard security does not imply security against selective-opening. In: Pointcheval, D., Johansson, T. (eds.) EURO-CRYPT 2012. LNCS, vol. 7237, pp. 645–662. Springer, Heidelberg (2012). doi:10.1007/978-3-642-29011-4_38

2. Böhl, F., Hofheinz, D., Kraschewski, D.: On definitions of selective opening security. In: Fischlin, M., Buchmann, J., Manulis, M. (eds.) PKC 2012. LNCS, vol. 7293, pp. 522–539. Springer, Heidelberg (2012). doi:10.1007/978-3-642-30057-8_31

3. Bellare, M., Hofheinz, D., Yilek, S.: Possibility and impossibility results for encryption and commitment secure under selective opening. In: Joux, A. (ed.) EURO-CRYPT 2009. LNCS, vol. 5479, pp. 1–35. Springer, Heidelberg (2009). doi:10.1007/978-3-642-01001-9_1

4. Bellare, M., Yilek, S.: Encryption schemes secure under selective opening attack. IACR Cryptology ePrint Archive 2009/101 (2009)

5. Choi, S.G., Dachman-Soled, D., Malkin, T., Wee, H.: Improved non-committing encryption with applications to adaptively secure protocols. In: Matsui, M. (ed.) ASIACRYPT 2009. LNCS, vol. 5912, pp. 287–302. Springer, Heidelberg (2009). doi:10.1007/978-3-642-10366-7_17

6. Canetti, R., Feige, U., Goldreich, O., Naor, M.: Adaptively secure multi-party computation. In: 28th ACM STOC, pp. 639–648. ACM, New York, May 1996

7. Canetti, R., Halevi, S., Katz, J.: Adaptively-secure, non-interactive public-key encryption. In: Kilian, J. (ed.) TCC 2005. LNCS, vol. 3378, pp. 150–168. Springer, Heidelberg (2005). doi:10.1007/978-3-540-30576-7_9

8. Canetti, R., Poburinnaya, O., Raykova, M.: Optimal-rate non-committing encryption in a CRS model. IACR Cryptology ePrint Archive 2016/511 (2016)

9. Cramer, R., Shoup, V.: Universal hash proofs and a paradigm for adaptive chosen ciphertext secure public-key encryption. In: Knudsen, L.R. (ed.) EUROCRYPT 2002. LNCS, vol. 2332, pp. 45–64. Springer, Heidelberg (2002). doi:10.1007/3-540-46035-7_4

10. Dachman-Soled, D., Dov Gordon, S., Liu, F.-H., O'Neill, A., Zhou, H.-S.: Leakage-resilient public-key encryption from obfuscation. In: Cheng, C.-M., Chung, K.-M., Persiano, G., Yang, B.-Y. (eds.) PKC 2016. LNCS, vol. 9615, pp. 101–128. Springer, Heidelberg (2016). doi:10.1007/978-3-662-49387-8_5

11. Damgård, I., Nielsen, J.B.: Improved non-committing encryption schemes based on a general complexity assumption. In: Bellare, M. (ed.) CRYPTO 2000. LNCS, vol. 1880, pp. 432–450. Springer, Heidelberg (2000). doi:10.1007/3-540-44598-6_27

12. Freeman, D.M., Goldreich, O., Kiltz, E., Rosen, A., Segev, G.: More constructions of lossy and correlation-secure trapdoor functions. In: Nguyen, P.Q., Pointcheval, D. (eds.) PKC 2010. LNCS, vol. 6056, pp. 279–295. Springer, heidelberg (2010). doi:10.1007/978-3-642-13013-7_17

13. Fehr, S., Hofheinz, D., Kiltz, E., Wee, H.: Encryption schemes secure against chosen-ciphertext selective opening attacks. In: Gilbert, H. (ed.) EUROCRYPT 2010. LNCS, vol. 6110, pp. 381–402. Springer, Heidelberg (2010). doi:10.1007/978-3-642-13190-5_20

14. Fuchsbauer, G., Heuer, F., Kiltz, E., Pietrzak, K.: Standard security does imply security against selective opening for markov distributions. In: Kushilevitz, E., Malkin, T. (eds.) TCC 2016. LNCS, vol. 9562, pp. 282–305. Springer, Heidelberg (2016). doi:10.1007/978-3-662-49096-9_12

15. Garg, S., Gentry, C., Halevi, S., Raykova, M., Sahai, A., Waters, B.: Candidate indistinguishability obfuscation and functional encryption for all circuits. In: 54th FOCS, pp. 40–49. IEEE Computer Society Press, October 2013

16. Håstad, J., Impagliazzo, R., Levin, L.A., Luby, M.: A pseudorandom generator from any one-way function. SIAM J. Comput. **28**(4), 1364–1396 (1999)

17. Heuer, F., Jager, T., Kiltz, E., Schäge, S.: On the selective opening security of practical public-key encryption schemes. In: Katz, J. (ed.) PKC 2015. LNCS, vol. 9020, pp. 27–51. Springer, Heidelberg (2015). doi:10.1007/978-3-662-46447-2_2

18. Huang, Z., Liu, S., Qin, B., Chen, K.: Fixing the Sender-equivocable encryption scheme in eurocrypt 2010. In: INCOS, pp. 366–372 (2013)

19. Hemenway, B., Libert, B., Ostrovsky, R., Vergnaud, D.: Lossy encryption: constructions from general assumptions and efficient selective opening chosen ciphertext security. In: Lee, D.H., Wang, X. (eds.) ASIACRYPT 2011. LNCS, vol. 7073, pp. 70–88. Springer, Heidelberg (2011). doi:10.1007/978-3-642-25385-0_4

20. Hemenway, B., Ostrovsky, R., Rosen, A.: Non-committing encryption from Φ-hiding. In: Dodis, Y., Nielsen, J.B. (eds.) TCC 2015. LNCS, vol. 9014, pp. 591–608. Springer, Heidelberg (2015). doi:10.1007/978-3-662-46494-6_24

21. Hazay, C., Patra, A., Warinschi, B.: Selective opening security for receivers. In: Iwata, T., Cheon, J.H. (eds.) ASIACRYPT 2015. LNCS, vol. 9452, pp. 443–469. Springer, Heidelberg (2015). doi:10.1007/978-3-662-48797-6_19. IACR Cryptology ePrint Archive 2015/860

22. Hofheinz, D., Rupp, A.: Standard versus selective opening security: separation and equivalence results. In: Lindell, Y. (ed.) TCC 2014. LNCS, vol. 8349, pp. 591–615. Springer, Heidelberg (2014). doi:10.1007/978-3-642-54242-8_25

23. Hofheinz, D., Rao, V., Wichs, D.: Standard security does not imply indistinguishability under selective opening. IACR Cryptology ePrint Archive 2015/792 (2015)

24. Lai, J., Deng, R.H., Liu, S., Weng, J., Zhao, Y.: Identity-based encryption secure against selective opening chosen-ciphertext attack. In: Nguyen, P.Q., Oswald, E. (eds.) EUROCRYPT 2014. LNCS, vol. 8441, pp. 77–92. Springer, Heidelberg (2014). doi:10.1007/978-3-642-55220-5_5

25. Liu, S., Paterson, K.G.: Simulation-based selective opening CCA security for PKE from key encapsulation mechanisms. In: Katz, J. (ed.) PKC 2015. LNCS, vol. 9020, pp. 3–26. Springer, Heidelberg (2015). doi:10.1007/978-3-662-46447-2_1

26. Liu, S., Zhang, F., Chen, K.: Public-key encryption scheme with selective opening chosen-ciphertext security based on the decisional Diffie-Hellman assumption. Concurr. Comput.: Pract. Exp. **26**(8), 1506–1519 (2014)

27. Mol, P., Yilek, S.: Chosen-ciphertext security from slightly lossy trapdoor functions. In: Nguyen, P.Q., Pointcheval, D. (eds.) PKC 2010. LNCS, vol. 6056, pp. 296–311. Springer, Heidelberg (2010). doi:10.1007/978-3-642-13013-7_18

28. Matsuda, T., Hanaoka, G.: Constructing and understanding chosen ciphertext security via puncturable key encapsulation mechanisms. In: Dodis, Y., Nielsen, J.B. (eds.) TCC 2015. LNCS, vol. 9014, pp. 561–590. Springer, Heidelberg (2015). doi:10.1007/978-3-662-46494-6_23

29. Nielsen, J.B.: Separating random oracle proofs from complexity theoretic proofs: the non-committing encryption case. In: Yung, M. (ed.) CRYPTO 2002. LNCS, vol. 2442, pp. 111–126. Springer, Heidelberg (2002). doi:10.1007/3-540-45708-9_8

30. Ostrovsky, R., Rao, V., Visconti, I.: On selective-opening attacks against encryption schemes. In: Abdalla, M., Prisco, R. (eds.) SCN 2014. LNCS, vol. 8642, pp. 578–597. Springer, Heidelberg (2014). doi:10.1007/978-3-319-10879-7_33

31. Peikert, C., Waters, B.: Lossy trapdoor functions and their applications. In: STOC, pp. 187–196 (2008)

32. Rosen, A., Segev, G.: Chosen-ciphertext security via correlated products. In: Reingold, O. (ed.) TCC 2009. LNCS, vol. 5444, pp. 419–436. Springer, Heidelberg (2009). doi:10.1007/978-3-642-00457-5_25

33. Sahai, A., Waters, B.: How to use indistinguishability obfuscation: deniable encryption, and more. In: Shmoys, D.B. (ed.) 46th ACM STOC, pp. 475-484. ACM Press, May/June 2014

34. Wee, H.: Efficient chosen-ciphertext security via extractable hash proofs. In: Rabin, T. (ed.) CRYPTO 2010. LNCS, vol. 6223, pp. 314–332. Springer, Heidelberg (2010). doi:10.1007/978-3-642-14623-7_17

Format Preserving Sets: On Diffusion Layers of Format Preserving Encryption Schemes

Kishan Chand Gupta[1], Sumit Kumar Pandey[2($\boxtimes$)], and Indranil Ghosh Ray[3]

[1] Applied Statistics Unit, Indian Statistical Institute,
203, B.T. Road, Kolkata 700108, India
kishan@isical.ac.in
[2] School of Physical and Mathematical Sciences,
Nanyang Technological University, Singapore, Singapore
emailpandey@gmail.com
[3] Department of Electrical and Electronic Engineering,
City University, London, UK
indranilgray@gmail.com

Abstract. Format preserving encryption refers to a set of techniques for encrypting data such that the ciphertext has the same format as the plaintext. Here, we consider the design of diffusion layers only which can be defined by, in general, a linear transformation. In this paper, we study and explore the format preserving diffusion layers, in particular, the relationship between the $n \times n$ diffusion matrix M over the field $\mathbb{F}_q$ and the format preserving set $\mathbb{S} \subseteq \mathbb{F}_q$ such that whenever $\mathbf{v} \in \mathbb{S}^n$, $M\mathbf{v} \in \mathbb{S}^n$. It is proved in this paper that if such a set $\mathbb{S}$ with respect to a certain type of matrix M contains $\bar{0} \in \mathbb{F}_q$, then it is always a vector space over the smallest field containing entries of M. Moreover, some more interesting results are found when this condition, $\bar{0} \in \mathbb{S}$, is relaxed. We illustrate our results by a credit card example where plaintext and ciphertext both come from the set $\{0, \cdots, 9\}$. We further show that only certain type of 4×4 matrices over the field $\mathbb{F}_{2^4}$ can be constructed which yield a format preserving set of cardinality 10 which is suited for our credit card example. However, to the best of our knowledge, such matrices do not have any cryptographic significance. Thus, it is impossible to construct any cryptographically significant 4×4 matrices over the field $\mathbb{F}_{2^4}$ in the diffusion layer which yields a format preserving set of cardinality 10.

Keywords: Diffusion layer · Format preserving encryption · Format preserving set

1 Introduction

In recent past, a massive leakage of private information like credit card numbers, social security numbers and so many more private and confidential data give rise to the importance of a newly emerging field of research in applied cryptography,

© Springer International Publishing AG 2016
O. Dunkelman and S.K. Sanadhya (Eds.): INDOCRYPT 2016, LNCS 10095, pp. 411–428, 2016.
DOI: 10.1007/978-3-319-49890-4_23

called *format preserving encryption,* or in short FPE. In cases like credit card based transactions, problems associated with integrating encryption into existing applications further demand the ciphers to be well-defined data models which is not possible using general block ciphers. FPE makes it possible to integrate the data level encryption into the legacy business application frameworks which were previously difficult or impossible to address.

Cryptographic literature contains good solutions for FPE [3,5,7,12,13,15, 18]. In recent papers [2,5,18], different FPE based constructions were proposed. An FPE scheme is used to transform data of a predefined specific format into a ciphertext of identical format. Extensive use of financial servers and also some recent attacks on these servers urge the need of secure, efficient and robust FPE schemes. In 2002 paper [5], authors proposed a practical approach of building an FPE construction. There are three popular approaches of designing FPE based scheme, namely *prefix ciphering, cyclic walking* and a *Feistel* based construction.

In 2008, Terence Spies [21] submitted a proposal to NIST, named FFSEM, which combines cycle walking and an AES-based balanced Feistel network. In 2010 paper [2,4], authors proposed a Feistel based design called FFX (Format-preserving, Feistel-based encryption). FFX is based on ten rounds of iteration involving at least ten invocation of pseudorandom permutation (AES). In practical applications like credit card numbers or social security numbers, generally 12 to 36 rounds of AES invocations are required. A database with millions of entries and an FPE scheme which needs 12 to 36 AES invocation per entry is a fairly inefficient system. The other known schemes are BPS [6] and VFPE [20].

Consider a block cipher which uses substitution-permutation network (SPN). Let this block cipher contains l number of blocks, each block having size m. An lm-bit input can be divided into l blocks each containing m bits. In FPE, the plaintext and the ciphertext both follow the same specified format. For example, in the encryption of credit card numbers, the desired format for both input and output may be digits 0 to 9. In such case, the total number of inputs in one block of the block cipher which is 2^m cannot be equal to 10. Therefore, some of the inputs (in this example, $2^m - 10$ inputs) are not considered in the encryption. The natural way to achieve it is to map the set $\{0, \cdots, 9\}$ to $\{0,1\}^m$ injectively (one-one). Let $X = \{0, \cdots, 9\}$, $Y = \{0,1\}^m$ and $\phi : X \to Y$ be an injective map. To preserve the format of plaintext and its corresponding ciphertext, one way to encrypt an element from X^l, say $X_1||X_2|| \cdots ||X_l$ where $X_i \in \{0, \cdots, 9\}$ for all $i = 1, \cdots, l$, is the following -

1. First encode the element $X_1||X_2|| \cdots ||X_l$ using the map ϕ. Let the output be an element from Y^l, say, $Y_1||Y_2|| \cdots ||Y_l = \phi(X_1)||\phi(X_2)|| \cdots ||\phi(X_l)$.
2. Then encrypt the encoded element using the encryption algorithm $\mathcal{E}$ of the block cipher. Let the ciphertext be $\bar{Y}_1||\bar{Y}_2|| \cdots ||\bar{Y}_l$.
3. Finally, decode the ciphertext $\bar{Y}_1||\bar{Y}_2|| \cdots ||\bar{Y}_l$ using ϕ^{-1}. After decoding, we get $\phi^{-1}(\bar{Y}_1)||\phi^{-1}(\bar{Y}_2)|| \cdots ||\phi^{-1}(\bar{Y}_l)$.

For consistency of the encryption, it is easy to check that $\bar{Y}_i$ must belong to $\phi(X)$ for all $i = 1, \cdots, l$. Decryption can also be defined in a similar manner discussed above.

The method discussed above seems obvious and straightforward solution to the format preserving encryption. However, it must be noted that this method is only a solution, not the solution and thus we do not claim that the above method is the only way to achieve the format preserving encryption. Furthermore, this method provides a solution for format preserving encryption if and only if $\mathcal{E}(\phi(X)^l) = \phi(X)^l$. Nevertheless, this condition does not seem trivial to achieve.

Now, the whole problem boils down to construct a block cipher and a map ϕ such that $\mathcal{E}(\phi(X)^l) = \phi(X)^l$. In the example discussed above, the set X contains the digits 0 to 9. In general, we consider X as any arbitrary set unless specified otherwise. Further, we assume that constructing encoding map ϕ is easy. It is trivial to see that if $|X| = 2^m$, constructing our desired block cipher also is easy (any block cipher with block size m will suffice), but what if $|X| \neq 2^{m_1}$ for any $m_1 > 0$? When $|X| \neq 2^{m_1}$, it is not trivial to achieve the goal and thus becomes interesting to see the possibility of those block ciphers which can be used in format-preserving encryption using the method discussed above only.

A block cipher can be viewed as a pseudo-random permutation which maps $\{0, 1\}^{lm}$ to $\{0, 1\}^{lm}$. Our problem demands a pseudo-random permutation which maps $\phi(X)^l$ to $\phi(X)^l$. In the example of SPN-based block cipher, the main problem can be divided into two smaller problems - (i) Can an S-box be constructed which maps $\phi(X)$ to $\phi(X)$ and (ii) Can an $n \times n$ matrix be constructed which given any input vector from $\phi(X)^n$ outputs a vector from $\phi(X)^n$ only. Keeping aside the security properties of S-box for a moment, constructing such an S-box is easy which satisfies condition (i). The only remaining problem is to construct a matrix in the diffusion layer whose both input and output vectors belong to $\phi(X)^n$. If such matrices are possible to construct, our desired block cipher can also be constructed (not sure about security properties!!). However, the impossibility of constructing any such matrix does not rule out the possibility of constructing our desired block cipher.

This paper considers not only block ciphers but, as a whole, those cryptographic primitives whose diffusion layer can be defined by, in general, a linear transformation. In diffusion layers of many block ciphers or hash functions, some matrices are used to provide the diffusion. If the matrix is MDS, it gives optimal branch number, i.e. optimal diffusion [1,8–11]. In the diffusion layer, input and output to an $n \times n$ matrix are mn-bit $\{0, 1\}$ strings. These mn-bit strings are divided into n tuples each consisting m-bit $\{0, 1\}$ string. Mathematically, operations required for matrix multiplications are addition and multiplication and hence entries of input vectors, output vectors and matrix are defined as elements of a subset of a ring containing 2^m elements. In many cases, these entries are from a special type of ring which is the field $\mathbb{F}_{2^m}$.

Our Contribution: Let p be a prime number and $q = p^s$ for some $s > 0$. Consider an $n \times n$ matrix, M, whose entries are from the field $\mathbb{F}_q$ (not necessarily binary field only). Let X be any set and ϕ be an injective map from X to $\mathbb{F}_q$, i.e. $\phi : X \to \mathbb{F}_q$. We say $\phi(X)$ to be a format preserving set with respect to the matrix M if $M\mathbf{v} \in \phi(X)^n$ for all $\mathbf{v} \in \phi(X)^n$. The question is - for a given field

$\mathbb{F}_q$ and a matrix M whose entries are from $\mathbb{F}_q$, what are the possible cardinalities of a format preserving set $\phi(X)$? In response to this question, this paper shows that if M has at least one row which contains at least two non-zero entries, then there does not exist any format preserving set of cardinality 10 with respect to 4×4 matrix over the field $\mathbb{F}_{2^4}$. Not only that, for such matrices and for any prime characteristic p, if $\bar{0} \in \phi(X)$, then $\phi(X)$ must be a vector space over the smallest field containing entries of M and therefore $|\phi(X)|$ must be $p^{m'}$ for some $m' \geq 1$ which cannot be 10. But, if each row of matrix M contains at most one non-zero entry, then any cardinality of $\phi(X)$ is possible. However, to the best of our knowledge, such matrices do not have any cryptographic significance and thus useful only for theoretical completeness.

Organisation of the Paper: The paper is organised as follows: In Sect. 2 we provide definitions and preliminaries. Section 3 covers the main results. In Sect. 4, we discuss the possibility or impossibility of constructing a format preserving set of cardinality 10 with respect to a 4×4 matrix over the field $\mathbb{F}_{2^4}$. Section 5 concludes the paper and mentions some of future works in this direction.

2   Definition and Preliminaries

A formal definition of format preserving definition has been given in [2], but, in this paper, for the sake of simplicity, we do not present that definition here. However, in the simplest form, a format preserving encryption is a function $E : K \times X \to X$ where K is called key space and X is called domain. Informally speaking, the plaintext and ciphertext both follow the same format.

Definition 1. *A non-empty set G together with a binary operation, $\cdot$, is said to form a group if it satisfies the following four conditions:*

1. *$a, b \in G$ implies that $a \cdot b \in G$ (closed);*
2. *$a, b, c \in G$ implies $(a \cdot b) \cdot c = a \cdot (b \cdot c)$ (associative);*
3. *There exists an element $e \in G$ such that $a \cdot e = e \cdot a = a$ for all $a \in G$ (existence of identity element);*
4. *For every element $a \in G$, there exists an element $b \in G$ such that $a \cdot b = b \cdot a = e$. (existence of inverse).*

If $a \cdot b = b \cdot a$ for all $a, b \in G$ (commutative), then the group G is said to be abelian. For more details about group, refer [14].

For notational convenience, we denote $a \cdot b$ as simply ab. Let S be a subset of a group G. Then the subgroup of G generated by S, denoted by $< S >$, is defined to be the intersection of all subgroups of G containing S. The subgroup $< S >$ is the smallest subgroup of G which contains S. Equivalently,

$$< S >= \{b_1^{\lambda_1} b_2^{\lambda_2} \cdots b_r^{\lambda_r} \mid r \geq 0, b_i \in S, \lambda_i \in \{-1, 1\}\}$$

with the convention that if $r = 0$, the product over the empty list is e. If G is finite, $\lambda_i \neq -1$ (-1 in the exponent not required). Let $S = \{g_1, g_2, \cdots, g_k\}$. If G is a finite abelian group, then

$$< S > = \{g_1^{\lambda_1} g_2^{\lambda_2} \cdots g_k^{\lambda_k} \mid \lambda_i \geq 0\}.$$

Let $h \in G$. By hS, we mean $\{hg_1, hg_2, \cdots, hg_k\}$.

Definition 2. *A field is a non-empty set $\mathbb{F}$ together with two binary operations, addition (+) and multiplication ($\cdot$) which satisfies the following conditions:*

1. *$\mathbb{F}$ must be an abelian group under addition (+);*
2. *The set of non-zero elements $\mathbb{F}^* = \mathbb{F} \setminus \{\bar{0}\}$ ($\bar{0}$ is the additive identity of $\mathbb{F}$) must be an abelian group under multiplication ($\cdot$);*
3. *For every element $a, b, c \in \mathbb{F}$, $c \cdot (a + b) = c \cdot a + c \cdot b$ (distributive law).*

Here also, for notational convenience, we denote $a \cdot b$ as simply ab. The multiplicative identity is denoted by $\bar{1}$.

Let $a \in \mathbb{F}$ and $Z \subseteq \mathbb{F}$. If $Z = \{z_1, z_2, \cdots, z_k\}$, then $aZ = \{az_1, az_2, \cdots, az_k\}$ and $a + Z = \{a + z_1, a + z_2, \ldots, a + z_k\}$. The smallest field containing the subset Z is the intersection of all subfields of $\mathbb{F}$ which contain Z. We denote this smallest field as $SF(Z)$. Fields may have finite or infinite cardinalities. In this paper, we consider only finite fields, i.e. fields having finite cardinalities.

The characteristic of a field $\mathbb{F}$ is the least positive integer n such that $n \cdot \bar{1} = \bar{0}$. If such n exists, we say that the characteristic of the field $\mathbb{F}$, denoted by $char(\mathbb{F})$, is n else 0. It can be shown that n must be a prime number and thus every finite field must have prime characteristic.

Let p be a prime number. Any finite field with characteristic p will have $q = p^e$ number of elements for some integer $e > 0$. A field having q number of elements is denoted by $\mathbb{F}_q$ and the set of non-zero elements of $\mathbb{F}_q$ by $\mathbb{F}_q^*$. Any subfield of the field $\mathbb{F}_q$ contains $p^{e'}$ number of elements where e' divides e. Conversely, if e' is a positive divisor of e, then there exists exactly one subfield having $p^{e'}$ number of elements. For more details about finite fields, see [17].

Definition 3. *A non-empty set V is said to be a vector space over a field $\mathbb{F}$ if it satisfies the following conditions:*

1. *V is an abelian group under addition (+);*
2. *For every $\alpha \in \mathbb{F}$ and for every $\mathbf{v} \in V$, there is defined an element $\alpha\mathbf{v}$ which belongs to V;*
3. *$\alpha(\mathbf{v} + \mathbf{w}) = \alpha\mathbf{v} + \alpha\mathbf{w}$ for every $\alpha \in \mathbb{F}$ and for every $\mathbf{v}, \mathbf{w} \in V$;*
4. *$(\alpha + \beta)\mathbf{v} = \alpha\mathbf{v} + \beta\mathbf{v}$ for every $\alpha, \beta \in \mathbb{F}$ and for every $\mathbf{v} \in V$;*
5. *$\alpha(\beta\mathbf{v}) = (\alpha\beta)\mathbf{v}$;*
6. *$1\mathbf{v} = \mathbf{v}$ for all $\mathbf{v} \in V$.*

For more details about vector space, see [16]. If V is finite, then V is finite-dimensional too over $\mathbb{F}$, but converse need not be true. If V is a d-dimensional vector space over a finite field $\mathbb{F}$, then the cardinality of V is the cardinality of $\mathbb{F}$

raised to the d-th power, i.e. $|V| = |\mathbb{F}|^d$. If $\mathbb{F}_{q'}$ is a subfield of the field $\mathbb{F}_q$, then $\mathbb{F}_q$ is a vector space over $\mathbb{F}_{q'}$ having dimension e/e'.

In this paper, we assume $V = \mathbb{F}_q^n$ and $\mathbb{F} = \mathbb{F}_q$ or some subfield of $\mathbb{F}_q$. Any vector can be denoted as a horizontal array or a vertical array. A vector denoted horizontally is called a row vector whereas a vector denoted vertically is called a column vector. Throughout this paper, we assume that the vector is n-dimensional and thus has n entries. If these entries are from a set Z, we say $\mathbf{v} \in Z^n$ (abuse of notation!). The transpose of a row vector is a column vector. Let $\mathbf{v} = [v_1 \ v_2 \ \cdots \ v_n]$ be a row vector. Then $\mathbf{v}^T = [v_1 \ v_2 \ \cdots \ v_n]^T$ is a column vector.

An $m \times n$ matrix is a rectangular array having m rows and n columns. If entries of a matrix A are from a set Z, then we denote it as $A(Z)$. We simply write A if entries of the matrix A are evident from the context. The transpose of a matrix A is denoted as A^T. If the matrix has all entries $\bar{0}$, we call it null matrix denoted as $\mathbf{O}_{m \times n}$. If an $n \times n$ matrix has diagonal entries $\bar{1}$ and rest $\bar{0}$, we call it identity matrix denoted as $I_{n \times n}$. For more about matrices and their operations, refer [19].

Definition 4. *A non-empty set $\mathbb{S} \subseteq \mathbb{F}_q$ is said to be a format preserving set with respect to an $n \times n$ matrix $M(\mathbb{F}_q)$ if $M\mathbf{v} \in \mathbb{S}^n$ for all $\mathbf{v} \in \mathbb{S}^n$.*

If $\mathbb{S}$ is a format preserving set with respect to M, we write $\mathbb{S}$ is FPS wrt M. All other notations used in this paper are standard notations. Any undefined terms have usual standard definitions.

3   Our Results

Let $\mathbb{S} \subseteq \mathbb{F}_q$. We denote the $(i,j)^{\text{th}}$-entry of the matrix $M(\mathbb{F}_q)$ by $m_{i,j}$ where $1 \leq i, j \leq n$. We divide our results into three different subsections. Subsection 3.1 covers the case when $\bar{0} \in \mathbb{S}$ while Subsect. 3.2 considers the case when $\bar{0}$ may or may not belong to $\mathbb{S}$. Subsection 3.3 deals with some conditions which ensure that $\bar{0} \in \mathbb{S}$.

Lemma 1. *Let M be $\mathbf{O}_{n \times n}$ (null matrix). If $\mathbb{S}$ is FPS wrt M, then $\bar{0} \in \mathbb{S}$.*

Proof. It's obvious.   □

From the definition of format preserving set, any set $\mathbb{S}$ will be an FPS wrt null matrix M if and only if $\bar{0} \in \mathbb{S}$. From Lemma 1, it is evident that if $\mathbb{S}$ is an FPS wrt M, then $\bar{0} \in \mathbb{S}$. Conversely, assume $\bar{0} \in \mathbb{S}$. Take any $\mathbf{v} \in \mathbb{S}^n$. Then $M\mathbf{v} = [\bar{0} \ \bar{0} \ \cdots \bar{0}] \in \mathbb{S}^n$. Therefore, $M\mathbf{v} \in \mathbb{S}^n$ for all $\mathbf{v} \in \mathbb{S}^n$ and thus $\mathbb{S}$ is a format preserving set with respect to the null matrix M.

To the best of our knowledge, null matrix M does not play any significant role in the design of diffusion layer of cryptographic primitives like block ciphers and hash functions, however, Lemma 1 is essential for the theoretical completeness of the results.

Lemma 2. *Let $s \in \mathbb{F}_q^*$ and $\mathbb{S}' = s\mathbb{S}$. Then, $\mathbb{S}$ is an FPS wrt M if and only if $\mathbb{S}'$ is an FPS wrt M.*

Proof. Let $\mathbf{v} \in \mathbb{S}^n$ and $\mathbf{v}' = s\mathbf{v} \in \mathbb{S}'^n$.

Suppose $\mathbb{S}$ is an FPS wrt M. Then, $M\mathbf{v} \in \mathbb{S}^n$ for all $\mathbf{v} \in \mathbb{S}^n$. Consider $M\mathbf{v}' = M(s\mathbf{v}) = s(M\mathbf{v})$. Since $M\mathbf{v} \in \mathbb{S}^n$, therefore $M\mathbf{v}' = s(M\mathbf{v}) \in \mathbb{S}'^n$ for all $\mathbf{v}' \in \mathbb{S}'^n$.

Conversely, assume $\mathbb{S}'$ be an FPS wrt M. Therefore, $M\mathbf{v}' \in \mathbb{S}'^n$ for all $\mathbf{v}' \in \mathbb{S}'^n$. Consider $M\mathbf{v} = M(s^{-1}\mathbf{v}') = s^{-1}M(\mathbf{v}')$ (s^{-1} exists because $s \in \mathbb{F}_q^*$). Since $M\mathbf{v}' \in \mathbb{S}'^n$, therefore $M\mathbf{v} = s^{-1}(M\mathbf{v}') \in \mathbb{S}^n$ for all $\mathbf{v} \in \mathbb{S}^n$. Hence, the lemma. $\qquad\square$

Let $s' \in \mathbb{S}'$ be a non-zero element. Then, there exists $s \in \mathbb{F}_q^*$ such that $ss' = \bar{1}$. Let $\mathbb{S} = s\mathbb{S}'$. It is clear that $\bar{1} \in \mathbb{S}$. From Lemma 2, $\mathbb{S}$ is an FPS wrt M if and only if $\mathbb{S}'$ is an FPS wrt M.

Lemma 3. *If $\mathbb{S}$ is an FPS wrt M, then $\mathbb{S}$ is an FPS wrt M^k also for all $k \geq 1$.*

Proof. We prove it by induction. Let $\mathbf{v} \in \mathbb{S}^n$. Since $\mathbb{S}$ is an FPS wrt M, the vector $\mathbf{v}^{(1)} = M\mathbf{v}$ also belongs to $\mathbb{S}^n$ for all $\mathbf{v} \in \mathbb{S}^n$. Assume $\mathbf{v}^{(r-1)} = M^{r-1}\mathbf{v} \in \mathbb{S}^n$ for some $k = r - 1$ and for all $\mathbf{v} \in \mathbb{S}^n$. Now, we show that $\mathbf{v}^{(r)} = M^{(r)}\mathbf{v}$ also belongs to $\mathbb{S}^n$.

Take any $\mathbf{v} \in \mathbb{S}^n$. Consider $M^{(r)}\mathbf{v} = M(M^{(r-1)}\mathbf{v}) = M\mathbf{v}^{(r-1)}$. We assumed that $\mathbf{v}^{(r-1)} \in \mathbb{S}^n$. Since $\mathbb{S}$ is an FPS wrt M, therefore $\mathbf{v}^{(r)} = M\mathbf{v}^{(r-1)}$ also belongs to $\mathbb{S}^n$. Since it is true for all $\mathbf{v} \in \mathbb{S}^n$, hence the lemma. $\qquad\square$

3.1   What if $\bar{0} \in \mathbb{S}$?

In this section, we assume that $\bar{0} \in \mathbb{S}$ and then explore the complete algebraic structure of $\mathbb{S}$ with respect to the matrix M.

Lemma 4. *Let $\bar{0} \in \mathbb{S}$. Suppose $s \in \mathbb{S}$. If $\mathbb{S}$ is an FPS wrt M, then $sm_{i,j} \in \mathbb{S}$ for all $1 \leq i, j \leq n$.*

Proof. Let $\mathbf{v}^{(j)} = [\bar{0}\ \bar{0}\ \cdots\ s\ \cdots\ \bar{0}\ \bar{0}]^T$ where s is at the jth position of the vector $\mathbf{v}^{(j)}$ and rest $\bar{0}$. Then $M\mathbf{v}^{(j)} = [sm_{0,j}\ sm_{1,j}\ \cdots\ sm_{n,j}]^T \in \mathbb{S}^n$. Therefore, $sm_{i,j} \in \mathbb{S}$ for all $i = 1, \cdots, n$. Now, consider $M\mathbf{v}^{(j)}$ for $j = 1, \cdots, n$. Thus $sm_{i,j} \in \mathbb{S}$ for $1 \leq i, j \leq n$. Hence, the lemma. $\qquad\square$

Note that Lemma 4 does not assume that $\bar{1} \in \mathbb{S}$. If we assume that $\bar{1} \in \mathbb{S}$, then $m_{i,j} \in \mathbb{S}$ for all $1 \leq i, j \leq n$. The justification of assuming $\bar{1} \in \mathbb{S}$ comes from Lemma 2. From Lemma 4, we get the next corollary.

Let $Z = \{m_{i,j} \mid m_{i,j} \neq \bar{0}\}$. Using the fact that $< Z >$ is a subgroup of the multiplicative group $\mathbb{F}_q^*$, we get the following corollary.

Corollary 1. *Let $\bar{0} \in \mathbb{S}$. Suppose $s \in \mathbb{S}$. If $\mathbb{S}$ is an FPS wrt M, then $s < Z > \subseteq \mathbb{S}$.*

Using the above Corollary 1, we get the following theorem which characterises the structure of format preserving set $\mathbb{S}$ with respect to such matrices M whose each row contains at most one non-zero entry.

Theorem 1. *Let $\bar{0} \in \mathbb{S}$. Suppose each row of M contains at most one non-zero entry. Then, $\mathbb{S}$ is an FPS wrt M if and only if there exists a set $H \subseteq \mathbb{F}_q^*$ such that $\mathbb{S} = \bigcup_{s \in H} s < Z > \cup \{\bar{0}\}$.*

Proof. Let $\mathbb{S}$ be an FPS wrt M. Take H to be an empty set. Choose $s_1 \neq \bar{0}$ from the set $\mathbb{S}$. If there is no non-zero element in $\mathbb{S}$, then $\mathbb{S} = \{\bar{0}\}$ only, otherwise from Corollary 1, $s_1 < Z > \subseteq \mathbb{S}$. Consider $\mathbb{S}_1 = \mathbb{S} \backslash s_1 < Z >$ and add s_1 into H. Repeat this process until we are left with $\bar{0}$ only. Finally, we get $\mathbb{S} = \bigcup_{s \in H} s < Z > \cup \{\bar{0}\}$.

Conversely, let $\mathbb{S} = \bigcup_{s \in H} s < Z > \cup \{\bar{0}\}$. Assume $s^{(i)} \in H$ and $\alpha_i \in < Z >$ for all $i = 1, \cdots, n$. Consider a vector $\mathbf{v} = [s^{(1)} \alpha_1 \,|\, \bar{0}\; s^{(2)} \alpha_2 \,|\, \bar{0} \;\cdots\; s^{(n)} \alpha_n \,|\, \bar{0}]^T$. By the term $s^{(i)} \alpha_i \,|\, \bar{0}$, we mean either $s^{(i)} \alpha_i$ or $\bar{0}$. It is easy to see that $\mathbf{v} \in \mathbb{S}^n$. We assume that each row of M has at most one non-zero entry. Without loss of generality, we may assume that each row has exactly one non-zero entry. Suppose M has non-zero entries in columns $j_1, j_2, \cdots, j_n$ corresponding to rows $1, 2, \cdots, n$. Then, $M\mathbf{v} = [s^{(j_1)} \alpha_{j_1} m_{1,j_1} \,|\, \bar{0}\; s^{(j_2)} \alpha_{j_2} m_{2,j_2} \,|\, \bar{0} \;\cdots\; s^{(j_n)} \alpha_{j_n} m_{n,j_n} \,|\, \bar{0}] \in \mathbb{S}^n$ because from Corollary 1, $s^{(j_i)} \alpha_{j_i} m_{i,j_i} \in \mathbb{S}$ for all $i = 1, \cdots, n$. Therefore, $\mathbb{S}$ is an FPS wrt M. Hence, the lemma. $\qquad\square$

Theorem 1 gives a nice characterisation for a certain type of of format preserving sets (we assumed $\bar{0} \in \mathbb{S}$) with respect to those matrices whose each row contains at most one non-zero entry. In-fact, this result can be used to count the number of elements in $\mathbb{S}$ with respect to such type of matrices. Consider the set $\bigcup_{s \in H} s < Z >$. Take two elements, say $s_1 \alpha_1$ and $s_2 \alpha_2$, from this set. The equality $s_1 \alpha_1 = s_2 \alpha_2$ implies $s_1 s_2^{-1} = \alpha_2 \alpha_1^{-1} = \alpha_3$ for some $\alpha_3 \in < Z >$ and thus $s_1 = s_2 \alpha_3 \in s_2 < Z >$. If so, then $s_1 < Z > = s_2 < Z >$. Therefore, it can be concluded that either $s_1 < Z > = s_2 < Z >$ or $s_1 < Z > \cap s_2 < Z > = \phi$ (empty set) for any two elements $s_1, s_2 \in H$. We assume that if $s_1, s_2 \in H$ such that $s_1 \neq s_2$, then $s_1 < Z > \cap s_2 < Z > = \phi$. Let $|H| = k$ and $| < Z > | = d$. Since $< Z >$ is a subgroup of $\mathbb{F}_q^*$, $| < Z > | = d$ divides $q - 1$. Then, the total number of elements in $\mathbb{S}$ will be $dk + 1$.

Consider the example of credit card. In this example, $|\mathbb{S}| = dk + 1 = 10$. So, $dk = 9$. Possible values of (d, k) are $(1, 9), (3, 3)$ and $(9, 1)$. Take $char(\mathbb{F}_q) = p = 2$, i.e. binary field. Some examples for these possible cases are as follows:

Case 1: $(d, k) = (1, 9)$. If $d = 1$, $< Z > = \{\bar{1}\}$ which implies $Z = \{\bar{1}\}$. And thus, each row of M has at most one non-zero entry and that non-zero element is $\bar{1}$. Since $k = 9$, choose any 9 elements from $\mathbb{F}_q^*$. To get 9 distinct non-zero values, $q \geq 2^4$.

Case 2: $(d, k) = (3, 3)$. In this case, $3 \mid q - 1$, i.e., $3 \mid 2^e - 1$ implies that e must be a positive even number. Furthermore, as the subgroup of a cyclic group is cyclic, $< Z >$ must be a cyclic group because $< Z >$ is the subgroup of a cyclic group $\mathbb{F}_q^*$. Let $< Z > = \{\bar{1}, \alpha, \alpha^2\}$ for some $\alpha \in \mathbb{F}_q^*$ whose order is 3. Thus, each row of M has at most one non-zero entry and that non-zero entry must be either $\bar{1}$, α, or α^2. To be $< Z >= \{\bar{1}, \alpha, \alpha^2\}$, it is required that at least one row must contain either α or α^2 as a non-zero element. Now, consider the quotient group $(\mathbb{F}_q^* / < Z >) = \{\Gamma_1, \Gamma_2, \cdots, \Gamma_z\}$ where $z = (q-1)/3$. Choose any three cosets and then choose one element from each coset. Let these elements be $\{s_1, s_2, s_3\}$. Take $H = \{s_1, s_2, s_3\}$.

Case 3: $(d, k) = (9, 1)$. In this case, $9 \mid q - 1$ or $9 \mid 2^e - 1$ implies that e must be a multiple of 6. Let $< Z >= \{\bar{1}, \beta, \beta^2, \beta^3, \cdots, \beta^8\}$ for some $\beta \in \mathbb{F}_q^*$ whose order is 9. Thus, each row of M has at most one non-zero entry and that non-zero entry must be from $< Z >$. To be $< Z >= \{\bar{1}, \beta, \beta^2, \beta^3, \cdots, \beta^8\}$, it is required that at least one row must contain one element from the set $\{\beta, \beta^2, \beta^4, \beta^5, \beta^7, \beta^8\}$ as a non-zero element. Now, choose any element, say s, from $\mathbb{F}_q^*$. Take $H = \{s\}$.

For the credit card example over binary field, we showed how to construct the matrix M using $< Z >$ and the desired format preserving set $\mathbb{S}$ with respect to M using $< Z >$ and $< H >$. The idea of constructing $\mathbb{S}$ and M was based upon the Theorem 1. Although Theorem 1, as per our best knowledge, do not provide any cryptographically significant matrices which might be used in diffusion layer, it has undoubtedly a theoretical significance.

The left case is when M has at least one row which has at least two non-zero entries. This case is covered by the next lemma.

Lemma 5. *Let $\bar{0} \in \mathbb{S}$. Suppose M has at least one row which contains at least two non-zero entries. Suppose $s_1, s_2 \in \mathbb{S}$. If $\mathbb{S}$ is an FPS wrt M, then $s_1 + s_2 \in \mathbb{S}$.*

Proof. From Corollary 1, $s_1 < Z > \subseteq \mathbb{S}$ and $s_2 < Z > \subseteq \mathbb{S}$. Let i^{th} row of the matrix M contains at least two non-zero entries, say m_{i,j_1} and m_{i,j_2}, at column positions j_1 and j_2. Let $\mathbf{v} = [\bar{0} \cdots s_1 m_{i,j_1}^{q-2} \bar{0} \cdots \bar{0} s_2 m_{i,j_2}^{q-2} \bar{0} \cdots \bar{0}]^T$ where $s_1 m_{i,j_1}^{q-2}$ is at the j_1^{th}, $s_2 m_{i,j_2}^{q-2}$ is at the j_2^{th} position of the vector $\mathbf{v}$ and rest $\bar{0}$. Since $s_1 < Z >$ and $s_2 < Z >$ both are subsets of $\mathbb{S}$, therefore $s_1 m_{i,j_1}^{q-2}$ and $s_2 m_{i,j_2}^{q-2}$ both belong to $\mathbb{S}$ and thus $\mathbf{v} \in \mathbb{S}^n$. As $\mathbb{S}$ is an FPS wrt M, the vector $M\mathbf{v} \in \mathbb{S}^n$. The ith entry of the vector $M\mathbf{v}$ will be $s_1 m_{i,j_1}^{q-1} + s_2 m_{i,j_2}^{q-1}$ which is equal to $s_1 + s_2$ due to the fact that $m_{i,j_1}^{q-1} = m_{i,j_2}^{q-1} = \bar{1}$ in $\mathbb{F}_q^*$. Hence, the lemma. $\square$

We now define a new set

$$\mathbb{K} = \{k_1 \alpha_1 + k_2 \alpha_2 + \cdots + k_r \alpha_r \mid r \geq 0, k_i \geq 1, \alpha_i \in < Z >\}$$

with the convention that if $r = 0$, the sum over the empty list is $\bar{0}$. The set $\mathbb{K}$ is in-fact the smallest field containing entries of the matrix M, or containing entries of Z because Z contains all non-zero entries of M. The only difference lies when M has at least one entry which is $\bar{0}$. In such case, Z does not contain all entries of M, however, the smallest field containing Z has $\bar{0}$ and hence it becomes equal to the smallest field containing entries of M. Let $SF(Z)$ and $SF(M)$ be the smallest field containing entries of Z and M respectively. The next three lemmas will show the relation between $SF(Z)$, $SF(M)$ and $\mathbb{K}$.

Lemma 6. $\mathbb{K} = SF(Z)$.

Proof. $SF(Z)$ contains Z and therefore contains $< Z >$ too. Thus $< Z > \subseteq SF(Z)$. Take any $\alpha \in Z$. Since $Z \subseteq SF(Z)$, therefore $\alpha \in SF(Z)$. As $SF(Z)$ is a field, so $\alpha + \alpha + \cdots + \alpha$ (k times addition), i.e. $k\alpha$ belongs to $SF(Z)$. Take any element from $\mathbb{K}$. If that element is $\bar{0}$, it belongs to $SF(Z)$ too. Otherwise, the element will be of the form of $k_1\alpha_1 + k_2\alpha_2 + \cdots + k_r\alpha_r$ for some $r \geq 1$ and $k_i \geq 1$. All elements $k_i\alpha_i \in SF(Z)$. Since $SF(Z)$ is a field, hence $k_1\alpha_1 + k_2\alpha_2 + \cdots + k_r\alpha_r$ also belongs to $SF(Z)$. Thus $\mathbb{K} \subseteq SF(Z)$.

To complete the proof, we need to show that $\mathbb{K}$ is a field. To show it, we must prove - (a) $\mathbb{K}$ is an abelian group under addition, (b) $\mathbb{K} \setminus \{\bar{0}\}$ is an abelian group under multiplication and (c) follows distributive law (see definition and preliminaries). Using the fact that (i) $\bar{0} \in \mathbb{K}$ (ii) $\bar{1} \in < Z >$, so $\bar{1} \in \mathbb{K}$ and (iii) $\alpha_1\alpha_2 \in < Z >$, so $\alpha_1\alpha_2 \in \mathbb{K}$, it can be easily proved that $\mathbb{K}$ is a field.

Since $SF(Z)$ is the smallest field containing Z, therefore $SF(Z) \subseteq \mathbb{K}$. Thus $\mathbb{K} = SF(Z)$. □

Lemma 7. $\mathbb{K} = SF(Z) = SF(M)$.

Proof. Using similar arguments as in Lemma 6, it can be shown that $\mathbb{K} = SF(M)$. Therefore $\mathbb{K} = SF(Z) = SF(M)$. □

Theorem 2. *Let $\bar{0} \in \mathbb{S}$. Suppose M has at least one row which contains at least two non-zero entries. Then, $\mathbb{S}$ is an FPS wrt M if and only if $\mathbb{S}$ is a vector space over the field $SF(M)$.*

Proof. Suppose $\mathbb{S}$ is a format preserving set. To show that $\mathbb{S}$ is a vector space over $SF(M)$, we need the following (see definition and preliminaries):

1. $\mathbb{S}$ is an abelian group under addition $(+)$ - (a) If $s_1, s_2 \in \mathbb{S}$, then $s_1 + s_2 \in \mathbb{S}$ (From Lemma 5), (b) associativity comes from the fact that $\mathbb{S} \subseteq \mathbb{F}_q$, (c) $\bar{0}$ is the additive identity (we assumed that $\bar{0} \in \mathbb{S}$) (d) For every $s \in \mathbb{S}$, its inverse $(p-1)s \in \mathbb{S}$ (because of Lemma 5 and the fact that $char(\mathbb{F}_q) = p$) and (e) commutativity comes from the fact that $\mathbb{S} \subseteq \mathbb{F}_q$.
2. Take any $\alpha \in SF(M)$. If $\alpha = \bar{0}$, then $\alpha s = \bar{0} \in \mathbb{S}$ for any $s \in \mathbb{S}$. Suppose $\alpha \neq \bar{0}$, say $\alpha = k_1\alpha_1 + k_2\alpha_2 + \cdots + k_r\alpha_r$ (from Lemmas 6 and 7) where $\alpha_i \in < Z >$, $r \geq 1$ and $k_i \geq 1$ for all $i = 1, \cdots, r$. Take $s \in \mathbb{S}$. Then $\alpha s = k_1(\alpha_1 s) + k_2(\alpha_2 s) + \cdots + k_r(\alpha_r s)$. From Corollary 1, $\alpha_i s \in \mathbb{S}$ for all $i = 1, \cdots, r$. From Lemma 5, it can be concluded that $k_i(\alpha_i s) \in \mathbb{S}$ and from the same lemma again, $\sum_{i=1}^{r} k_i(\alpha_i s) \in \mathbb{S}$.

Rest conditions trivially come from the fact that $\mathbb{S}$ and $SF(M)$ both are subsets of $\mathbb{F}_q$.

Conversely, let $\mathbb{S}$ be a vector space over $SF(M)$. Since $\mathbb{S} \subseteq \mathbb{F}_q$, therefore $\mathbb{S}$ is finite and hence finite dimensional. Let $\{\gamma_1, \gamma_2, \cdots, \gamma_d\}$ be the basis of $\mathbb{S}$. Let $s_r = \sum_{j=1}^{d} \alpha_j^{(r)}\gamma_j$ for $r = 1, \cdots, n$. Consider the vector $\mathbf{v} = [s_1 \ s_2 \ \cdots \ s_n]^T$. Take $M\mathbf{v} = M[s_1 \ s_2 \ \cdots \ s_n]^T$. Then the ith element of the vector $M\mathbf{v}$ will be $\sum_{r=1}^{n} m_{i,r}s_r = \sum_{r=1}^{n} m_{i,r}(\sum_{j=1}^{d} \alpha_j^{(r)}\gamma_j) = \sum_{j=1}^{d} \gamma_j(\sum_{r=1}^{n} m_{i,r}\alpha_j^{(r)})$. Since $\alpha_j^{(r)}$,

$m_{i,r} \in SF(M)$, therefore $\alpha_j^{(r)} m_{i,r} \in SF(M)$. So, $\sum_{j=1}^d \gamma_j (\sum_{r=1}^n m_{i,r} \alpha_j^{(r)}) \in \mathbb{S}$ (because $\mathbb{S}$ is a vector space over $SF(M)$). Thus, the ith element of the vector $M\mathbf{v}$ belongs to $\mathbb{S}$ and hence $M\mathbf{v} \in \mathbb{S}^n$. Therefore, $\mathbb{S}$ is an FPS wrt M. Hence, the theorem.   □

Suppose $\mathbb{S}$ is a format preserving set with respect to a matrix M which has at least one row that contains at least two non-zero entries. Then from Theorem 2, $|\mathbb{S}| = |SF(M)|^d$ where d is the dimension of the vector space $\mathbb{S}$ over the field $SF(M)$. Let $char(\mathbb{F}_q) = p$ and $|SF(M)| = p^{m'}$ for some $m' \geq 1$. Then, $|\mathbb{S}| = p^{m'd}$. Therefore, in our credit card example, in such case, it is impossible to get a format preserving set whose cardinality is 10.

Now, we consider the case when $\bar{0}$ may not belong to $\mathbb{S}$.

3.2   What if $\bar{0}$ May or May not Belong to $\mathbb{S}$?

In the last subsection, we assumed that $\bar{0} \in \mathbb{S}$ and then showed the complete algebraic structure of $\mathbb{S}$. In-fact, the broader question is what happens if $\bar{0}$ may or may not belong to $\mathbb{S}$?

For matrices M and M^k for $k \geq 1$, let $\mathrm{m}_i = \sum_{j=1}^n m_{i,j}$ and $\mathrm{m}_i^{(k)} = \sum_{j=1}^n m_{i,j}^{(k)}$ where $m_{i,j}^{(k)}$ is the $(i,j)^{\text{th}}$ entry of the matrix M^k. When $k = 1$, $m_{i,j} = m_{i,j}^{(k)}$ and $\mathrm{m}_i = \mathrm{m}_i^{(k)}$ and hence can be used interchangeably. If any of $\mathrm{m}_i^{(k)} = \bar{0}$ for any $k \geq 1$, we obtain the following lemma.

Lemma 8. Let $\mathrm{m}_i^{(k)} = \bar{0}$ for some $i \in \{1, \cdots, n\}$ and for any $k \geq 1$. If $\mathbb{S}$ is an FPS wrt M, then $\bar{0} \in \mathbb{S}$.

Proof. Suppose $s \in \mathbb{S}$. Let $\mathbf{v} = [s \; s \; \cdots \; s]^T \in \mathbb{S}^n$. From Lemma 3, $\mathbb{S}$ is an FPS wrt M^k also for $k \geq 1$. Therefore i^{th} element of $M^k \mathbf{v}$ will be $s\mathrm{m}_i^{(k)} = \bar{0} \in \mathbb{S}$. Hence the lemma.   □

Let $R = \{\mathrm{m}_i \mid \mathrm{m}_i \neq \bar{0}\}$ and $R^{(k)} = \{\mathrm{m}_i^{(k)} \mid \mathrm{m}_i^{(k)} \neq \bar{0}\}$. It is easy to see that when $k = 1$, R and $R^{(k)}$ are same and hence can be used interchangeably. Furthermore, it is easy to observe that $< R >$ and $< R^{(k)} >$ are subgroups of $\mathbb{F}_q^*$.

Lemma 9. Let $s \in \mathbb{S}$. If $\mathbb{S}$ is an FPS wrt M, then $s < R^{(k)} > \subseteq \mathbb{S}$ for all $k \geq 1$.

Proof. Consider the vector $\mathbf{v} = [s \; s \; \cdots \; s]^T$. Because $\mathbb{S}$ is an FPS wrt M^k also for any $k \geq 1$ (from Lemma 3), $M^k \mathbf{v} = [s\mathrm{m}_1^{(k)} \; s\mathrm{m}_2^{(k)} \; \cdots \; s\mathrm{m}_n^{(k)}]^T \in \mathbb{S}^n$. Thus $s\mathrm{m}_i^{(k)} \in \mathbb{S}$ for all $i = 1, \cdots, n$. Now, we show that $s(\mathrm{m}_{i_1}^{(k)\lambda_1} \mathrm{m}_{i_2}^{(k)\lambda_2} \cdots \mathrm{m}_{i_r}^{(k)\lambda_r}) \in \mathbb{S}$ where $\mathrm{m}_{i_t}^{(k)} \in R^{(k)}$, $r \geq 1$ and $\lambda_t \geq 0$ for all $t = 1, \cdots, r$.

We prove it by induction. It is assumed that $s \in \mathbb{S}$ and shown that $s\mathrm{m}_i^{(k)} \in \mathbb{S}$ for all $i = 1, \cdots, n$. Lets assume that $s(\mathrm{m}_{i_1}^{(k)e_1} \mathrm{m}_{i_2}^{(k)e_2} \cdots \mathrm{m}_{i_r}^{(k)e_r}) \in \mathbb{S}$ for some $e_t = \lambda_t \geq 0$ for all $t = 1, \cdots, r$. Now, we show $s(\mathrm{m}_{i_1}^{(k)e_1} \mathrm{m}_{i_2}^{(k)e_2} \cdots \mathrm{m}_{i_j}^{(k)e_j+1} \cdots \mathrm{m}_{i_r}^{(k)e_r}) \in \mathbb{S}$ for any $j \in \{1, \cdots, r\}$. Let $s' = s(\mathrm{m}_{i_1}^{(k)e_1} \mathrm{m}_{i_2}^{(k)e_2} \cdots \mathrm{m}_{i_j}^{(k)e_j} \cdots \mathrm{m}_{i_r}^{(k)e_r})$. Then $s'\mathrm{m}_{i_j}^{(k)} \in \mathbb{S}$ for any $j \in \{1, \cdots, r\}$ (proved in the first paragraph). Hence the lemma.   □

Next theorem characterises the format preserving set $\mathbb{S}$ with respect to any matrix M using Lemma 9. Although this characterisation does not provide a complete picture of the possibility of cardinalities of $\mathbb{S}$, still it allows to eliminate some candidates.

Theorem 3. *Let $k \geq 1$. If $\mathbb{S}$ is an FPS wrt M, then there exists a set $H \subseteq \mathbb{F}_q$ such that $\mathbb{S} = \bigcup_{s \in H} s < R^{(k)} >$.*

Proof. Take H to be an empty set. Choose s_1 from the set $\mathbb{S}$. From Lemma 9, $s_1 < R^{(k)} > \subseteq \mathbb{S}$. Consider $\mathbb{S}_1 = \mathbb{S} \setminus s_1 < R^{(k)} >$ and add s_1 into H. Repeat this process until we are left with empty set only. Finally, we get $\mathbb{S} = \bigcup_{s \in H} s < R^{(k)} >$. Hence the theorem.   □

3.3   When $\bar{0}$ May Belong to $\mathbb{S}$?

This section explores some relationship between the format preserving set $\mathbb{S}$ and the matrix M which ensure that $\bar{0} \in \mathbb{S}$. These results become significant in the sense that if $\bar{0} \in \mathbb{S}$, then from Subsect. 3.1, we can completely identify the algebraic structure of $\mathbb{S}$ with respect to M.

Lemma 10. *Let $m_{i,j} \in < R > \cup \{\bar{0}\}$ for some $1 \leq i \leq n$ and for all $j = 1, \cdots, n$. Let the characteristic of the underlying field be p. Suppose the ith row has $l \geq 1$ number of non-zero entries where $l \not\equiv 1 \mod p$. If $\mathbb{S}$ is an FPS wrt M, then $\bar{0} \in \mathbb{S}$.*

Proof. Being a subgroup of the cyclic group $\mathbb{F}_q^*$, $< R >$ also is a cyclic group. Let $\beta \in \mathbb{F}_q^*$ be the generator of this group, i.e. $< R >=< \beta >$. Then, entries of the row i will be either $\bar{0}$ or β^{i_j} for some $i_j \geq 0$. Let the ith row of the matrix M be $[m_{i,1} \; m_{i,2} \; \cdots \; m_{i,n}]$ where $m_{i,j}$ is either β^{i_j} or $\bar{0}$. For every $\beta^{i_j} \in < R >$, there exists $\beta^{k_j} \in < R >$ such that $\beta^{i_j} \beta^{k_j} = \bar{1}$. Consider the vector $\mathbf{v}^{(1)} = [\bar{m}_{i,1} \; \bar{m}_{i,2} \; \cdots \; \bar{m}_{i,n}]^T$ where $\bar{m}_{i,j} = \beta^{k_j}$ when $m_{i,j} = \beta^{i_j}$ else $\bar{m}_{i,j} = \bar{1}$. It is easy to see that $\mathbf{v}^{(1)} \in \mathbb{S}^n$.

Now, consider the vector $M\mathbf{v}^{(1)}$. The ith entry of the vector $M\mathbf{v}^{(1)}$ will be $l\bar{1} = (l-1)\bar{1} + \bar{1}$. Since $\mathbb{S}$ is an FPS wrt M, therefore $l\bar{1} \in \mathbb{S}$. From Lemma 9, $(l\bar{1}) < R > \subseteq \mathbb{S}$. Let j_1 be the first column entry in the ith row which has non-zero entry. Take the vector $\mathbf{v}^{(2)} = [\bar{\bar{m}}_{i,1} \; \bar{\bar{m}}_{i,2} \; \cdots \; \bar{\bar{m}}_{i,n}]^T$ where $\bar{\bar{m}}_{i,j} = \bar{1}$ if $m_{i,j} = \bar{0}$ else $\bar{\bar{m}}_{i,j} = (l\bar{1})\beta^{k_{j_1}}$ if $j = j_1$ otherwise $\bar{\bar{m}}_{i,j} = \beta^{k_j}$. Now, consider $M\mathbf{v}^{(2)}$. The ith entry of the vector $M\mathbf{v}^{(2)}$ will be $2(l-1)\bar{1} + \bar{1} \in \mathbb{S}$.

In a similar manner, take the vector $\mathbf{v}^{(3)} = [\bar{\bar{\bar{m}}}_{i,1} \; \bar{\bar{\bar{m}}}_{i,2} \; \cdots \; \bar{\bar{\bar{m}}}_{i,n}]$ where $\bar{\bar{\bar{m}}}_{i,j} = \bar{1}$ if $m_{i,j} = \bar{0}$ else $\bar{\bar{\bar{m}}}_{i,j} = (2l-1)\bar{1}\beta^{k_{j_1}}$ if $j = j_1$ otherwise $\bar{\bar{\bar{m}}}_{i,j} = \beta^{k_j}$. Now, consider $M\mathbf{v}^{(3)}$. The ith entry of the vector $M\mathbf{v}^{(3)}$ will be $3(l-1)\bar{1} + \bar{1} \in \mathbb{S}$.

Repeating the process $e - 1$ times in a similar fashion, we get $e(l-1)\bar{1} + \bar{1} \in \mathbb{S}$ for all $e \geq 0$. In a field of characteristic p, if $l - 1 \not\equiv 0 \mod p$, there exists an $1 \leq e < p$ such that $e(l-1)\bar{1} + \bar{1} = \bar{0}$. The value of e is in-fact $-(l-1)^{-1} \mod p$. Thus $\bar{0} \in \mathbb{S}$. Hence the lemma.   □

Similarly, from Lemmas 3 and 10, we get the following theorem.

Theorem 4. *Suppose $k_1, k_2 \geq 1$ and $s \in \mathbb{S}$. Let $m_{i,j}^{(k_1)} \in s < R^{(k_2)} > \cup \{\bar{0}\}$ for some $1 \leq i \leq n$ and for all $j = 1, \cdots, n$. Let the characteristic of the underlying field be p. Suppose the ith row has $l \geq 1$ number of non-zero entries where $l \not\equiv 1 \mod p$. If $\mathbb{S}$ is an FPS wrt M, then $\bar{0} \in \mathbb{S}$.*

Lemma 11. *Let $\bar{1} \in \mathbb{S}$. Suppose $r \geq 1$. If $\mathbb{S}$ is an FPS wrt M, then $(m_i - \bar{1})$ $(\sum_{l=0}^{r} m_{i,j}^{l}) + \bar{1} \in \mathbb{S}$ for all $1 \leq i, j \leq n$.*

Proof. Recall that $m_i = \sum_{j=1}^{n} m_{i,j}$. Consider $\mathbf{v} = [\bar{1}\ \bar{1}\ \cdots\ m_i\ \cdots\ \bar{1}]^T \in \mathbb{S}^n$ (from Lemma 9) where m_i is at the jth position of the vector $\mathbf{v}$ and rest $\bar{1}$. The ith element of the vector $M\mathbf{v}$ will be $(m_i - \bar{1})(\bar{1} + m_{i,j}) + \bar{1} \in \mathbb{S}$. Now, we show that $(m_i - \bar{1})(\sum_{l=0}^{r} m_{i,j}^{l}) + \bar{1} \in \mathbb{S}$ for any $r \geq 1$. We prove it by induction.

For $r = 1$, we have shown that $(m_i - \bar{1})(\bar{1} + m_{i,j}) + \bar{1} \in \mathbb{S}$. Assume that $(m_i - \bar{1})(\sum_{l=0}^{r} m_{i,j}^{l}) + \bar{1} \in \mathbb{S}$ for some $r = r_1$. Now, we show that its true for $r = r_1 + 1$ also. Consider the vector $\mathbf{v} = [\bar{1}\ \bar{1}\ \cdots\ (m_i - \bar{1})(\sum_{l=0}^{r_1} m_{i,j}^{l}) + \bar{1}\ \cdots\ \bar{1}]^T \in \mathbb{S}^n$ where $(m_i - \bar{1})(\sum_{l=0}^{r_1} m_{i,j}^{l}) + \bar{1}$ is at the jth position of the vector $\mathbf{v}$ and rest $\bar{1}$. Then the ith element of the vector $M\mathbf{v}$ will be $m_{i,j}((m_i - \bar{1})(\sum_{l=0}^{r_1} m_{i,j}^{l}) + \bar{1}) + m_i - m_{i,j} = (m_i - \bar{1})(\sum_{l=0}^{r_1+1} m_{i,j}^{l}) + \bar{1} \in \mathbb{S}$. Hence the lemma. □

Lemma 12. *Let $s \neq \bar{1}$ and $\{\bar{1}, s\} \subseteq \mathbb{S}$. Suppose $r \geq 1$ and $m_i = \bar{1}$ for some $i \in \{1, \cdots, n\}$. If $\mathbb{S}$ is an FPS wrt M, then $m_{i,j}^{r}(s - \bar{1}) + \bar{1} \in \mathbb{S}$ for all $j = 1, \cdots, n$.*

Proof. Consider the vector $\mathbf{v} = [\bar{1}\ \bar{1}\ \cdots\ s\ \cdots\ \bar{1}]^T \in \mathbb{S}^n$ where s is at the jth position of the vector $\mathbf{v}$ and rest $\bar{1}$. The ith element of the vector $M\mathbf{v}$ will be $(s - \bar{1})m_{i,j} + \bar{1} \in \mathbb{S}$ (because $\mathbb{S}$ is a format preserving set). Now, we show that $(s - \bar{1})m_{i,j}^{r} + \bar{1} \in \mathbb{S}$ for all $r \geq 1$. We prove it by induction.

For $r = 1$, we have shown that $(s - \bar{1})m_{i,j} + \bar{1} \in \mathbb{S}$. Assume that $(s - \bar{1})m_{i,j}^{r} + \bar{1} \in \mathbb{S}$ for some $r = r_1 > 1$. Now, we show that its true for $r = r_1 + 1$ also. Consider the vector $\mathbf{v} = [\bar{1}\ \bar{1}\ \cdots\ (s - \bar{1})m_{i,j}^{r_1} + \bar{1}\ \cdots\ \bar{1}]^T \in \mathbb{S}^n$ where $(s - \bar{1})m_{i,j}^{r_1} + \bar{1}$ is at the jth position of the vector $\mathbf{v}$ and rest $\bar{1}$. Then the ith element of the vector $\mathbf{v}$ will be $\bar{1} - m_{i,j} + m_{i,j}((s - \bar{1})m_{i,j}^{r_1} + \bar{1}) = (s - \bar{1})m_{i,j}^{r_1+1} + \bar{1} \in \mathbb{S}$. Hence the lemma. □

Using Lemmas 11 and 12, we get the next theorem.

Theorem 5. *Let $s \neq \bar{1}$ and $\{\bar{1}, s\} \subseteq \mathbb{S}$. Suppose the ith row of the matrix M has $l \geq 1$ number of non-zero entries where $l \not\equiv 1 \mod p$. For some $j \in \{1, \cdots, n\}$, let there be an element $m_{i,j}$ such that $SF(M)^* = < m_{i,j} >$. If $\mathbb{S}$ is an FPS wrt M, then $\bar{0} \in \mathbb{S}$.*

Proof. If $m_{i,j} = \bar{1}$, then $SF(M) = \mathbb{F}_2$. In such case, all entries of M will be either $\bar{0}$ or $\bar{1}$. If i^{th} row of the matrix has $l \not\equiv 1 \mod 2$ number of non-zero entries, i.e., even number of $\bar{1}$s, then $m_i = \bar{0}$ which further implies $\bar{0} \in \mathbb{S}$ (from Lemma 8).

We assume that $m_{i,j} \neq \bar{1}$ and divide it into three cases - (a) when $m_i \notin \{\bar{1}, m_{i,j}\}$, (b) when $m_i = \bar{1}$ and (c) when $m_i = m_{i,j}$. Consider these following cases:

(a) From Lemma 11, $(m_i - \bar{1})(\sum_{l=0}^{r} m_{i,j}^l) + \bar{1} = (m_i - \bar{1})(m_{i,j} - \bar{1})^{-1}(m_{i,j}^{r+1} - \bar{1})$
$+ \bar{1} \in \mathbb{S}$ for all $r \geq 1$. Since $< m_{i,j} > = SF(M)^*$, so for $r \geq 1$, $(m_{i,j}^{r+1} - \bar{1})$
varies over all the elements of the field $SF(M)$ except $-\bar{1}$. In this case,
$m_i \notin \{\bar{1}, m_{i,j}\}$ and $m_{i,j} \neq \bar{1}$, therefore there exists $r = r_1 \geq 1$, such that
$(m_i - \bar{1})(m_{i,j} - \bar{1})^{-1}(m_{i,j}^{r+1} - \bar{1}) = -\bar{1}$ and thus $\bar{0} \in \mathbb{S}$.

(b) From Lemma 12, $m_{i,j}^r(s - \bar{1}) + \bar{1} \in \mathbb{S}$ for all $r \geq 1$. Since $SF(M)^* = < m_{i,j} >$,
so $m_{i,j}^r$ varies over all the elements of the field $SF(M)^*$. As $s \neq \bar{1}$, there
exists some $r = r_1 \geq 1$ such that $m_{i,j}^r(s - \bar{1}) + \bar{1} = \bar{0} \in \mathbb{S}$.

(c) If $< m_i > = SF(M)^*$, then $m_{i,l} \in < m_i > \cup \{\bar{0}\} \subseteq < R > \cup \{\bar{0}\}$ for all
$l = 1, \cdots, n$. From Lemma 10, we can conclude that $\bar{0} \in \mathbb{S}$.

Hence the theorem.   □

4   Credit Card Example over the Field $\mathbb{F}_{2^4}$

In the credit card example, we fixed our requirement to be $|\mathbb{S}| = 10$. In Sect. 3,
the case when $\bar{0} \in \mathbb{S}$ has been discussed and that's why, in this section, we do
not assume that $\bar{0} \in \mathbb{S}$. In this section, we discuss only for 4×4 matrices whose
entries are from the field $\mathbb{F}_{2^4}$.

From Theorem 3, there exists a subset $H \subseteq \mathbb{F}_{2^4}^*$ such that $\mathbb{S} = \cup_{s \in H} s < R >$.
Suppose $s_1, s_2 \in H$ such that $s_1 \neq s_2$. Since $< R >$ is the subgroup of $\mathbb{F}_{2^4}^*$, it can
be easily shown that either $s_1 < R > = s_2 < R >$ or $s_1 < R > \cap s_2 < R > = \phi$
(an empty set). Thus $| < R > |$ divides $|\mathbb{S}| = 10$. Moreover, $| < R > |$ divides
$|\mathbb{F}_{2^4}^*| = 15$. Therefore $| < R > |$ divides the greatest common divisor of 10 and
15 which is 5. So, the possible values of $| < R > |$ are 1 and 5.

The multiplicative group $\mathbb{F}_{2^4}^*$ is cyclic, therefore, its subgroup $< R >$ also
is cyclic. Let $< \gamma > = \mathbb{F}_{2^4}^*$. For $| < R > | = 1$, the subgroup $< R > = \{\bar{1}\}$,
whereas, for $| < R > | = 5$, the subgroup $< R > = \{\bar{1}, \gamma^3, \gamma^6, \gamma^9, \gamma^{12}\}$. Let
$\gamma^3 = \alpha$. Then $< R >$ will be either $\{\bar{1}\}$ or $\{\bar{1}, \alpha, \alpha^2, \alpha^3, \alpha^4\}$. Let $\beta = \gamma^5$. Then
$\mathbb{F}_{2^4}^* = < R > \cup \beta < R > \cup \beta^2 < R >$. For the case $| < R > | = 5$, there are
three possibilities - (a) $\mathbb{S} = < R > \cup \beta < R >$, (b) $\mathbb{S} = < R > \cup \beta^2 < R >$ and
(c) $\mathbb{S} = \beta < R > \cup \beta^2 < R >$.

A matrix can have either (a) all rows which contains at most one non-zero
entry or (b) at least one row which has at least two non-zero entries. We do not
consider those matrices which has at least one row whose all entries are $\bar{0}$ because
in such case, $\bar{0} \in \mathbb{S}$. Therefore, in case (a), we consider only those matrices whose
all rows have exactly one non-zero entry. Similarly, for case (b), there is no row
whose all entries are $\bar{0}$.

4.1   Case (a)

In this subsection, we provide the structure of 4×4 matrix M and the set $\mathbb{S}$
which is a format preserving set with respect to M. Let $m_{i,j_i} \neq \bar{0}$ for some
$j_i \in \{1, \cdots, 4\}$ and for all $i = 1, \cdots, 4$. Consider the following cases:

- When $< R >= \{\bar{1}\}$. In such case, $m_{i,j_i} = \bar{1}$ for all $i = 1, \cdots, 4$. Thus each row of M has exactly one non-zero entry whose value is $\bar{1}$. Furthermore, choose any 10 elements from $\mathbb{F}_{2^4}^*$. Let these elements be $\{s_1, s_2, \cdots, s_{10}\}$. Then $\mathbb{S} = \{s_1, s_2, \cdots, s_{10}\}$.
- When $< R >= \{\bar{1}, \alpha, \alpha^2, \alpha^3, \alpha^4\}$. Since, $| < R > | = 5$, a prime number, hence, $\alpha, \alpha^2, \alpha^3$ and α^4 all are generators of $< R >$. Therefore $m_{i,j_i} \in \{\bar{1}, \alpha, \alpha^2, \alpha^3, \alpha^4\}$ for all $i = 1, \cdots, 4$ with the condition that at least one of $m_{i,j_i} \in \{\alpha, \alpha^2, \alpha^3, \alpha^4\}$. Furthermore, $\mathbb{S} =< R > \cup \beta < R >$ or $\mathbb{S} =< R > \cup \beta^2 < R >$ or $\mathbb{S} = \beta < R > \cup \beta^2 < R >$.

4.2   Case (b)

This subsection shows the impossibility of the existence of our desired matrix M. We assume that the matrix M has at least one row, say i^{th}, which has $l \geq 2$ number of non-zero entries. Moreover, no row contains all entries whose values are $\bar{0}$. Now, consider the following cases:

- When $< R >= \{\bar{1}\}$. In such case $m_i = \bar{1}$ for all $i = 1, \cdots, 4$. As $|\mathbb{S}| = 10$, there exists an $s \in \mathbb{S}$ such that $s \neq \bar{1}$. From Theorem 5, in case of $l = 2$ and 4, if $< m_{i,j} >= \mathbb{F}_{2^4}^*$ for some $j \in \{1, \cdots, 4\}$, then $\bar{0} \in \mathbb{S}$. Therefore, we consider $m_{i,j} \in \{\bar{0}, \bar{1}, \alpha, \alpha^2, \alpha^3, \alpha^4, \beta, \beta^2\}$ for all $j = 1, \cdots, 4$ because all other elements of $\mathbb{F}_{2^4}^*$ are the generators of $\mathbb{F}_{2^4}^*$. Consider those $m_{i,j}$'s which are not zero. Then, non-zero $m_{i,j}$s belong to $\{\bar{1}, \alpha, \alpha^2, \alpha^3, \alpha^4, \beta, \beta^2\}$ only. Consider the following cases -
  - $l = 2$. Two non-zero $m_{i,j}$s can be either $\{\alpha^{r_1}, \alpha^{r_2}\}$ or $\{\alpha^{r_1}, \beta\}$ or $\{\alpha^{r_1}, \beta^2\}$ or $\{\beta, \beta^2\}$ for some $1 \leq r_1, r_2 \leq 5$. The only possible candidate is $\{\beta, \beta^2\}$ because none other than $\{\beta, \beta^2\}$ will have sum $\bar{1}$. Suppose $\delta \in \mathbb{F}_{2^4}$. Consider the set $H_\delta = \{\delta, \beta\delta, \beta^2\delta\}$. It is easy to verify that $\bar{0} \in H_\delta$ if and only if $\delta = \bar{0}$. If $\delta \neq \bar{0}$, the set H_δ will have all distinct elements. Suppose $\delta_1, \delta_2 \in \mathbb{F}_{2^4}^*$ such that $\delta_1 \neq \delta_2$. It is easy to verify then that either $H_{\delta_1} = H_{\delta_2}$ or $H_{\delta_1} \cap H_{\delta_2} = \phi$. Therefore, there exists a set D such that $\mathbb{F}_{2^4}^* = \cup_{\delta \in D} H_\delta$.
  Let $\beta^{a_1}\delta$ and $\beta^{a_2}\delta$ be two distinct elements from the set H_δ where $a_1 \neq a_2 \bmod 3$ and $\delta \neq 0$. If $\beta^{a_1}\delta$ and $\beta^{a_2}\delta$ both belong to the set $\mathbb{S}$, then $(\beta^{a_1+1} + \beta^{a_2+2})\delta \in \mathbb{S}$ (because $[\beta \ \beta^2] \cdot [\beta^{a_1}\delta \ \beta^{a_2}\delta]^T \in \mathbb{S}$). Therefore, if two distinct elements from H_δ belong to the set $\mathbb{S}$, then $\bar{0} \in \mathbb{S}$. For $\bar{0} \notin \mathbb{S}$, there can be at most $15/3 = 5$ elements in $\mathbb{S}$. Thus $|\mathbb{S}| \neq 10$, a contradiction.
  - $l = 4$. Let $R_i = \{m_{i,1}, m_{i,2}, m_{i,3}, m_{i,4}\}$. Consider these following cases -
    * The sum of four elements from the set $\{\bar{1}, \alpha, \alpha^2, \alpha^3, \alpha^4\}$ can be $\bar{1}$ only when those four elements are $\alpha, \alpha^2, \alpha^3$ and α^4. Let $R_i = \{\alpha, \alpha^2, \alpha^3, \alpha^4\}$. Suppose $\delta \in \mathbb{F}_{2^4}$. Consider $H_\delta = \{\delta, \alpha\delta, \alpha^2\delta, \alpha^3\delta, \alpha^4\delta\}$. It is easy to verify that $\bar{0} \in H_\delta$ if and only if $\delta = \bar{0}$. If $\delta \neq \bar{0}$, the set H_δ will have all distinct elements. Suppose $\delta_1, \delta_2 \in \mathbb{F}_{2^4}^*$ such that $\delta_1 \neq \delta_2$. It is easy to verify then that either $H_{\delta_1} = H_{\delta_2}$ or $H_{\delta_1} \cap H_{\delta_2} = \phi$. Therefore, there exists a set D such that $\mathbb{F}_{2^4}^* = \cup_{\delta \in D} H_\delta$.

If any 4 distinct elements from the set H_δ belong to $\mathbb{S}$, then there exists a vector $\mathbf{v} \in \mathbb{S}^4$ such that $\bar{0}$ becomes an element in the vector $M\mathbf{v}$. Therefore, there can be at most 3 elements from the set H_δ which may belong to the set $\mathbb{S}$. For $\bar{0} \notin \mathbb{S}$, there can be at most $(15/5)*3 = 9$ elements in the set $\mathbb{S}$, a contradiction.

* If $R_i \subseteq \{1, \beta, \beta^2\} \cup \{\bar{0}\}$, then, for $m_i = \bar{1}$, only 4 non-zero possible values are $\beta, \beta^2, \beta^a, \beta^a$ for some $a = 0, 1, 2$. Such case can be dealt in a similar manner as done in the case for $l = 2$ above. Thus, it can be shown that $\bar{0} \in \mathbb{S}$.

* If $R_i \cap \{\alpha, \alpha^2, \alpha^3, \alpha^4\} \neq \phi$ and $R_i \cap \{\beta, \beta^2\} \neq \phi$ both, then there exists column indices $1 \leq j_1 \neq j_2 \leq n$ such that $m_{i,j_1} = \alpha^{a_1}$ and $m_{i,j_2} = \beta^{a_2}$ for some $a_1 = 1, \cdots, 4$ and $a_2 = 1, 2$. If some $s - \bar{1} \in \{\alpha, \alpha^2, \alpha^3, \alpha^4\}$, then from Lemma 12, there exists $r \geq 1$ such that $m_{i,j_1}^r (s - \bar{1}) + \bar{1} = \bar{0} \in \mathbb{S}$. Similarly, when $s - \bar{1} \in \{\beta, \beta^2\}$, then from Lemma 12 again, there exists $r \geq 1$ such that $m_{i,j_2}^r (s - \bar{1}) + \bar{1} = \bar{0} \in \mathbb{S}$. Thus, in this case, $s - \bar{1} \notin \{\alpha, \alpha^2, \alpha^3, \alpha^4, \beta, \beta^2\}$. Thus $\mathbb{S}$ can have at most $15 - 6 = 9$ elements, a contradiction.

• $l = 3$. If, for some $1 \leq j_1 \leq 4$, $m_{i,j_1} \in R_i$ such that $< m_{i,j_1} > = \mathbb{F}_{2^4}^*$, then from Lemma 12, $\bar{0} \in \mathbb{S}$. Therefore, we assume that $R_i \subseteq \{\bar{1}, \alpha, \alpha^2, \alpha^3, \alpha^4, \beta, \beta^2\} \cup \{\bar{0}\}$. Consider these following cases-

* If $R_i \subseteq \{\bar{1}, \alpha, \alpha^2, \alpha^3, \alpha^4\} \cup \{\bar{0}\}$, then only possible non-zero values in R_i which make $m_i = \bar{1}$ are $1, \alpha^a, \alpha^a$ where $a \in \{1, \cdots, 5\}$. If $s_1, s_2 \in \mathbb{S}$, then $\bar{1} + \alpha^a(s_1 + s_2) \in \mathbb{S}$. For $\bar{0} \notin \mathbb{S}$, it is required that s_1 and $-(\alpha^{-a} + s_1)$ both should not belong to $\mathbb{S}$. Vary s_1 over all elements of $\mathbb{F}_{2^4}^*$; $-(\alpha^{-a} + s_1)$ will vary from all elements of $\mathbb{F}_{2^4}$ except $-\alpha^{-a}$. There will be exactly one non-zero value s_1 for which $-(\alpha^{-a} + s_1)$ becomes $\bar{0}$. Thus, there can be at most 8 elements in $\mathbb{S}$, a contradiction.

* Similar case occurs if $R_i \subseteq \{\bar{1}, \beta, \beta^2\} \cup \bar{0}$.

* Therefore we assume that $R_i \cap \{\alpha, \alpha^2, \alpha^3, \alpha^4\} \neq \phi$ and $R_i \cap \{\beta, \beta^2\} \neq \phi$ both. For such case, similar argument holds which has been discussed in the case for $l = 4$ above.

– When $< R > = \{\bar{1}, \alpha, \alpha^2, \alpha^3, \alpha^4\}$. In such case, without loss of generality, we may assume that $\mathbb{S} = < R > \cup \beta < R >$. Moreover, $m_i \in < R >$ for all $i = 1, \cdots, 4$ but all $m_i \neq \bar{1}$. From Theorem 5, in case of $l = 2$ and 4, if $< m_{i,j} > = \mathbb{F}_{2^4}^*$ for some $j \in \{1, \cdots, 4\}$, then $\bar{0} \in \mathbb{S}$. Therefore, we consider $m_{i,j} \in \{\bar{0}, \bar{1}, \alpha, \alpha^2, \alpha^3, \alpha^4, \beta, \beta^2\}$ for all $j = 1, \cdots, 4$. Consider the following cases:

• $l = 2$. Two non-zero $m_{i,j}$s can be either (a) $\{\alpha^{r_1}, \alpha^{r_2}\}$ or (b) $\{\alpha^{r_1}, \beta\}$ or (c) $\{\alpha^{r_1}, \beta^2\}$ or (d) $\{\beta, \beta^2\}$ for some $1 \leq r_1, r_2 \leq 5$. Choose $s_1 = \alpha^{5-r_1}$, $s_2 = \alpha^{5-r_2}$ for (a), $s_1 = \alpha^{5-r_1}, s_2 = \bar{1}$ for (b), $s_1 = \alpha^{5-r_1}, s_2 = \beta$ for (c) and $s_1 = \beta, s_2 = \bar{1}$ for (d). All four choices of s_1 and s_2 belong to $\mathbb{S}$ and for each such choices, $\bar{0} \in \mathbb{S}$ in (a), (c), (d) and $\beta^2 \in \mathbb{S}$ in (b), a contradiction.

• $l = 4$. Any four non-zero values from the set $\{\bar{1}, \alpha, \alpha^2, \alpha^3, \alpha^4, \beta, \beta^2\}$ will yield either $\bar{0} \in \mathbb{S}$ or $\beta^2 \in \mathbb{S}$, a contradiction.

- $l = 3$, In this case, we cannot apply Theorem 5, therefore $m_{i,j} \in < R >$ $\cup\ \beta < R > \cup\ \beta^2 < R > \cup\ \bar{0}$ for all $j = 1, \cdots, 4$. If $m_{i,j} \in < R > \cup\ \{\bar{0}\}$ for all $j = 1, \cdots, 4$, then $\mathrm{m}_i \notin < R >$, a contradiction. Thus we assume that there exists at least one $j_1 \in \{1, \cdots, 4\}$ such that $m_{i,j_1} \in \beta < R > \cup\ \beta^2 < R >$. Since $\mathbb{S} = < R > \cup\ \beta < R >$, it can be shown that either $\bar{0} \in \mathbb{S}$ or $\beta^2 \in \mathbb{S}$, a contradiction.

Thus, we conclude that if a 4×4 matrix M over the field $\mathbb{F}_{2^4}$ has a row which contains at least two non-zero entries, then there does not exist any format preserving set $\mathbb{S}$ with respect to the matrix M such that $|\mathbb{S}| = 10$.

5   Conclusion and Future Work

This paper discusses the algebraic structure of the format preserving set $\mathbb{S}$ with respect to the matrix M over the field $\mathbb{F}_q$. It is shown that if the matrix M has a row which contains at least two non-zero entries and $\bar{0} \in \mathbb{S}$, then $\mathbb{S}$ becomes a vector space over the smallest field containing entries of M. Therefore, in a field of characteristic p, for such matrices M, $|\mathbb{S}| = p^{m'}$ for some $m' \geq 1$. But, this paper does not provide the complete algebraic structure of format preserving set as it is unknown what happens when $\bar{0}$ may not belong to $\mathbb{S}$? In this direction, we obtain some more interesting results which can be used to find out the possibility or impossibility of the algebraic structure of format preserving set $\mathbb{S}$ with respect to M. Using these results, it is shown that if a 4×4 matrix M over the field $\mathbb{F}_{2^4}$ has a row which contains at least two non-zero entries, then it is impossible to construct a format preserving set whose cardinality is 10.

But, if each row of the matrix M has at most one non-zero entry, then a format preserving set $\mathbb{S}$ of any given cardinality can be constructed. Although, to the best of our knowledge, such matrices do not have any cryptographic significance, these results are useful in providing the theoretical completeness.

Future Work: This paper does not provide the complete structure of format preserving set $\mathbb{S}$ with respect to M when the condition, $\bar{0} \in \mathbb{S}$ is relaxed. Therefore, it would be interesting to explore the complete structure of $\mathbb{S}$ with respect to any matrix M. Furthermore, this paper considers that $\mathbb{S}$ is a subset of some field $\mathbb{F}_q$ and entries of the matrix M also are from the same field. It would be worth to explore what happens if instead of the field $\mathbb{F}_q$, the set $\mathbb{S}$ is a subset of some ring $\mathcal{R}$ and entries of the matrix M also are from the same ring.

References

1. Augot, D., Finiasz, M.: Direct construction of recursive MDS diffusion layers using shortened BCH codes. In: Cid, C., Rechberger, C. (eds.) FSE 2014. LNCS, vol. 8540, pp. 3–17. Springer, Heidelberg (2015). doi:10.1007/978-3-662-46706-0_1
2. Bellare, M., Ristenpart, T., Rogaway, P., Stegers, T.: Format-preserving encryption. In: Jacobson, M.J., Rijmen, V., Safavi-Naini, R. (eds.) SAC 2009. LNCS, vol. 5867, pp. 295–312. Springer, Heidelberg (2009). doi:10.1007/978-3-642-05445-7_19

3. Bellare, M., Rogaway, P.: On the construction of variable-input-length ciphers. In: Knudsen, L. (ed.) FSE 1999. LNCS, vol. 1636, pp. 231–244. Springer, Heidelberg (1999). doi:10.1007/3-540-48519-8_17

4. Bellare, M., Rogaway, P., Spies, T.: The FFX mode of operation for format-preserving encryption (2010). http://csrc.nist.gov/groups/ST/toolkit/BCM/documents/proposedmodes/ffx/ffx-spec.pdf

5. Black, J., Rogaway, P.: Ciphers with arbitrary finite domains. In: Preneel, B. (ed.) CT-RSA 2002. LNCS, vol. 2271, pp. 114–130. Springer, Heidelberg (2002). doi:10.1007/3-540-45760-7_9

6. Brier, E., Peyrin, T., Stern, J.: BPS: A Format-Preserving Encryption Proposal (2010). http://csrc.nist.gov/groups/ST/toolkit/BCM/documents/proposedmodes/bps/bps-spec.pdf

7. Chang, D., Kumar, A., Sanadhya, S.K.: SPF: a new family of efficient format-preserving encryption algorithms. In: Preprint

8. Daemen, J., Rijmen, V.: The Design of Rijndael: AES-The Advanced Encryption Standard. Springer, Berlin (2002)

9. Gupta, K.C., Ray, I.G.: On constructions of involutory MDS matrices. In: Youssef, A., Nitaj, A., Hassanien, A.E. (eds.) AFRICACRYPT 2013. LNCS, vol. 7918, pp. 43–60. Springer, Heidelberg (2013). doi:10.1007/978-3-642-38553-7_3

10. Gupta, K.C., Ray, I.G.: On constructions of MDS matrices from companion matrices for lightweight cryptography. In: Cuzzocrea, A., Kittl, C., Simos, D.E., Weippl, E., Xu, L. (eds.) CD-ARES 2013. LNCS, vol. 8128, pp. 29–43. Springer, Heidelberg (2013). doi:10.1007/978-3-642-40588-4_3

11. Gupta, K.C., Ray, I.G.: On constructions of circulant MDS matrices for lightweight cryptography. In: Huang, X., Zhou, J. (eds.) ISPEC 2014. LNCS, vol. 8434, pp. 564–576. Springer, Heidelberg (2014). doi:10.1007/978-3-319-06320-1_41

12. Halevi, S., Rogaway, P.: A tweakable enciphering mode. In: Boneh, D. (ed.) CRYPTO 2003. LNCS, vol. 2729, pp. 482–499. Springer, Heidelberg (2003). doi:10.1007/978-3-540-45146-4_28

13. Halevi, S., Rogaway, P.: A parallelizable enciphering mode. In: Okamoto, T. (ed.) CT-RSA 2004. LNCS, vol. 2964, pp. 292–304. Springer, Heidelberg (2004). doi:10.1007/978-3-540-24660-2_23

14. Herstein, I.N.: Topics in Algebra. Wiley, Hoboken (1975)

15. Hoang, V.T., Rogaway, P.: On generalized feistel networks. In: Rabin, T. (ed.) CRYPTO 2010. LNCS, vol. 6223, pp. 613–630. Springer, Heidelberg (2010). doi:10.1007/978-3-642-14623-7_33

16. Hoffman, K.M., Kunze, R.: Linear Algebra. Prentice-Hall, Upper Saddle River (1971)

17. Lidl, R., Niederreiter, H.: Finite Fields. Cambridge University Press, Cambridge (2008)

18. Morris, B., Rogaway, P., Stegers, T.: How to encipher messages on a small domain. In: Halevi, S. (ed.) CRYPTO 2009. LNCS, vol. 5677, pp. 286–302. Springer, Heidelberg (2009). doi:10.1007/978-3-642-03356-8_17

19. Rao, A.R., Bhimasankaram, P.: Linear algebra, vol. 19 of texts and readings in mathematics. Hindustan Book Agency, New Delhi. Technical report, ISBN 81-85931-26-7 (2000)

20. Sheets, J., Wagner, K.R.: VISA Format Preserving Encryption (2011). http://csrc.nist.gov/groups/ST/toolkit/BCM/documents/proposedmodes/vfpe/vfpe-spec.pdf

21. Terence Spies. Feistel Finite Set Encryption Mode (2008). http://csrc.nist.gov/groups/ST/toolkit/BCM/documents/proposedmodes/ffsem/ffsem-spec.pdf

Author Index

Printed in the United States
By Bookmasters